The **Crest** *of* *the* **Peacock**

The **Crest** of *the* **Peacock**

Non-European Roots of Mathematics THIRD EDITION

George Gheverghese Joseph

Princeton University Press | Princeton & Oxford

Copyright © 2011 by
Princeton University Press
Published by Princeton University Press,
41 William Street, Princeton,
New Jersey 08540
In the United Kingdom:
Princeton University Press,
6 Oxford Street, Woodstock,
Oxfordshire OX20 1TW
First published by I. B. Tauris 1991
Published in Penguin Books 1992
Reprinted with minor revisions 1993
Reprinted with revisions 1994
Reprinted with additional material 2000
press.princeton.edu

Library of Congress Cataloging-in-Publication Data

Joseph, George Gheverghese.
    The crest of the peacock : non-European roots of
mathematics / George Gheverghese Joseph. — [3rd ed.].
    Originally published: London ; New York : I.B. Tauris,
1991.
    Includes bibliographical references and index.
    ISBN 978-0-691-13526-7 (pbk. : alk. paper)
1. Mathematics Ancient. 2. Mathematics—History.
I. Title.
    QA22.J67   2001
    510.9—dc22   2010015214

British Library of Cataloging-in-Publication Data
is available

This book has been composed in Minion
Printed on acid-free paper ∞
Printed in the United States of America
10 9 8 7 6 5 4 3 2 1

To the memory of my parents and elder sister,
Adangapuram Gheverghese Joseph Panicker,
Sara Joseph, and Mary Jacob

Like the crest of a peacock, like the gem
on the head of a snake, so is mathematics
at the head of all knowledge.

—*Vedanga Jyotisa* (C. 500 BC)

Contents

Preface to the Third Edition

It is now almost twenty years since the first edition of this book came out as a hardback. Four reprints, two editions, and a number of translations later, the book is badly in need of a revamp, owing to new theories and evidence as well as comments, suggestions, and criticisms that have come from so many different parts of the world. It was particularly fortuitous that while I was preparing the new edition of the book there appeared *The Mathematics of Egypt, Mesopotamia, China, India, and Islam: A Sourcebook* (edited by Victor Katz), on the history of mathematics of the named cultures, published in 2007. My debt to this book will become evident from the references and acknowledgments that follow.

The readership of this book in the past has included mainly teachers and the general public, with the technical content of the mathematical material being accessible to anyone having a reasonable precalculus background. And since it is a similar readership that this edition addresses, the demands on the reader have been kept to a level not different from those of the earlier editions. However, it is hoped that the new edition will also attract greater interest among the historians of mathematics. Toward that end, and for other readers who wish to pursue their interests further, this edition contains a major innovation: the introduction of endnotes for each chapter. These endnotes will hopefully serve different objectives: to provide references for those who wish to pursue their own reading on specific subjects, to qualify and elaborate on points made in the main text, to respond to comments and criticisms on earlier editions, and occasionally to make connections between different traditions and their "ways of doing mathematics." It is hoped that the introduction of these endnotes will not disturb or distract the flow of the narrative in the main text. Yet another innovation with this edition is a revised and enlarged reference list. Because of substantial additions, it was felt necessary to regroup the entries according to the mathematics of particular cultures, with a separate category for general items. It is hoped that these innovations will make this book a more effective resource for students and teachers of mathematics while remaining accessible to general readers.

While researching for this edition, I also came across another book, *A History of Mathematics: From Mesopotamia to Modernity*, by Luke Hodgkin. In its introduction, there is a section on "Eurocentrism," which I found both thought provoking and persuasive. Incidentally, Hodgkin's book is the first history of mathematics I have come across that acknowledges the pervasiveness and durability of the Eurocentric version of history. The quotation below from his book encapsulates his view:

> It would appear that the argument set out by Joseph [in the *Crest of the Peacock*] has not been won yet. . . . For what [Eurocentrism] might mean in mathematics, we should go back to Joseph who, at the time he began his project (in the 1980s), had a strong, passionate and undeniable point. . . . his book is important: it is the only book in the history of mathematics written from a strong personal conviction, and it is valuable for that reason alone. It stands as the single most influential work in changing attitudes to non-European mathematics. The sources, such as Neugebauer on the Egyptians and Babylonians, or Youschkevitch on the Islamic tradition, may have been available for some time before, but Joseph drew their findings into a forceful argument which since (like Kuhn's work) its main thrust is easy to follow has made many converts. (pp. 12–13)

A number of "mainstream" historians of mathematics have in recent years taken up the task of casting a wider net in writing history and considering seriously the contributions of not only the ancient Egyptian and Mesopotamian civilizations but also the Chinese, Indian, and Islamic civilizations. There are substantial and growing communities of mathematics historians of all these civilizations, a number of whom are engaged in the task of making new evidence accessible to everyone. Nevertheless, it is argued that change in historical perceptions is slow, and that a significant part of the new studies in the history of mathematics has failed to reach the broader community.

A central theme of the earlier editions, seen by some as their principal strength and by others as either an irritating irrelevance or even a fatal weakness, is their critique of the widespread acceptance of the hegemony of a Western version of mathematics, following from the assumption that mathematics was largely a Greek and European creation. I have argued elsewhere that in the past the two tactics used to propagate this view were (1) omission and appropriation and (2) exclusion by definition. The first is examined in some detail in chapter 1. The second needs clarification and elaboration here.

A Eurocentric approach to the history of mathematics is intimately connected with the dominant view of mathematics, both as a sociohistorical practice and as an intellectual activity. Despite evidence to the contrary, a number of earlier histories viewed mathematics as a deductive system, ideally proceeding from axiomatic foundations and revealing, by the "necessary" unfolding of its pure abstract forms, the eternal/universal laws of the "mind."

The concept of mathematics found outside the Graeco-European praxis was very different. The aim was not to build an imposing edifice on a few self-evident axioms but to validate a result by any suitable method. Some of the most impressive work in Indian and Chinese mathematics examined in later chapters, such as the summations of mathematical series, or the use of Pascal's triangle in solving higher-order numerical equations, or the derivations of infinite series, or "proofs" of the so-called Pythagorean theorem, involve computations and visual demonstrations that were not formulated with reference to any formal deductive system. The view that mathematics is a system of axiomatic/deductive truths inherited from the Greeks, and enthroned by Descartes, has traditionally been accompanied by the following cluster of values that reflect the social context in which it originated:

1. An *idealist* rejection of any practical, material(ist) basis for mathematics: hence the tendency to view mathematics as value-free and detached from social and political concerns

2. An *elitist* perspective that sees mathematical work as the exclusive preserve of a high-minded and almost priestly caste, removed from mundane preoccupations and operating in a superior intellectual sphere

Mathematical traditions outside Europe did not generally conform to this cluster of values and have therefore been dismissed on the grounds that they were dictated by utilitarian concerns with little notion of rigor, especially relating to proof. Any attempt at excavation and restoration of non-European mathematics is a multifaceted task: confront historical bias, question the social and political values shaping the mathematics (and the writing of the history of mathematics), and search for different ways of "knowing" or establishing mathematical truths in various traditions. I have *written* elsewhere (Joseph 1994b, 1997a) on the same subject in relation to the Indian tradition. (Documentation can be found in the "India" and "General" sections of the reference list at the end of this book.) Because of the centrality of the issue of "proof" in judging the quality of mathematics outside the European tradition, we will be returning to this subject at different points in this book.

The responses of some critics to the earlier versions of this book have helped to confirm my belief that words or labels in common use need careful scrutiny. It has been pointed out that terms such as "Classical," "Dark Ages," and "Renaissance" are peculiarly European concepts of little relevance to the rest of the world. Also, words such as "ancient," "medieval," and "modern" are of doubtful provenance when applied to other histories. It is, however, in the labeling of geographical areas that the distortions could, potentially, take on grotesque proportions.

Consider the term "Europe." A Eurasian peninsula has been elevated to the status of a continent, equal in importance, if not superior, to the rest of the continent combined. The Mercator projection may have contributed to this perception, with its visually distorted image exaggerating the northward bounds to make "Europe" look larger than the whole of Africa, and enormous compared with the other Eurasian peninsula, India. The special status accorded to Europe in the standard histories of the world has strengthened the notion, shared by many Europeans and their overseas descendants, that they played a starring role in the Eurasian theater of world history. The resulting categories such as European/non-European, West/East, Europe/Asia have tended to reinforce this Eurocentric illusion.

It is precisely because of this tendency to enhance Europe through the use of labels that any project involving the writing of a balanced history should carefully address the question of labels. Unfortunately, these labels have existed for so long that they have acquired legitimacy through usage. Fairly early in writing *The Crest of the Peacock*, I was faced with the problem of finding a subtitle for the book. In the hardback first edition, the publisher took the decision out of my hands and provided the subtitle *Non-Western Roots of Mathematics*. The paperback versions replaced it with *Non-European Roots of Mathematics,* which had at least the benefit of confronting the widely held view that mathematics began in Greece—a part of Europe. However, both subtitles fail to avoid the implication that a large part of the world's population is defined by *not* being something.

A further problem was in the choice of labels to identify different mathematical traditions. Labeling traditions by names of regions such as India and China, or of historical cultures such as the Islamic cultures between AD 800 and 1500 or Greek/Hellenistic cultures between 600 BC and AD 300, has both spatial and temporal justification. If we use the linguistic marker, descriptions of Greek, Chinese, and Arabic mathematics have legitimacy, although there appears to be a certain inconsistency implied in the first, and incompleteness implied in the last, of the descriptive terms. In terms of content, historical legitimacy poses yet another problem. In earlier editions, chapter 8 on ancient Indian mathematics contained sections on the so-called Vedic multiplication and on the "mathematics inherent in a meditation device, the *sriyantra."* For this edition, it was decided to omit those sections, the first being of doubtful historical authenticity and the second because of the dubious assumption that it contained "hidden geometry."

Labeling has yet another dimension. When Alexander appointed his general Ptolemy I to rule Egypt, he also appointed Seleucus I to rule Mesopotamia. Historians who tended to downplay the African influence called the science of Ptolemaic Egypt "Greek," while they continued to label the science and mathematics of Seleucid period as "Mesopotamian." For example, Otto Neugebauer (1962, p. 97) writes: "Early Mesopotamian astronomy appeared to be crude and merely qualitative, quite similar to its contemporary, Egyptian astronomy.... Only the last three centuries BC [have] furnished us with texts [from the former] ... fully comparable to the corresponding Greek systems (of the latter)."

Outstanding among these so-called Greek systems were those of Alexandria, which Neugebauer labels "Greek" rather than "Egyptian," unlike the description of the Seleucid astronomy as "Mesopotamian," although the Mesopotamia of that time was also under Greek rule. In this edition we will continue to retain the label "Greek" to describe all mathematical works written in Greek, irrespective of their geographical origins. However, a distinction is drawn between "Greek" and "Hellenistic" (or "Alexandrian") in providing a geographical reference. Thus, Alexandrian mathematics would be firmly situated in Egyptian mathematics.

The label "Arabic mathematics" is a clumsy and problematic construction that neglects the rich source of non-Arabic mathematical texts, especially those written in Pahlavi, Syriac, and Hebrew. In place of "Arabic" other labels are available, such as "Middle Eastern" or "Semitic," among others. However, the terms "Middle Eastern" and "Semitic" were quickly rejected for their geographical and conceptual imprecision. So it was decided, despite the limitations as a descriptive label, to use the term "Islamic" to refer to a civilization that contained a number of other ethnic and religious groups. This is a marked departure from the use of the term "Arab" as a descriptive label in the earlier editions. It is now recognized that the term "Arab" is too restrictive and imprecise, given that at its height the civilization referred to included "non-Arab" lands such as present-day Iran, Turkey, Afghanistan, and Pakistan, all of which have distinctive Islamic cultures. By the same token, it should be emphasized that medieval Islamic civilization included non-Muslim populations such as communities of Christians, Jews, and Zoroastrians.

There have been reservations expressed about the trajectories introduced in chapter 1. Now, one of the main purposes of the three trajectories introduced in that chapter was to bring out differing *perspectives* on the origins of mathematics. No claims are made for completeness or balance in the marshaling of historical evidence underlying the discussion. Sicily, as a staging post in the spread of mathematical ideas from the East, was of less consequence than Spain, despite the appearance to the contrary in figure 1.3. Knowledge regarding the role of Jund-i-shapur as a center of learning and scholarship remains speculative compared with the roles of Baghdad or Toledo. In the case of the alternative trajectory (see figure 1.3), the role of centers in the Maghreb from the beginning of the twelfth century to the end of the fifteenth century AD, discussed in the works of Djebbar (1981, 1985, 1990, 1997), was ignored in the earlier editions. A brief treatment of the subject is now incorporated in chapter 11 of this book. However, it should be remembered that the main thrust of the argument in the case of the "alternative" trajectory was how mathematics spread into Europe through the intervention of the Islamic scientists. The role of the Maghreb in this dissemination process is, as yet, unclear.

Since the first edition of this book, we are not any closer to gathering further definitive evidence of *direct* transmission of mathematical knowledge to Europe after the Islamic encounter. But direct written evidence is not the only evidence taken into account in establishing transmissions.

A number of historians of mathematics now admit that there were some remarkable mathematical ideas emanating from China, India, and the Islamic world, but many would argue that Europeans in later centuries were unaware of this work and carried out their exploration of new ideas independent of these earlier efforts. It is interesting, in this context, that a strong point of contention has been the somewhat tentative suggestion, in the first edition, of possible transmission of seminal ideas in modern mathematics from India, China, and the Islamic world.

With hindsight, the author regrets not having expressed these ideas of transmission more clearly and forcefully. The area of greatest promise of new discoveries relates to the transmission from the Islamic world to Europe. Cumulating circumstantial evidence now strongly supports the thesis that significant ideas—notably in algebra, trigonometry, non-Euclidean geometry, number theory, and combinatorics—were transmitted from Islamdom to Christendom, through Arabic and Hebrew texts, to contribute toward the development of modern mathematics. The absence of a tradition of attribution during this period makes our task of tracing the transmission more difficult. However, there are some intriguing possibilities raised by Katz (2007) worth further exploration.

In twelfth-century Muslim Spain, a book by Jabir ibn Aflah on spherical geometry was translated into Latin and Hebrew. A method of solving triangles on the surface of a sphere, discussed in Jabir's book, appeared in a book by Regiomontanus (1436–1476), an influential European mathematician of his time. The Italian algebraist Cardona (1501–1576) noted the close similarity between the passages in the two books. The trigonometry contained in Jabir's book also makes an appearance in Copernicus's seminal text, De Revolutionibus. In the same text, according to Saliba (1994, 2007), Copernicus resolved the problem of the "equant" with the help of two mathematical theorems discovered by the Islamic scientists Nasir al-Din al-Tusi (1201–1274) and Muayyad al-Din al-Urdi (d. 1266) and named after them as the Tusi Couple and the Urdi Lemma respectively. In chapter 11 (Islamic mathematics), other examples are quoted, notably the influence of Nasir al-Din al-Tusi's geometry on the Italian Girolamo Saccheri's (1667–1733) attempt to prove the parallel postulate, and Thabit ibn Qurra's (826–901) formula for finding amicable numbers, which was proposed and used by the French mathematician Pierre Fermat (1601–1665), who may have come across a Latin translation of an Arabic or Hebrew text from the twelfth to the fourteenth centuries containing ibn Qurra's formula.

Katz also mentions another example of a possible transmission, starting from a Hebrew text of Levi ben Gerson (1288–1344), containing the basic formulas for finding permutations and combinations that reappear in one of Cardano's manuscripts, and then an almost carbon copy in Mersenne's (1588–1648) classic book on music theory. It is interesting, in this context, that almost fifteen hundred years before Mersenne's book, the basic formulas are found in Indian mathematics through the work of the Jaina school (c. 300 BC), with the link between combinatorics and music theory explored by Pingala around the same time.* A discussion of the Indian work is found in chapter 8.

In our discussion of the transmission thesis in chapter 1, as exemplified by the second and third (or alternative) trajectories, we have to an extent subscribed to a view that Saliba (2007, pp. 3–25) describes as a "classical narrative." The narrative starts with the assumption that initially the Islamic civilization was a desert civilization, which began to develop its scientific thought when it came into contact with other, more ancient civilizations— mainly the Graeco-Hellenistic in the West and to a lesser extent the Persian (and by extension the Indian) in the East. An active appropriation of the sciences of these cultures took place, and translations of many texts emanating from these cultures were undertaken during the early period of the Abbasid caliphates (AD 750–900), which would usher in a golden period of Islamic mathematics and science. But from around the eleventh to the twelfth centuries, this great enterprise, jeopardized externally by the Mongol threat and internally by the conservative religious forces as exemplified by the work of the Islamic theologian al-Ghazali (d. 1111), was gradually abandoned. But before it was lost forever, Europe woke up from its slumber and set in motion a translation movement that resulted in the start of the scientific revolution there. The European dependence on Islamic science was, however, short-lived, for soon European Renaissance thinkers found a way of bypassing the Arabs and reconnecting themselves with their Graeco-Roman legacy, where (according to the classical narrative) all science and philosophy began. Saliba's detailed critique, which points to

---

*Even more remarkable is the little-known mathematical debt owed to the ancient Chinese for the "single most important development in Western European music in the last 400 years: the invention of Equal Temperament" (Goodall 2000, p. 111). A short discussion of the Chinese work on the subject will be found in chapter 7. I am grateful to Keekok Lee for bringing this to my attention.

both the inconsistencies and distortions that underpin this narrative, is a useful corrective to oversimplification of what is a complex story.

An area signposted as worthy of further study in the second edition is the possible transmission westward of the remarkable work in Kerala, South India. A connecting link that needs further exploration is the one through the Islamic world. We know that the works of ibn al-Haytham (fl. AD 1000), the great Islamic scientist, particularly on geometric series (discussed in chapter 11), was studied at certain *madrassahs* (Muslim "schools") in close proximity to the epicenters of the Kerala School of Mathematics and Astronomy, notably at Ponnani and near present-day Kannur. So was it possible that, through the medium of the Islamic scholars, some of this work moved west? Or even, possibly, that influences (technological or otherwise) from Islamic astronomy reached Kerala, notably in the work of Paramesvara (c. 1360–1450) in his long-term observations of eclipses? An investigation to establish these transmissions would require an extensive study at the various centers of Muslim learning in Kerala and elsewhere.

However, a more important and better-known connection is the role of the Jesuits and the Portuguese: there is evidence that Matteo Ricci, the Jesuit astronomer and mathematician who is generally credited with bringing European sciences to the Chinese, spent almost two years in Cochin, South India, after being ordained in Goa in 1580. During that time he was in correspondence with the rector of Collegio Romano, the primary institution for the education of those who wished to become Jesuits. The Jesuits of that time were not merely priests but also scholars who were very knowledgeable in science and mathematics. In fact, if you wanted to be trained as a mathematician in Italy at that time, you could not do better than go to a Jesuit seminary. For a number of Jesuits who followed Ricci, Cochin was a staging post on the way to China. Cochin was only seventy kilometers from the largest repository of astronomical manuscripts in Trissur, from where, two hundred years later, Whish and Heyne, two of the earliest Europeans who reported on the work in Kerala, obtained their manuscripts. The Jesuits were expected to submit regularly a report to their headquarters in Rome, and it is a reasonable conjecture that some of the reports may have contained appendixes of a technical nature that would then be passed on by Rome to those who would understand them, including notable mathematicians. Materials gathered by the Jesuits were scattered all over Europe: at Pisa, where Galileo, Cavalieri, and Wallis spent time; at Padua, where James Gregory was engaged on mathematical studies; and at

Paris, where Mersenne, through his correspondence with Galileo, Wallis, Fermat, and Pascal, acted as an agent for the transmission of mathematical ideas. We will examine some of these links more closely in chapter 10.

Now, in studying the mathematics of any ancient culture, three related questions arise:

1. *What* was the content of the mathematics known to that culture?

2. *How* was that mathematics thought about and discussed?

3. *Who* was doing the mathematics?

In relation to (1) and (2) there are further subquestions relating to how the information about the mathematics is available to us. If the information requires translation from another language, long dead and esoteric, such a translation can be either "user-friendly," in that the purpose is to make the ancient mathematics familiar and easily comprehensible, *or* "alienating," whereby the translation that results is "literal" in that it tries to be as faithful as possible to the structure, vocabulary, and syntax of the original. It is evident from the chapters that follow it is content rather than presentation that is seen as important in this book, so that the texts may be interpreted legitimately in our own terms. Indeed, the content may often be seen as independent of presentation. However, this approach has its dangers in that we may unintentionally distort ancient mathematical concepts and procedures by imposing modern concepts and symbolic packages. It is hoped that a consciousness of this danger may help to make such qualifications as necessary in the discussions that follow.*

In preparing this edition, I have been particularly fortunate in receiving constructive comments from a number of scholars who have, depending on their expertise, read different sections of the book. In particular, I wish to thank Glen Van Brummelen, Jöran Friberg, Takao Hayashi, Victor Katz, and four other reviewers for their detailed and careful scrutiny of the man-

---

*This difference in approach highlighted here may be summed up as "historicism" versus "presentism." The former asserts that works from the past can be interpreted only in the context of that past culture, while the latter attempts to understand such works on our own terms in the present. Whether one inclines to historicism or presentism in one's own interpretation depends to an extent on whether one sees present-day mathematics as having evolved from older mathematics, so that the older mathematics has been absorbed into our own, or whether one sees different mathematical traditions as being to a significant extent incommensurable.

uscript. I greatly appreciate the time and effort they put into their work. Glen Van Brummelen also provided most generously a copy of his manuscript, now published as *The Mathematics of the Heavens and the Earth: The Early History of Trigonometry*, which has helped me immeasurably in composing the sections on trigonometry in various chapters of the book. Finally, at the stage of preparing the manuscript for publication, Vickie Kearn's role has been invaluable. Not only has she provided meticulous editorial assistance in spotting ambiguities, omissions, and inconsistencies in the text, but at various places she also suggested changes that have significantly improved the presentation. More important than anything else, she has been extremely supportive in guiding the product to its final stages. However, I assume all responsibility for any weaknesses in the final product. Finally, there is one person who has consistently encouraged me in all my endeavors and provided the necessary confidence and fortitude to complete the task, and that has been my wife, Leela. She has also been a wonderful traveling companion on the trips we have made together to different parts of the world to promote the ideas in this book. To her, my deepest gratitude and love.

George Gheverghese Joseph
University of Manchester
JULY 2009

Preface to the First Edition

In 1987 I visited the birthplace of the Indian mathematician Srinivasa Ramanujan. Exactly a hundred years had passed since his birth. Ramanujan was born in a small town called Erode in southern India. At his death, aged thirty-two, he was recognized by some as a natural genius, the like of whom could be found only by going back two centuries to Euler and Gauss. Among his contemporaries, particularly his close collaborator G. H. Hardy, there was a sense of disappointment—the feeling that Ramanujan's ignorance of modern mathematics, his strange ways of "doing" mathematics, and his premature death had diminished his achievements and therefore his influence on the future of the subject. Yet today few mathematicians would accept this assessment. In 1976, George Andrews, an American mathematician, was rummaging through some of Ramanujan's papers in a library at Cambridge University and came across 130 pages of scrap paper filled with notes representing Ramanujan's work during the last year of his life in Madras. This is what Richard Askey, a collaborator of Andrews, had to say about what has come to be known as Ramanujan's "Lost Notebook":

> The work of that one year, while he was dying [and obviously in considerable pain a lot of the time, according to his wife], was the equivalent of a lifetime of work of a very great mathematician. What he accomplished was unbelievable. If it were in a novel, nobody would believe it. (My comment in brackets.)

The riches contained in the "Lost Notebook" and his earlier works are being mined with increasing success and excitement by mathematicians today. He is believed to have contributed to the creation of one of the most revolutionary concepts of recent theoretical physics—superstring theory in cosmology. A 1914 paper of Ramanujan, "Modular Equations and Approximations to π," was used to program a computer some years ago to evaluate π to a level of accuracy (to millions of digits) never attained previously. But one should not make excessive claims for the practical applications of Ramanujan's work. As Kanigel (1991, pp. 349–50) states,

> What makes Ramanujan's work so seductive is not the prospect of use in the solution of real-world problems, but its richness, beauty, mystery— its sheer mathematical loveliness.

However, for me the most intriguing aspect of Ramanujan's mathematics work remains his method. Here was someone poorly educated in modern mathematics and isolated for most of his life from work going on at the frontiers of the subject, yet who produced work of a quality and durability that is increasingly tending to overshadow that of some of his more prominent contemporaries, including Hardy. Ramanujan's style of doing mathematics was very different from that of the conventional mathematician trained in the deductive axiomatic method of proof. From the accounts of his wife and close associates he made extensive use of a slate on which he was always jotting down and erasing what his wife described as "sums," and then transferring some of the final results into his Notebook when he was satisfied with his conclusions. He felt no strong compulsion to prove that the results were true—what mattered were the results themselves. This has provided a growing number of mathematicians with a singular task: to prove the results that Ramanujan simply stated. And from the endeavors of these mathematicians have emerged a number of subdisciplines, promoting gatherings and collaborations among their practitioners, whose approach stands in stark contrast to that taken by the original inspirer.

In writing this book, I found the life and work of Ramanujan instructive because it raises a number of interesting questions. First (and this is a question that is rarely addressed by historians of mathematics, one for which there can in any case never be a fully satisfactory answer), how far did cultural influences determine Ramanujan's choice of subjects or his methods? It is interesting in this context that Ramanujan came from the Ayyangar Brahmans of Tamil Nadu in southern India, a group that enjoyed a high social status for their traditional learning and religious observances. Given this background, Ramanujan's tendency to credit his discoveries to the intervention of the family goddess, Namagiri, is understandable, though it must have been a source of embarrassment to some of his admirers, both in India and in the West. But it is perfectly consistent with a culture that saw mathematics in part as an instrument of divine intervention and astrological prediction. The Western mathematical temperament finds it difficult to come to terms with the speculative, extrarational, and intuitive elements in Ramanujan's makeup.

At another level, the example of Ramanujan is a sure indication that the highest level of mathematical achievement is well within the scope of those educated and brought up in traditions and environments far removed from Western society. However, a second question, an interesting and indeed a central one, is raised by Ramanujan's work: is it possible to identify any features in his own culture that were conducive to creative work in mathematics? Any attempt to answer this question should delve deeply into the role of Ayyangars as custodians of traditional knowledge of astronomy and mathematics. Ramanujan's mother was a well-known local astrologer, and it is likely that his first exposure to mathematics, and in particular his special interest in the theory of numbers, came about through his mother's astrology. Mathematics and numbers had a special significance within the Brahmanical tradition as extrarational instruments for controlling fate and nature.

Ramanujan's work also raises questions about what constitutes mathematics. Is there a need to conform to a particular method of presentation before something is recognizable as mathematics? His notebooks contain many jottings that do not conform to a conventional view of mathematical results, since there is no attempt at any demonstration or examination of the theory behind these results. Yet a number of mathematicians not only have found these jottings sufficiently worthwhile to devote years of their time to proving theorems Ramanujan knew to be true, but may even have gained more from the very act of deriving the formulas than the knowledge of the formulas themselves. This is quite consistent with both the Indian and Chinese traditions, where great mathematicians merely state the results, leaving their students to provide oral demonstrations or written commentaries. The students are thus encouraged to allow both their critical and their creative faculties to develop at the same time.

An author is not expected to explain why he writes a book. But the motives are often quite revealing. If I am to explain why I have spent the last three years on this book, I would think my being a product of four different heritages is relevant. I was born in Kerala, southern India, and spent the first nine years of my life there and in the town of Madurai, the cultural capital of the neighboring state of Tamil Nadu. My early awareness of the sheer diversity of Indian culture was helped by living close to the famous Meenakshi Temple at Madurai, a great center of pilgrimage, dance, music, and religious festivals. This exposure during my formative years contributed to my Indian heritage. I come from a family of Syrian Orthodox Christians that traces its descent directly to one of the families (the Sankarapuri) who were converted by Christ's disciple St. Thomas in about AD 50. This is my Middle Eastern Christian heritage. My family moved to Mombasa in Kenya, where I was brought up in that rich mixture of African and Arab influences that makes the distinctive Swahili culture. My African heritage is a result of the time I spent there, first at school and then at work, both in Mombasa and in the neighboring country of Tanzania. The period I have spent in Britain, at the University of Leicester, where I did my first degree, and at the University of Manchester, where I continued my postgraduate studies and subsequently worked, now accounts for more than half my life. This is my Western heritage. To keep a balance between my four heritages and not allow any one to take over permanently is important to me. Hence my travels abroad, which have taken me to East and Central Africa, to India, to Papua New Guinea, and to South and Southeast Asia. And hence, in a different way, the driving passion behind this book, which emphasizes the global nature of mathematical pursuits and creations.

In writing this book, I am indebted to many who have over the years patiently and skillfully translated and interpreted the original sources of mathematics from different cultures so that they are now more accessible and comprehensible. I owe them more than I can acknowledge merely through entries in the bibliography at the back of this book. They have often had to work in environments that are not particularly sympathetic to their efforts and have rarely received sufficient academic recognition. In a number of cases their attempts at collecting and transliterating ancient manuscripts show a desperate sense of urgency, as the storage and preservation of these documents often leave much to be desired.

During the time I have been working on this book, several people have given me advice, constructive criticism, and encouragement. Burjor Avari and I have shared a long and close association, which has taken the form, among other things, of a study of the nature and consequences of a Eurocentric view of the history of knowledge. Our collaboration in this area is clearly reflected in some of the ideas found in the first chapter. In particular, the historic backgrounds to a number of chapters have benefited from his criticisms. David Nelson read the whole manuscript carefully and suggested a number of changes to improve the clarity and balance of the book. I found his comments so useful and persuasive that I have tried in almost all cases to respond to them. Even so, I am conscious of having fallen short of the thoroughness that his detailed comments deserve. To Bill Farebrother I am grateful for having gone through the manuscript at various stages, making useful criticisms of the style and mathematical presentation. I should also like to acknowledge my debt to others in the Department of Econometrics and Social Statistics, University of Manchester, who not only tolerated my project (as removed as it was from the usual concerns of the department) but in some cases went through chapters and provided constructive responses. Finally, at the stage of preparing the manuscript for publication, John Woodruff's role has been invaluable. Not only has he provided meticulous editorial assistance in spotting ambiguities, omissions, and inconsistencies in the text and bibliography, but at various places he has suggested changes that have significantly improved the presentations. It is appropriate, given my insufficient response to some of the advice offered me, that I exclude all those mentioned above from responsibility for any errors of fact and interpretation present in this book.

George Gheverghese Joseph
University of Manchester
1991

The Crest of
the Peacock

Chapter One

The History of Mathematics: Alternative Perspectives

A Justification for This Book

An interest in history marks us for life. How we see ourselves and others is shaped by the history we absorb, not only in the classroom but also from the Internet, films, newspapers, television programs, novels, and even strip cartoons. From the time we first become aware of the past, it can fire our imagination and excite our curiosity: we ask questions and then seek answers from history. As our knowledge develops, differences in historical perspectives emerge. And, to the extent that different views of the past affect our perception of ourselves and of the outside world, history becomes an important point of reference in understanding the clash of cultures and of ideas. Not surprisingly, rulers throughout history have recognized that to control the past is to master the present and thereby consolidate their power.

During the last four hundred years, Europe and its cultural dependencies[1] have played a dominant role in world affairs. This is all too often reflected in the character of some of the historical writing produced by Europeans in the past. Where other people appear, they do so in a transitory fashion whenever Europe has chanced in their direction. Thus the history of the Africans or the indigenous peoples of the Americas often appears to begin only after their encounter with Europe.

An important aspect of this Eurocentric approach to history is the manner in which the history and potentialities of non-European societies were represented, particularly with respect to their creation and development of science and technology. The progress of Europe during the last four hundred years is often inextricably—or even causally—linked with the rapid growth of science and technology during the period. In the minds of some, scientific progress becomes a uniquely European phenomenon, which can be emulated by other nations only if they follow a specifically European path of scientific and social development.

Such a representation of societies outside the European cultural milieu raises a number of issues that are worth exploring, however briefly. First, these societies, many of them still in the grip of an intellectual dependence that is the legacy of European political domination, should ask themselves some questions. Was their indigenous scientific and technological base innovative and self-sufficient during their precolonial period? Case studies of India, China, and parts of Africa, contained, for example, in the work of Dharampal (1971), Needham (1954), and Van Sertima (1983) and summarized by Teresi (2002), seem to indicate the existence of scientific creativity and technological achievements long before the incursions of Europe into these areas. If this is so, we need to understand the dynamics of precolonial science and technology in these and other societies and to identify the material conditions that gave rise to these developments. This is essential if we are to see why modern science did not develop in these societies, only in Europe, and to find meaningful ways of adapting to present-day requirements the indigenous and technological forms that still remain.[2]

Second, there is the wider issue of who "makes" science and technology. In a material and nonelitist sense, each society, impelled by the pressures and demands of its environment, has found it necessary to create a scientific base to cater to its material requirements. Perceptions of what constitute the particular requirements of a society would vary according to time and place, but it would be wrong to argue that the capacity to "make" science and technology is a prerogative of one culture alone.

Third, if one attributes all significant historical developments in science and technology to Europe, then the rest of the world can impinge only marginally, either as an unchanging residual experience to be contrasted with the dynamism and creativity of Europe, or as a rationale for the creation of academic disciplines congealed in subjects such as development studies, anthropology, and oriental studies. These subjects in turn served as the basis from which more elaborate Eurocentric theories of social development and history were developed and tested.[3]

One of the more heartening aspects of academic research in the last four or five decades is that the shaky foundations of these "adjunct" disciplines are being increasingly exposed by scholars, a number of whom originate from countries that provide the subject matter of these disciplines. "Subversive" analyses aimed at nothing less than the unpackaging of prevailing Eurocentric paradigms became the major preoccupation of many of these scholars. Syed Husain Alatas (1976) studied intellectual dependence and imitative thinking among social scientists in developing countries. The growing movement toward promoting a form of indigenous anthropology that sees its primary task as questioning, redefining, and if necessary rejecting particular concepts that grew out of colonial experience in Western anthropology is thoroughly examined by Fahim (1982). Edward Said (1978) has brilliantly described the motives and methods of the so-called orientalists who set out to construct a fictitious entity called "the Orient" and then ascribe to it qualities that are a mixture of the exotic, the mysterious, and the otherworldly. The rationale for such constructs is being examined in terms of the recent history of Europe's relations with the rest of the world.

In a similar vein, and in the earlier editions of this book, it was the intention to show that the standard treatment of the history of non-European mathematics exhibited a deep-rooted historiographical bias in the selection and interpretation of facts, and that mathematical activity outside Europe has as a consequence been ignored, devalued, or distorted. It is interesting in this context that since the first edition of this book there has been a growing recognition of the mathematics outside the European and Greek traditions, especially in the mainstream teaching of the history of mathematics. The Eurocentric argument has shifted its ground and now questions both the nature of the European debt to other mathematical traditions and the existence and quality of proofs and demonstrations in traditions outside Europe. A brief discussion of the shifting ground of Eurocentrism in the history of mathematics is found in the preface to this edition.[4]

The Development of Mathematical Knowledge

A concise and meaningful definition of mathematics is difficult. In the context of this book, the following aspects of the subject are highlighted. Modern mathematics has developed into a worldwide language with a particular kind of logical structure. It contains a body of knowledge relating to number and space, and prescribes a set of methods for reaching conclusions about the physical world. And it is an intellectual activity which calls for both intuition and imagination in deriving "proofs" and reaching conclusions. Often it rewards the creator with a strong sense of aesthetic satisfaction.

The "Classical" Eurocentric Trajectory

Most histories of mathematics that have had a great influence on later work were written in the late nineteenth or early twentieth century. During that period, two contrasting developments were taking place that had an impact on both the content and the balance of these books, especially those produced in Britain and the United States. Exciting discoveries of ancient mathematics on papyri in Egypt and clay tablets in Mesopotamia pushed back the origins of written mathematical records by at least fifteen hundred years. But a far stronger and countervailing influence was the culmination of European domination in the shape of political control of vast tracts of Africa and Asia. Out of this domination arose the ideology of European superiority that permeated a wide range of social and economic activities, with traces to be found in histories of science that emphasized the unique role of Europe in providing the soil and spirit for scientific discovery. The contributions of the colonized peoples were ignored or devalued as part of the rationale for subjugation and dominance. And the development of mathematics before the Greeks—notably in Egypt and Mesopotamia—suffered a similar fate, dismissed as of little importance to the later history of the subject. In his book *Black Athena* (1987), Martin Bernal has shown how respect for ancient Egyptian science and civilization, shared by ancient Greece and pre-nineteenth-century Europe alike, was gradually eroded, leading eventually to a Eurocentric model with Greece as the source and Europe as the inheritor and guardian of the Greek heritage.

Figure 1.1 presents the "classical" Eurocentric view of how mathematics developed over the ages. This development is seen as taking place in two sections, separated by a period of stagnation lasting for over a thousand years: Greece (from about 600 BC to AD 400), and post-Renaissance Europe from the sixteenth century to the present day. The intervening period of inactivity was the "Dark Ages"—a convenient label that expressed both post-Renaissance Europe's prejudices about its immediate past and the intellectual self-confidence of those who saw themselves as the true inheritors of the "Greek miracle" of two thousand years earlier.

Two passages, one by a well-known historian of mathematics writing at the turn of the century and the other by a more recent writer whose books are still referred to on both sides of the Atlantic, show the durability of this Eurocentric view and its imperviousness to new evidence and sources:

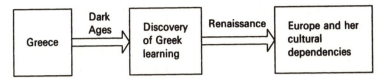

FIGURE 1.1: The "classical" Eurocentric trajectory

The history of mathematics cannot with certainty be traced back to any school or period before that of the Ionian Greeks. (Rouse Ball 1908, p. 1)

[Mathematics] finally secured a new grip on life in the highly congenial soil of Greece and waxed strongly for a short period.... With the decline of Greek civilization the plant remained dormant for a thousand years ... when the plant was transported to Europe proper and once more imbedded in fertile soil. (Kline 1953, pp. 9–10)

The first statement is a reasonable summary of what was popularly known and accepted as the origins of mathematics at that time, except for the neglect of the early Indian mathematics contained in the *Sulbasutras* (The Rules of the Cord), belonging to the period between 800 and 500 BC, which would make it at least as old as the earliest-known Greek mathematics. Thibaut's translations of these works, made around 1875, were known to historians of mathematics at the turn of the century. The mathematics contained in the *Sulbasutras* is discussed in chapter 8.

The second statement, however, ignores a considerable body of research evidence pointing to the development of mathematics in Mesopotamia, Egypt, China, pre-Columbian America, India, and the Islamic world that had come to light in the intervening period. Subsequent chapters will bear testimony to the volume and quality of the mathematics developed in these areas. But in both these quotations mathematics is perceived as an exclusive product of European civilization. And that is the central message of the Eurocentric trajectory depicted in figure 1.1.

This comforting rationale for European dominance became increasingly untenable for a number of reasons. First, there is the full acknowledgment given by the ancient Greeks themselves of the intellectual debt they owed the Egyptians. There are scattered references from Herodotus (c. 450 BC) to Proclus (c. AD 400) of the knowledge acquired from the Egyptians in fields such as astronomy, mathematics, and surveying, while other commentators even considered the priests of Memphis to be the true founders of science.

To Aristotle (c. 350 BC), Egypt was the cradle of mathematics. His teacher, Eudoxus, one of the notable mathematicians of the time, had studied in Egypt before teaching in Greece. Even earlier, Thales (d. 546 BC), the legendary founder of Greek mathematics, and Pythagoras (c. 500 BC), one of the earliest and greatest of Greek mathematicians, were reported to have traveled widely in Egypt and Mesopotamia and learned much of their mathematics from these areas. Some sources even credit Pythagoras with having traveled as far as India in search of knowledge, which could explain some of the parallels between Indian and Pythagorean philosophy and religion.[5]

A second reason why the trajectory depicted in figure 1.1 was found to be wanting arose from the combined efforts of archaeologists, translators, and interpreters, who between them unearthed evidence of a high level of mathematics practiced in Mesopotamia and in Egypt at the beginning of the second millennium, providing further confirmation of Greek reports. In particular, the Mesopotamians had invented a place-value number system, knew different methods of solving what today would be described as quadratic equations (methods that would not be improved upon until the sixteenth century AD), and understood (but had not proved) the relationship between the sides of a right-angled triangle that came to be known as the Pythagorean theorem.[6] Indeed, as we shall see in later chapters, this theorem was stated and demonstrated in different forms all over the world.

A four-thousand-year-old clay tablet, kept in a Berlin museum, gives the value of $n^3 + n^2$ for $n = 1, 2, \ldots, 10, 20, 30, 40, 50$, from which it has been surmised that the Mesopotamians may have used these values in solving cubic equations after reducing them to the form $x^3 + x^2 = c$. A remarkable solution in Egyptian geometry found in the Moscow Papyrus from the Middle Kingdom (c. 2000–1800 BC) follows from the correct use of the formula for the volume of a truncated square pyramid. These examples and other milestones will be discussed in the relevant chapters of this book.

The neglect of the Islamic contribution to the development of European intellectual life in general and mathematics in particular is another serious drawback of the "classical" view. The course of European cultural history and the history of European thought are inseparably tied up with the activities of Islamic scholars during the Middle Ages and their seminal contributions to mathematics, the natural sciences, medicine, and philosophy.[7]

In particular, we owe to the Islamic world in the field of mathematics the bringing together of the technique of measurement, evolved from its Egyptian roots to its final form in the hands of the Alexandrians, and the remarkable instrument of computation (our number system) that originated in India. These strands were supplemented by a systematic and consistent language of calculation that came to be known by its Arabic name, algebra. An acknowledgment of this debt in more recent books contrasts sharply with the neglect of other Islamic contributions to science.[8]

Finally, in discussing the Greek contribution, there is a need to recognize the differences between the Classical period of Greek civilization (i.e., from about 600 to 300 BC) and the post-Alexandrian period (i.e., from about 300 BC to AD 400). In early European scholarship, the Greeks of the ancient world were perceived as an ethnically homogeneous group, originating from areas that were mainly within the geographical boundaries of present-day Greece. It was part of the Eurocentric mythology that from the mainland of Europe had emerged a group of people who had created, virtually out of nothing, the most impressive civilization of ancient times. And from that civilization had emerged not only the cherished institutions of present-day Western culture but also the mainspring of modern science. The reality, however, is different and more complex. The term "Greek," when applied to times before the appearance of Alexander (356–323 BC), really refers to a number of independent city-states, often at war with one another but exhibiting close ethnic or cultural affinities and, above all, sharing a common language. The conquests of Alexander changed the situation dramatically, for at his death his empire was divided among his generals, who established separate dynasties. The two notable dynasties from the point of view of mathematics were the Ptolemaic dynasty of Egypt and the Seleucid dynasty, which ruled over territories that included the earlier sites of the Mesopotamian civilization. The most famous center of learning and trade became Alexandria in Egypt, established in 332 BC and named after the conqueror. From its foundation, one of its most striking features was its cosmopolitanism—part Egyptian, part Greek, with a liberal sprinkling of Jews, Persians, Phoenicians, and Babylonians, and even attracting scholars and traders from as far away as India. A lively contact was maintained with the Seleucid dynasty. Alexandria thus became the meeting place for ideas and different traditions. The character of Greek mathematics began to change slowly, mainly as a result of continuing cross-fertilization between different mathematical traditions, notably the algebraic and empirical basis of Mesopotamian and Egyptian mathematics interacting with the geometric and antiempirical traditions of Greek mathematics. And from this mixture came some of the greatest mathematicians of antiquity, notably Archimedes and Diophantus. It is therefore important to recognize the Alexandrian dimension to Greek mathematics while noting that Greek intellectual and cultural traditions served as the main inspiration and the Greek language as the medium of instruction and writing in Alexandria. In a later chapter, based on some innovative work of Friberg (2005, 2007), we will examine the close and hitherto unexamined links between Egyptian, Mesopotamian, and Greek mathematics.

A Modified Eurocentric Trajectory

Figure 1.2 takes on board some of the objections raised about the "classical" Eurocentric trajectory. The figure acknowledges that there is growing awareness of the existence of mathematics before the Greeks, and of their debt to earlier mathematical traditions, notably those of Mesopotamia and Egypt. But this awareness was until recently tempered by a dismissive rejection of their importance in relation to Greek mathematics: the "scrawling of children just learning to write as opposed to great literature" (Kline 1962, p. 14).

The differences in character of the Greek contribution before and after Alexander are also recognized to a limited extent in figure 1.2 by the separation of Greece from the Hellenistic world (in which the Ptolemaic and Seleucid dynasties became the crucial instruments of mathematical creation). There is also some acknowledgment of the "Arabs" but mainly as custodians of Greek learning during the Dark Ages in Europe. The role of the Islamic world as transmitter and creator of knowledge is often ignored; so are the contributions of other civilizations—notably China and India—which have been perceived either as borrowers from Greek sources or as having made only minor contributions that played an insignificant role in mainstream mathematical development (i.e., the development eventually culminating in modern mathematics).

Figure 1.2 is therefore still a flawed representation of how mathematics developed: it contains a series of biases and remains quite impervious to new evidence and arguments. Until a couple of decades ago, and with minor modifications, it was the model to which a number of books on the history of mathematics conformed. But this has changed even during the twenty-odd years that this book has been in print. "Mainsteam" histories of mathematics are casting a wider net by seriously considering the contributions not only

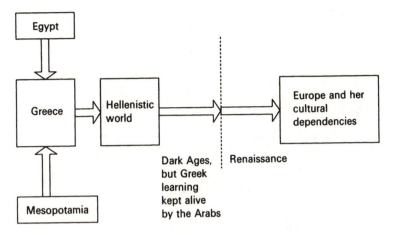

FIGURE 1.2: A modified Eurocentric trajectory

of ancient Egyptian and Mesopotamian civilizations; they are punctilious in incorporating as well the contributions of Chinese, Indian, and Islamic civilizations. The recent sourcebook for the histories of mathematics of the five civilizations edited by Katz (2007) is a testimony to this change.

It is interesting that a similar Eurocentric bias had existed in other disciplines as well: for example, diffusion theories in anthropology and social geography implied that "civilization" has spread from the center ("greater" Europe) to the periphery (the rest of the world). And the theories of globalization or evolution developed in recent years within some Marxist and neo-Marxist frameworks were characterized by a similar type of Eurocentrism. In all such conceptual schemes, the development of Europe is seen as a precedent for the way in which the rest of the world will follow—a trajectory whose spirit is not dissimilar to the one suggested by figures 1.1 and 1.2.

An Alternative Trajectory for the Dark Ages

If we are to construct an unbiased alternative to figures 1.1 and 1.2, our guiding principle should be to recognize that different cultures in different periods of history have contributed to the world's stock of mathematical knowledge. Figure 1.3 presents such a trajectory of mathematical development but confines itself to the period between the fifth and fifteenth centuries AD—the period represented by the arrow labeled in figures 1.1 and 1.2 as the "Dark Ages" in Europe. The choice of this trajectory as an illustration is deliberate: it serves to highlight the variety of mathematical

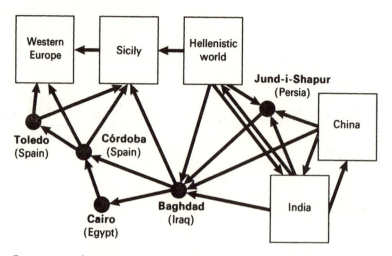

Figure 1.3: An alternative trajectory for the "Dark Ages"

activity and exchange between a number of cultural areas that went on while Europe was in deep slumber. A trajectory for the fifteenth century onward would show that mathematical cross-fertilization and creativity were more or less confined to countries within Europe until the emergence of the truly international character of modern mathematics during the twentieth century.

The role of the Islamic civilization is brought out in figure 1.3. Scientific knowledge that originated in India, China, and the Hellenistic world was sought out by Islamic scholars and then translated, refined, synthesized, and augmented at different centers of learning, starting at Jund-i-Shapur[9] in Persia around the sixth century (even before the coming of Islam) and then moving to Baghdad, Cairo, and finally to Toledo and Córdoba in Spain, from where this knowledge spread into western Europe. Considerable resources were made available to the scholars through the benevolent patronage of the caliphs, the Abbasids (the rulers of the eastern Arab empire, with its capital at Baghdad) and the Umayyads (the rulers of the western Arab empire, with its capital first at Damascus and later at Córdoba).

The role of the Abbasid caliphate was particularly important for the future development of mathematics. The caliphs, notably al-Mansur (754–775), Harun al-Rashid (786–809), and al-Mamun (809–833), were in the forefront of promoting the study of astronomy and mathematics in

Baghdad. Indian scientists were invited to Baghdad. When Plato's Academy was closed in 529, some of its scholars found refuge at Jund-i-Shapur, which a century later became part of the Islamic world. Greek manuscripts from the Byzantine empire, the translations of the Syriac schools of Antioch and Damascus, and the remains of the Alexandrian library in the hands of the Nestorian Christians at Edessa were all eagerly sought by Islamic scholars, aided by the rulers who had control over or access to men and materials from the Byzantine empire, Persia, Egypt, Mesopotamia, and places as far east as India and China.

Caliph al-Mansur built at Baghdad a Bait al-Hikma (House of Wisdom), which contained a large library for the manuscripts that had been collected from various sources; an observatory that became a meeting place of Indian, Babylonian, Hellenistic, and probably Chinese astronomical traditions; and a university where scientific research continued apace.[10] A notable member of the institution, Muhammad ibn Musa al-Khwarizmi (fl. AD 825), wrote two books that were of crucial importance to the future development of mathematics. One of them, the Arabic text of which is extant, is titled *Hisab al-jabr w'al-muqabala* (which may be loosely translated as Calculation by Reunion and Reduction). The title refers to the two main operations in solving equations: "reunion," the transfer of negative terms from one side of the equation to the other, and "reduction," the merging of like terms on the same side into a single term.[11] In the twelfth century the book was translated into Latin under the title *Liber algebrae et almucabola*, thus giving a name to a central area of mathematics. A traditional meaning of the Arabic word *jabr* is "the setting of a broken bone" (and hence "reunion" in the title of al-Khwarizmi's book). Some decades ago it was not an uncommon sight on Spanish streets to come across a sign advertising "Algebrista y Sangrador" (i.e., someone dedicated to setting dislocated bones) at the entrance of barbers' shops.[12]

Al-Khwarizmi wrote a second book, of which only a Latin translation is extant: *Algorithmi de numero indorum*, which explained the Indian number system. While al-Khwarizmi was at pains to point out the Indian origin of this number system, subsequent translations of the book attributed not only the book but the numerals to the author. Hence, in Europe any scheme using these numerals came to be known as an "algorism" or, later, "algorithm" (a corruption of the name al-Khwarizmi) and the numerals themselves as Arabic numerals.

Figure 1.3 shows the importance of two areas of southern Europe in the transmission of mathematical knowledge to western Europe. Spain and

Sicily were the nearest points of contact with Islamic science and had been under Arab hegemony, Córdoba succeeding Cairo as the center of learning during the ninth and tenth centuries. Scholars from different parts of western Europe congregated in Córdoba and Toledo in search of ancient and contemporary knowledge. It is reported that Gherardo of Cremona (c. 1114–1187) went to Toledo, after its recapture by the Christians, in search of Ptolemy's *Almagest*, an astronomical work of great importance produced in Alexandria during the second century AD. He was so taken by the intellectual activity there that he stayed for twenty years, during which time he was reported to have copied or translated eighty-seven manuscripts of Islamic science or Greek classics, which were later disseminated across western Europe. Gherardo was just one of a number of European scholars, including Plato of Tivoli, Adelard of Bath, and Robert of Chester, who flocked to Spain in search of knowledge.[13]

The main message of figure 1.3 is that it is dangerous to characterize the history of mathematics solely in terms of European developments. The darkness that was supposed to have descended over Europe for a thousand years before the illumination that came with the Renaissance did not interrupt mathematical activity elsewhere. Indeed, as we shall see in later chapters, the period saw not only a mathematical renaissance in the Islamic world but also high points of Indian and Chinese mathematics.

Mathematical Signposts and Transmissions across the Ages

Alternative trajectories to the ones shown in figures 1.1 and 1.2 should highlight the following three features of the plurality of mathematical development:

1. The global nature of mathematical pursuits of one kind or another

2. The possibility of independent mathematical development within each cultural tradition followed or not followed by cross-fertilization

3. The crucial importance of diverse transmissions of mathematics across cultures, culminating in the creation of the unified discipline of modern mathematics

However, to construct a feasible diagram we must limit the number of geographical areas of mathematical activity we wish to include. Selection inevitably introduces an element of arbitrariness, for some areas that may merit inclusion are excluded, while certain inclusions may be controversial. Two considerations have influenced the choice of the cultural areas represented in figure 1.4. First, a judgment was made, on the basis of existing evidence, as to which places saw significant developments in mathematics. Second, an assessment of the nature and direction of the transmission of mathematical knowledge also helped to identify the areas of interest.

On the basis of these two criteria, ancient Egypt and Mesopotamia, Greece (and the Hellenistic world), India, China, the Islamic (or "Arab") world, and Europe were selected as being important in the historical development of mathematics. For one cultural area, the application of the two selection criteria produced conflicting results: from existing evidence, the Maya of Central America were isolated from other centers of mathematical activity, yet their achievements in numeration and calendar construction were quite remarkable by any standards. I therefore decided to include the Maya in figure 1.4, and to examine their contributions briefly in chapter 2.

The limited scope of this book and the application of the above criteria make it impossible to examine the mathematical experience of Africa, Korea, and Japan in greater detail. However, chapter 2 contains a discussion of the Ishango bone and the Yoruba numerals, and chapter 3 a detailed examination of Egyptian mathematics, all of which were products of Africa. Further information on the mathematical traditions of Korea and Japan is available in the second of the two chapters on Chinese mathematics (chapter 7), since these traditions were both heavily influenced by China.

Figure 1.4, together with its detailed legend, emphasizes the following features of mathematical activity through the ages:

1. The continuity of mathematical traditions until the last few centuries in most of the selected cultural areas

2. The extent of transmissions between different cultural areas that were geographically or otherwise separated from one another

3. The relative ineffectiveness of cultural barriers (or "filters") in inhibiting the transmission of mathematical knowledge (In a number of other areas of human knowledge, notably in philosophy and the arts, the barriers are often insurmountable unless filters can be devised to make foreign "products" more palatable.)

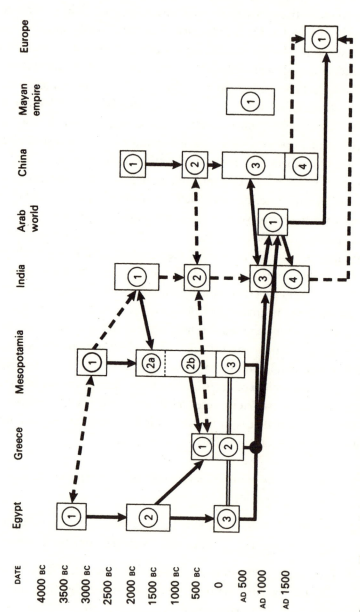

FIGURE 1.4: The spread of mathematical ideas down the ages

EGYPT

1 *Predynastic period:* Appearance of the earliest forms of writing and hieroglyphic numerals

2 *Middle Kingdom to New Kingdom:* Egyptian mathematics mainly contained in the Moscow and Ahmes papyri

3 *Greek and Roman period:* Flowering of mathematics at Alexandria

GREECE

1 *Classical period:* Beginnings of deductive geometry and number theory

2 *Hellenistic period:* Growing synthesis of classical, Egyptian, and Babylonian mathematics

MESOPOTAMIA

1 *Sumerian period:* Beginnings of cuneiform numerals

2a *First Babylonian dynasty:* Early algebra, commerical arithmetic, and geometry from clay tablets

2b *New Babylonian and Persian periods:* Mathematics and astronomy

3 *Seleucid dynasty:* Hellenistic mathematics

INDIA

1 *Harappan period:* Protomathematics from bricks, baths, etc.

2 *Vedic period:* Ritual geometry

3 *Classical period:* Indian numerals, computing algorithms, algebra, and trigonometry

4 *Medieval period:* Kerala mathematics

ARAB WORLD

1 Preservation and synthesis of mathematical traditions from different areas, laying the foundations of modern mathematics

CHINA

1 *River valley civilization:* Beginnings of practical mathematics

2 *Shang and Zhou dynasties:* Rod numerals, *Zhou Bi* (the earliest extant mathematics textbook)

3 *Han to Tang dynasties:* "Arithmetic in Nine Sections" (the most important text in Chinese mathematics)

4 *Song to Ming dynasties:* Golden age of Chinese mathematics

MAYAN EMPIRE

1 Construction of a highly accurate calendar and the development of a place-value number system (base 20) with zero

EUROPE

1 Development of modern mathematics, building on mathematics from other sources

Hellenistic cultural areas (Egypt, Greece, Mesopotamia)

Confirmed lines of transmission (two-way)

(I) Harappan culture *and* first Babylonian dynasty

(II) Han to Tang dynasties *and* classical period (India)

Confirmed lines of transmission (one-way)

(I) Middle Kingdom to New Kingdom (Egypt) *to* classical period (Greece)

(II) New Babylonian and Persian periods *to* classical period (Greece)

(III) Hellenistic cultural areas *to* classical period (India)

(IV) Hellenistic cultural areas *to* Arab world

(V) Classical period (India) *to* Arab world

(VI) Arab world *to* Europe and India

Unconfirmed or tentative lines of transmission (two-way)

(I) Predynastic period (Egypt) *and* Sumerian period (Mesopotamia)

(II) Classical period (Greece) *and* Vedic period (India)

(III) Vedic period (India) *and* Shang and Zhou dynasties (China)

Unconfirmed or tentative lines of transmission (one-way)

(I) Sumerian culture to Harappan culture

(II) Song to Ming dynasties to Europe

(III) Kerala mathematics to Europe

In both Egypt and Mesopotamia there existed well-developed written number systems as early as the third millennium BC. The peculiar character of the Egyptian hieroglyphic numerals led to the creation of special types of algorithms for basic arithmetic operations. Both these developments and subsequent work in the area of algebra and geometry, especially during the period between 1800 and 1600 BC, will form the subject matter of chapter 3. Figure 1.4 brings out another impressive aspect of Egyptian mathematics—the continuity of a tradition for over three thousand years, culminating in the great period of Alexandrian mathematics around the beginning of the Christian era. We shall not examine the content and personalities of this mature phase of Egyptian mathematics in any detail, since its coverage in standard histories of mathematics is more than adequate. There is, however, a widespread tendency in many of these texts to view Alexandrian mathematics as a mere extension of Greek mathematics, in spite of the distinctive character of the mathematics of Archimedes, Heron, Diophantus, and Pappus, to mention a few notable names of the Alexandrian period.

The other early contributor to mathematics was the civilization that grew around the twin rivers, the Tigris and the Euphrates, in Mesopotamia. There mathematical activity flourished, given impetus by the establishment of a place-value sexagesimal (i.e., base 60) system of numerals, which must surely rank as one of the most significant developments in the history of mathematics. However, the golden period of mathematics in this area (or at least the period for which considerable written evidence exists) came during the First Babylonian period (c. 1800–1600 BC), which saw not only the introduction of further refinements to the existing numeral system but also the development of an algebra more advanced than that in use in Egypt. The period is so important that the mathematics that developed in Mesopotamia is often simply referred to as Babylonian mathematics. As with Egypt, the next period of significant advance followed Alexander's conquest and the establishment of the Seleucid dynasty. Babylonian mathematics (a term that will be used interchangeably with Mesopotamian hereafter to describe the mathematics of this cultural area) is discussed in chapter 4.

There is growing evidence of mathematical links between Egypt and Mesopotamia before the Hellenistic period, which we would expect, given their proximity and the records we have of their economic and political contacts. Earlier, Parker (1972) had examined the evidence for a spread of

Mesopotamian algebra and geometry to Egypt. He pointed out that certain parallel developments in both geometry and algebra provided at least some support for links between the two cultural areas. This has now been reinforced by Friberg (2005), who examined "Egyptian mathematics against an up-to-date background in the history of Mesopotamian mathematics." We will discuss Friberg's work in greater detail in chapter 5. However, given that there is more evidence than hitherto believed, we represent the contacts between Egypt and Mesopotamia by a two-headed arrow in figure 1.4.

There is also evidence of the great debt that Greece owed to Egypt and Mesopotamia for its earlier mathematics and astronomy. We have mentioned the acknowledgment of this debt by the Greeks themselves, who believed that mathematics originated in Egypt. The travels of the early Greek mathematicians such as Thales, Pythagoras, and Eudoxus to Egypt and Mesopotamia in search of knowledge have been attested to both by their contemporaries and by later historians writing on the period. The period of greatest Egyptian influence on the Greeks may have been the first half of the first millennium BC. The Greek colonies scattered across the Mediterranean provided a wide channel of interchange. It is at the time of their heyday that we hear of Anaximander of Miletus (610–546 BC) introducing the gnomon (a geometric shape of both mathematical and astronomical significance)[14] from Babylon. During the same period, contacts with the Greeks were maintained through the campaigns of the Assyrian king Sargon II (722–705 BC), and later through Ashurbanipal's occupation of Egypt and his meeting with Gyges of Lydia toward the middle of the seventh century BC. Even when Assyria ceased to exist, the Jewish captivity played a significant part in disseminating Babylonian learning. This was followed by the Persian invasion of Greece at the beginning of the fifth century and the final defeat of the Persians at the end of the fifth century. Thus continuous contacts were maintained throughout a period in which Greek mathematics was still in its infancy, as the foundations were being laid for the flowering of Greek creativity in a couple of centuries. In the next five hundred years, the pupil would learn and develop sufficiently to teach the teachers.

Adding to these historical conjectures, there is now stronger evidence of links between the mathematical traditions of Egypt, Mesopotamia, and Greece. In a recent book Friberg (2007b) has argued as a sequel to his earlier thesis (Friberg 2005) of "unexpected links between Egyptian and Babylonian mathematics" that there are "amazing traces of a Babylonian origin in Greek mathematics."[15] These "traces" (discussed in chapter 5) are found in the fact that several of the famous Greek mathematicians showed an easy familiarity with what Friberg describes as Babylonian "metric algebra," that is, a characteristic approach that combines geometry, metrology, and the solution of quadratic equations.

The transmissions to Greece from the two areas are shown in figure 1.4 by the arrows from 2 in Egypt and 2b in Mesopotamia to 1 in Greece. All three areas then became part of the Hellenistic world, and during the period between the third century BC and the third century AD, and partly due to the interaction between the three mathematical traditions, there emerged one of the most creative periods in mathematics. We usually associate this period with names such as Euclid, Archimedes, Apollonius and Diophantus. But if Friberg's thesis is sustained, there was a 'non-Euclidean lower level' of mathematics present in these traditions. These links are represented by the double lines between 3 in Egypt, 2 in Greece and 3 in Mesopotamia.

The geographical location of India made it throughout history an important meeting place of nations and cultures. This enabled India from the very beginning to play an important role in the transmission and diffusion of ideas. The traffic was often two-way, with Indian ideas and achievements traveling abroad as easily as those from outside entered. Archaeological evidence shows both cultural and commercial contacts between Mesopotamia and the Indus Valley. While there is no direct evidence of mathematical exchange between the two cultural areas, certain astronomical calculations of the longest and shortest day included in the *Vedanga Jyotisa*, the oldest extant Indian astronomical/astrological text, as well as the list of twenty-eight *nakshatras* found in the early Vedic texts, have close parallels with those used in Mesopotamia. And hence the tentative link, shown by broken lines in figure 1.4, between 1 in Mesopotamia and 1 in India.[16]

The relative seclusion that India had enjoyed for centuries was broken by the invasion of the Persians under Darius around 513 BC. In the ensuing six centuries, except for a century and a half of security under the Mauryan dynasty, India was subjected to incursions by the Greeks, the Sakas, the Pahlavas, and the Kusanas. Despite the turbulence, the period offered an opportunity for a close and productive contact between India and the West. Beginning with the appearance of the vast Persian empire, which touched Greece at one extremity and India at the other, tributes from Greece and from the frontier hills of India found their way to the same imperial treasure houses at Ecbatana or Susa. Soldiers from Mesopotamia, the Greek cities of Asia Minor, and India served in the same armies. The word *indoi* for Indians began to appear in Greek literature. Certain interesting parallels between Indian and Pythagorean philosophy have already been pointed out. Indeed, according to some Greek sources, Pythagoras had ventured as far afield as India in his search for knowledge.

By the time Ptolemaic Egypt and Rome's Eastern empire had established themselves just before the beginning of the Christian era, Indian civilization was already well developed, having founded three great religions—Hinduism, Buddhism, and Jainism—and expressed in writing some subtle currents of religious thought and speculation as well as fundamental theories in science and medicine. There are scattered references to Indian science in literary sources from countries to the west of India after the time of Alexander. The Greeks had a high regard for Indian "gymnosophists" (i.e., philosophers) and Indian medicine. Indeed, there are various expressions of nervousness about the Indian use of poison in warfare. In a letter to his pupil Alexander in India, Aristotle warns of the danger posed by intimacy with a "poison-maiden," who had been fed on poison from her infancy so that she could kill merely by her embrace!

There is little doubt that the Mesopotamian influence on Indian astronomy continued into the Hellenistic period, when the astronomy and mathematics of the Ptolemaic and Seleucid dynasties became important forces in Indian science, readily detectable in the corpus of astronomical works known as *Siddhantas*, written around the beginning of the Christian era. Evidence of such contacts (especially in the field of medicine) has been found in places such as Jund-i-Shapur in Persia dating from between AD 300 and 600. As mentioned earlier, Jund-i-Shapur was an important meeting place of scholars from a number of different areas, including Indians and, later, Greeks who sought refuge there with the demise of Alexandria as a center of learning and the closure of Plato's Academy. All such contacts are shown in figure 1.4 by lines linking 2 in India to 1 in Greece and 3 in India to the Hellenistic cultural areas.

By the second half of the first millennium AD, the most important contacts for the future development of mathematics were those between India and the Islamic world. This is shown by the arrow from 3 in India to 1 in the "Arab" world. As we saw in figure 1.3, the other major influence on the Islamic world was from the Greek cultural areas, and the nature of these influences has been discussed in some detail. As far as Indian influence via the Islamic world on the future development of mathematics is concerned, it is possible to identify three main areas:

1. The spread of Indian numerals and their associated algorithms, first to the Islamic world and later to Europe

2. The spread of Indian trigonometry,[17] especially the use of the sine function

3. The solutions of equations in general, and of indeterminate equations in particular[18]

These contributions will be discussed in chapters 8–10, which deal with Indian mathematics.

We have already looked briefly at the contributions of Islamic scholars as producers, transmitters, and custodians of mathematical learning. Their role as teachers of mathematics to Europe is not sufficiently acknowledged. The arrow from 1 in the "Arab" world to 1 in Europe represents the crucial role of the Islamic world in the creation and spread of mathematics, which culminated in the birth of modern mathematics. These contributions will be discussed in the final chapter of this book.

Figure 1.4 shows another important cross-cultural contact, between India and China. There is very fragmentary evidence (as shown by the broken line between 2 in India and 2 in China) of contacts between the two countries before the spread of Buddhism into China. After this, from around the first century AD, India became the center for pilgrimage of Chinese Buddhists, opening the way for a scientific and cultural exchange that lasted for several centuries. In a catalogue of publications during the Sui dynasty (c. 600), there appear Chinese translations of Indian works on astronomy, mathematics, and medicine. Records of the Tang dynasty indicate that from 600 onward Indian astronomers were hired by the Astronomical Board of Changan to teach the principles of Indian astronomy. The solution of indeterminate equations, using the method of *kuttaka* in India and of *qiuyishu* in China, was an abiding passion in both countries. The nature and direction of transmission of mathematical ideas between the two areas is a complex but interesting problem, one to which we shall return in later chapters. The two-headed arrow linking 3 in India with 3 in China is a recognition of the existence of such transmission. Also, there is some evidence of a direct transmission of mathematical (and astronomical) ideas between China and the Islamic world, around the beginning of the second millennium AD.[19] Numerical methods of solving equations of higher order such as quadratics and cubics, which attracted the interest of later Islamic mathematicians, notably al-Kashi (c. 1400), may have been influenced by Chinese work in this area. There is every likelihood that some of the important trigonometric concepts introduced into Chinese mathematics around this period may have an Islamic origin.

There are broken lines of transmission in figure 1.4 that need some explanation. One of the conjectures posed and elaborated in chapter 10 is the possibility that mathematics from medieval India, particularly from the southern state of Kerala, may have had an impact on European mathematics of the sixteenth and seventeenth centuries. While this cannot be substantiated at present by existing direct evidence, the circumstantial evidence has become much stronger as a result of some recent archival research. The fact remains that around the beginning of the fifteenth century Madhava of Kerala derived infinite series for p and for certain trigonometric functions, thereby contributing to the beginnings of mathematical analysis about 250 years before European mathematicians such as Leibniz, Newton, and Gregory were to arrive at the same results from their work on infinitesimal calculus. The possibility of medieval Indian mathematics influencing Europe is indicated by the arrow linking 4 in India with 1 in Europe.

During the medieval period in India, especially after the establishment of Mughal rule in North India, the Arab and Persian mathematical sources became better known there. From about the fifteenth century onward there were two independent mathematical developments taking place, one Sanskrit-derived and constituting the mainstream tradition of Indian mathematics, then best exemplified in the work of Kerala mathematicians in the South, and the other based in a number of Muslim schools (or *madrassahs*) located mainly in the North. We recognize this transmission by constructing an arrow linking 1 in the Arab world to 4 in India. A discussion of the flourishing mathematical tradition introduced into India during the medieval times, where the sources were Persian and Arabic texts, will be found in chapter 9.

The medieval period also saw a considerable transfer of technology and products from China to Europe, which has been thoroughly investigated by Lach (1965) and Needham (1954). The fifteenth and sixteenth centuries witnessed the culmination of a westward flow of technology from China that had started as early as the first century AD. It included, from the list given by Needham (1954, pp. 240–41), the square-pallet chain pump, metallurgical blowing engines operated by water power, the wheelbarrow, the sailing carriage, the wagon mill, the crossbow, the technique of deep drilling, the so-called Cardan suspension for a compass, the segmental arch-bridge, canal lock-gates, numerous inventions in ship construction (including watertight compartments and aerodynamically efficient sails), gunpowder, the magnetic compass for navigation, paper and printing, and porcelain. The conjecture here is that with the transfer of technology went certain mathematical ideas, including different algorithms for extracting square and cube roots, the "Chinese remainder theorem," solutions of cubic and higher-order equations by what is known as Horner's method, and indeterminate analysis. Such a transmission from China need not have been a direct one but may have taken place through India and the Islamic world. We shall return to the question of influences and transmission from China to the rest of the world in chapter 7.

During the first half of the first millennium of the Christian era, the Central American Mayan civilization attained great heights in a number of different fields including art, sculpture, architecture, mathematics, and astronomy. In the field of numeration, the Maya shared in two fundamental discoveries: the principle of place value and the use of zero. Present evidence indicates that the principle of place value was discovered independently four times in the history of mathematics. At the beginning of the second millennium BC, the Mesopotamians were working with a place-value notational system to base 60. Around the beginning of the Christian era, the Chinese were using positional principles in their rod numeral computations. Between the third and fifth centuries AD, Indian mathematicians and astronomers were using a place-value decimal system of numeration that would eventually be adopted by the whole world. And finally, the Maya—apparently cut off from the rest of the world—had developed a positional number system to base 20. As regards zero, there are only two original instances of its modern use in a number system: by the Maya and by the Indians around the beginning of the Christian era.

But mathematics is not the only area in which the Maya surprise us. With the most rudimentary instruments at their disposal they undertook astronomical observations and calendar construction with a precision that went beyond anything available in Europe at that time. They had accurate estimates of the duration of solar, lunar, and planetary movements. They estimated the synodic period of Venus (i.e., the time between one appearance at a given point in the sky and its next appearance at that point) to be 584 days, which is an underestimate of 0.08 days. They achieved these discoveries with no knowledge of glass or, consequently, of any sort of optical device. Neither did they apparently have any device for measuring the passage of time, such as clocks or sandglasses, without which it would now seem impossible to produce astronomical data.

Figure 1.5 shows the geographical areas whose mathematics form the subject matter of this book. I am conscious of not having examined in sufficient detail the mathematical pursuits of other groups, notably the Africans south of the Sahara, the Amerindians of North America, and the indigenous Australasians, although the topics treated in chapter 2 should go some way in making up for this neglect.[20] Much research still needs to be done on mathematical activities in these areas, despite some promising work on ethnomathematics in recent years, notably by Gerdes (1995, 1999, 2002) and Zaslavsky (1973a) on African mathematics.[21]

Since the publication of the first edition of this book in 1991, there has been an increase in interest in ethnomathematics, or the study of mathematical concepts in their cultural context, often within socially cohesive and small-scale indigenous groups. Within the definition of mathematics given earlier, the emphasis is on how structures and systems of ideas involving number, pattern, logic, and spatial configuration arose in different cultures. This view has had to contend with the strongly entrenched notion that mathematics, having originated in some primitive unformed state, advanced in a linear direction to the current state of modern mathematics and will continue to grow in that direction. A mathematical system that emerges in a culture removed from this "mainstream" would then be perceived as a mere distraction of little relevance to the ideas and activities supported by modern mathematics.

A telling criticism of the first edition of *The Crest of the Peacock* is that it implicitly subscribed to this "linear" view, being "epistemologically based on the idea of direct literal translations of non-western mathematics to the western tradition" (Eglash 1997, p. 79). In response to this criticism and in subsequent editions, the coverage has been extended to include areas in the Pacific and elaborate further on the mathematical activities in the African and American continents.

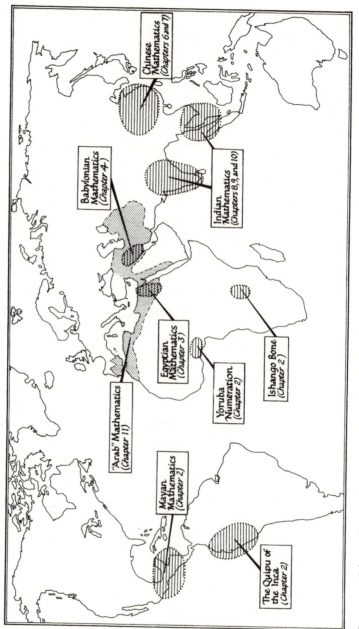

Chinese Mathematics (Chapters 6 and 7)

Indian Mathematics (Chapters 8, 9, and 10)

Babylonian Mathematics (Chapter 4)

Egyptian Mathematics (Chapter 3)

Yoruba Numeration (Chapter 2)

Ishango Bone (Chapter 2)

"Arab" Mathematics (Chapter 11)

Mayan Mathematics (Chapter 2)

The Quipu of the Inca (Chapter 2)

FIGURE 1.5: Cultures whose mathematics form the subject of this book

Notes

1. The term "cultural dependencies" is used here to describe those countries—notably the United States, Canada, Australia, and New Zealand—which are inhabited mainly by populations of European origin or which have historical and cultural roots similar to those of European peoples. For the sake of brevity, the term "Europe" is used hereafter to include these cultural dependencies as well.

2. This is a variation on the famous Needham question (named after Joseph Needham, the well-known twentieth-century British scientist and sinologist): Why did modern science develop in Europe when China with its momentous inventions like printing and gunpowder seemed so much better placed to achieve it? A similar question may be asked substituting instead of China the names of India or the Islamic world. For further discussion, see Bala (2006) and Bala and Joseph (2007).

3. See Brohman (1995a, 1995b) for further details.

4. The shift has occurred not only in the history of mathematics. The traditional Eurocentric world history presupposed the existence of an imaginary line of "civilizational apartheid" between the European and the non-European world whereby the former had single-handedly propelled the whole world from tradition into modernity while the latter remained stagnant. In recent years, spurred by a non-Eurocentric global history focusing on the historical resource portfolios (i.e., ideas, institutions, and technologies) diffused from the East across to the West, one discerns the emergence of what may be described as a neo-Eurocentric approach: one that acknowledges the borrowing of non-Western resources in the rise of the West but recasts Europe as "cosmopolitan, tolerant, open to others ideas, and highly adaptive insofar as it put all these non-Western sources together in a unique way to produce modernity." I am grateful to John Hobson for making this point in a private communication. It follows logically from his book *The Eastern Origins of Western Civilisation* (2004).

5. These parallels include (a) a belief in the transmigration of souls; (b) the theory of four elements constituting matter; (c) the reasons for not eating beans; (d) the structure of the religio-philosophical character of the Pythagorean fraternity, which resembled Buddhist monastic orders; and (e) the contents of the mystical speculations of the Pythagorean schools, which bear a striking resemblance to the Hindu *Upanishads*. According to Greek tradition, Pythagoras, Thales, Empedocles, Anaxagoras, Democritus, and others undertook journeys to the East to study philosophy and science. While it is far-fetched to assume that all these individuals reached India, there is a strong possibility that some of them became aware of Indian thought and science through Persia.

6. It is interesting to note that the terminology used in modern mathematics has a mixed origin consisting mainly of Greek, Latin, and modern European languages. The terms used in both Egyptian and Mesopotamian texts date back to the period before

the Greeks. Given the nature and scope of this book, we will continue to use modern terminology and avoid literal translations of the technical terms given in the ancient texts. Thus, for example, we use the modern term "triangle" (three angles) rather than the Babylonian term translated as "wedge" (three sides). The concept of an "angle" came only with the Greeks. A right-angled triangle in Old Babylonian mathematics had no angle connotation and was literally transliterated as one of two triangles into which a rectangle was divided by the longer diagonal. Similarly, although we use a modern term such as a "square" in the presentation of the ancient mathematics of Egypt and Mesopotamia, it should be noted that the corresponding term in these texts is "equal side" (or "same side").

7. In terms of historical, geographical, as well as intellectual proximity, Islamic science could be regarded as the most immediate predecessor of modern Western science. Some of the more recent studies (Bala 2006; Saliba 2007) show the existence of epistemological links between the two sciences. The "mathematicization" of nature, the centrality of the empirical method in scientific methodology, and the rationality of scientific discourses are features of Islamic science inherited by founders of modern Western science.

8. They include (a) an early description of pulmonary circulation of the blood, by ibn al-Nafis, usually attributed to Harvey, though there are records of an even earlier explanation in China; (b) the first known statement about the refraction of light, by ibn al-Hayatham, usually attributed to Newton; (c) the first known scientific discussion of gravity, by al-Khazin, again attributed to Newton; (d) the first clear statement of the idea of evolution, by ibn Miskawayh, usually attributed to Darwin; and (e) the first exposition of the rationale underlying the "scientific method," found in the works of ibn Sina, ibn al-Hayatham, and al-Biruni but usually credited to Roger Bacon. A general discussion of the Western debt to the Middle East is given by Savory (1976), while detailed references to specific contributions of Islamic science are given by Gillespie (1969–).

9. Jund-i-Shapur (or Guneshahpuhr) was founded around AD 260 by Shahpuhr I (241–272) to settle Roman prisoners captured in the war against Valerian and was located in Khuzistan in southwestern Iran. Early settlers included Roman engineers and physicians, and doubtless others who may have been acquainted with Greek, Egyptian, and Mesopotamian mathematics. The Christian bishop Demetrianus from Antioch founded a bishopric there, and during the fifth and sixth centuries Nestorianism was the only form of Christianity permitted in Iran. This intolerance contrasted with the openness and tolerance exhibited toward other religious immigrants, for when Zeno closed the School of the Persians in Edessa (AD 489), its intellectual and spiritual center moved to Persian Nisibis, where the exiles re-created their famous seat of learning. The Medical School of Jund-i-Shapur was founded on Greek medical knowledge (itself from Egyptian and Babylonian) by these Nestorian physicians. In the realms of philosophy, it is often forgotten that the Sasanian king Khusro I welcomed the major seven Neoplatonist Greek philosophers who fled Athens in 529 when the Academy there was

closed on the orders of the Byzantine Justinian. Some of these scholars worked for some time at Jund-i-Shapur but became homesick; Khusro negotiated their safe conduct and pardon for their return to Athens. Indeed, it was said of the enlightened Khusro that he was "a disciple of Plato seated on the Persian throne." The Jund-i-Shapur Medical School remained a center of excellence right through to Islamic times and indeed well past the mid-ninth century. While there are no extant records relating to mathematical activities in Jund-i-Shapur, we have evidence to indicate that during the reign of Shahpuhr I and later Khusro I, translations into Middle Persian (Pahlavi) were made in Iran from Greek and Sanskrit texts. It is more than likely that these included texts in astronomy, mathematics, and other sciences. After the downfall of the Sasanians, the Islamic regimes of the caliphs were by turns favorable or otherwise to the ancient learning enshrined at Jund-i-Shapur. Either way, Islamic knowledge was vastly increased through such deep and enduring exchanges.

10. This familiar story (or even some believe a caricature) about the role played by the House of Wisdom is now being reassessed. For further details see Gutas (1998) and Saliba (2007). See also endnote 2 of chapter 11.

11. But see the comment and reference given in endnote 24 of chapter 11 for further clarification.

12. A Spanish dictionary gives the following meanings: *álgebra*. 1. f. Parte de las matemáticas en la cual las operaciones aritméticas son generalizadas empleando números, letras y signos. 2. f. desus. Arte de restituir a su lugar los huesos dislocados (translation: the art of restoring broken bones to their correct positions).

13. For further details of these transmissions, see Zaimeche (2003, p. 10).

14. *Gnomon* is an ancient Greek word meaning "indicator" or "that which reveals." There are references to the gnomon in other traditions, for example, the seminal Chinese text *Nine Chapters on the Mathematical Art,* and it was referred to earlier by the Duke of Zhou (eleventh century BC). "Gnomon" also refers to the triangular part of a sundial that casts the shadow.

15. In the concluding paragraph Friberg (2005, p. 270) writes: "The observation that Greek *ostraca* [i.e., limestone chippings and pottery used as writing material] and papyri with Euclidean mathematics existed side by side with demotic and Greek papyri with Babylonian style mathematics is important for the reason that this surprising circumstance is an indication that when the Greeks themselves claimed that they got their mathematics from Egypt, they can really have meant that they got their mathematical inspiration from Egyptian texts with the mathematics of the Babylonian type. To make this thought more explicit would be a natural continuation of the present investigation." Friberg (2007) is the continuation of the investigation alluded to and provides the material for the Greek links with the two earlier civilizations.

16. In the case of Indian astronomy and the mathematics associated with it, the early influences from Mesopotamia came through the mediation of the Greeks. Probably in the fifth century BC, India acquired Babylonian astronomical period relations and arithmetic (e.g., representing continuously changing quantities with "zigzag" functions). Around the early centuries AD, the Babylonian arithmetical procedures were combined with Greek geometrical methods to determine solar and lunar positions, as reported in the Indian astronomical treatises *Romaka-siddhanta* and *Paulisa-siddhanta*. For further details, see Pingree (1981).

17. Since this is the first time we use the term "trigonometry," a word of caution is necessary. Trigonometry (meaning "triangle measurement") is a relatively modern term dating back to the sixteenth century. While today we have difficulty disentangling the concept of trigonometry from the ratio of sides in a right-angled triangle, for a long period of history the concept related only to circles and their arcs. And this was particularly so for the Greeks and the Indians. It was a search for a measure of the angle (or the inclination) of one line to another, an interest (and ability) to estimate the lengths of line segments, and a "systematic ability to convert back and forth between measures of angles and of lengths" that gave rise to modern trigonometry. I am grateful to Van Brummelen (2008) for this insight.

18. An example of an indeterminate equation in two unknowns (x and y) is $3x + 4y = 50$, which has a number of positive whole-number (or integer) solutions for (x, y). For example, $x = 14$, $y = 2$ satisfies the equation, as do the solution sets $(10, 5)$, $(6, 8)$ and $(2, 11)$.

19. An exchange of astronomical knowledge took place between the Islamic world and the Yuan dynasty in China in the latter part of the thirteenth century, when both territories were part of the Mongol empire. A few Chinese astronomers were employed at the observatory in Maragha (set up by Hulegu Khan in 1258) and probably helped in the construction of the Chinese-Uighur calendar (a type of a lunisolar calendar or a calendar whose date indicates both the phase of the moon and the time of the solar year). This calendar was widely used in Iran from the late thirteenth century onward. There were at least ten Islamic astronomers working in the Islamic Astronomical Bureau in Beijing founded by the first Mongol emperor of China, Kublai Khan, in 1271. At this bureau, continuous observations were made and a *zij* (or astronomical handbook with tables) was compiled in Persian. This work was then translated into Chinese during the early Ming dynasty (1383) and, together with Kushayar's influential Islamic text, *Introduction to Astrology*, served for a number of years as important sources for further research and study by Chinese scholars. For further details, see van Dalen (2002).

20. It could be argued that in the examples discussed in chapter 2 there is undue emphasis on the role of number systems and insufficient attention paid to what Gerdes (1995) describes as "frozen geometry." These would include geometric or logical relationships embedded in diverse activities such as basket weaving, knitting, and sand

drawings highlighted by scholars such as Gerdes (1999) and Harris (1997). The problem in including such ethnomathematical activities is partly one of determining their historical origins and partly one of deciding what are to be included/excluded given the scope of this book.

21. The burgeoning study of African mathematics in recent years has highlighted a variety of mathematics that goes under the blanket term "ethnomathematics." Rambane and Mashige (2007, 184–85) have constructed the following list, with references to those who have worked in these areas.

1. *Oral mathematics*. The mathematical knowledge that is transmitted orally from one generation to another.

2. *Oppressed mathematics*. The mathematical elements in daily life that remain unrecognized by the dominant (colonial and neocolonial) ideologies (Gerdes 1985b).

3. *Indigenous mathematics*. A mathematical curriculum that uses everyday indigenous mathematics as the starting point. The origin of this concept is found in Gay and Cole (1967), who criticized the teaching of Kpelle children in Liberia in Western-oriented schools "things that have no point or meaning within their culture."

4. *Sociomathematics of Africa*. "The applications of mathematics in the lives of African people, and, conversely, the influence that African institutions had upon the evolution of their mathematics" (Zaslavsky 1973b, 1991).

5. *Informal mathematics*. Mathematics that is transmitted and learned outside the formal system of education, sometimes referred to as "street mathematics" (Posner 1982; Nunes et al. 1993).

6. *Nonstandard mathematics*. A distinctive mathematics beyond the standard form, found outside the school and university (Gerdes 1985b).

7. *Hidden* or *frozen mathematics*. Mathematics that has to be unfrozen from "hidden" or frozen objects or techniques, such as basket making, weaving, or traditional architecture (Gerdes 1985b).

Chapter Two

Mathematics from Bones, Strings, and Standing Stones

It is taking an unnecessarily restrictive view of the history of mathematics to confine our study to written evidence. Mathematics initially arose from a need to count and record numbers. As far as we know there has never been a society without some form of counting or tallying (i.e., matching a collection of objects with some easily handled set of markers, whether it be stones, knots, or inscriptions such as notches on wood or bone). If we define mathematics as any activity that arises out of, or directly generates, concepts relating to numbers or spatial configurations together with some form of logic, we can then legitimately include in our study protomathematics, which existed when no written records were available.

Beginnings: The Ishango Bone

High in the mountains of central equatorial Africa, on the borders of Uganda and Congo, lies Lake Rutanzige (Edward), one of the furthest sources of the river Nile. It is a small lake by African standards, about eighty kilometers long and fifty wide. Though the area is remote and sparsely populated today, about twenty-five thousand years ago by the shores of the lake lived a small community that fished, gathered food, or grew crops, depending on the season of the year. The settlement had a relatively short life span of a few hundred years before being buried in a volcanic eruption. These Neolithic people have come to be known as the Ishango, after the place where their remains were found. There exists today a small village by that name.

Archaeological excavations at Ishango have unearthed human remains and tools for fishing, hunting, and food production (including grinding and pounding stones for grain). Harpoon heads made from bone may have

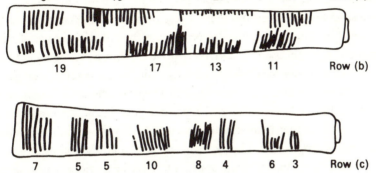

FIGURE 2.1: The Ishango bone (Courtesy of Dr. J. de Heinzelin)

served as prototypes for tools discovered as far away as northern Sudan and West Africa. However, the most interesting find, from our point of view, is a bone tool handle (figure 2.1) which is now at the Musée d'Histoire Naturelle in Brussels.[1] The original bone may have petrified or undergone chemical change through the action of water and other elements. What remains is a dark brown object on which some markings are clearly visible. At one end is a sharp, firmly fixed piece of quartz which may have been used for engraving, tattooing, or even writing of some kind.

The markings on the Ishango bone, as it is called, consist of series of notches arranged in three distinct rows. The asymmetrical grouping of these notches, as shown in figure 2.1, would make it unlikely that they were put there merely for decorative purposes. Row (a) contains four groups of notches with 9, 19, 21, and 11 markings. In row (b) there are also four groups, of 19, 17, 13, and 11 markings. Row (c) has eight groups of notches in the following order: 7, 5, 5, 10, 8, 4, 6, 3. The last two groups (6, 3) are spaced closer together, as are (8, 4) and (5, 5, 10), suggesting a deliberate arrangement in distinct subgroups.

If these groups of notches were not decorative, why were they put there? An obvious explanation is that they were simply tally marks. Permanent records of counts maintained by scratches on stones, knots on strings, or notches on sticks or bones have been found all over the world, some going back to the very early history of human habitation. During an excavation of a cave in the Lebembo Mountains on the borders of Swaziland in southern Africa, a small section of the fibula of a baboon was discovered, with 29 clearly visible notches, dating to about 35,000 BC.[2] This is one of the earliest artifacts we have that provide evidence of a numerical recording device. An interesting feature of this bone is its resemblance to the "calendar sticks" still used by some inhabitants of Namibia to record the passage of time. From about five thousand years later we have the shin-bone of a young wolf, found in Czechoslovakia, which contains 57 deeply cut notches arranged in S-shaped groups. It was probably a record kept by a hunter of the number of kills to his credit. Such artifacts represent a distinct advance, a first step toward constructing a numeration system, whereby the counting of objects in groups is supplemented by permanent records of these counts.

However, the Ishango bone appears to have been more than a simple tally. Certain underlying numerical patterns may be observed within each of the rows marked (a) to (c) in figure 2.1. The markings on rows (a) and (b) each add up to 60: $9 + 19 + 21 + 11 = 60$, and $19 + 17 + 13 + 11 = 60$, respectively. Row (b) contains the prime numbers between 10 and 20. Row (a) is quite consistent with a numeration system based on 10, since the notches are grouped as $20 + 1$, $20 - 1$, $10 + 1$, and $10 - 1$. Finally, row (c), where subgroups (5, 5, 10), (8, 4), and (6, 3) are clearly demarcated, has been interpreted as showing some appreciation of the concept of duplication or multiplying by 2.

De Heinzelin (1962), the archaeologist who helped to excavate the Ishango bone, wrote that it "may represent an arithmetical game of some sort, devised by a people who had a number system based on 10 as well as a knowledge of duplication and of prime numbers" (p. 111). Further, from the existing evidence of the transmission of Ishango tools, notably harpoon heads, northward up to the frontiers of Egypt, de Heinzelin considered the possibility that the Ishango numeration system may have traveled as far as Egypt and influenced the development of its number system, the earliest decimal-based system in the world.

The African origins of Egyptian civilization are well attested to by archaeological and early written evidence. Herodotus wrote of the Egyptian people and culture having strong African roots, coming from the lands of the "long-lived Ethiopians," which meant in those days the vast tract of inner Africa inhabited by black people. However, de Heinzelin's speculations about the state of mathematical knowledge of the Ishango, based as they are on the evidence of a single bone, seem far-fetched. A single bone with suggestive markings raises interesting possibilities of a highly developed sense of arithmetical awareness; it does not provide conclusive evidence.

There is, however, another answer, more firmly rooted in the cultural environment, to the puzzle of the Ishango bone. Rather than attribute the development of a numeration system to a small group of Neolithic settlers living in relative isolation on the shores of a lake, apparently cut off from other traceable settlements of any size and permanence, a more plausible hypothesis is that the bone markings constitute a system of sequential notation—for example, a record of different phases of the moon. Whether this is a convincing explanation would depend in part on establishing the importance of lunar observations in the Ishango culture, and in part on how closely the series of notches on the bone matches the number of days contained in successive phases of the moon.

Archaeological evidence of seasonal changes in the habitat and activity of the Ishango highlights how important it was to maintain an accurate lunar calendar. At the beginning of the dry season, the Ishango moved down to the lake from the hills and valleys that formed their habitat during the rains. For those who were permanently settled along the shores of the lake, the onset of the dry season brought animals and birds to the lake in search of water. Now assume, for the sake of argument, that migration took place around the full moon or a few days before the full moon. About six months later the rainy season would begin, and the water levels of the lake would rise. Between the beginning of the dry season and the onset of the rainy season, there might be festivities that coincided with particular phases of the moon. And such events might very well be what is recorded notationally on the bone. Activities such as gathering and processing of nuts and seeds, or hunting, both of which archaeological evidence suggests were important in the Ishango economy, could be incorporated sequentially into the lunar calendar represented by the Ishango bone. Similarly, religious rituals associated with seasonal and other festivities could be recorded on the bone. Such a scenario is still conjectural, but consistent with what we know of present-day peoples who still follow the hunter-gatherer lifestyle of the Ishango.

A cursory examination of the pattern of notches on the Ishango bone shows no obvious regularity that one can associate with lunar phenomena. Two of the rows add up to 60, so that each of these rows may be said to represent two lunar months. The third row contains only 48 notches, which would account for only a month and a half. But a mere count of the notches would ignore the possible significance of the different sizes and shapes of the markings as well as the sequencing of the subgroups demarcated on the bone.

Marshack (1972) carried out a detailed microscopic examination of the Ishango bone and found markings of different indentations, shapes, and sizes. He concluded that there was evidence of a close fit between different phases of the moon and the sequential notation contained on the bone, once the additional markings—visible only through the microscope—were taken into account. Also, the different engravings represented by markings of various shapes and sizes may have been a calendar of events of a ceremonial or ritual nature.

These conjectures about the Ishango bone highlight three important aspects of protomathematics. First, the close link between mathematics and astronomy has a long history and is tied up with the need felt even by early humans to record the passage of time, out of curiosity as well as practical necessity. Second, there is no reason to believe that early humans' capacity to reason and conceptualize was any different from that of their modern counterparts. What has changed dramatically over the years is the nature of the facts and relationships with which human beings have had to operate. Thus the creation of a complicated system of sequential notation based on a lunar calendar was well within the capacity of prehistoric humans, whose desire to keep track of the passage of time and changes in seasons was translated into observations of the changing aspect of the moon. Finally, in the absence of records, conjectures about the mathematical pursuits of early human beings have to be examined in the light of their plausibility, the existence of convincing alternative explanations, and the quality of evidence available. A single bone may well collapse under the heavy weight of conjectures piled upon it.[3]

The notches of the Ishango bone open up other interesting conjectures. The epoch of the Ishango bone was around the same period when women were supposedly the temporal and spiritual leaders of their clans. Since a woman's menstrual cycle mimics the phases of the moon, would it be too fanciful to argue that the markings on the bone represent an early calendar of events of a ceremonial or ritual nature superimposed on a record of a lunar/menstrual cycle constructed by a woman? After all, among the Siaui of the Solomon Islands in the Pacific, a menstruating female is described as "going to the moon."[4]

There is yet another interpretation.[5] The Ishango bone may have been a precursor of writing. In that case, writing originated not in drawing figures or in attempting to record speech but in storing numerical information. The rows of notches became "graphically isomorphic" to the recorder's counting numbers. In this and subsequent chapters, there are illustrations of counting systems with forms of recording in which the iconic origin of the dash or stroke is the human digit (i.e., the finger or toe). These strokes are graphically isomorphic with the corresponding words used for counting. It makes little difference whether we "read" the sign pictorially, as standing for so many fingers held up, or in "script," as standing for a certain numeral.

Counting Systems and Numeration: The Pacific Dimension

The study of worldwide systems of numeration as they occur in natural languages has had a checkered career. A rich source of information on "exotic" languages and customs from the published literature of the explorers, administrators, and missionaries forms the core of the data now available on non-European counting systems. Initiated originally by those interested in linguistics and anthropology, the subject is now of only marginal interest to these groups. Yet no growth of interest on the part of historians or philosophers of mathematics has matched the waning interest on the part of the linguist or the ethnographer.

There are various reasons for this lack of interest. For the Western philosopher, the study of natural-language number systems seems to have little or no relevance to understanding the nature of number—an abstract philosophical concept derived from ancient Greeks and independent of linguistics and cultural vagaries. Historians of mathematics tend to concentrate on the origins of written numbers because of their uneasiness about straying into territories where culture and language interpose. In any case, they prefer to work with written records, even if these records are mainly confined to those in European languages. The occurrence of ideas relating to numbers as existing among "primitive" tribes, if considered at all, is mentioned in a dismissive fashion.

This lack of interest in societies outside the usual ambit of historians and philosophers has had some unfortunate consequences. Societies that lack a tradition of historical documentation constitute the majority of world languages. Hence, a great part of humanity's numeration and counting practices is ignored. Also, those societies whose numerical practices are considered in historical writing tend to fall into a small range of counting types, leading to an overwhelming bias in discussions on systems that concentrate on base 10. Allowance may sometimes be made for irregularities introduced by vestiges of base 12 (English and German) or base 20 (French and Dutch). An unfortunate consequence of this concentration is the susceptibility to the reductionist fallacy that sees humans' response to the need for enumeration of their world as being unvarying across time and culture.

A renewed interest, since the 1960s, in documenting mathematical ideas from non-Western societies is reflected in histories of numbers, notably the works of Flegg (1983), Ifrah (1985), Menninger (1969), and Schmandt-Besserat (1999). However, part of the material on which these studies were based came from nineteenth-century reports of the agents of European colonization. Regions such as Melanesia, Polynesia, Micronesia, and Australasia, which account for more than one-quarter of the existing world languages, have been neglected. This situation has now changed dramatically with the monumental twenty-one-volume "Counting Systems of Papua New Guinea and Oceania" by Glendon Lean (1996). The study, which remains unpublished,[6] is a valuable guide to natural-language counting systems used by nine hundred out of twelve hundred linguistic groups in Papua New Guinea, Irian Jaya, Solomon Islands, Vanuata, New Caledonia, and parts of Polynesia and Micronesia. The study relating to Papua New Guinea and Irian Jaya, the other half of the island of New Guinea, is especially useful since nowhere in the world is the diversity of cultures and languages so marked as on this island. This study will form the main basis of the discussion here.

To present a coherent framework for a discussion of the large database available from Lean's study, a system of classifying the number systems is required. The classification discussed later in this chapter under the heading "Emergence of Written Number Systems" is found in the older literature and is built around the descriptive term "base." This classification is unsatisfactory for various reasons. It is particularly inadequate for examining "mixed" systems of counting, which may, for example, have elements of base 2, base 5, and base 10 at the same time.[7] Also, Lean's study indicates

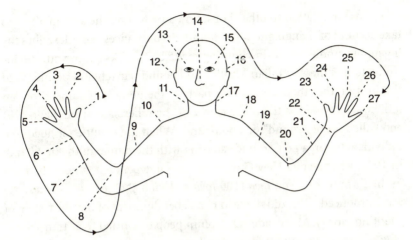

FIGURE 2.2: "Body numbers" based on Saxe's 1982 drawing

that groups within Papua New Guinea, sometimes in close proximity to one another, differ in significant ways in their counting practices according to the importance that they attribute to the enumeration of objects as well as the type of objects counted, and the circumstances in which the counting takes place.

Laycock (1975) introduced a clearer fourfold classification of number systems. The first consists of a "body part" tally system whereby number representation is based on the body and entails counting body parts according to a conventionally defined order. Consider the Oksapmin, a group found in the West Sepik Province of Papua New Guinea, as an example. Saxe (1982) describes their counting system in the following terms: starting with the thumb of a hand, counting proceeds along the fingers of that hand, so that the little finger would be 5; it further proceeds down that hand, so that 7 is the forearm along the upper periphery of the body; to the face, with 12 being the nearest ear to the hand that has been counted; to the nearest eye (13); the nose (14); the other eye (15); the other ear (16); and down along the other side of the body to the little finger on the other hand (27). If necessary, one could repeat the counting process again to reach 54 on the second count. A specific number is referred to by its associated body part. (See figure 2.2)

Such a system of number representation has been applied to a range of diverse activities, whether counting a set of objects (e.g., number of pigs), measuring the length of an object (e.g., a bow), or establishing the location of a house in relation to other houses on a path. And the system alters to take account of changing socioeconomic circumstances. Saxe describes the ingenious modifications in counting made by the Oksapmin with the introduction of the Australian "shilling and pound" currency (1 pound = 20 shillings): rather than using all 27 parts of the body for enumeration, an individual stopped at the inner elbow of the other side of the hand (20) and called it one "round" or one pound. When the count continued, the individual began a new "round" starting with the thumb of the first hand. In 1975, when Papua New Guinea became independent, a new currency in the form of *kina* and *toea* (100 *toea* = 1 *kina*, 200 *toea* = 2-*kina* note) was introduced.[8] This did not lead to an abandonment of the old system of counting money. Many older Oksapmin people resorted to a "translation" of *kina* and *toea* into pounds and shillings, and then continued the count by referring to a 10-*toea* coin as one shilling and one 2-*kina* note as one pound.

A second type of number system, according to Laycock, is one that has two to four discrete number-words and a matching "base" to carry out the tally. For example, in the case of an indigenous Australian group, the Cumulgal, counting proceeds as *urapon* (1), *ukasar* (2), *ukasar-urapon* (3), *ukasar-ukasar* (4), *ukasar-ukasar-urapon* (5), indicating counting by twos (or a base 2 system).

A third type is the "quinary-vigesimal" system, which has a 5, 20 cyclic pattern and may employ fingers and toes as an aid to tallying. For example, in the Melanesian language Sesake, counting proceeds as *sekai* (1), *dua* (2), *dolu* (3), *pati* (4), *lima* (5), *la-tesa* (6), *la-dua* (7), . . . , *dua-lima* (10), . . . , *dua-lima dua* (20). The system has a cycle of 5, in which numbers six or seven use the roots for words one or two respectively, and a superordinate cycle of 20, in which 20 is two fives twice.

The final type is a decimal system, normally with no reference to "body parts" and having six to ten discrete number-words. In the Micronesian language Kiribati, counting proceeds as *tenuana* (1), *uoua* (2), *tenua* (3), *aua* (4), *nimua* (5), *onoua* (6), *itiua* (7), *wanua* (8), *ruainua* (9), *tebwina* (10), *tebwina-ma-tenuana* (11), . . . , *vabui* (20), *vabui-ma-tenuana* (21), . . . , *tebubua* (100), . . . , *tenga* (1000), ... , *tebina-tenga* (10,000), . . . , *tebubuna-tenga* (100,000). Here we have a straightforward base 10 counting system with no association to body parts. In Kiribiti, number-words also vary according to the object being counted. Thus the number-word for nine is *ruaman* when counting animals, *ruakai* when counting plants, *ruai* when counting knives, *ruakora* when counting baskets, and *rauawa* when counting boats.

Numeration relating to time has not been a significant influence on the development of counting systems in the Pacific. Time is usually reckoned in units determined by lunar and seasonal cycles. Precision of reckoning is not very important, and in only two activities would reckoning time be significant: agriculture and ceremonial events. The Mae Enga, a group in New Guinea, have a lunar "calendar of events" consisting of twelve *kana* (or "garden" months) with which they monitor agriculture and other activities. The name of each *kana* is indicative of the activity undertaken that month. A month is allocated for each activity, such as planting of specific crops, harvesting, preparing the "garden" for planting, as well as undertaking trading trips or engaging in fighting. It is not clear whether the Mae Enga ever kept a record of the passage of time as measured by their calendar, such as the Ishango bone from central Africa.

The recording of ceremonial events does not require anything as elaborate as a calendar. Among the Kewa, the body-part system of counting is sufficient to track the occurrence of ceremonial dances. According to Johnson (1997, p. 659), if a cycle of dances were to begin in eleven months time, this would be counted as *komane roba summa* (elbow). If the next occurrence of the dance were three months later, this would be shown as *pesame roba suma* (shoulder). The third occurrence, six or seven months later, would be counted as *rigame robasuna* (between the eyes). This sequence of dances would continue until the climax of the feast, when pigs were killed. This happened on the month represented by the division of the wrist and the finger. A number of these festivals required a strong numerical sense on the part of the participants, since a person's prestige was measured by the quantity of pigs, shells, or any other "currency" given away as gifts.

The Yoruba Counting System: The African Dimension

The origins of the Yoruba people of southwestern Nigeria are lost in the mists of time.[9] Oral traditions indicate that they came from the east, and certain similarities between the customs and practices of ancient Egyptians and those of the Yoruba would support this. These include similarities in religious practices and institutions, in particular, the carved idols used in worship, the shape and design of sacrificial altars, and the role of a powerful priesthood.

Their more recent history began with the foundation of the Oyo state around the early centuries of the second millennium AD. Commercial and other contacts with the north provided an important stimulus to scientific and cultural activity in the region. Later centuries saw the establishment of the vast Benin empire, independent of the Oyo kingdom, both of which were finally dissolved by the British at the end of the nineteenth century.

The Yoruba system of numeration is essentially a base 20 counting system, its most unusual feature being a heavy reliance on subtraction. The subtraction principle operates in the following way. As in our system, there are different names for the numbers one (*okan*) to ten (*eewa*). The numbers eleven (*ookanla*) to fourteen (*eerinla*) are expressed as compound words that may be translated as "one more than ten" to "four more than ten." But once fifteen (*aarundinlogun*) is reached the convention changes, so that fifteen to nineteen (*ookandinlogun*) are expressed as "twenty less five" to "twenty less one," respectively, where twenty is known as *oogun*. Similarly, the numbers twenty-one to twenty-four are expressed as additions to twenty, and twenty-five to twenty-nine as deductions from thirty (*ogbon*). At thirty-five (*aarundinlogun*), however, there is a change in the way the first multiple of twenty is referred to: forty is expressed as "two twenties" (*ogoji*), while higher multiples are named *ogota* (three twenties), *ogerin* (four twenties), and so on to "ten twenties," for which a new word, *igba*, is used. It is in the naming of some of the intermediate numbers that the subtraction principle comes into its own. To take a few examples, the following numbers are given names that indicate the decomposition shown on the right:

$$45 = (20 \times 3) - 10 - 5,$$

$$50 = (20 \times 3) - 10,$$

$$108 = (20 \times 6) - 10 - 2,$$

$$300 = 20 \times (20 - 5),$$

$$318 = 400 - (20 \times 4) - 2,$$

$$525 = (200 \times 3) - (20 \times 4) + 5.$$

All the numbers from 200 to 2,000 (except those that can be directly related to 400, or *irinwo*) are reckoned as multiples of 200. From the name *egbewa* for two thousand, compound names are constructed for numbers in excess of this figure using subtraction and addition wherever appropriate, in ways similar to those shown in the above examples.

The origin of this unusual counting system is uncertain. One conjecture is that it grew out of the widespread practice of using cowrie shells for counting and computation. A description of the cowrie-shell counting procedure given by Mann in 1887 is interesting. From a bag containing a large number of shells, the counter draws four lots of 5 to make 20. Five 20s are then combined to form a single pile of 100. The merging of two piles of 100 shells gives the next important unit of Yoruba numeration, 200. As a direct result of counting in 5s, the subtraction principle comes into operation: taking 525 as an illustration, we begin with three piles of *igba* (200), remove four smaller piles of *oogun* (20), and then add 5 (*aarun*) cowrie shells to make up the necessary number.

This amazingly complicated system of numeration, in which the expression of certain numbers involves considerable feats of arithmetical manipulation, runs counter to the widespread view that indigenous African mathematics is primitive and unsophisticated. But does it have any intrinsic merit for computation? As an example of a calculation that exploits Yoruba numeration to the full, consider the multiplication 19 × 17. The cowry calculator begins with twenty piles of 20 shells each. From each pile, 1 shell is removed (−20). Then three of the piles, now containing 19 shells each, are also removed. The three piles are adjusted by taking 2 shells from one of them and adding 1 each to the other two piles to bring them back to 20: $-20 \times 2 - (20 - 3)$. At the end of these operations, we have

$$400 - 20 - (20 \times 2) - (20 - 3) = 323.$$

While the Yoruba system shows what is possible in arithmetic without a written number system, it is clearly impractical for more difficult multiplications. It is a cumbersome method requiring a good deal of recall and mental arithmetic. Its peculiar characteristics, the base 20 and the subtraction principle of reckoning, seem to have had only a limited impact on other counting systems, even within West Africa.

Further Reflections on African Mathematics

Most books on the history of mathematics ignore Africa, especially areas south of the Sahara. After a cursory treatment of Egyptian mathematics, which in any case is not often considered African, Africa disappears from the reckoning. The publication of Claudia Zaslavsky's *Africa Counts: Number and Pattern in African Cultures* (1973a) seems to have made little impression on the dominant Eurocentrism of the practitioners. Yet a recognition and evaluation of Africa's contribution to mathematics is important, and not only for reasons of restoring a historical balance to the subject. Consideration of African mathematics reminds us of what is often forgotten: mathematics is a pancultural phenomenon that manifests itself in a number of different ways, in counting and numeration systems, in games and leisure pursuits, in art and design, in record keeping and metrology. Our very definition of mathematics has to be broadened to include activities such as counting, locating, measuring, designing, playing, explaining, classifying, sorting. Further, a search for the origins and nature of African mathematics takes us on an adventure of establishing interdisciplinary connections: between mathematics and cosmology; between mathematics and philosophy; between mathematics and technology; between mathematics and linguistics; and between mathematics and cross-cultural psychology. Gerdes and Djebbar (2007) has a list of references to recent studies that attempt to establish these connections.

In this book, we examine in a piecemeal and scattered fashion the mathematics of the African continent. Apart from Egyptian mathematics, examined in chapter 3, there are brief discussions of the Ishango bone, the Yoruba system of numeration, and a mention in passing of the counting system of the Zulus. The geometry of African art and design, the mathematical "ingredients" of games and puzzles, and the implicit mathematics of certain aspects of African astronomy are ignored. Some mention of the type of work done in this area over the last forty years may be instructive. Ascher (1988) discusses the cultural background and mathematical properties of the continuous graphs drawn by the Booshong and Tshokwe who live on the borders of the area adjoining Angola, Zaire, and Zambia. In a general discussion of archaeastronomy, Aveni (1981) draws parallels between cultures in the tropics that appear to have adopted a horizon and zenith approach to the sky, as opposed to the approach with the celestial pole (now Polaris) and the ecliptic/celestial equator, which is more familiar to those from the temperate climates. As a result, navigators tended to use stars on the horizon instead of compass directions. The use of the star compass seemed a characteristic of cultures as geographically far removed as the inhabitants of the Caroline Islands in the Pacific, the Maya of Central America, the Mursi of Ethiopia, and the Bambara of the Sudan. Crowe (1971, 1975, 1982a) discerns complex geometrical patterns from the designs on the smoking pipes of Begho (Ghana), on the textiles and wood carvings of the Bakuba (Tanzania) and in Benin art. Eglash (1995) considers an example of fractals (i.e., scaling in street branching, recursive rectangular enclosures, circles of circular dwellings, etc.) in the layout of settlements of the Mokoulek in Cameroon. Gerdes (1990, 1991, 1995 and 1999) has written voluminously on the "frozen" geometry found in the material artifacts and games of different groups from southern Africa, and his 1995 publication provides a good overview of the burgeoning literature in this area in recent years. Zaslavsky's classic text, originally written in 1973, with a new edition that came out in 1999, still remains a valuable reference to the geometry implicit in African art and design.

There are two other aspects of African mathematics that are often neglected.

1. The enforced diaspora of the Africans resulting from the slave trade was destructive of existing mathematical traditions, and yet games such as "mancala" (a board game based solely on strategy akin to mathematical reasoning, and found today with minor variations in many parts of Africa and elsewhere) were taken in their earlier forms to the Caribbean and the American continents.

2. Skills in drawing and design, and a rich tradition of mental arithmetic, were also taken over in slave ships.

African Diaspora Mathematics: The Case of Thomas Fuller

In discussions of people with extraordinary powers of mental calculation, there is occasionally a mention of Thomas Fuller, an African, shipped to America in 1724 as a slave at the age of fourteen. He was born somewhere between present-day Liberia and Benin. Late in his life, his remarkable powers of calculation made him an example for the abolitionists to demonstrate blacks are not mentally inferior to whites. After his death, Fuller became a source of interest for psychics and psychologists; the latter, even when denying mental abilities of blacks, supported the notion of Fuller as an idiot savant. This was not borne out by those who met him. They remarked on Fuller's general self-taught intelligence and decried a system that prevented him from attaining formal education.

On his death in 1790, an obituary contained the following passage:

Though he could never read or write, he had perfectly acquired the art of enumeration. . . . He could multiply seven into itself, that product by seven, and the products, so produced, by seven, for seven times. He could give the number of months, days, weeks, hours, minutes, and seconds in any period of time that any person chose to mention, allowing in his calculation for all leap years that happened in the time; he could give the number of poles, yards, feet, inches, and barley-corns in any distance, say the diameter of the earth's orbit; and in every calculation he would produce the true answer in less time than ninety-nine men out of a hundred would produce with their pens. . . . He drew just conclusions from facts, surprisingly so, for his (limited) opportunities. . . . Had his opportunity been equal to those of thousands of his fellow-men . . . even a Newton himself, need not have been ashamed to acknowledge him a Brother in Science. (*Columbian Centennial*, December 29, 1790)

Our new understanding of the ethnomathematics of his birthplace allows us to claim that when Thomas Fuller arrived in 1724 Virginia, he had already developed his calculation abilities based on his indigenous traditions. The existing evidence for this claim is not conclusive. However, Bardot's 1732 account of the numerical abilities of the inhabitants of Fida (on the coast of Benin) may be of relevance:

The Fidasians are so expert in keeping their accounts, that they easily reckon as exact, and as quick by memory, as we can do with pen and ink. . . . [This] very much facilitates the trade the Europeans have with them.

Thomas Clarkson backed this up in 1788, writing:

It is astonishing with what facility the African brokers reckon up the exchange of European goods for slaves. One of these brokers has perhaps ten slaves to sell, and for each of these he demands ten different articles. He reduces them immediately by the head to bars, coppers ounces . . . and immediately strikes the balance. The European, on the other hand, takes his pen, and with great deliberation, and with all the advantage of the arithmetic and letters, begins to estimate also. He is so unfortunate often, as to make a mistake; but he no sooner errs, than he is detected by this man of inferior capacity, whom he can neither deceive in the name or quality of his goods, nor in the balance of his account.

Gerdes and Fauvel (1990) wrote an interesting account of Thomas Fuller (1710–1790). Shirley (1988) shows the survival of a numerate tradition, placing high value on mental calculations, among a mainly illiterate population, from whose ancestors captive slaves such as Fuller were transported to the American continents.

Native Americans and Their Mathematics

It is difficult to estimate precisely how many Native Americans were present on the two continents when the Europeans "discovered" them in 1492. An estimate (Denevan 1992, p. 244) puts the number at fifty-four million. The question as to where they came from has fascinated scholars since Columbus's time. It is now generally agreed that the distribution of their population in the Americas, together with their physical appearance and the underlying unity in some aspects of their cultures across the Americas, implies a common Asian origin. An interesting lacuna in all Native American cultures may help to establish a lower bound for the date of their migration from Asia to the Americas. In no culture in the pre-Columbian Americas is there any evidence of the use of the wheel for transport. A probable date for the first use of the wheel for this purpose in central Eurasia is the beginning of the sixth millennium BC. Further, the early Native Americans and the Asians had certain technologies in common: these included the smelting of bronze, the casting of gold, silver, and copper, as well as the arts of weaving and dyeing. These facts together would indicate that the migration was likely to have occurred not long before 12,000 BC.

There is yet another singular feature in the early history of the Native Americans: of all the Americans, the Maya of Central America seem to exhibit in their arts, folklore, and myths a clearer historical memory of their Asian origins. Among them one finds sculptures of elephants, a species not found on the American continents. The Mayan is perceived as being among the most "advanced" of all Native American cultures in terms of its mathematics, astronomy, and technology, and is one of the few cultures that possessed a written language. This leads to the conjecture that the Mayan may have been the prototype of the culture that was transferred from Asia to America during the migration. Other groups, as they roamed across the Americas, splintered away and forgot their origins, whereas the Maya remained closer to their roots.

Knotted Strings from South America

Knots as Aids to Memory

Human memory is remarkable for both its capacity and its complexity. It can store an incredible amount of information, but as a storage device it is often unreliable and not particularly well organized. Therefore from early times all types of mnemonic devices, including notches and knots, have been used as aids to memory. Compared with writing, the use of knots is a clumsy device, though for a preliterate culture it would have had its advantages. Knots were easy to use, convenient to carry around, and they had familiar associations with everyday pursuits such as sewing or fishing. In fact, Niles (2007) reports that recent work by anthropologists has concentrated on the users and uses of knots and how the codes implied in them functioned to validate authority. The knots served one primary purpose: to record and preserve information.

There are a host of anecdotes and legends about knots used for recording the passage of time. To take just one story: at the turn of the twentieth century a German, Karl Weule, reported a conversation he had had with an old inhabitant of the Makonde Plateau in East Africa. At the beginning of a journey the old man would present his wife a piece of bark string with eleven knots. She would be asked to untie a knot each day. The first knot represented the day of his departure, the next three knots the period of his journey, the fifth knot the day he reached his destination, the sixth and seventh knots the days he spent conducting his business, and the next three knots the period of his return journey. So when she had untied the tenth knot, she would know that he was returning home the next day.

It is important not to confuse the purpose of these simple mnemonic knots with that of the *quipu*, to which we shall now turn. Such confusion may arise from failure to distinguish a straightforward numerical magnitude represented by tally marks or knots from the ordinal representation possible on a *quipu* or with a written number system. The fact that the *quipu* cannot be manipulated for calculations, while a written number system can, does not affect the argument. Tying knots in a cord to show a certain numerical quantity is no different from writing the same number on a piece of paper using some widely accepted symbols. This point will become clearer as we proceed

The Inca *Quipu*: Appearance and History

Quipu is a Quechua (the language of the Incas) word meaning "knot." A *quipu* resembles a mop that has seen better days. It consists of a collection of cords, often dyed in one or more colors and containing knots of different types but not, apparently, arranged in any systematic fashion. *Quipus*, of which there are about four hundred authenticated examples, are to be found in the museums of western Europe and the Americas. They were initially thought of as primitive artifacts with little aesthetic appeal. About fifty of these objects have now been carefully studied, the credit for unraveling part of their mystery going to Leyland Locke (1912, 1923). From a close study of statements made by Spanish chroniclers of the sixteenth century and a detailed examination of some of the *quipus*, Locke concluded that the *quipu* was basically a device for recording numbers in a decimal base system.

At its height during the last decades of the fifteenth century AD, the Inca empire occupied an area that today would include all of Peru and parts of Bolivia, Chile, Ecuador, and Argentina. In this vast and difficult terrain lived a culturally diverse population of about six million. It was a well-organized society, cooperative in character, its material culture the creation of a number of different groups that the Inca state was able to organize and control during its short 150-year period of dominance. Yet, despite the level of their material culture, the Incas seem to have lacked the three widely accepted basics of early civilizations: the wheel, beasts of burden, and a written language. Yet the high level of organization required the keeping of detailed accounts and records. In the absence of a system of writing, they used *quipus*.

There is, of course, no contemporary written evidence on the nature and uses of a *quipu* from the society that used it. There are, however, the chronicles of Spanish soldiers, priests, and administrators. The most reliable and unbiased of these chroniclers was a soldier, Cieza de León, who began keeping a record in 1547—fifteen years after the Spanish conquest—and stopped writing three years later. It provides a fascinating account, both of the flora and fauna of the vast territory and of the society there.

There was one aspect of the former Inca state that Cieza found impressive. Across the imperial highways, many of them more substantial than the Roman roads of his native land, were to be found small post houses, the staging posts for runners who carried messages across the difficult mountainous terrain, impossible for any animal to negotiate. These trained runners, called *chasquis*, were stationed in pairs at intervals of about a mile along the highways. Running at top speed and handing their *quipus* on from one runner to the next, as in a relay race, they could transmit a message to the imperial capital Cuzco from three hundred miles away in twenty-four hours. Given the terrain, Cieza noted that this method of carrying messages was superior to using horses and mules, with which he was more familiar.

In two words, Cieza summed up the strengths of the former Inca empire and its ability later to withstand to some degree the havoc brought about by Spanish plunder: order and organization. And the essential prerequisite for maintaining good order and efficient organization was the existence of detailed and up-to-date information (or government statistics, as we would describe such information today) that the state could call upon whenever necessary. Records of all such information were kept on *quipus*.

A whole inventory of resources that included agricultural produce, livestock, and weaponry—as well as people—was maintained and updated regularly by a group of special officials known as *quipucamayus* (*quipu* keepers). Each district under the rule of the Incas had its own specially trained *quipucamayu*, and larger villages had as many as thirty. For information on the role and status of the *quipucamayus* we have a set of remarkable drawings by one Guaman Pomade Ayala, a Peruvian, which form part of a 1,179-page letter to the king of Spain sent in about 1600, some eighty years after the Spanish conquest. Apart from being one of the most searing indictments of Spanish rule, it contains a series of illustrations in which the Inca bureaucracy figures prominently. Seven of these drawings show people carrying *quipus*; two of them are reproduced in figure 2.3.

The inscription in figure 2.3a indicates that the figure holding the *quipu* is none other than the secretary to the Inca (emperor) and his Council. Figure 2.3b shows the chief treasurer to the Inca. There is little doubt that "*quipu* literacy" was widespread among government officials, of whom the *quipucamayus* were important members enjoying high social status.

Figure 2.3b contains another interesting feature, apart from the blank (i.e., unknotted) *quipu* held by the Inca's treasurer. At the bottom left of the drawing is a rectangle divided into twenty cells, in each of which there is a systematic arrangement of small circles and dots probably representing seeds, stones, or similar objects. The Inca abacus, as it has been nicknamed, may have been the device on which computations were worked out before the results were recorded on the *quipu*. We shall return to an explanation of how computations may have been carried out on this "abacus" in a later section. But we begin by looking at how a *quipu* was constructed and used for storing numerical data.

The Construction and Interpretation of a *Quipu*

A *quipu* is constructed by joining together different types of cord. Each cord is at least two-ply, with one end looped and the other tapered and tied with a small knot. Four different kinds of cord can be distinguished. The first type, which is thicker than the rest, is termed the *main cord*. From it are attached like a fringe a number of other cords, most of which hang down and are known as the *pendant cords*, but a few have knotted ends that are directed upward, the *top cords*. In some *quipus* there may be an additional cord whose looped end is connected to the looped end of the main cord and tightened, which explains its name—the *dangle end* cord. To any of these cords suspended from the main cord there may be attached *subsidiary cords*. And this process of attachment may be carried further, so that a subsidiary may be connected to a subsidiary of a subsidiary, and so on. Also, it is possible that some of the pendant cords may be drawn together by means of a single top cord to form a distinct group. What we have after the process is complete is a blank *quipu*, rather like the one in figure 2.3b, which apparently has no top cords. A blank *quipu* can have as few as three cords or as many as two thousand.

There is a further dimension to the construction of a *quipu*—color. The predominant colors of the cords in the *quipus* that have survived are dull white and varying shades of brown. It is not clear whether the small differences in the shades of brown are simply a reflection of the age of the *quipus* rather than real color differences. However, early chroniclers of the Inca culture refer to the use of symbolic color representation for different things: white for silver, yellow for gold, red for soldiers, and so on. The symbolic use of colors is common in many societies. The use of red and green in traffic lights, for example, conveys a meaning that cuts across cultural barriers. Even in those societies where red is not traditionally associated with danger, its appearance in a traffic signal is sufficient to produce a rapid response, much more readily than if the warning were in the form of printed words. And in any case, color being more recognizable than print over a longer distance would clinch the argument for its adoption. But the use of color codes to distinguish between mathematical quantities or operations is unusual in modern mathematics (though not in modern bookkeeping). Yet, as we shall see later, the ancient Egyptians used red ink to represent "auxiliaries," which they calculated as part of arithmetical operations with fractions; the Chinese distinguished between positive and negative numbers by using red and black rods, respectively; and the Indians called algebraic unknowns by the names of different colors. In a *quipu*, color was used primarily to distinguish between different attributes. Each *quipu* had a color coding system to relate some of the cords to one

SECRETARIO:DELINGAICÕZEJO
INCAPQVIPOCNINCAPAC
APOCONARCAMACHICVININQVIPOC:

apolliuyac poma

secretario

(a)

FIGURE 2.3: Two Inca officials holding *quipus*. An "Inca abacus" can be seen in (b) at the bottom left. (Poma de Ayala 1936, pp. 358 and 360. Reproduced with permission.)

CÕTADOR·MAÏOR·ÏTEЗORERO
TAVANTIN·SVÏO·QVÏPOC
CVRACA·CON DOR·CHAVA

con tador ykgouro   con tador

(b)

another and at the same time to distinguish them from other cords. The range and subtlety of color coding was extended by using different combinations of colored yarns.

There have been suggestions that the colors had some numerical significance, but we cannot be certain. What we do know is that numerical representation on a *quipu* was achieved by means of knots. Contemporary records clearly indicate that the Incas used a decimal system of numeration. According to Garcilaso de la Vega (b. 1539), whose mother was the niece of the last king of the Incas (Inca Huayna Capac) and whose father was a Spaniard, the knots indicated a system of notation by position:

> According to their position, the knots signified units, tens, hundreds, thousands, ten thousands and, exceptionally, hundred thousands, and they are all as well aligned on their different cords as the figures that an accountant sets down, column by column, in his ledger.

On each cord except the main cord, clusters of knots were used to represent a certain number. A number shown on one of the pendant cords could be read by counting the number of knots in the cluster of knots closest to the main cord, which represented the highest-value digit, and proceeding along the cord to the next cluster of knots, representing the next positional digit (i.e., the next-lowest power of 10) as far as the "units" cluster, at the other end of the pendant cord. To distinguish the units cluster of knots from the other clusters representing higher positional digits, a different knot was used. Generally a long knot with four turns indicated the units position unless a 1 occurred in the units position, in which case a figure-of-eight (or Flemish) knot was used instead. For all other positions single (or short) knots were used. The absence of a knot indicated zero in any of the positions.

An illustration will be useful at this point. Figure 2.4 shows how the numbers 1,351, 258, and 807 may be represented on a *quipu*, with L, S, and F denoting long, single, and Flemish knots respectively. The left-hand pendant cord contains four knot clusters, of one single knot (1S), three single knots (3S), five single knots (5S), and one Flemish knot (1F), reading downward from the main cord. This may be read as

$$(1 \times 1,000) + (3 \times 100) + (5 \times 10) + (1 \times 1) = 1,351.$$

In a similar manner, the other two pendant cords may be read as

$$(2 \times 100) + (5 \times 10) + (8 \times 1) = 258,$$
$$(8 \times 100) + (0 \times 10) + (7 \times 1) = 807.$$

The spacing of the knot clusters is crucial here. For example, the pendant cord on the right is read as representing 807 because of the considerable space without any knots that exists between the cluster of eight knots (or eight 100s, shown as 8S in figure 2.4) and the cluster of seven knots (or

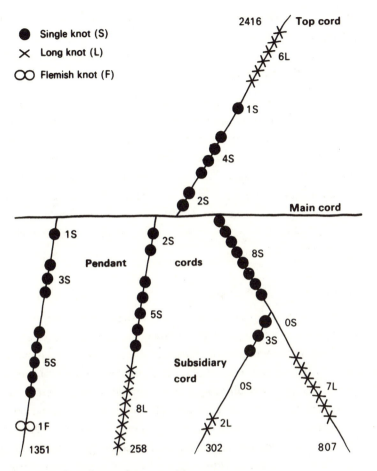

Single knot (S)

Long knot (L)

Flemish knot (F)

2416   Top cord

6L

1S

4S

2S

Main cord

1S

2S

8S

Pendant   cords

3S

5S

5S   0S

3S

Subsidiary

cord   0S   7L

5S

8L

1F   2L

1351   258   302   807

FIGURE 2.4: Recording numbers on a *quipu*

seven 1s, shown as 7L). If there were the usual space between the two clusters of knots, this pendant cord would read as 87.

The knot clusters on a top cord usually represent the sum of the numbers of the pendant cords. So in figure 2.4 the knots on the top cord may be interpreted as

$$(2 \times 1,000) + (4 \times 100) + (1 \times 10) + (6 \times 1) = 2,416 = 1,351 + 258 + 807.$$

The same principles apply to interpreting numbers on subsidiary cords, of which there is only one (with knots representing 302) in figure 2.4. This example is only a simple illustration of one way of using a *quipu*. There are other ways of forming cord groups, by using different colors or by distinguishing between different subgroups of pendant cords to extend the versatility of the *quipu*.

As an illustration of another use[10] of the *quipu*, we have the report of the early chronicler Garcilaso de la Vega:

> The ordinary judges gave a monthly account of the sentences they im-posed to their superiors, and they in turn reported to their immediate superiors, and so on finally to the Inca or those of his Supreme Council. The method of making these reports was by means of knots, made of various colors, where knots of such and such colors denote that such and such crimes had been punished. Smaller threads attached to thicker cords were of different colors to signify the precise nature of the punish-ment that had been inflicted. By such a device was information stored in the absence of writing.

The Mathematics of the *Quipu*

The *quipu* served as a device for storing ordered information, cross-referenced and summed within and between categories. One of the few real-life examples known to us is a *quipu* that was used to record data from a household census of an Andean population in 1567 (Ascher and Ascher 1981; Murra 1968). We shall look at this example in some detail, for it serves to bring out clearly the versatility of the *quipu* as a recording device.

Data for the Andean population of Lupaqa are given for seven prov-inces whose households were classified into two ethnic groups (Alasaa and Maasaa). Each of the two groups is further divided into two subgroups (Uru and Aymara). However, for two of the seven provinces the only in-formation available is the total number of Uru and Aymara households. How was this information fitted into a logical structure, involving cross-categorization and summation, so that it could be recorded on a *quipu*?

We can see from the above information that the household census data contains 26 independent items of information consisting of:

1. The populations of the five provinces for which complete information is available, divided into Alasaa and Maasaa groups and further sub-divided into Uru and Aymara (making a total of 20 items of information)

2. The populations of the two provinces for which the only information available is the number of Uru and Aymara households in each (4 items of information)

3. The total population of households in the two provinces for which information is incomplete (2 items of information).

From the same data it is possible to obtain 20 derived items of information:

1. The grand total of all households: 1 item of information

2. The total number of Uru and Aymara households: 2 items of information

3. The number of households in each province: 7 items of information

4. The number of Alasaa and Maasaa households in each province: 10 items of information

These are the 26 + 20 = 46 items of information—a mixture of given and derived values, and partial and total summations—represented on this household census *quipu*.

The simplest way would be to represent each item of information along the main cord on a pendant cord, equally spaced, using different colors to distinguish categories. But this is a most uneconomical method of formatting information, for it takes no account of the relationships that exist among a number of these items. A more efficient construction, but not optimal in any sense, is to proceed as follows. The information is arranged in seven groups, each group having seven pendant cords. The first four groups relate to the number of Uru and Aymara households in Alasaa and Maasaa for the seven provinces. The first eight pendant cords are blank (i.e., unknotted) since they represent the two provinces for which information is not available separately for Alasaa and Maasaa. The other twenty pendant cords have all the relevant information in the form of clusters of knots. Partial sums and total sums—the 20 items of derived information listed above—can be shown by proper positioning of top cords.

This is one of a number of possible arrangements of cords on a *quipu*. The more the *quipucamayu* considers the pattern of distribution by taking account of the relative sizes and positions of different cords, the better the logical structure of the final representation. Cord placement, color coding, and number representation are the basic constructional features, repeated and recombined to define a format and convey a logical structure. This search for a coherent numerical/logical structure is mathematical thinking.

An Inca Abacus?

The *quipu* could not have been used as a calculating device. While results of summations and other simple arithmetical operations were recorded on the *quipu*, the computations were worked out elsewhere. How did the Incas carry out these calculations? The clue may lie in a passage from a book written by Father Jose de Acosta, a Spanish priest, who lived in Peru from 1571 to 1586:[11]

> To see them use another kind of *quipu* with maize kernels, is a perfect joy. In order to carry out a very difficult computation for which an able computer would require paper and pen, these Indians make use of their kernels. They place one here, three somewhere else and eight, I know not where. They move one kernel here and there and the fact is that they are able to complete their computation without making the smallest mistake. As a matter of fact, they are better at practical arithmetic than we are with pen and ink. Whether this is not ingenious and whether these people are wild animals, let those judge who will! What I consider as certain is that in what they undertake to do they are superior to us. (de Acosta 1596)

Is the priest here describing a form of counting board (*yupanu*) similar in appearance to Poma's drawing shown in figure 2.3b and reproduced in figure 2.5a? There can of course be no conclusive answer. But Wassen (1931), who was among the first to describe and interpret Guarnan Poma's drawings, had an interesting explanation. He interpreted the row values of figure 2.5a, from bottom to top, as successive powers of 10. More controversial is his explanation of the column values of the counting board: that, from left to right, they represent the values 1, 5, 15, and 30.[12] According to this interpretation, the number represented by the dark circles on the counting board, worked out in figure 2.5b, is

$$47 + 21(10) + 20(100) + 36(1,000) + 37(10,000) = 408,257.$$

This is determined in the following way. In the bottom row, there is one black dot in the column marked 30, three in the column marked 5, and 2 in the column marked 1. This gives you $30 + 15 + 2$, or 47. The other numbers in the equation above are derived in a similar manner.

There is no other evidence to substantiate this idiosyncratic interpretation of the columns of the counting board. Indeed, it would appear a

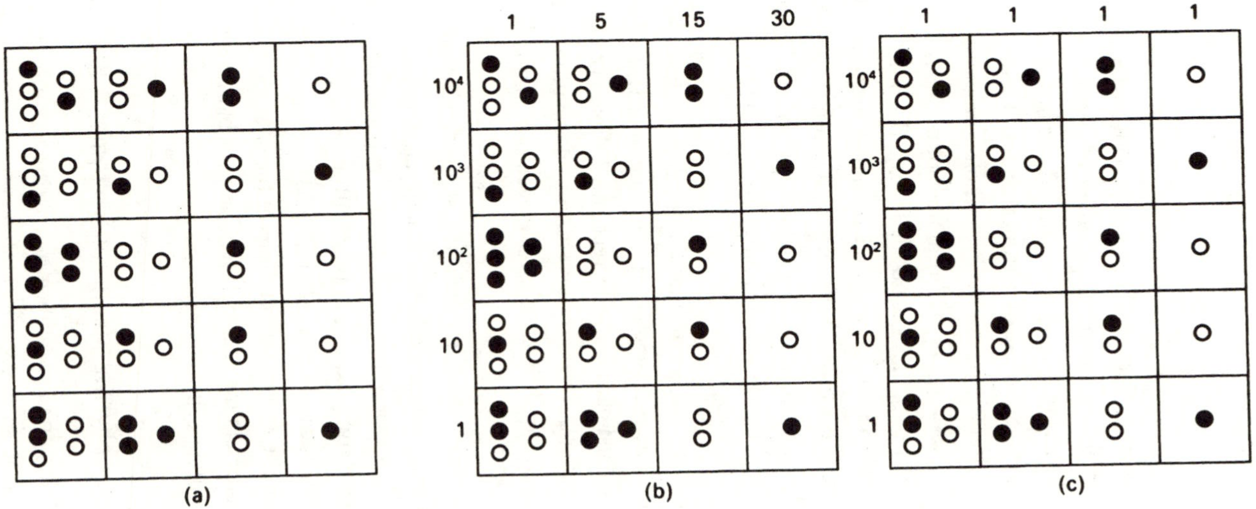

FIGURE 2.5: The Inca abacus: recording numbers

strange choice given the use of a decimal base. A more plausible explanation would be that all column values are equal to 1, as in figure 2.5c, so that the number represented is

$$6 + 3(10) + 6(100) + 3(1,000) + 5(10,000) = 53,636.$$

How might the Incas have used this counting board for computations? Addition and subtraction present few problems. We can only conjecture as to how multiplications may have been carried out before the results were recorded on a *quipu*. Suppose an astronomer-priest were faced with the need to multiply 116 days (to the nearest whole number, the synodic period of the planet Mercury) by 52 (a number of considerable astronomical significance to both the Maya and the Incas). The multiplication could have proceeded as in figure 2.6. Figure 2.6a shows the number 1,160 (i.e., 116×10). A successive process of repeatedly adding 1,160 to itself five times will give the result 5,800 (i.e., $1,160 \times 5$), as shown in figure 2.6b. To complete the multiplication we need only add twice 116 to 5,800 to obtain 6,032, shown in figure 2.6c).[13]

It is not part of my argument that the Incas used this precise method of multiplication (i.e., the *yupanu*, or counting board), or even that this representation on the board is the correct one.[14] What is clear, however, is that before a *quipu* could be used for storing information, some calculations had to be made, and these may have been done on a device like the Inca abacus.

The decimal system found in *quipus* is just a way of recording numbers before writing words. But sometimes humans use a system for arithmetic and a different system for putting their results on paper, papyrus, or cords. There are other intriguing aspects of Inca arithmetic that merit further investigation. Did the Inca use a base 4 or base 5 system of numeration? How would one explain the absence of fractions in Inca numeration? Does their likely use of remainders instead indicate that they had one of the earliest examples of a modular arithmetic system?

The Emergence of Written Number Systems: A Digression

It is possible to view the appearance of a written number (or numeral) system as a culmination of earlier developments. First was the recognition of the distinction between more and less (a capacity we share with certain other animals). From this developed first simple counting, then

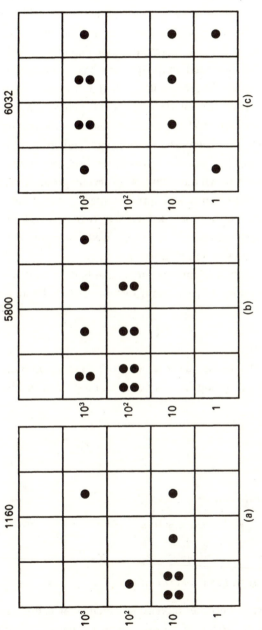

FIGURE 2.6: The Inca abacus: possible multiplication procedure

the different methods of recording the counts as tally marks, of which the Ishango bone is one example. This progression continued with the emergence of more and more complex means of recording information, culminating in the construction of devices such as the *quipu*. Before the appearance of such devices, there must have emerged an efficient system of spoken numbers founded on the idea of a base to enable numbers to be arranged into convenient groups. It was then only a matter of time before a system of symbols was invented to represent different numbers.

There is ample historical and anthropological evidence indicating that a variety of bases have been used over the ages and around the world. The numbers 2, 3, and 4 may have served as the earliest and simplest bases. Anthropologists have drawn our attention to certain groups in central Africa who operated until recently with a rudimentary binary base, so that their spoken numbers would proceed thus: one, two, two and one, two twos, many. Similar systems using base 3 or 4 have been reported to be used by remote communities in South America.

The sheer variety of counting systems that have existed at some time is brought out in figure 2.7. The main systems included counting by twos, fives, and tens.

Counting by Twos

A typical example, discussed earlier, was found among an indigenous Australian group, the Gumulgal. Their counting proceeded as follows:

1 = *urapon*

2 = *ukasar*

3 = *ukasar-urapon*

4 = *ukasar-ukasar*

5 = *ukasar-ukasar-urapon*

6 = *ukasar-ukasar-ukasar*

7 = *ukasar-ukasar-ukasar-urapon*

Clearly, this system becomes increasingly inefficient as the number-words become longer. A version of the two-count system modified to take account of longer word-numbers uses special words for three and four, so six and eight become "twice three" and "twice four." Both versions of the

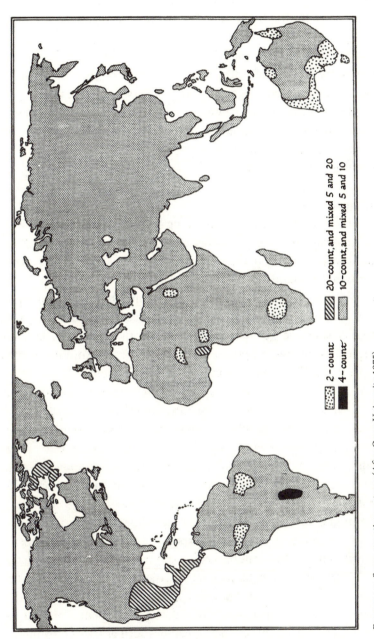

FIGURE 2.7: Some counting systems (After Open University 1975)

two-count system were found to coexist in adjacent areas in Africa, in southern Australia, and in South America, probably indicating a form of evolution from the less to the more efficient version.

Counting by Fives

This suggests the use of finger counting, and proved a more efficient method since it avoided the repetitions of the two-count systems. It was also a more productive procedure since it could be more easily extended to include the use of "two hands," "three hands," and so on for increasing multiples of five. It was only a matter of time before this counting system gave way to counting by tens.

Counting by Tens

It is clear from figure 2.7 that this counting system has been the most widespread, probably because it is associated with the use of fingers on both hands. It was, according to the accounts of the Spaniards, the base used by the Incas in their numeration. The etymology of number-words in the ten-count system may bring out its close association with finger counting. As an illustration, the meanings of the words for the numbers from one to ten in the Zulu language are given in table 2.1. Note the use of the subtraction principle in forming the words for eight and nine: the word for nine, for example, means "leave out one finger" (from ten fingers).

Two other bases have been either popular or mathematically important. The vigesimal scale (i.e., base 20) had its most celebrated development as a number system during the first millennium of the present era among the Maya of Central America. There have also been other base 20 number systems, of which the Yoruba system from West Africa, discussed earlier, is one of the better-known examples. Base 20 systems may have originated from finger and toe counting among early societies.

The origins of the sexagesimal scale (i.e., base 60), first developed in Mesopotamia about five thousand years ago, cannot be directly traced to human physiology. This scale has certain computational advantages arising from the number of integral fractional parts, and survives in certain time and angular measurements even today. There was also another scale (duodecimal, or base 12), which must have enjoyed some popularity in the past because it survives in astronomical quantities such as (twice) the number of hours in a day or the approximate number of lunar months in a year. Other remnants of its usage are found in British units of measurement (12

TABLE 2.1: ZULU WORDS FOR ONE TO TEN

Number	Zulu word	Meaning
1	Nyi	State of being alone
2	Bili	Raise a separate finger
3	Tatu	To pick up
4	Ne	?
5	Hlanu	(All fingers) together
6	Tatisitupa	Take the right thumb
7	Ikombile	Point with forefinger of right hand
8	Shiya' ngalombi/e	Leave out two fingers
9	Shiya'ngalolunye	Leave out one finger
10	Shumi	Make all (fingers) stand

inches = 1 foot), old money (12 pence = 1 shilling), and also in terms such as "dozen" and "gross" (12 dozen).[15] It had some powerful advocates in the past. So convinced was Charles XII of Sweden (1682–1718) of the superiority of this scale over the decimal that he tried—though unsuccessfully—to ban the latter.

Constructing a Written Number System

Once a base (say b) has been chosen, one of the simpler ways of constructing a number system is to introduce separate symbols to represent b^0, b^1, $b^2, \ldots$, whereby these symbols, repeated if necessary, may be used additively to represent any number. Possibly the earliest and certainly the best-known example of such a system with $b = 10$ is the Egyptian hieroglyphic number system, dating back to 3500 BC. (We shall be discussing the principles underlying the construction and arithmetical operations with these numerals in the next chapter.) The Aztecs of Central America later developed a system of numerals similar in principle to the Egyptian number system.

The Aztecs were a people who migrated to Mexico from the north in the early thirteenth century AD and founded a large tribute-based empire, ruled from their capital city Tenochtitlán, which reached the height of its power in the fifteenth and early sixteenth centuries. The empire's prosperity was founded on a highly centralized agricultural system in which existing land was intensively cultivated, irrigation systems were built, and swampland was reclaimed. The staple crop, maize, figures prominently in the Aztec number system.

This number system was vigesimal (base 20) and used four different symbols. The unit symbol was a "blob" representing a maize seedpod; the symbol for 20 was a flag, commonly used to mark land boundaries; 400 was represented by a schematic maize plant; and the symbol for 8,000 is thought to be a "maize dolly," similar to the decorative figures traditionally woven from straw in some European countries. The four symbols were

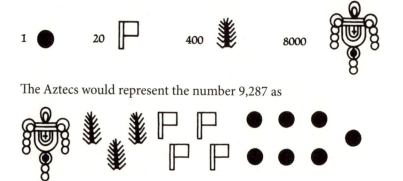

The Aztecs would represent the number 9,287 as

They also developed an intricate system of counting in which the bases depended on the type of objects being counted. Cloths or tortillas would be counted in twenties, while round objects such as eggs or oranges would be counted in tens.

Often an additive grouping system may evolve into a "ciphered" number system. Here new symbols are introduced not only for the powers of the base b but also for

$$1, 2, 3, \ldots, b - 1; 2b, 3b, \ldots, b(b - 1); b^2, 2b^2, 3b^2, \ldots, b^2(b - 1); \ldots$$

While such a system calls for a greater effort to memorize many more symbols, the representation of numbers is obviously more compact and computation more efficient than with a simple additive system. Examples of such systems include later Egyptian numerals (i.e., hieratic and demotic), Ionic (or Greek) alphabetic numerals, early Arabic numerals, and the Indian Brahmi numerals. Some of these number systems will be discussed in later chapters.

There are a few instances of a simple additive grouping system developing into a multiplicative system. In such a system, after a base b has been selected, separate sets of symbols are used for

$$1, 2, \ldots, (b - 1); b, b^2, b^3, \ldots.$$

To represent any number greater than b, the symbols of both sets are used together. One of the best-known of the multiplicative number systems is the standard Chinese system, discussed in chapter 6.

It is likely that a positional or place-value number system such as ours, which evolved from the Indian number system, may have its origin in an earlier multiplicative or even a ciphered number system. In a positional system, after the base b has been selected, any number can be represented by b different symbols. For example, our decimal numeral system requires symbols (or digits) to represent the numbers from 0 to 9: $0, 1, 2, \ldots, 9$. With these symbols we can represent any integer N uniquely as

$$N = c_n b^n + c_{n-1} b^{n-1} + \ldots + c_2 b^2 + c_1 b + c_0,$$

where $c_0, c_1, \ldots, c_{n-1}, c_n$ represent the basic symbols (in our number system, $0, 1, 2, \ldots, 9$). (Note that $0 \leq c_i < b$ for $i = 0, 1, 2, \ldots, n$) For example,

$$1,385 = (1 \times 10^3) + (3 \times 10^2) + (8 \times 10) + (5 \times 1).$$

This is a number system that is both economical in representation, given the requirement of only ten symbols to write any number, and less taxing on memory than a ciphered system. But its chief advantage is its immense computational efficiency when it comes to working with paper and pencil. It is not that arithmetic was impossible in nonpositional number systems, as we shall soon see when we discuss arithmetical operations with Egyptian numerals, but rather that computations were often cumbersome or relied on a mechanical device such as an abacus or a counting board, or even cowrie shells. The crucial advantage of our number system is that it gave birth to an arithmetic that could be done by people of average ability, not just an elite.

There are historical records of only three other number systems that were based on the positional principle. Predating all other systems was the Mesopotamian, which must have evolved during the third millennium BC. A sexagesimal scale was employed, with a simple collection of the correct number of symbols employed to write numbers less than 60. Numbers in excess of 60 were written according to the positional principle, though the absence of a symbol for zero until the early Hellenistic period limited the usefulness of the system for computational and representational purposes. The Chinese rod numeral system is essentially a base 10 system. The numbers $1, 2, \ldots, 9$ are represented by rods whose orientation and location determine the place value of the number represented and whose color shows whether the quantity is positive or negative. We shall be discussing this system in detail in chapter 6. The third positional system was the Mayan, essentially a vigesimal ($b = 20$) system incorporating a symbol for zero. It is to the details of this system that we now turn.

Mayan Numeration

There is an area of Central America whose history and culture were shaped by the Mayan civilization. On the eve of the Spanish conquest of 1519–1520, it occupied over 300,000 square kilometers (approximately the area of the British Isles) and covered present-day Belize, central and southern Mexico, Guatemala, El Salvador, and parts of Honduras. Some cultural similarities helped to unite this vast territory: hieroglyphic writing, lip ornaments, positional numeration, and a calendar built around a year consisting of eighteen months of twenty days and a final month of five extra days. The main agency for the spread of this distinctive culture and retention of cultural links between different regions of Central America was the remarkable civilization of the Maya, which reached its classical phase between the third and tenth centuries AD.

Evidence relating to pre-Columbian Mayan civilization comes from following main sources:

1. Hieroglyphic inscriptions cover the stelae (upright pillars or slabs) found scattered around the region. They were constructed by the Maya every twenty years and span at least fifteen centuries. These stone monuments generally recorded the exact day of their erection, the principal events of the previous twenty years, and the names of prominent nobles and priests.

2. The walls of some Mayan ruins and caves contain paintings and hieroglyphs that provide valuable evidence of not only their everyday life but also their scientific activities.

3. A few manuscripts escaped the destruction of the Spanish conquerors. The most notable are five screen-fold books called codices, namely the Dresden Codex, the Paris (or Peresianus) Codex, the Madrix Codex, the Grolier Codex, and the Troano-Cortesian Codex. These books were made of durable paper from the fiber of the plant *Agave americana* and then covered with size (a gelatinous solution used to glaze paper) before the hieroglyphs were recorded in various colors. There are also thousands of ceramic vessels scattered around in various museums across the world that may supply useful information.

Many hieroglyphs remain undeciphered even today, though some notable investigative work toward the end of the nineteenth century shed some

light on part of the mysteries, especially those inscriptions relating to astronomical or calendar data. It was in the course of this work that the remarkable achievement of the Maya in the field of numeration was discovered.

Mayan Numerals

The Mayan system of number notation was one of the most economical
systems ever devised. In the form that was used mainly by the priests for
calendar computation from as early as 400 BC, it required only three symbols. A dot was used for 1, and a bar for 5; these symbols are thought to
represent a pebble and a stick. Larger numbers were represented by a combination of these symbols up to 19. To write 20, the Maya introduced a
symbol for zero that resembled a snail's shell. A few examples are

$$\bullet \; \vdots\!|||\; | \; \oplus \; = (1 \times 7{,}200) + (18 \times 360) + (5 \times 20) + 0 = 13{,}780.$$

The Dresden Codex, an important source of written evidence on Mayan
numerals, contains a representation of the "snake numbers," which were of
great significance in Mayan cosmology; it is reproduced here in figure 2.8.
There are two sets of numerals represented on the coils: one represented by
solid black circles and bars and the other represented by open circles and
bars. Reading from bottom to top, the Mayan numerals can be transliterated as in table 2.2. In the table, a departure from a strictly vigesimal system occurs at the second number group, where instead of $b^2 = 20^2 = 400$
we have $18b = 360$. Subsequent groups are of the form $18b^i$, where $i = 2$,

TABLE 2.2: NUMBERS REPRESENTED IN FIGURE 2.8

Black number			Red (white) number		
$4 \times 18(20)^4$	=	11,520,000	$4 \times 18(20)^4$	=	11,520,000
$6 \times 18(20)^3$	=	864,000	$6 \times 18(20)^3$	=	864,000
$9 \times 18(20)^2$	=	64,800	$1 \times 18(20)^2$	=	7,200
$15 \times 18(20)$	=	5,400	$9 \times 18(20)$	=	3,240
12×20	=	240	15×20	=	300
19×1	=	19	0×1	=	
		12,454,459			12,394,740

3, ... This anomaly reduces efficiency in arithmetical calculation in that the Mayan zero does not work as an operator, as it should in a true place-value system. For example, one of the most useful facilities with our number system is the ability to multiply a given number by 10 by adding a zero to the end of it. An addition of a Mayan zero to the end of a number would not in general multiply the number by 20, because of the mixed base system employed.

What is the reason for the presence of this curious irregularity in Mayan numeration? To understand the anomaly, we need to appreciate the social context in which the number system was used. As far as we know, this form of writing numbers was used only by a tiny elite, the priests who were responsible for carrying out astronomical calculations and constructing calendars, and the exigencies of these tasks are what lie at the root of the explanation. In other words, the rationale behind the Mayan numeration system was not its effectiveness as a system for calculation but the calendrical requirement of counting the days of eighteen months, each of twenty days.

Before we examine the Mayan calendars, it is worth noting that the Maya had an alternative notation, shown in figure 2.9, which often occurred in inscriptions alongside the "dot and dash" numerals. This was the "head variant" system, relying on a series of distinct anthropomorphic deity-head glyphs to represent zero and 1 to 19. The heads are those of the thirteen deities of the Superior World, with six variants, whose significance is revealed in the next section.

FIGURE 2.8: Mayan "snake numbers" (After Spinden 1924, p. 27)

Mayan Calendars

The Maya had three kinds of calendar. The first, known as the *tzolkin* or "sacred year," was specially devised for carrying out certain religious rituals. It contained 260 days, in twenty cycles of thirteen days each. Superimposed

FIGURE 2.9: Mayan "head variant" numerials (Closs 1986, p. 335. Reproduced with permission)

on each of the cycles was an unchanging series of twenty days, each of which was considered a god to whom prayers and other supplications were to be made. For example, the first of these was known as Imix, which represented the god associated with a crocodile or a water lily. One of the most auspicious days was the last of the twenty days (or gods), known as Ahau and associated with the sun. To complicate matters still further, each of the twenty days was in turn assigned a number from 1 to 13. So once the fourteenth day of a series of twenty was reached, it was allocated the number 1 in a new cycle; thus the twentieth day became 7 Ahau. This procedure for assigning the basic twenty days to the thirteen numbers continued indefinitely. Now, after the date 1 Imix, 260 days would have to pass before it recurred (as there are $13 \times 20 = 260$ possible combinations of the twenty basic days and the first thirteen numbers). Thus a particular day in the religious year of 260 days could be indicated uniquely by adding to the hieroglyph associated with one of the twenty basic days a number corresponding to it from the series of thirteen numbers. Each of the thirteen numbers could represent one of the thirteen gods of the Superior World or one of the thirteen gods of the Inferior World. Even today, among certain descendants of the Maya in Guatemala, a child takes the name and persona of the god associated with its date of birth.

This religious calendar was of limited utility to people like farmers who needed to keep track of the passing seasons for their livelihood. For them there was a second calendar, a true solar calendar known variously as the civil, secular, or vague calendar. This calendar had 360 days, grouped into eighteen monthly periods of twenty days and an extra "month" consisting of five days. The regular months were known as *uinals*, and the additional period was called *uayeb* (which means a period without a name; it was shown by a hieroglyph that represented chaos, corruption, and disorder). Anyone born during this most unlucky period of the civil year was supposed to have been cursed for life.

There was yet a third calendar, known as the *tun*, mainly used for "long" counts and with a unit of 360 days. This calendar was also based on a vigesimal system but the third order, as we saw earlier, was irregular since it consisted of $18 \times 20 = 360$ *kins* (days). The calendar took the following form:

$$20 \textit{ kins } = 1 \textit{ uinal or } 20 \text{ days,}$$

$$18 \textit{ uinals } = 1 \textit{ tun or } 20 \times 18 = 360 \text{ days,}$$

$$20 \textit{ tuns } = 1 \textit{ katun or } 20^2 \times 18 = 7{,}200 \text{ days,}$$

$$20 \textit{ katuns } = 1 \textit{ baktun or } 20^3 \times 18 = 144{,}000 \text{ days,}$$

$$20 \textit{ baktuns } = 1 \textit{ piktun or } 20^4 \times 18 = 2{,}880{,}000 \text{ days,}$$

$$20 \; piktuns \;\; = \;\; 1 \; calabtun \;\; \text{or} \;\; 20^5 \times 18 = 57,600,000 \; \text{days},$$

$$20 \; calabtuns \;\; = \;\; 1 \; kinchiltun \;\; \text{or} \;\; 20^6 \times 18 = 1,152,000,000 \; \text{days},$$

$$20 \; kinchiltuns \;\; = \;\; 1 \; alautin \;\; \text{or} \;\; 20^7 \times 18 = 23,040,000,000 \; \text{days}.$$

For each of these units there was a special head-variant hieroglyph, the head taking one of various forms—man, animal, bird, deity, or some mythological creature. These hieroglyphs were accompanied by the bars and dots standing for the numerals we discussed earlier. On a stela at Quirigua in Guatemala, shown in figure 2.10, is inscribed the date on which it was built, using a calendar system of "long counts."[16] The number represented in figure 2.10 reads:

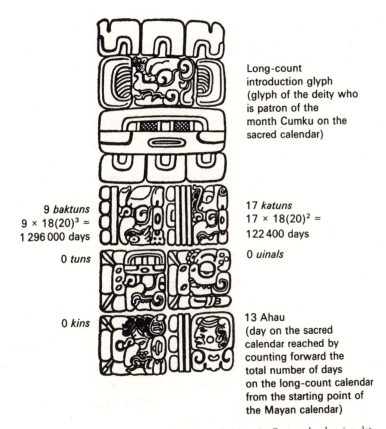

Long-count
introduction glyph
(glyph of the deity who
is patron of the
month Cumku on the
sacred calendar)

9 baktuns
$9 \times 18(20)^3 =$
1 296 000 days

17 katuns
$17 \times 18(20)^2 =$
122 400 days

0 tuns

0 uinals

0 kins

13 Ahau
(day on the sacred
calendar reached by
counting forward the
total number of days
on the long-count calendar
from the starting point of
the Mayan calendar)

FIGURE 2.10: The upper section of the stela at Quirigua in Guatemala, showing dating according to the "long count" calendar (Midonick 1965, p. 143. Reproduced with permission)

9 *baktuns*   17 *katuns*

0 *tuns*   0 *uinals*

0 *kins*

This corresponds to

$$[9 \times 18(20)^3] + [17 \times 18(20)^2] = 1,418,400 \text{ days.}$$

One of the oldest stelae found in Ires Zapotes bears the numbers 8 + $[16 \times 20] + [6 \times 18(20)^2] + [16 \times (20)^3] + [7 \times 18(20)^4]$, which corresponds to 31 BC (Closs 1986, p. 327).[17]

The presence of the irregularity in the operation of the place-value system arose from the need to make all three calendars compatible with one another. Given the central importance in Mayan culture of the measurement of time, the curious anomaly in the third place-value position becomes more comprehensible. But this inconsistency inhibited the development of further arithmetical operations, particularly those involving fractions. Yet one of the more amazing aspects of Mayan astronomy was the high degree of accuracy that was obtained without ever working with fractions, rational or decimal. The Mayan estimate of the duration of a solar year (i.e., the time that it takes the sun to travel from one vernal equinox to the next), expressed in modern terms, is 365.242 days; the currently accepted value is 365.242198 days.[18] A similar degree of accuracy was obtained for the average duration of a lunar month. According to the Mayan astronomers, 149 lunar months lasted 4,400 days. This is equivalent to an average lunar month of 29.5302 days—with all our present-day knowledge and technology, the figure we get is 29.53059 days!

Social Context of Mayan Mathematics

The best account of the Mayan culture around the time of the Spanish conquest comes from a Franciscan priest, Diego de Landa, who recorded the history and traditions of the Mayan people around 1566. Piecing together these different strands of evidence, Closs (1995) and others have constructed an account of the social context in which mathematical sciences flourished during the period between the fourth and tenth centuries AD. This note is based on their findings.

At the top of the educational pyramid was a hereditary leader who was both a high priest (*Ahaucan*) and a Mayan noble. Under him were the master scribes, priests as well as teachers and writers ("engaged in teaching their sciences as well as in writing books about them"). Mathematics was recognized as such an important discipline that depictions of scribes who were adept at that discipline appear in the iconography of Mayan artists. Their mathematical identity was signified in the manner in which they were depicted: either with the Maya bar and dot numerals coming out of their mouths, or with a number scroll being carried under their armpit. Closs (1995, pp. 8–9) discusses a Mayan vase that portrays two seated scribes with opened codices bound in jaguar skin. One of the scribes has, under his armpit, a scroll containing numerical data. The location of the scroll with numerical data under the armpit would indicate that it was a status symbol. It supports Landau's observation that not all the scribes understood the *katuns* (calendrical computation) and those who did enjoyed special prestige. There is an interesting illustration on another Mayan vase of about the same period: at the center of the illustration is seated a supernatural figure with the ears and hooves of a deer, attended by a number of human figures, including a kneeling scribe-mathematician from whose armpit emanates a scroll containing the sequence of numbers 13, 1, 3, 3, 4, 5, 6, 7, 8, and 9. At the top right-hand corner of this illustration there is the small figure of a scribe who looks female, with a number scroll under her armpit indicating that she is a mathematician and, possibly, the one who painted the scene and wrote the text on the vase. She is described as *Ah T'sib* (the scribe). Preceding this text is a glyph that has not been deciphered but which could be her name. Once the name is deciphered, and if the scribe is female, we may have the name of one of the earliest-known female mathematicians. The existence of female mathematician-scribes among the Maya is further supported by another depiction found on a ceramic vase examined by Closs (1992). The text on this vessel contains the statement of the parentage of the scribe in question: "Lady Scribe Sky, Lady Jaguar Lord, the Scribe." Closs (1995, p. 10) adds: "Not only does she carry the scribal title at the end of her name phrase but she incorporates it into one of her proper names, an indication of the importance she herself places on that reality."

• • •

In this chapter we have traveled three continents in search of protomathematics. The intellectual bases of the examples given are mathematical in that they all consist of manipulating numerical systems in order to create some form of number record. Objections to considering these examples as mathematical are mainly on the following grounds.

1. The nature and scope of the mathematical ideas contained in these examples are perceived as fairly trivial, or unimportant in the long-term development of mainstream mathematics. The first part of this argument, dismissing something as trivial, is to ignore the socio-economic environment in which it developed—a reflection of an ahistorical bias that cannot be taken too seriously. The second part, the charge of insignificance, is valid only if one perceives mathematical development as an essentially linear or autonomous process. It is not, as we have sought to show in the previous chapter.

2. A more serious point relates devices such as *quipus*. While *quipus* were important to the Inca state, they were little more than mnemonics and therefore, it is argued, hardly fruitful in generating mathematical concepts or algorithms. How valid is this argument? From our earlier discussion, it is clear that a *quipu* is more than a set of knots to jog the memory: it is a unique device in which numerical, logical, and spatial relationships are brought together for the purpose of recording information and showing correlations between data. As such it throws an interesting light on the nature of different types of categorizations and summations as well as providing elementary exercises in combinatorials. A *quipu* to whose top cord are attached pendant cords from which are suspended subsidiary cords resembles an inverted tree. Questions of mathematical interest then arise: both specific ones, such as how many paths are possible along a particular tree, and general ones, such as how many trees can be constructed with a given number of cords. In modern mathematics the concept of a "tree" is first found in the work of Gustav Kirchhoff, who in 1847 applied it to the study of electrical networks. Arthur Cayley used the same concept in his study of chemical isomers in 1857. From these attempts to answer questions that were essentially similar to those that confronted the makers of the *quipu* emerged a field of modern mathematics known as graph theory. Gerdes (1988a) has another illustration: traditional Angolan sand drawings, wherein the skill lies in constructing figures without retracing a line or removing one's finger from the sand. It was a similar problem about the bridges of Konigsberg that led the Swiss mathematician Leonhard Euler to achieve his breakthrough in graph theory around the middle of the eighteenth century.

3. There is also the argument that the term "mathematics" should be used only for the study of numerical and spatial concepts for their

own sake, rather than for their applications. This is a highly restrictive view of mathematics, often attributed to the Greeks—that mathematics devoid of a utilitarian bent is in some sense a nobler or better mathematics.[19] Where this attitude has percolated into the mathematics curriculum in schools and colleges, it engenders a sense of remoteness and irrelevance associated with the subject in many who study it, and an ingrained elitism in many who teach it.

4. Finally, the most substantive criticism of protomathematics points to a danger that inferences about mathematical activities of the past, especially from artifacts, may be unsound because of a natural tendency to attribute to the ancients our modern modes of thought and knowledge. This is a legitimate concern, and in the course of this chapter I have pointed out the need for caution and for a search for corroborative evidence before any definitive conclusions are drawn.

Notes

1. Since the earlier editions of this book, a second Ishango bone has been found lying in a drawer in the Musée unnoticed for about fifty years. Exhaustive tests indicate that the two bones are 22,000 years old, and not 8,500 nor 11,000 as previously thought. For further details about the latest discovery, see Huylebrouck 2008.

2. For further details, see Bogoshi et al. 1987.

3. Although we shall not be discussing in this book the inferences that have been made by Thom (1967), van der Waerden (1983), and others on the mathematical attainments of the constructors of megalithic monuments, such as Stonehenge in England, it is worth sounding the same note of caution: it is extremely unlikely that the Neolithic lifestyle of the builders of these monuments would have generated the demands or supplied the resources required for developing the "advanced" mathematics attributed to them by such writers.

4. This possibility prompted Zaslavsky (1991) to ask: "Who but a woman keeping track of her cycles would need a lunar calendar?" This led her to conclude: "[African] women were undoubtedly the first mathematicians!" (p. 4).

5. The latest conjecture about the Ishango bone rejects it as a "prime number" recorder or a "moon-month" calendar. Pletser and Huylebrouck (1999) sees the Ishango bone as a "slide ruler." They propose that the counting methods in present-day Africa as well as de Heinzelin's archaeological evidence about the relationship between Egypt, West Africa, and Ishango provide circumstantial evidence supporting their hypothesis.

6. However, a keynote address by Alan Bishop (1995) gives a useful summary of Glendon Lean's work on counting systems.

7. This point will be taken up when we discuss the numeration systems in Mesopotamia in Chapter 4.

8. The names *kina* and *toea* come from the *kina* shell, which was traditionally used as currency in the Papuan region of the country.

9. Africa is a storehouse of different counting practices. Lagercrantz (1973) provides an interesting survey of African counting practices using strings, counting sticks, and tattoos on the body. He draws attention to ancient practices that involved counting by lines drawn on the ground or painted on doors and walls. Zaslavsky (1973a) points to the practice among the Fulani of northern Nigeria, who indicate the fact that they have 100 animals by placing two sticks on the ground to form a V. Sticks forming a cross (X) show 50; and with horizontal sticks indicating tens and vertical sticks units, 23 is denoted by = |||, a written system is already in place. Going farther south, the Bashongo of Congo showed 10 by drawing lines in the sand with three fingers of one hand, completing three groups of lines and then one line: ||| ||| ||| | (Mubumbila 1988). The Bambala count to 5, but the Bangongo, Bohindu, and Sungu to 4. Recent reports show base 10 or 12 were far from being the only ones: as linguists showed, local languages still reveal the use of bases 2, 5, 6, 12, 16, 20, or even 32. Some people do not say "seventeen or seven and teen" but "three less than twenty" or "five and twelve," or use expressions such as "three less than two twenties" or "thirty-two plus five" instead of thirty-seven. For further details and references, see the publications mentioned above.

10. In a recent paper, there is discussion of the discovery of a *quipu* used to store astronomical data. For further details, see Laurencich-Minelli and Magli (2008).

11. In the book, Acosta points to similarities between the customs of the Incas and those of the peoples of Siberia, but then proceeds to make the incorrect inference that the Incas came from Asia about two thousand years before the Spanish conquest of Mexico.

12. In fairness to Wassen, his choice of the values 1, 5, 15, and 30 are not entirely random; they come out from the number of slots in each square in the board.

13. Note that the assumption here is that the three parts of figure 2.6 record the three different stages of the multiplication process. Thus, to find the product of 116 and 52,

$116 \times 10 = 1,160$ (figure 2.6a);

$1,160 \times 5 = 5,800$ (figure 2.6b);

$5,800 + (2 \times 116) = 6,032$ (figure 2.6c).

I am grateful to Ramón Glez-Regueral of Madrid for an alternative interpretation of the passage quoted earlier from Acosta, as well as the drawing of the abacus given in figure

2.3b. In the drawing (if we count only the black dots in each cell of the abacus), the series of numbers given may be arranged as a Fibonacci series of the form 0, 1, 1, 2, 3, 5, 8, 13, 21, ... Let me quote from a personal communication from Ramón Glez-Regueral:

> Don't you see here a trace of Fibonacci before Fibonacci? It has been mathematically proven that Fibonacci, Lucas and Golden Mean number systems are possible. All of them use a binary alphabet (0, 1).... The Inca minds had not yet been "contaminated" by any pre-conceived algorithmic ideas. Just as you observed (in the *Crest of the Peacock*) the mental algorithms developed in the most varied manner by ancient civilizations, and then adopted generation after generation. Maybe the Inca brain jingled some little Fibonacci bells.... Maybe the Fibonacci approach could help in solving the Inca Abacus riddle.

14. There are opposing opinions on whether the *yupanu* was a calculating device. Ascher and Ascher (1997) argue that on the basis of Poma's drawing, "the interpretation of the configuration and the meaning of the unfilled and filled holes (is) highly speculative." On the other hand, Pareja (1986) maintains that it is a calculating device. From the existing information, it is a difficult task to decide whether the *yupanu* provides fresh insight into the computational methods of the Incas.

15. Other historical examples of duodecimal systems include the British units of length: 12 lines = 1 inch, 12 inches = 1 foot, 6 feet = 1 fathom; and the French (ancien régime): 12 *lignes* = 1 *pouce*, 12 *pouces* = 1 *pied*, 6 *pieds* = 1 *toise*, augmented in geodesy by 12 *pieds* = 1 *regle* (*de Perou*).

16. A recent item in the newspapers states that on December 21, 2012, the "long count" calendar of the Mayan people clicks over to year zero, marking the end of a five-thousand-year era. This has led to a minor panic in the Netherlands, where, despite the country's "enlightened" image, thousands of Dutch people are convinced the date coincides with a world catastrophe.

17. According to the calendar of "long counts," this corresponds to a total of 22,507,528 days. The start date of the Mayan "long count" calendars expressed in terms of our calendar was August 13 in the year 3114 BC. Their calendar had an end date of December 21 of AD 2012, when there was supposedly some enormous catastrophe that might even mark the end of the world.

18. Or more precisely, as Kelley (2000) notes, the duration of the solar year at the height of the Mayan civilization was 365.24258 days; the Maya gave dates indicating that twenty-nine periods of 18,980 days equaled 1,508 Mayan years of 365 days, which corresponded to 1,507 solar years. With the Mayan calculation, it would take a magnitude of 70,000 years to accumulate one full day of error!

19. This is brought out well in a passage from Plutarch's description of the Roman siege of Syracuse in 214 BC, at which Archimedes lost his life. After stating that Eudoxus and

Archytas had shown considerable ingenuity in devising mechanical demonstrations of some of the more difficult geometric theorems, Plutarch adds: "Plato was indignant at these developments, and attacked both men for having corrupted and destroyed the ideal purity of geometry. He complained that they had caused [geometry] to forsake the realm of disembodied and abstract thought for that of material objects and to employ instruments which required much base and manual labor" (quoted by Fauvel and Gray 1987, p. 173).

Chapter Three

The Beginnings of Written Mathematics: Egypt

The Urban Revolution and Its African Origins

In the previous chapter we began our examination of early evidence of mathematical activity with an artifact found in the middle of Africa. For the next stage of our journey we remain on the same continent but move north to Egypt. Egypt is generally recognized as the homeland of one of the four early civilizations that grew up along the great river valleys of Africa and Asia over five thousand years ago, the other three being in Mesopotamia, India, and China. Egyptian civilization did not emerge out of the blue as a full-blown civilization without any African roots. This is supported by evidence of large concentrations of agricultural implements carbon-dated to around 13,000 BC, found during the UNESCO-led operations to salvage the ancient monuments of Nubia.

Although there are no tangible traces of the origins of these Neolithic communities, recent archaeological discoveries indicate that they may have belonged to groups from the once-fertile Sahara region who were forced to migrate, initially to the areas south and east, as the desert spread. So, just as Egypt was a "gift of the Nile" (in the words of Herodotus), the culture and people of Egypt were at least initially a "gift" of the heartlands of Africa, the inhabitants of which were referred to at times as "Ethiopians." This is borne out by the historian Diodorus Siculus, who wrote around 50 BC that the Egyptians "are colonists sent out by the Ethiopians. . . . And the large part of the customs of the Egyptians . . . are Ethiopian, the colonists still preserving their ancient manners" (Davidson 1987, p. 7).[1]

It is important that the African roots of the Egyptian civilization be emphasized so as to counter the still deeply entrenched view that the ancient Egyptians were racially, linguistically, and even geographically separated from Africa.[2] The work during the last fifty years, well summarized by Bernal (1987) and Davidson (1987), lays bare the flimsy scholarship and ideological bias of those who persist in regarding ancient Egypt as a separate entity, plucked out of Africa and replanted in the middle of the Mediterranean Sea.

What were the origins of the urban revolution that transformed Egypt into one of the great ancient civilizations? It is not possible to give a definitive answer. All we can do is surmise that the gradual development of effective methods of flood control, irrigation, and marsh drainage contributed to a significant increase in agricultural yield. But each of these innovations required organization. An irrigation system calls for digging canals and constructing reservoirs and dams. Marsh drainage and flood control require substantial cooperation among what may have been quite scattered settlements. Would it be too fanciful to conjecture that, before the emergence of the highly centralized government of pharaonic Egypt, a form of *ujamaa* (self-help communities)[3] may have come into existence as an institutional backup for these agricultural innovations? This may eventually have led to the establishment of administrative centers that grew into cities.[4]

Between 3500 and 3000 BC the separate agricultural communities along the banks of the Nile were gradually united, first to form two kingdoms—Upper and Lower Egypt—which were then brought together, in about 3100 BC, as a single unit by a legendary figure called Menes, who came from Nubia (part of present-day Sudan). Menes was believed to have founded a long line of pharaohs, thirty-two dynasties in all, who ruled over a stable but relatively isolated society for the next three thousand years. With the discovery of the Narmer Palette (dating back to the thirty-first century BC), some archaelogists have raised the possibility that Pharaoh Narmer predates Menes, which would then cast doubts on the traditional accounts. However, there are others who believe that Narmer and Menes are in fact the same person.

It is worth remembering that up to 1350 BC the territory of Egypt covered not only the Nile Valley but also parts of modern Israel and Syria. Control over such a wide expanse of land required an efficient and extensive administrative system. Censuses had to be taken, taxes collected, and large armies maintained. Agricultural requirements included not only drainage, irrigation, and flood control but also the parceling out of scarce arable land among the peasantry and the construction of silos for storing grain and other produce. Herodotus, the Greek historian who lived in the fifth century BC, wrote that

Sesostris [Pharaoh Ramses II, c. 1300 BC] divided the land into lots and gave a square piece of equal size, from the produce of which he exacted an annual tax. [If] any man's holding was damaged by the encroachment of the river. . . . The King . . . would send inspectors to measure the extent of the loss, in order that he might pay in future a fair proportion of the tax at which his property had been assessed. Perhaps this was the way in which geometry was invented, and passed afterwards into Greece. (Herodotus 1984, p. 169)

He also tells of the obliteration of the boundaries of these divisions by the overflowing Nile, regularly requiring the services of surveyors known as *harpedonaptai* (literally "rope stretchers"). Their skills must have impressed the Greeks, for Democritus (c. 410 BC) wrote that "no one surpasses me in the construction of lines with proofs, not even the so-called rope-stretchers among the Egyptians." One can only suppose that "lines with proofs" in this context refers to constructing lines with the help of a ruler and a compass.

There were other pursuits requiring practical arithmetic and mensuration. As the Egyptian civilization matured, there evolved financial and commercial practices demanding numerical facility. The construction of calendars and the creation of a standard system of weights and measures were also products of an evolving numerate culture serviced by a growing class of scribes and clerks. And the high point of this practical culture is well exemplified in the construction of ancient Egypt's longest-lasting and best-known legacy—the pyramids.

Sources of Egyptian Mathematics

Time has been less kind to Egyptian mathematical sources recorded on papyri than to the hard clay tablets from Mesopotamia. The exceptional nature of the climate and the topography along the Nile made the Egyptian civilization one of the more agreeable and peaceful of the ancient world. In this it contrasted sharply with its Mesopotamian neighbors, who not only had a harsher natural environment to contend with but were often at the mercy of invaders from surrounding lands. Yet the very dryness of most of Mesopotamia, as well as the unavailability of any natural writing material, resulted in the creation of a writing medium that has stood the test of time far better than the Egyptian papyrus. However, it must be remembered that papyrus is quite a bit more durable than the palm leaves, bark, or bamboo used as writing materials by the ancient Chinese and Indians. It is interesting in this context to note that, owing to climatic conditions, almost all the papyri that survive are from Egypt and, even among these papyri, the ones that are best preserved belong to certain favored texts. It would therefore follow that basing one's impressions of ancient Egypt on these records could result in a skewed image of the society of that time.

There are two major sources and a few minor ones on early Egyptian mathematics. Most minor sources relate to the mathematics of a later period, the Hellenistic (332 BC to 30 BC) or Roman (30 BC to AD 395) periods of Egyptian history. The most important major source is the Ahmes (or Ahmose) Papyrus, named after the scribe who copied it around 1650 BC from an older document. It is also known as the Rhind Mathematical Papyrus, after the British collector who acquired it in 1858 and subsequently donated it to the British Museum. (Since we know in this instance who penned the document, it would be more proper to name it after the writer than the collector.) The second major source is the Moscow Papyrus, written in about 1850 BC; it was brought to Russia in the middle of the last century, finding its way to the Museum of Fine Arts in Moscow. Between them, the Ahmes and Moscow papyri contain a collection of 112 problems with solutions. At the time of the receipt of the Ahmes Papyrus by the British Museum in 1864, it was highly brittle with sections missing. A fortunate discovery of the missing fragments in the possession of the New York Historical Association in 1922 helped to restore it to its original form, although the two parts still remain with their separate owners.[5]

Other sources include the Egyptian Mathematical Leather Roll, from the same period as the Ahmes Papyrus, which is a table text consisting of twenty-six decompositions into unit fractions; the Berlin Papyrus, which contains two problems involving what we would describe today as simultaneous equations, one of second degree; the Reisner Papyri containing administrative texts from around 1900 BC, consisting of accounts of building construction and carpentry, including a list of workers arranged in groups needed for these activities; the Lahun mathematical fragments, formerly known as the Kahun Papyrus, also from around 1800 BC, containing six scattered mathematical fragments, all of which have now been deciphered;[6] and the Cairo Wooden Boards from the Middle Kingdom period. From a later period, there are the two ostraca texts (i.e., texts written on tiles/

potteries) from the New Kingdom and demotic texts from the Greek and Roman periods. The latter consists of one large papyrus, the Cairo Papyrus, plus six smaller texts plus several ostraca.

There is a third group of Egyptian mathematical texts that come from the last few centuries of the first millennium BC and the first half of the first millennium AD, all of which are written in Greek. A small subsection of texts in this group containing six ostraca, one papyrus roll, and three papyrus fragments are in some way related to Euclid's *Elements*. However, the majority in this group show little or no sign of having been influenced by Greek mathematics. Friberg (2005, p. vii) describes the manuscripts from this group as "non-Euclidean" mathematical texts.[7] And they constitute important evidence, as we shall see later, for tracing possible links between Egyptian, Babylonian, and Greek mathematics.

Ahmes tells us that his material is derived from an earlier document belonging to the Middle Kingdom (2025–1773 BC). There is even the possibility that this knowledge may ultimately have been derived from Imhotep (c. 2650 BC), the legendary architect and physician to Pharaoh Zoser of the Third Dynasty. The opening sentence claims that the Papyrus contains "rules for enquiring into nature, and for knowing all that exists, [every] mystery, . . . every secret." While an examination of the Ahmes Papyrus does not bear this out, it remains, with the tables and eighty-seven problems and their solutions, the most comprehensive source of early Egyptian mathematics, and it was more likely than not a teacher's manual. The Moscow Papyrus was composed (or copied) by a less competent scribe, who remains unknown. It shows little order in the arrangement of topics covered, which are not very different from those in the Ahmes Papyrus. It contains twenty-five problems, among them two notable results of Egyptian mathematics: the formula for the volume of a truncated square pyramid (or frustum), and a remarkable solution to the problem of finding what some interpreters consider to be the curved surface area of a hemisphere. Before looking in detail at the mathematics in these and other sources, we begin the next section with a discussion of the Egyptian system of numeration.

Three types of source materials on Egyptian mathematics can be distinguished: table texts, problem texts, and administrative texts. These texts were the product of a group of scribes, with their clearly defined hierarchy. An interesting glimpse into professional rivalry is shown in the Papyrus Anatasi from the New Kingdom in which one scribe taunts another:[8]

You come here and [try] to impress me with your official status as "the scribe and commander of a work gang." Your arrogance and boastful behavior will be shown up by (how you tackle) the following problem: "A ramp, 730 cubits [long] and 55 cubits wide, must be built, with 120 compartments filled with reeds and beams." It should be at a height of 60 cubits at its peak, 30 cubits in the middle, a slope of 15 cubits with a base of 5 cubits. The quantity of bricks required can be obtained from the troop commander." The scribes are all assembled but no one knows how to solve the problem. They put their faith in you and say: "You are a clever scribe, my friend! Solve [the problem] quickly for your name is famous. . . . Let it not be said: 'There is something he does not know.' Give us the quantity of bricks required. Behold, its measurements are before you; each of its compartments is 30 cubits [long] and 7 cubits [wide]."

It would seem that the problem set deals with four situations that require different calculations: (1) calculating the number of bricks needed to build a ramp; (2) calculating the number of persons needed to move an obelisk; (3) calculating the number of persons needed to erect a colossal statue; and (4) calculating the rations of a group of soldiers of a given size. It is not known whether the "arrogant" scribe solved the problem. However, for a modern reader, the data provided are insufficient to solve the problem, and hence a variety of interpretations have been suggested.[10]

It is clear in this instance that the task set for the scribe was a problem in practical mathematics. A number of other problems had little connection with real life. The teacher scribes were simply showing their student scribes how to apply certain procedures correctly. A scribe was either an instructor or an accountant. If he was an instructor, he was expected to teach "advanced" calculations to his students. If he was an accountant, he had to work out labor requirements, food rations, land allocation, grain distributions, and similar matters for his employers, who were either government officials or wealthy private individuals. It is usually an accountant scribe who is depicted on wall frescoes walking a few paces behind his master.

Number Recording among the Egyptians

From the beginning of the third millennium BC, there are records of names of persons and places as well as those of commodities and their quantities. An example of this is a mace head containing a list of tributes received by the pharaoh Narmer: 120,000 men, 400,000 oxen, and 1,422,000 goats.[11]

To record such large numbers would require a system of numerals that allowed counting to continue almost indefinitely by the introduction of a new symbol wherever necessary.

There is an impression, fostered (no doubt inadvertently) by many textbooks on the history of mathematics, that only one scheme of numeration was used in ancient Egypt: the hieroglyphic. This impression is quite consistent with a view of Egyptian civilization as stable and unchanging, with mathematics primitive yet sufficient to serve the economic and technological needs of the time. The truth is very different from this view. It is possible to distinguish three different notational systems—hieroglyphic (pictorial), hieratic (symbolic), and demotic (from the Greek word meaning "popular")—the first two of which made their appearance quite early in Egyptian history. The hieratic notation was employed in both the Ahmes and the Moscow papyri. It evolved into a script written with ink and a reed pen or other implements on papyrus, ostracon (tile/pottery), leather, or wood, changing from what earlier resembled the hieroglyphic script to a more cursive and variable style suiting the handwriting of the individual scribe. The demotic variant was a popular adaptation of the hieratic notation and became important during the Greek and Roman periods of Egyptian history.

The hieroglyphic system of writing was a pictorial script in which each character represented an object, some easily recognizable. Special symbols were used to represent each power of 10 from 1 to 10^7. Thus a unit was commonly written as a single vertical stroke, though when rendered in detail it resembled a short piece of rope. The symbol for 10 was in the shape of a horseshoe. One hundred was a coil of rope. The pictograph for 1,000 resembled a lotus flower, though the plant sign formed the initial *khaa*, the beginning of the Egyptian word "to measure." Ten thousand was shaped like a crooked finger which probably had some obscure phonetic or allegorical connotation. The stylized tadpole for 100,000 may have been a general symbol of large numbers. One million was shown by a figure with arms upraised, representing the god Heh supporting the sky. On rare occasions, 10 million was represented by the rising sun and possibly associated with Ra, the sun-god, one of the more powerful of the Egyptian deities.[12]

Thus the earliest Egyptian number system was based on the following symbols:

1	10	10^2	10^3	10^4	10^5	10^6	10^7

Any reasonably large number can be written using the above symbols additively, for example:

$$12,013 = 3 + 1(10) + 2(10^2) + 1(10^4) = \;\text{⌁}$$

No difficulties arose from not having a zero or placeholder in this number system. It is of little consequence in what order the hieroglyphs appeared, though the practice was generally to arrange them from right to left in descending order of magnitude, as in the example above. While there were exceptions regarding the orientation of the number symbols in the case of the hieroglyphic numbers, the hieratic was invariably written from right to left.

Addition and subtraction posed few problems. In adding two numbers, one made a collection of each set of symbols that appeared in both numbers, replacing them with the next higher symbol as necessary. Subtraction was merely the reversal of the process for addition, with decomposition achieved by replacing a larger hieroglyph with ten of the next-lower symbol.

The absence of zero is a shortcoming of Egyptian numeration that is often referred to in histories of mathematics. It is clear that an absence of zero as a placeholder is perfectly consistent with a number system such as the Egyptian system. However, in two other senses it may be argued, as Lumpkin (2002, pp. 161–67) has done, that the concept of zero was present in Egyptian mathematics. First, there is zero as a number. Scharff (1922, pp. 58–59) contains a monthly balance sheet of the accounts of a traveling royal party, dating back to around 1770 BC, which shows the expenditure and the income allocated for each type of good in a separate column. The balance of zero, recorded in the case of four goods, is shown by the *nfr* symbol that corresponds to the Egyptian word for "good," "complete," or "beautiful." It is interesting, in this context, that the concept of zero has a positive association in other cultures as well, such as in India *(sunya)* and among the Maya (the shell symbol).

The same *nfr* symbol appears in a series of drawings of some Old Kingdom constructions. For example, in the construction of Meidun Pyramid, it appears as a ground reference point for integral values of cubits given as "above zero" (going up) and "below zero" (going down). There are other examples of these number lines at pyramid sites, known and referred to by Egyptologists early in the century, including Borchardt, Petrie, and Reiner, but not mentioned by historians of mathematics, not even Gillings (1972), who played such an important role in revealing the treasures of Egyptian mathematics to a wider public. About fifteen hundred years after Ahmes, in a deed from Edfu, there is a use of the "zero concept as a replacement to a magnitude in geometry," according to Boyer (1968, p. 18). Perhaps there are other examples waiting to be found in Egypt.

The hieratic representation was similar to the hieroglyphic system in that it was additive and based on powers of 10. But it was far more economical, as a number of identical hieroglyphs were replaced with fewer symbols, or just one symbol. For example, the number 57 was written in hieroglyphic notation as

$$|\,|\,|\,|\,|\,\cap_{\cap}\cap_{\cap}\cap$$

But the same number would be written in hieratic notation as $\rightarrow\!\!\!\!\angle$ where $\rightarrow$ and $\angle$ represent 7 and 50 respectively. It is clear that the idea of a ciphered number system, which we discussed in the previous chapter, is already present here.

While the hieratic notation was no doubt more taxing on memory, its economy, speed, conciseness, and greater suitability for writing with pen and ink must have been the main reasons for its fairly early adoption in ancient Egypt. For example, to represent the number 999 would take altogether twenty-seven symbols in hieroglyphics, compared with three number signs in hieratic representation! And from the point of view of the history of mathematics, the hieratic notation may have inspired, at least in its formative stages, the development of the alphabetic Greek number system around the middle of the first millennium BC. Over the years the hieratic was replaced by an "abnormal hieratic"[13] and demotic, while the hieroglyphic script remained in use throughout.

From as early as the First Dynasty in the Archaic period (3000–2686 BC), thin sheets of whitish "paper" were produced from the interior of the stem of a reedlike plant that grew in the swamps along the banks of the Nile. Fresh stems were cut, the hard outer parts removed, and the soft inner pith was laid out and beaten until it formed into sheets, the natural juice of the plant acting as the adhesive. Once dried in the sun, the writing surface was scraped smooth and gummed into rolls, of which the longest known measures over 40 meters. On these rolls the Egyptians wrote with a brushlike pen, using for ink either a black substance made from soot or a red substance made from ocher.

Egyptian Arithmetic

The Method of Duplation and Mediation

One of the great merits of the Egyptian method of multiplication or division is that it requires prior knowledge of only addition and the 2-times table. A few simple examples will illustrate how the Egyptians would have done their multiplication and division. Only in the first example will the operation be explained in terms of both the hieroglyphic and present-day notation.

EXAMPLE 3.1 Multiply 17 by 13.

Solution

The scribe had first to decide which of the two numbers was the multiplicand—the one he would multiply by the other. Suppose he chose 17. He would proceed by successively multiplying 17 by 2 (i.e., continuing to double each result) and stopping before he got to a number on the left-hand side of the "translated" version below that exceeded the multiplier, $13 (= 1 + 4 + 8)$:

		$\rightarrow 1$	17
		2	34
		$\rightarrow 4$	68
		$\rightarrow 8$	136
		$1 + 4 + 8 = 13$	$17 + 68 + 136 = 221$

The hieroglyph ☐, resembling a papyrus roll, meant "total." The numbers to be added to obtain the multiplier 13 are arrowed.

If this method is to be used for the multiplication of any two integers, the following rule must apply: Every integer can be expressed as the sum of integral powers of 2. Thus

$$15 = 2^0 + 2^1 + 2^2 + 2^3;$$
$$23 = 2^0 + 2^1 + 2^2 + 2^4.$$

It is not known whether the Egyptians were aware of this general rule, though the confidence with which they approached all forms of multiplication by this process suggests that they had an inkling.

This ancient method of multiplication provides the foundation for Egyptian calculation. It was widely used, with some modifications, by the Greeks and continued well into the Middle Ages in Europe. In a modern variation of this method, still popular among rural communities in Russia, Ethiopia, and the Near East, there are no multiplication tables, and the ability to double and halve numbers (and to distinguish odd from even) is all that is required.

EXAMPLE 3.2 Multiply 225 by 17.

Solution

→	225	17
	112	34
	56	68
	28	136
	14	272
→	7	544
→	3	1,088
→	1	2,176

Inspect the left-hand column for odd numbers or "potent" terms (ancient lore in many societies imputed "potency" to odd numbers). Add the corresponding terms in the right-hand column to get the answer.

$$17 + 544 + 1,088 + 2,176 = 3,825.$$

This method, known in the West as the "Russian peasant method," works by expressing the multiplicand, 225, as the sum of integral powers of 2:

$$225 = 1 + 32 + 64 + 128 = 1(2^0) + 0(2^1) + 0(2^2) + 0(2^3) + 0(2^4) + 1(2^5) + 1(2^6) + 1(2^7).$$

Continued . . .

Continued . . .

Adding the results of multiplying each of these components by 17 gives

$(17 \times 2^0) + (17 \times 2^5) + (17 \times 2^6) + (17 \times 2^7) = 17 + 544 + 1{,}088 + 2{,}176 = 3{,}825.$

In Egyptian arithmetic, the process of division was closely related to the method of multiplication. In the Ahmes Papyrus a division x/y is introduced by the words "reckon with y so as to obtain x." So an Egyptian scribe, rather than thinking of "dividing 696 by 29," would say to himself, "Starting with 29, how many times should I add it to itself to get 696?" The procedure he would set up to solve this problem would be similar to a multiplication exercise:

EXAMPLE 3.3 Divide 696 by 29.

Solution

1		29
2		58
4		116
8	→	232
16	→	464

$16 + 8 = 24 \quad 232 + 464 = 696$

The scribe would stop at 16, for the next doubling would take him past the dividend, 696. Some quick mental arithmetic on the numbers in the right-hand column shows the sum of 232 and 464 would give the exact value of the dividend 696. Taking the sum of the corresponding numbers in the left-hand column gives the answer $16 + 8 = 24$.

Where a scribe was faced with the problem of not being able to get any combination of the numbers in the right-hand column to add up to the value of the dividend, fractions had to be introduced. And here the Egyptians faced constraints arising directly from their system of numeration: their method of writing numerals did not allow any unambiguous way of expressing fractions. But the way they tackled the problem was quite ingenious.

Egyptian Representation of Fractions

Nowadays, we would write a noninteger number as either a fraction (2/7) or a decimal (0.285714). From very early times, the ancient Egyptians (as far as we can tell from surviving documents) wrote nonintegers as a sum of unit fractions. So that while a number like 1/7 in hieroglyphic notation consisted of seven vertical lines crowned by the hieroglyphic sign for mouth, the number 2/7 was written as a sum of unit fractions or 2/7 = 1/4 + 1/28. Note that the same unit fraction was not used twice in one representation (i.e., 2/7 = 1/7 + 1/7 was not allowed); and there is no known explanation for the adoption of this convention. This system of representing fractions became known as the Egyptian system.

There have been different views as to why Egyptians followed this style of fractional representation. A traditional view, widely accepted even today, is that the representation reflected the notational, or conceptual, limitation of their number system. To many of us brought up on fixed-base representations of numbers, it is difficult to imagine a situation that favored the Egyptian usage. One could, of course, contrast the "exactness" of the Egyptian representation with the "approximate" nature of fixed-base expansions of the neighboring Mesopotamians, whose system is discussed in chapter 4. We could marvel at the way that Egyptian scribes performed complicated arithmetical operations. They would take 16 + 1/56 + 1/679 + 1/776, find 2/3 of it as 10 + 2/3 + 1/84 + 1/1,358 + 1/4,074 + 1/1,164, then add 1/2 of it and 1/7 of it and show that it all adds up to 37!

The Egyptian preference for "exactness" has some interesting historical parallels with the preference of their students, the Greeks, for geometry over symbolic arithmetic. The Egyptian style may have contributed to the Pythagorean number mysticism and the number theory that grew out of it. For example, the notion of a "perfect number," which is equal to the sum of its proper divisors, would now seem to many as esoteric. But perfect numbers have certain practical uses in working with Egyptian fractions. And the subject provides, even today, a source of mathematical puzzles and problems in abstract number theory.

The puzzle still remains: what practical purpose was served by the Egyptian unit fractions? One reason for expanding rational fractions is to facilitate easy comparison of different quantities. For example, if you had to choose between being paid 1/7 of a bushel of corn or 13/89 of a bushel, which should you take? In terms of the Babylonian arithmetic the two quantities could be expressed in base 60 as

$$\frac{1}{7} \approx \frac{8}{60} + \frac{34}{60^2} + \frac{17}{60^3},$$

$$\frac{13}{89} \approx \frac{8}{60} + \frac{45}{60^2} + \frac{50}{60^3}.$$

It is clear from inspection that while the first terms on the right of both quantities are identical, the second term of 13/89 is larger than the second term of 1/7.

An estimate of the relative magnitude of the two fractions using the Egyptian approach is cumbersome. For example, the decomposition of 13/89 is impossibly complicated, as shown below:

$$\frac{13}{89} = \frac{(1+4+8)}{89} = \frac{1}{89} + \left(\frac{1}{30} + \frac{1}{178} + \frac{1}{267} + \frac{1}{445}\right) + \left(\frac{1}{15} + \frac{1}{89} + \frac{2}{267} + \frac{2}{445}\right)$$

$$= \frac{1}{15} + \frac{1}{30} + \frac{1}{60} + \frac{1}{178} + \frac{1}{267} + \frac{1}{356} + \frac{1}{445} + \frac{1}{534} + \frac{1}{890} + \frac{2}{267} + \frac{2}{445}.$$

Note that the last two terms in the expansion for 13/89 can be converted to unit fractions if we had a $2/n$ table extending up to $n = 445$. (The $2/n$ table given in the Ahmes Papyrus and shown in table 3.1 later in this chapter provides only odd values of n up to 101). At some point, the quantities have to be reduced to expansions that have a common denominator for a comparison to be made.

Operations with Unit Fractions

Operating with unit fractions is a singular feature of Egyptian mathematics and is absent from almost every other mathematical tradition. A substantial proportion of surviving ancient Egyptian calculations make use of such operations—of the eighty-seven problems in the Ahmes Papyrus, only six do not. Two reasons may be suggested for this great emphasis on fractions. In a society that did not use money, where transactions were carried out in kind, there was a need for accurate calculations with fractions, particularly in practical problems such as division of food, parceling out land, and mixing different ingredients for beer or bread. We shall see later that a number of problems in the Ahmes Papyrus deal with such practical concerns.

A second reason arose from the peculiar character of Egyptian arithmetic. The process of halving in division often led to fractions. Consider how the Egyptians solved the following problem (no. 25) from the Ahmes Papyrus.

EXAMPLE 3.4 Divide 16 by 3.

Solution

1	→	3
2		6
4	→	12
2/3		2
1/3	→	1

$1 + 4 + 1/3 = 5 + 1/3$ 16

As $12 + 3 = 15$ falls one short of 16, the Egyptian scribe would proceed by working out 2/3 of 1 and then halving the result (i.e., $1/2 \times 2/3 = 1/3$). These steps are shown on the left. Now, $3 + 12 + 1 = 16$. The sum of the corresponding figures in the left-hand column gives the answer $5\frac{1}{3}$.

Two important features of Egyptian calculations with fractions are highlighted here:

1. Perverse as it may seem to us today, to calculate a third of a number a scribe would first find two-thirds of that number and then halve the result. This was standard practice in all Egyptian computations.

2. Apart from two-thirds (represented by its own hieroglyph, either ⩗ or ⨁), Egyptian mathematics had no compound fractions: all fractions were decomposed into a sum of unit fractions (fractions such as 1/4 and 1/5).

To represent a unit fraction, the Egyptians used the symbol ⌒, meaning "part," with the denominator underneath. Thus 1/5 and 1/40 would appear as ⌒ and ⌒ respectively.

The 2/n Table: Its Construction

The dependence on unit fractions in arithmetical operations, together with the peculiar system of multiplication, led to a third aspect of Egyptian computation. Every multiplication and division involving unit fractions would invariably lead to the problem of how to double unit fractions. Now, doubling a unit fraction with an even denominator is a simple matter of halving the denominator. Thus doubling 1/2, 1/4, 1/6, and 1/8 yields 1, 1/2, 1/3, and 1/4. Doubling 1/3 raised no difficulty, for 2/3 had its own hieroglyphic or hieratic symbol. But it was in doubling unit fractions with other odd denominators that difficulties arose. For some reason unknown to us, it was not permissible in Egyptian computation to write 2 times 1/n as 1/n + 1/n. Thus the need arose for some form of ready reckoner that would provide the appropriate unit fractions that summed to 2/n, where n = 5, 7, 9, ...

At the beginning of the Ahmes Papyrus there is a table of decomposition of 2/n into unit fractions for all odd values of n from 3 to 101. In the papyrus, the decomposed unit fractions are marked in red ink. A few of its entries are given in table 3.1.

The usefulness of this table for computations cannot be overemphasized: it may quite legitimately be compared in importance to the logarithmic tables that were used before the advent of electronic calculators. The table is interesting for a number of reasons. For one, it does not contain a single arithmetical error, in spite of the long and highly involved calculations that its construction must have entailed; it may be a final corrected version of a number of earlier attempts that have not survived.

There is an even more remarkable aspect to this table. With the help of a computer, it has been worked out that there are about twenty-eight thousand different combinations of unit fraction sums that can be generated for 2/n, n = 3, 5, . . . , 101. The constructor of this table arrived at a particular subset of fifty unit-fraction expressions, one for each value of n. According to Gillings (1972), it is possible to discern certain guidelines for the sets of values chosen. There is

1. a preference for small denominators, and none greater than 900;

2. a preference for combinations with only a few unit fractions (no expression contains more than four);

3. a preference for even numbers as denominators, especially as the denominator of the first unit-fraction in each expression, even

TABLE 3.1: SOME ENTRIES FROM THE AHMES PAPYRUS 2/N TABLE

2/n	Unit fractions
2/5	1/3 + 1/15
2/7	1/4 + 1/28
2/9	1/6 + 1/18
2/15	1/10 + 1/30
2/17	1/12 + 1/51 + 1/68
2/19	1/12 + 1/76 + 1/114
2/45	1/30 + 1/90
2/47	1/30 + 1/141 + 1/470
2/49	1/28 + 1/196
2/51	1/34 + 1/102
2/55	1/30 + 1/330
2/57	1/38 + 1/114
2/59	1/36 + 1/236 + 1/531
2/95	1/60 + 1/380 + 1/570
2/97	1/56 + 1/679 + 1/776
2/99	1/66 + 1/198
2/101	1/101 + 1/202 + 1/303 + 1/606

though they are large or might increase the number of terms in the expression.

To take an example, according to Gillings's calculations the fraction 2/17 can be decomposed into unit-fraction summations in just one way if there are two unit-fraction terms, 11 ways with three unit-fraction terms, and 467 ways with four unit-fraction terms. Table 3.1 shows that the constructor opted for one of the three-unit-fraction groups, $2/17 = 1/12 + 1/51 + 1/68$, rather than the solitary two-unit-fraction group, $2/17 = 1/9 + 1/153$. It would seem that criteria (1) and (3) prevailed in this instance.[14]

Multiplication and Division with Unit Fractions

The main purpose of constructing the table was to use it for multiplication and division. Let us consider one example of each to illustrate its use. First, multiplication.

EXAMPLE 3.5

Multiply $1\frac{8}{15}$ by $30\frac{1}{3}$ (or $1 + \frac{1}{3} + \frac{1}{5}$ by $30 + \frac{1}{3}$).

Solution

	1	$1 + 1/3 + 1/5$
→	2	$2 + 2/3 + 2/5 = 2 + 2/3 + 1/3 + 1/15$
→	4	$6 + 2/15 = 6 + 1/10 + 1/30$
→	8	$12 + 1/5 + 1/15$
→	16	$24 + 2/5 + 2/15 = 24 + 1/3 + 1/15 + 1/10 + 1/30$
	2/3	$2/3 + 2/9 + 2/15 = 2/3 + 1/6 + 1/18 + 1/10 + 1/30$
→	1/3	$1/3 + 1/12 + 1/36 + 1/20 + 1/60$

$2 + 4 + 8 + 16 + 1/3 = 30 + 1/3$ $46 + 1/5 + 1/10 + 1/12 + 1/15 + 1/30 + 1/36$

The product of the two numbers using modern multiplication would be $46\frac{23}{45}$, which is exactly equivalent to the Egyptian result given in the last row. In the course of multiplication we have taken the unit fraction terms of 2/5, 2/15, and 2/9 from table 3.1. And because Egyptian multiplication was based on doubling, only table 3.1 was required. The sheer labor and tedium of this form of multiplication should not make us forget how modest is the "tool kit" required. The ability to double, halve, and work with the fraction "two-thirds," together with the $2/n$ table, is sufficient.

To illustrate division with fractions, we take one of the more difficult problems of its kind from the Ahmes Papyrus, problem 33, which may be restated in modern language as follows.

EXAMPLE 3.6 The sum of a certain quantity together with its two-thirds, its half, and its one-seventh becomes 37. What is the quantity?

Continued . . .

Continued...

Solution

In the language of modern algebra, this problem is solved by setting up an equation of the first degree in one unknown. Let the quantity be x. The problem is then to solve

$$\left(1 + \frac{2}{3} + \frac{1}{2} + \frac{1}{7}\right)x = 37$$

to give

$$x = 37 \div \left(1 + \frac{2}{3} + \frac{1}{2} + \frac{1}{7}\right) = 16\frac{2}{97}.$$

The problem restated: Divide 37 by $(1 + 2/3 + 1/2 + 1/7)$.

1	1 + 2/3 + 1/2 + 1/7	
2	4 + 1/3 + 1/4 + 1/28 (2/7 = 1/4 + 1/28 from the 2/n table)	
4	8 + 2/3 + 1/2 + 1/14	
8	18 + 1/3 + 1/7	
→ 16	36 + 2/3 + 1/4 + 1/28	

At this point in the procedure, two questions arise:

1. The right-hand side of the last row is close to 37, which is the dividend. What must be added to 2/3 + 1/4 + 1/28 to make up 1? With our present method, we find the answer, 1/21.

2. The next question is: By what must the divisor 1 + 2/3 + 1/2 + 1/7 be multiplied to get 1/21? The answer is 2/97, or in unit fractions 1/56 + 1/679 + 1/776, from table 3.1.

So the solution is

$$37 \div \left(1 + \frac{2}{3} + \frac{1}{2} + \frac{1}{7}\right) = 16 + \frac{1}{56} + \frac{1}{679} + \frac{1}{776} = 16\frac{2}{97}.$$

Egyptian Division: The Use of "Red Auxiliaries"

The real question remains: How would the Egyptians, working within the constraints of their arithmetic, have dealt with the problems raised by questions 1 and 2 in example 3.6? A study of some of the problems in the Ahmes Papyrus provides us with the answer. Problems 21 to 23 are commonly known as "problems in completion," since they are expressed as

Complete 2/3 1/15 to 1.   (Problem 21)

Complete 1/4 1/8 1/10 1/35 1/45 to 3.   (Problem 23)

These problems are similar to the one in question 1 above, which may also be expressed in this way:

Complete 2/3 1/4 1/28 to 1.

The Egyptians adopted a method of solution that is analogous (but not equivalent) to the present-day method of least common denominator. First they took the denominator of the smallest unit-fraction as a reference number, and then they multiplied each of the fractions by this number to obtain "red auxiliaries" (so named because the scribe wrote these numbers in red ink). They proceeded to calculate by how much the sum of these auxiliaries fell short of the reference number. This shortfall quantity was then expressed as a fraction of the reference number to obtain the desired complement. If the shortfall quantity turned out to be an awkward fraction, a further search was made for a reference number that would result in more manageable auxiliaries. So, how was question 1 tackled the Egyptian way?

EXAMPLE 3.7 Complete 2/3 1/4 1/28 to 1.

Solution

$$\frac{2}{3} + \frac{1}{4} + \frac{1}{28} + \text{(some fraction)} = 1;$$
$$28 + \left(10 + \frac{1}{2}\right) + \left(1 + \frac{1}{2}\right) + 2 = 42.$$

The denominator of the smallest fraction, 28, is not a suitable reference number given the auxiliaries that result. Instead, 42 is chosen for ease of calculation because it is important that the sum of the auxiliaries belonging to the divisor in example 3.6 is an integer. Thus 42 is the lowest common multiple of the numbers 1, 3, 2, and 7. However, the reference number chosen in Egyptian computation was not necessarily the lowest common multiple. So what fraction(s) of 42 will give 2? The answer is 1/21.

Continued . . .

Continued . . .

The next step is to find by what fraction the divisor $1 + 2/3 + 1/2 + 1/7$ (from example 3.6) must be multiplied to get $1/21$. In other words, we have to divide $1/21$ by $1 + 2/3 + 1/2 + 1/7$:

→	1	21
→	2/3	14
→	1/2	10 + 1/2
→	1/7	3

$1 + 2/3 + 1/2 + 1/7$	$48 + 1/2$

Now, $1 \div (48 + 1/2) = 2/97 = 1/56 + 1/679 + 1/776$ (obtained from the $2/n$ table).

Hence $37 \div (1 + 2/3 + 1/2 + 1/7) = 16 + 1/56 + 1/679 + 1/776$.

We have not followed the scribe all the way in his solution to the problem, for the reason that at one stage his approach requires an addition of sixteen unit-fractions, the last six of which are $1/1,164$, $1/1,358$, $1/1,552$, $1/4,074$, $1/4,753$, and $1/5,432$! We can only assume that the scribe was either an incredible calculator or that he had a battery of tables that he could consult when called upon to add different combinations of unit fractions. However, the more likely but mundane explanation is that the scribe "cheated," since he knew what the answer should be! The fact remains: the Egyptians were inveterate table makers, and the summation table of unit fractions contained in the Leather Roll and the decomposition table of $2/n$ in the Ahmes Papyrus are prime examples.[15]

It is unlikely that the original problem (example 3.6) had any practical import. In an attempt probably to illustrate, for the benefit of trainee scribes, the solution of simple equations of this type, an unfortunate choice of numbers led to difficult sets of unit fractions with the attendant cumbersome operations, which the scribe accomplished without faltering. One is again struck by the mental agility of the scribes who could perform such feats with a minimum of mathematical tools to call upon. The use of red auxiliaries is further evidence of the high level of Egyptian achievement in computation, since they enabled any division, however complicated, to be performed.

Applications of Unit Fractions: Distribution of Loaves

As has been suggested, the exclusive use of unit fractions in Egyptian mathematics also had a practical rationale. This is brought out quite clearly in the first six problems of the Ahmes Papyrus, which are concerned with sharing out n loaves among ten men, where $n = 1, 2, 6, 7, 8, 9$. As an illustration let us consider problem 6, which relates to the division of 9 loaves among ten men. A present-day approach would be to work out the share of each man, i.e., 9/10 of a loaf, and then divide the loaves so that the first nine men would each get 9/10 cut from one of the 9 loaves. The last man, however, left with the 9 pieces of 1/10 remaining from each loaf, might well regard this method of distribution as less than satisfactory. The Egyptian method of division avoids such a difficulty. It consists of first looking up the decomposition table for $n/10$ and discovering that 9/10 = 2/3 + 1/5 + 1/30. The division would then proceed as shown in figure 3.1: seven men would each receive 3 pieces of bread, consisting of 2/3, 1/5, and 1/30 of a loaf. The other three men would each receive 4 pieces consisting of two 1/3 pieces, a single 1/5 piece, and a single 1/30 of a loaf. Justice is not only done, but seen to be done!

Applications of Unit Fractions: Remuneration of Temple Personnel

In a nonmonetary economy, payment for both goods and labor is made in kind. Often the choice of the goods that act as measures or standards

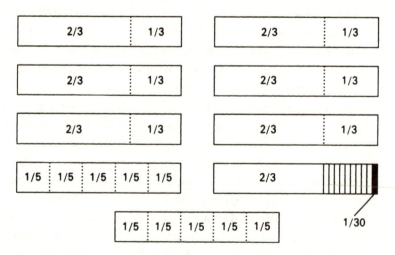

FIGURE 3.1: Problem 6 from the Ahmes Papyrus: sharing 9 loaves among 10 men (After Gillings 1962, p. 67)

of value provides interesting insights into the character of the society. In Egypt, bread and beer were the most common standards of value for exchange. A number of problems in the Ahmes Papyrus concern these goods, dealing with their distribution among a given number of workers, and also with strength (*pesu*) of different types of these two commodities. We shall be examining one of the *pesu* problems later in this chapter. But first we look at an example, brought to our attention by Gillings (1972), that sheds some interesting light.

Table 3.2, adapted from Gillings's book, is a record of payments to various temple personnel at Illahun around 2000 BC. The payments were made in loaves of bread and two different types of beer (referred to here as beer A and beer B). The temple employed 21 persons and had 70 loaves, 35 jugs of beer A, and $115\frac{1}{2}$ jugs of beer B available for distribution every day. The unit of distribution was taken to be 1/42 of a portion of each of these items, which worked out as 1 + 2/3 loaves of bread, 2/3 + 1/6 jug of beer A, and 2 + 2/3 + 1/10 jugs of beer B.

The table is interesting for a number of reasons. It contains an interesting example of an arithmetical error on the part of a scribe: the unit of distribution of beer B was wrongly worked out as 2 + 2/3 + 1/10 (the correct

TABLE 3.2: REMUNERATION OF THE PERSONNEL OF ILLAHUN TEMPLE (UNITS OF DISTRIBUTION PER PERSON)

	Number of portions received	Commodity		
Status of personnel		Bread (1 + 2/3 loaves)	Beer A (2/3 + 1/6 jugs)	Beer B* (2 + 1/2 + 1/4 jugs)
Temple director	10	16 + 2/3	8 + 1/3	27 + 1/2
Head reader	6	10	5	16 + 1/2
Usual reader	4	6 + 2/3	3 + 1/3	11
Head lay priest	3	5	2 + 1/2	8 + 1/4
Priests, various (7)	14	23 + 1/3	11 + 1/3	37 + 1/2
Temple scribe	1 + 1/3	2 + 1/6 + 1/18	1 + 1/9	3 + 2/3
Clerk	1	1 + 2/3	2/3 + 1/6	2 + 1/2 + 1/4
Other workers (8)	2 + 2/3	4 + 1/3 + 1/9	2 + 1/6 + 1/18	7 + 1/3
Totals	42	70	35	115 + 1/2

Note: Adapted from table 11.2 in Gillings (1972)

*The scribe made an error in working out the amount in one portion of beer b, 1/42 of 115 + 1/2, which he estimated as 2 + 2/3 + 1/10 instead of 2 + 1/2 + 1/4. This mistake has been rectified along the lines indicated by Gillings.

value is 2 + 1/2 + 1/4) . The scribe proceeded to use the incorrect figure in working out the shares of different personnel, but did not apparently check his calculations by adding all the shares. He wrote total as $115\frac{1}{2}$ jugs, which is what it should have been, whereas the table adds up to $114\frac{1}{2}$ jugs. Also, the table bears ample testimony to the facility with which the Egyptians could handle fractions. Given all the limitations of their number system, they proved to be extremely adept at computations. Further, the minute fractional division of both beer and bread suggests a highly developed system of weights and measures: it is intriguing how 2 + 1/6 + 1/18 jugs of beer A were shared equally among eight "other workers."

From the table we have some indication of the relative status of the personnel at the temple. At the top was the high priest, or temple director, often a member of the royal family. Among his duties was to pour out the drink-offering to the gods and to examine the purity of the sacrificial animals. It was only after he had "smelt" the blood and declared it pure that pieces of flesh could be laid on the table of offerings. Hence *Ue'b*, meaning "pure," was the name by which he was known. Perhaps more important than the *Ue'b*, from a ritual point of view, was the head reader (or reciter-priest), whose duty it was to recite from the holy books. Since magical powers were attributed to these texts, it was generally believed that the reciter-priest was a magician, making him in status and remuneration second only to the high priest. After him came other classes of the priesthood, the largest of which was known as the "servants of God." Some of them were prominent in civil life; others were appointed to serve particular gods. Their job included washing and dressing statues of assigned deities and making offerings of food and drink to them at certain times of the day. The scribes came quite low on the list, though this was not the case in other walks of life—most scribes, particularly those associated with the royal court, enjoyed considerable status and power.

Egyptian Algebra: The Beginnings of Rhetorical Algebra

It is sometimes claimed that Egyptian mathematics consisted of little more than applied arithmetic, and that one cannot therefore talk of Egyptian algebra or geometry. We shall come to the question of Egyptian geometry, but first we consider the existence or otherwise of an entity called Egyptian algebra. Algebra may be defined as a branch of mathematics of generalized arithmetical operations, often involving today the substitution of letters for numbers to express mathematical relationships.

The rules devised by mathematicians for solving problems about numbers of one kind or another may be classified into three types. In the early stages of mathematical development these rules were expressed verbally and consisted of detailed instructions, without the use of any mathematical symbols (such as $+$, $-$, $\div$, $\sqrt{}$), about what was to be done to obtain the solution to a problem. For this reason this approach is referred to as "rhetorical algebra." In time, the prose form of rhetorical algebra gave way to the use of abbreviations for recurring quantities and operations, heralding the appearance of "syncopated algebra." Traces of such algebra are to be found in the works of the Alexandrian mathematician Diophantus (c. AD 250), but it achieved its fullest development—as we shall see in later chapters—in the work of Indian and Islamic mathematicians during the first millennium AD. During the past five hundred years "symbolic algebra" has developed. In this type of algebra, with the aid of letters and signs of operation and relation ($+$, $-$, $\div$, $\sqrt{}$), problems are stated in such a form that the rules of solution may be applied consistently and systematically. The transformation from rhetorical to symbolic algebra marks one of the most important advances in mathematics. It had to await

1. the development of a positional number system, which allowed numbers to be expressed concisely and with which operations could be carried out efficiently;

2. the emergence of administrative and commercial practices which helped to speed the adoption, not only of such a number system, but also of symbols representing operators.

It is taking too narrow a view to equate the term "algebra" just with symbolic algebra. If one examines the hundred-odd problems in the existing Egyptian mathematical texts, of which most are found in the Ahmes Papyrus, one finds that they are framed in a manner that may be described as "rhetorical" and "algorithmic" or procedure-based. Further, in the case of examples from the Ahmes Papyrus, one can discern distinct stages in laying out a problem and its solution: statement of the problem, the procedure for its solution, and verification of the result. It is interesting to note that the examples in the Moscow Papyrus contain just the statement of the problem (or a diagram) and cryptic instructions for its solution.

As an illustration let us look at problem 72 of the Ahmes Papyrus, restated in modern terminology. It should be noted here that since the Egyptian system of rationing involved the two staple commodities of grain and beer, a frequent task set for the scribes was to record and calculate the amounts and types of these commodities that had to be allocated to various employees and beneficiaries.

EXAMPLE 3.8 100 loaves of *pesu* 10 are to be exchanged for a certain number of loaves of pesu 45. What is this certain number?

(Note: The word *pesu* (or *psw*) may be defined as a measure of the "weakness" of a commodity. Here it can be taken to be the ratio of the number of loaves produced to the amount of grain used in their production so that the higher the *pesu*, the weaker the bread.)

Solution

We would tackle the above problem today as one of simple proportions, obtaining the number of loaves as $45/10 \times 100 = 450$. The solution prescribed in the Egyptian text is quite involved. It is interesting from our point of view because it contains the germs of algebraic reasoning. Below are the Egyptian solution and a restatement of the same steps in modern symbolic terms.

EGYPTIAN EXPLANATION	MODERN EXPLANATION
	Let x and y be the loaves of p and q pesu, respectively. Find y if x, p, q are known.
1. Find excess of 45 over 10: result 35. Divide this 35 by 10: result $3 + 1/2$.	$(q - p)/p$
2. Multiply this $(3 + 1/2)$ by 100: result 350. Add 100 to 350: result 450.	$[(q - p)/p]x + x$
3. Then the exchange is 100 loaves of 10 *pesu* for 450 loaves of 45 *pesu*.	$y = [(q - p)/p]x + x = (q/p)x$

What is important here is not whether the scribe arrived at this method of solution by any thought process akin to ours, but that what we have here from four thousand years ago is a form of algebra, dependent on knowing that $y/x = q/p$ and $(y - x)/x = (q - p)/p$.

Solving Simple and Simultaneous Equations: The Egyptian Approach

To find topics that are represented in modern elementary algebra, we have to turn to problems 24–34 of the Ahmes Papyrus. One of these problems, problem 26, will serve as an illustration.

> EXAMPLE 3.9 A quantity, its 1/4 added to it so that 15 results. [I.e., a quantity and its quarter added become 15. What is the quantity?]
>
> *Solution*
>
> In terms of modern algebra, the solution is straightforward and involves finding the value of x, the unknown quantity, from an equation:
>
> $$x + \frac{1}{4}x = 15, \quad \text{so} \quad x = 12.$$
>
> The scribe, however, reasoned as follows: If the answer were 4, then 1 + 1/4 of 4 would be 5. The number that 5 must be multiplied by to get 15 is 3. If 3 is now multiplied by the assumed answer (which is clearly false), the correct answer will result: $4 \times 3 = 12$.

This problem belongs to a set of problems that are described as "quantity" or "number" problems and are basically concerned with showing how to determine an unknown quantity from a given relationship. The scribe was using the oldest and probably the most popular way of solving linear equations before the emergence of symbolic algebra—the method of false assumption (or false position). Variants of "quantity" problems of this kind included adding a multiple of an unknown quantity instead of a fraction of the unknown quantity. For example, problem 25 of the Moscow Papyrus asks for a method of calculating an unknown quantity such that twice that quantity together with the quantity itself adds up to 9. The instruction for its solution suggests assuming the quantity as 1 and that together with twice the assumed quantity gives 3. The number that 3 must be multiplied by to get 9 is 3. So the unknown quantity is 3. It is interesting to reflect that such an approach was still in common use in Europe and elsewhere until about a hundred years ago.

The Berlin Papyrus contains two problems that would appear to us today to involve second-degree simultaneous equations (i.e., equations with terms like x^2 and xy). It is badly mutilated in places, so the solution offered below is both conjectural and a reconstructed one.

EXAMPLE 3.10 It is said to thee [that] the area of a square of 100 [square cubits] is equal to that of two smaller squares. The side of one is 1/2 + 1/4 of the other. Let me know the sides of the two unknown squares.

Solution 1: The Symbolic Algebraic Approach

Let x and y be the sides of the two smaller squares. From the information given above, we can derive the following set of equations:

$$x^2 + y^2 = 100;$$
$$4x - 3y = 0.$$

The solution set, $x = 6$ and $y = 8$, is obtained by substituting $x = (3/4)y$ into $x^2 + y^2 = 100$.

Solution 2: The Egyptian Rhetorical Algebraic Approach

Take a square of side 1 cubit (i.e., a false value of y equal to 1 cubit). Then the other square will have side 1/2 + 1/4 cubits (i.e., $x = 1/2 + 1/4$). The areas of the squares are 1 and (1/2 + 1/16) square cubits respectively. Adding the areas of the two squares will give 1 + 1/2 + 1/16 square cubits. Take the square root of this sum: 1 + 1/4. Take the square root of 100 square cubits: 10. Divide 10 by 1 + 1/4. This gives 8 cubits, the side of one square. (So from the false assumption $y = 1$, we have deduced that $y = 8$.) At this point, the papyrus is so badly damaged that the rest of the solution has to he reconstructed. One can only assume that the side of the smaller square was calculated as 1/2 + 1/4 of the side of the larger square, which was 8 cubits. So the side of the smaller square is 6 cubits.

Geometric and Arithmetic Series

A series is the sum of a sequence of terms. The most common types are the arithmetic and geometric series. The terms of the former are an arithmetic progression, a sequence in which each term after the first (usually denoted by a) is obtained by adding a fixed number, called the common difference (usually denoted by d), to the preceding term. For example, 1, 3, 5, 7, 9, . . . is an arithmetic progression with $a = 1$ and $d = 2$. In a geometric progression, each term after the first (a) is formed from the preceding term by multiplying by a fixed number called the common ratio (usually denoted by r). For example, 1, 2, 4, 8, 16, ... is a geometric progression with $a = 1$ and $r = 2$.

The Egyptian method of multiplication leads naturally to an interest in such series, since it is based on operations with the basic geometric progression 1, 2, 4, 8, ... and an understanding that any multiplier may be expressed as the sum of elements of this sequence. It would follow that Egyptian interest would focus on finding rules that made it easier to add up certain elements of such sequences. Here is problem 79 from the Ahmes Papyrus.

EXAMPLE 3.11 The actual statement of the problem in the Ahmes Papyrus is uncharacteristically ambiguous. It presents the following information, and nothing else:

Houses	7		
Cats	49		
Mice	343	1	2,801
Emmer wheat	2,401[16]	2	5,602
Hekats of grain	16,807	4	11,204
Total	19,607	Total	19,607

There is no algorithm nor any instruction for solution, except for what can be inferred from the calculations. The presentation of this curious data has led to some interesting suggestions. It was first believed that the problem was merely a statement of the first five powers of 7, along with their sum; and that the words "houses," "cats," and so on were really a symbolic terminology for the second and third powers, and so on. Since no such terminology occurs elsewhere, this explanation is unconvincing. Moreover, it does not account for the other set of data on the right.

A more plausible interpretation is that we have here an example of a geometric series, where the first term (a) and common ratio (r) are both 7, which shows that the sum of the first five terms of the series is obtained as $7[1 + (7 + 49 + 343 + 2,401)] = 7 \times 2,801$. We now see that the second set of data in the problem is merely the multiplication of 7 by 2,801 in the Egyptian way.

A precursor to this Egyptian example of geometric progression may have been a mathematical text from the Old Babylonian period dealing with the same subject. It was discovered at Mari, a small kingdom in the northwest corner of Mesopotamia, which was conquered by Hammurabi in 1757 BC. A reconstruction of this example, which Friberg (2005, p. 5) describes as a "whimsical story," reads as follows: "There were 645,539 barleycorns, 9 barleycorns on each ear of barley, 9 ears of barley eaten by each ant, 9 ants swallowed by each bird, 9 birds caught by each of 99 men. How many were there altogether?" [Answer: 730,719 different items.]

In the next chapter, in the section dealing with geometric series in Mesopotamian mathematics, we return to this problem. But what is being strongly suggested here is the existence of links between the two mathematical traditions, long considered to have been independent of one another. We will return to this theme in chapter 5.

A detailed solution to another problem in the Ahmes Papyrus gives some support to the view that the Egyptians had an intuitive rule for summing n terms of an arithmetic progression. Problem 64 may be restated as follows:

EXAMPLE 3.12 Divide 10 hekats of barley among 10 men so that the common difference is one-eighth of a hekat of barley.

Solution

The solution of the problem as it appears in the Papyrus is given on the left-hand side. On the right-hand side the algorithm is stated symbolically.

EGYPTIAN METHOD	SYMBOLIC EXPRESSION
	Let a be the first term, f the last term, d the difference, n the number of terms, and S the sum of n terms.
1. Average value: $10/10 = 1$.	1. Average value of n terms $= S/n$.
2. Total number of common differences: $10 - 1 = 9$.	2. Number of common differences $= n - 1$

Continued...

Continued . . .

3. Find half the common difference: $1/2 \times 1/8 = 1/16$.

3. Half the common difference $= d/2$.

4. Multiply 9 by 1/16: $1/2 + 1/16$.

4. Multiply $n - 1$ by $d/2$: $(n - 1)d/2$.

5. Add this to the average value to get the largest share: $1 + 1/2 + 1/16$.

5. $f = S/n + (n - 1)d/2$.

6. Subtract the common difference (1/8) nine times to get the lowest share: $1/4 + 1/8 + 1/16$.

6. $a = f - (n - 1)d$.

7. Other shares are obtained by adding the common difference to each successive share, starting with the lowest. The total is 10 *hekats* of barley.

7. Now form $a, a + d, a + 2d \ldots, a + (n - 1)d$. So $S = an + (1/2) n(n - 1)d$, or $S/n = a + (1/2)(n - 1)d$.

The correspondence between the rhetorical algebra of the Egyptians and our symbolic algebra is quite close, though a word of caution is necessary here. It would not be reasonable to infer, on the basis of this correspondence, that the ancient Egyptians used anything like the algebraic reasoning on the right-hand side. It is more likely that they took a common-sense approach, listing the following sequence on the basis that the terms added to 10:

$$a, a + \frac{1}{8}, a + \frac{2}{8}, \ldots, a + \frac{9}{8}.$$

Each successive term gives the rising share of barley received by the 10 men.

Egyptian Geometry

The practical character of Egyptian geometry has led a number of commentators to question whether it can properly be described as geometry, but that is to take too restrictive a view. The word itself comes from two Greek words meaning "earth" and "measure," indicating that the subject had its origin in land surveying and other practical applications, and it was from the need to compute land areas and the volumes of granaries and pyramids that Egyptian geometry emerged with its peculiarly practical character. If there was any theoretical motivation, it was well hidden behind rules for computation.

If there is an abiding icon to characterize ancient Egypt, it would have to be the pyramid. The Ahmes Papyrus contains six problems referring to pyramids, in particular the relation between their base, height, and inclination. Consider problem 56, which is illustrated by a diagram of a pyramid whose numerical values of base and height are written next to the drawing. What is to be determined is the inclination of its walls (or the *seked* of the pyramid). A *seked* is measured by taking the number of "palms" that an inclined plane falls per cubit of height.[17]

EXAMPLE 3.13   A pyramid has base 360 cubits and height 250 cubits. Find its *seked*.

Solution

Take half of 360: result 180.

Divide 180 by 250 : result 1/2 + 1/5 + 1/50.

One cubit is 7 palms. So multiply the result by 7.

Therefore, $(1/2 + 1/5 + 1/50) \times 7 = 5 + 1/25 = 5\frac{1}{25}$ palms.

This is equivalent to the application of the formula

1 *seked* (in palms) $= \frac{1}{2}$ base (in cubits) divided by height (in cubits).

When it is asked what the three major achievements of Egyptian geometry were, there is general agreement on two—the approximation to the area of the circle, and the derivation of the rule for calculating the volume of a truncated pyramid—but some disagreement over the third: did they indeed find the correct formula for the surface area of a hemisphere?

The Area of a Circle: The Implicit Value for π

Problem 50 from the Ahmes Papyrus reads:

EXAMPLE 3.14 A circular field has diameter 9 *chet*. What is its area?

(Note: One *chet* is equal to 100 cubits, or approximately 50 meters. It is interesting that this problem is hardly a practical one, since the area works out to be about 16 hectares (0.16 square kilometers) and the circumference of the field is nearly a kilometer and a half!)

The Ahmes Solution

Subtract 1/9 of the diameter, namely 1 *chet*. The remainder is 8 *chet*. Multiply 8 by 8; it makes 64. Therefore, it contains 64 *cha-ta* (square *chet*) of land.

   In symbolic algebra this amounts to $A^E = [d - (d/9)]^2 = (8d/9)^2$, where d is the diameter and A^E is the area calculated the Egyptian way. Worked out the modern way (involving the ratio of circumference over the diameter, or π correct to three decimal places) the result is

$$A = \pi r^2 = (3.142)(4.5)^2 = 63.63,$$

which is close to the value estimated by the scribe.

The implicit estimate of π contained in the Egyptian method of calculating the area of a circle can be worked out quite easily by equating A with A^E:

$$\frac{\pi d}{4} = \left(\frac{8d}{9}\right)^2,$$

from which we get

$$\pi = 4\left(\frac{8}{9}\right)^2 = \frac{256}{81} = \left(\frac{16}{9}\right)^2 \approx 3.1605.$$

It is important to note that the Egyptian method is not based on recognizing the dependence of the circumference on the diameter (i.e., on the value of π).

   How was this rule derived by the Egyptians? Problem 48 may provide us with a clue. This is a problem expressed in the form of an "idealized" diagram, reproduced here as figure 3.2a, together with two calculations that can be identified as multiplications. It shows a square with four isosceles triangles in the corners. In the middle of the square is the normal hieratic symbol for 9, written as ϝↄ. The removal of the triangles, each of area 9/2

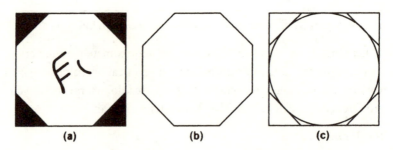

FIGURE 3.2: Problem from the Ahmes Papyrus: measuring a circle

square *chet*, would leave a regular octagon with side 3 *chet*, as in figure 3.2b. It is easily seen that the area of the octagon equals the area of the square minus the total area of the triangles cut off from the corners of the square:

$$A = 9^2 - 4\left(\frac{9}{2}\right) = 81 - 18 = 63.$$

This is nearly the value that is obtained by taking $d = 9$ in the expression $A^E = (8d/9)^2$. So the octagon is a reasonable approximation to the circle inscribed in the square, as illustrated by figure 3.2c.

There is something rather contrived and unconvincing about this explanation, for it assumes an algebraic mode of reasoning that is not immediately apparent in Egyptian mathematics. For a more plausible explanation, we turn to the geometric designs that were popular in ancient Egypt. A common motif found in burial chambers is the "snake curve," which looks like a snake coiled around itself several times. (It is found today in areas of Africa as far apart as Mozambique and Nigeria.) This spiral motif also appeared in the design of objects. We are told that at the time of Ramses III (c. 1200 BC) the royal bread was baked in the shape of a spiral. Sisal mats were shaped in the form of a snake curve, and are not uncommon objects in Africa even today.

If one of these spiral mats (whose presence in ancient Egypt is well attested to in the drawings of the period), of diameter 9 units, were uncoiled and a square formed of side 8 units, the close correspondence in the areas of the two shapes would be easy to establish experimentally.

Yet another explanation (Gerdes 1985a), also based on material evidence, is related to a board game that is still widely played in many parts of Africa. In ancient Egypt it was played on a board with three rows of

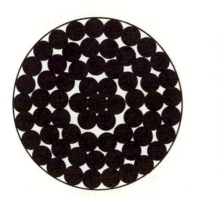

FIGURE 3.3: An alternative Egyptian way of measuring a circle, suggested by Gerdes (1985a, p. 267. Reproduced with permission)

fourteen hollows with counters that were, as they are today, round objects such as seeds, beans, or pebbles. Perhaps, then, the exigencies of the game prompted experimentation to find a square and a circle such that the same number of spherical counters could be packed into each as tightly as possible. Figure 3.3 shows that for both the circle of diameter 9 units and the square of side 8 units it is possible to pack in 64 small circles (or spherical counters). Thus the area of a circle with diameter 9 units is approximately equal to the area of a square of side 8 units, where the area is expressed as 64 small circles. An implicit value of π, given earlier, can easily be calculated from this approximate equivalence of the areas of the square and the circle.[18]

The Egyptian rule for obtaining the area of a circle is applied in a few examples from the Ahmes and Moscow papyri. In problem 41 of the Ahmes Papyrus, the volume of a granary with a circular base of diameter 9 cubits and height 10 cubits is calculated. From archaeological evidence, two types of granaries were commonly in use: one with a circular base that resembled a cone and one with a rectangular base. In the case of the circular one, the volume was calculated by determining the area of the base and multiplying this area by the height. The solution offered to problem 41 may be expressed symbolically as

$$\text{Volume} = A^{E}h = \left(\frac{8d}{9}\right)^{2}h = 640 \text{ cubic cubits.}$$

However, this is not the end of the calculations. A transformation from cubic cubits into more appropriate units for measuring large amounts of grain involves conversion into *hekats* (3/2 of 640 = 960), and then cubic *hekats* (1/20 of 960 = 48) gives the final answer.

Volume of a Truncated Square Pyramid

It is generally agreed that the Egyptian knowledge of the correct formula for the volume of a truncated square pyramid is the zenith of Egyptian geometry. Bell (1940) referred to this achievement as the "greatest Egyptian pyramid." A restatement of problem 14 of the Moscow Papyrus is as follows:

EXAMPLE 3.15 Example of calculating a truncated pyramid. You are told: a truncated pyramid of 6 cubits for vertical height by 4 cubits on the base by 2 cubits on the top. [Calculate the volume of this pyramid.]

Solution

The solution, as it appears in the Papyrus, is given on the left-hand side below; on the right-hand side the algorithm is stated in symbolic terms (see figure 3.4a).

EGYPTIAN METHOD

SYMBOLIC EXPRESSION

Let h be the vertical height, a and b the sides of the two squares which bound the solid above and below, and V the volume of the solid.

1. Square this 4: result 16.

1. Find the area of the square base: a^2.

2. Square this 2: result 4.

2. Find the area of the square top: b^2.

3. Take 4 twice: result 8.

3. Find the product of a and b: ab.

4. Add together this 16, this 8, and this 4: result 28.

4. Find $a^2 + ab + b^2$.

5. Take 1/3 of 6: result 2.

5. Find 1/3 of the height: $h/3$.

6. Take 28 twice: result 56.

6. $(a^2 + ab + b^2)h/3$.

7. Behold it is 56!

7. $V = (a^2 + ab + b^2)h/3$.

What has been found by you is correct.

Continued...

Continued . . .

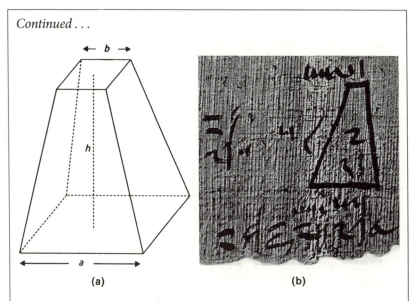

FIGURE 3.4: Problem 14 from the Moscow Papyrus: the truncated pyramid (a) in its modern version, and (b) the solution as given in the papyrus (Eves 1983, p. 42)

At the end of the calculations in the Moscow Papyrus is a drawing of the trapezoid (see figure 3.4b). Only this "idealized" drawing of the shape is given, without any Egyptian name for it. The sketch contains, exceptionally for the Moscow Papyrus, both the data and the results of the different arithmetical operations. The upper side of the drawing is labeled 2 and its square 4, the lower side is labeled 4 and its square 16, and its height labeled 6 and its third 2.

A number of attempts have been made to explain how the Egyptians may have arrived at the correct formula for the volume of a truncated pyramid. All explanations start with the assumption that they were aware of the formula for the volume of the complete pyramid, for otherwise it is difficult to explain the appearance of the factor 1/3 in the expression for the volume of a truncated pyramid. (The formula for the volume of the whole pyramid is in fact a special case of the more general formula for the truncated pyramid, since a substitution of $b = 0$ into the latter gives $V = a^2 h/3$. Putting $a = b$ gives $V = a^2 h$, the volume of a square prism).

There are three main explanations. The first suggests that the truncated pyramid was cut up into smaller and simpler solids whose volumes were then estimated before putting them back together again. It is with the last part of this explanation that difficulties arise, since the reduction of the sum of the volumes of all the component solids to the final formula would require a degree of algebraic knowledge and sophistication that few would concede to the Egyptians.

The second explanation is that the Egyptians had discovered empirically that the volume of a truncated pyramid can be obtained as the product of the height of the frustum, h, and the Heronian mean[19] of the areas of the bases, a^2 and b^2. The only evidence to support this viewpoint is provided by Heron (or Hero), an Alexandrian mathematician of the first century AD, whose work contains a useful synthesis of Egyptian, Greek, and Babylonian traditions. Book II of his *Metrica* has a detailed treatment of the volume mensuration of prisms, pyramids, cones, parallelepipeds, and other solids. The inference is that his method for estimating the volume of a truncated pyramid, using the "mean" named after him, derived directly from the Egyptian mathematical tradition.

Finally, there is the view that the volume was calculated as the difference between an original complete pyramid and a smaller one removed from its top. Gillings (1964) gives a detailed discussion of this explanation, which sounds the most plausible of the three, given the "concrete" approach to geometry that the Egyptians favored. Irrespective of how the Egyptians came to the discovery, the formula remains a lasting testimony to their mathematics.

The Area of a Curved Surface: A Semicylinder or a Hemisphere?

Problem 10 of the Moscow Papyrus is also concerned with calculation of areas. But unlike other examples of such calculations, there is no "idealized" drawing to help us to discern the shape of the figure in question. The problem reads as follows:

EXAMPLE 3.16 If you are told that an object [*nbet*] has diameter $4 + \frac{1}{2}$ and dimension (*adge*) [height?]. Let me know the area [of its surface].

Suggested Solution

1. Take 1/9 of 9 since the object is half a *nbet* [i.e., a hemisphere]: result 1.

Continued . . .

Continued ...

2. Take the remainder, which is 8.

3. Take 1/9 of 8: result 2/3 + 1/6 + 1/18.

4. Find the remainder of this 8 after [the subtraction of] 2/3 + 1/6 + 1/18: result 7 + 1/9.

5. Multiply 7 + 1/9 by 4 + 1/2: result 32.

Behold this is its area! You have found it correctly.

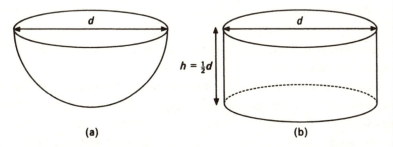

FIGURE 3.5: Problem 10 from the Moscow Papyrus. It may be asking for (a) the area of a hemisphere, or (b) the area of a semicylinder

A critical problem in interpretation is the shape of the object (*nbet*) under discussion. This term unfortunately appears only in this problem. In modern mathematical terms, the problem has been interpreted as *either* calculating the surface area of a hemisphere of diameter of $4\frac{1}{2}$ *or* finding the area of a curved semicylinder whose height is half its diameter (see figure 3.5a and b).

If the first interpretation is valid, the problem may be restated as

Find the surface area of a hemisphere of diameter $4\frac{1}{2}$

The suggested solution above may be expressed symbolically as

$$A = 2d \times \left(\frac{8}{9}\right) \times \left(\frac{8}{9}\right) \times d = 2d^2\left(\frac{8}{9}\right)^2 = 2\pi r^2,$$

where the Egyptian (implict) value of π was given earlier as 256/81. This is identical to the modern formula for the curved surface of a hemisphere, $A = \frac{1}{2}\pi d^2$, with a different value for π.

If this interpretation is valid, then here is an even more remarkable achievement than the application of the correct formula for the volume of a truncated pyramid, for the very idea of a curved surface (not simply one that can be obtained by rolling up a plane surface) is quite an advanced mathematical concept. It would predate the innovative work of Archimedes (c. 250 BC) by about fifteen hundred years.

However, doubts have been raised over this particular interpretation of the term *nbet*. Peet (1931) has argued that this term could also be taken in this context to mean a semicylinder (see figure 3.5b), in which case the suggested solution above, expressed symbolically, would become

$$A = 2\pi rh, \text{ where } \pi = \frac{256}{81} \text{ and } h = \frac{1}{2}d \text{ is the height.}$$

This would be the Egyptian counterpart of the modern formula for the area of the curved surface of a semicylinder.

Peet's translation and interpretation of the text has had its own critics. Gillings (1972) has argued that if the original interpretation of the basket as a hemisphere is accepted, the rule could have arisen from the empirical observation that, in weaving a hemispherical basket whose radius is approximately equal to its height, the quantity of material required to make a circular lid is approximately half that required for the basket itself. If so, then the rule was a matter of simple deduction, and the Egyptians' "greatest pyramid" remains the correct application of the formula for the volume of a truncated pyramid.

Other Exemplars of Egyptian Mathematics

Our discussion of problem texts of Egyptian mathematics has so far been confined to the contents of Ahmes and Moscow papyri. During the long period of the New Kingdom and subsequently (1550–650 BC) there are no extant mathematical texts, but a number of administrative texts containing applications of mathematics are available, notably those recorded on ostraca (pottery or tile). In the Valley of the Kings near Luxor, and particularly in the village of Deir-el-Medina, many documents of a legal and administrative nature have been found. These include data relating to the construction of tombs, such as estimation of volumes from quarrying tombs, and the numbers of different personnel employed to do so, their working schedules, and their remuneration. While these documents do illustrate the computing and organizational skills of the scribes, their mathematical content is quite limited.

The next corpus of extant mathematical documents appeared after Egypt came under Persian rule (525–404 BC and 343–332 BC) and then became part of the Hellenistic world (332 BC), finally ending as a province of the Roman empire (30 BC). The texts were recorded in the demotic script. It was during this period that Egyptian mathematics came under the increasing influence of Mesopotamia and Greece. We conclude our discussion of Egyptian mathematics with an example found in Parker (1972, pp. 35–37) taken from a text dated to around the fourth century BC. Similar problems exist in Mesopotamian mathematics and later in Chinese mathematics. Whether this indicates transmissions farther afield is an intriguing question for which no definitive evidence exists.

EXAMPLE 3.17 The foot of a pole [length] 10 cubits is moved outward so that its top is [resting] 8 cubits vertically. By how much has the top of the pole been lowered?

EGYPTIAN EXPLANATION

1. You should reckon 10, 10 times: result 100.

2. You should reckon 8, 8 times: result 64.

3. Take it from 100: result 36.

4. Reduce to its square root: result 6.

5. Take it from 10: remainder 4.

You shall say: "Four cubits is the answer."

MODERN EXPLANATION

Let a = length of the pole.

Let c = height of the pole once it is moved out.

Let b = the distance that the foot of the pole has moved out.

$c^2 = a^2 - b^2 = 10^2 - 8^2 = 36 \rightarrow c = 6$

$a - c = 4$.

It would seem that the Pythagorean theorem was applied in arriving at this solution.

Egyptian Mathematics: A General Assessment

Egyptian mathematics has been discussed here in more detail than in some of the past textbooks on the early history of mathematics. The treatment of Egyptian mathematics in many of these texts tends to be rather lopsided: Egyptian numeration is overemphasized, and consequently the rest of the mathematics receives less attention than it should. Where comparisons are made with contemporary or later mathematical traditions, the quality of Mesopotamian mathematics is stressed, but both the Egyptian and Mesopotamian contributions are judged to be meager, or—more charitably— seen merely as a prelude to the "Greek miracle." However, it should be pointed out that the amount of space devoted to Egyptian mathematics by some of the recent texts such as Burton (2005), Cooke (1997), Katz (1998), and Suzuki (2001) has increased considerably over the years.

Over several years, both on the Internet and in academic journals, there has been a lively debate on the historical relationship of Greek to Egyptian science in general, and mathematics in particular. The start of this debate in recent years may be traced to Martin Bernal's (1987, 1992, 1994) claim that Classical Greek culture was significantly influenced by ancient Egyptian civilization, partly through Egyptian colonization of parts of Greece. And this influence was also reflected in the debt that Greek science owed to its Egyptian counterpart. A debate that began between Bernal (1992, 1994) and Palter (1993) has spilled over to journals and popular magazines. Interesting interventions by Victor Katz and Beatrice Lumpkin have been published in a newsletter in July 1995. The discussion that follows has been influenced by their contributions.

The arguments marshaled by Katz relate to two claims made by Bernal and rejected by Palter:

1. There were scientific elements in Egyptian medicine, mathematics, and astronomy long before there was any Greek science at all.

2. Egyptian medicine, mathematics, and astronomy significantly influenced the development of corresponding Greek disciplines.

In relation to mathematics, we know from the evidence contained in this chapter that the Egyptians certainly knew how to solve various kinds of problems, from solving what we would now describe as linear equations to calculating the volumes and areas of different geometrical objects, including possibly the surface of a hemisphere. Lumpkin (2000) has summarized the nature and extent of the Egyptian contribution to science (and in particular mathematics long before the appearance of Greek science). What we are uncertain about is how the Egyptians discovered the methods they used in solving the problems. Presumably, at the very least, there was some "scientific" underpinning to their methods, although not necessarily based on reasoning from explicit axioms. On the other issue of whether

Egyptian mathematics influenced Greek mathematics, it may be possible to give a more definitive "yes." We have pointed out a similarity between the mathematical thinking of the Egyptians and the Greeks when it came to number theory. After all, many of the ancient Greek sources acknowledge this influence. As Katz (1995, pp. 10–11) stated:

> Not only is Pythagoras supposed to have studied in Egypt, but so is Thales, the supposed father of Greek geometry. Also, Oenopides. Herodotus, Heron of Alexandria, Diodorus Siculus, Strabo, Socrates (through Plato) and Aristotle—all say that geometry was first invented by the Egyptians and then passed on to the Greeks. The question always seems to be, in this regard, what we mean by geometry. If, by geometry, we mean an axiomatic treatment with theorems and proofs in the style of Euclid, then it is clear that this was a Greek invention. But mathematicians have always known that, in general, one does not discover theorems by the axiomatic method. One discovers theorems by experiment, by trial and error, by induction, etc. Only after the discovery is there a search for a rigorous "proof."

Srinivas Ramanujan's mathematics, discussed in the preface to the first edition of this book, is a good example of the discovery and the "proof" being undertaken by different persons and at different times. And it would seem clear that when the Greeks declare that the Egyptians invented (or discovered) geometry, it is the results that they have in mind and not necessarily the method of proof. Conceding that the Greeks learned various geometrical results from the Egyptians takes nothing away from Greek creativity. They were simply doing what mathematicians have always done, building on the results of their predecessors.

In chapter 1, we examined the issue of transmission between different mathematical traditions, concentrating mainly on the connections that arose during and after the Dark Ages in Europe. Some of the most promising recent research, notably that of Friberg (2005, 2007b), relates to discovering the three-way links that have been neglected so far: those between the Egyptian, Mesopotamian, and Greek mathematics. Past researchers have usually considered Egyptian and Mesopotamian mathematics as completely independent mathematical traditions despite their proximity to each other in time and space. Their differences, notably in the development of a decimal, nonpositional number system in Egypt compared with a sexagesimal positional number system in Mesopotamia, have been emphasized. Further, the persistence of a belief in the "Greek miracle" in some guise or other, alluded to in chapter 1, has underpinned the view of Greek mathematics as being unique and independent of the earlier mathematical traditions of Egypt and Mesopotamia. However, today there is far greater attention paid to the methodology of establishing transmission as well as a growing recognition that the cultural context in which a mathematical document arises is a crucial consideration as to its interpretation. In chapter 5, in the assessment of Egyptian and Mesopotamian mathematics, the linkages between these two mathematical traditions as well as the traces of these traditions in Greek mathematics will be examined in greater detail.

We have said little about the later phase of Egyptian mathematics, when Alexandria became the center of mathematical activity. It was the creative synthesis of Classical Greek mathematics, with its strong geometric and deductive tradition, and the algebraic and empirical traditions of Egypt and Mesopotamia that produced some of the greatest mathematics and astronomy of antiquity, best exemplified in the works of Archimedes, Ptolemy, Diophantus, Pappus, and Heron. We shall not take up the story of Hellenistic mathematics, which has been extensively explored in general histories, such as those by Boyer (1968), Eves (1983), Katz (1998), and Kline (1972), as well as in specialized works on Greek mathematics by Cuomo (2001), Fowler (1987), Heath (1921), van der Waerden (1961), and others.

Notes

1. Diodorus has been criticized as exhibiting "none of the critical faculties of the historian (but) merely setting down a number of unconnected details." But his English translator, Oldfather (1989), reminds us that of all the forty volumes of his *Universal History*, the first volume on Egypt (published in 1960) is the "fullest literary account of the history and customs of that country after Herodotus."

2. Davidson (1987, pp. 1–2), writing about public reactions to a television series that he presented on the history of the Africans, points out that what a number of viewers in Europe and North America found particularly difficult to accept was the "black" origins of the ancient Egyptians: "To affirm this, of course, is to offend nearly all established historiographical orthodoxy."

3. *Ujamaa* is a Swahili word meaning "brotherhood" that was used to describe a Tanzanian government initiative in the late 1960s and early 1970s to encourage scattered rural homesteads to form villages, which would then serve both as pools of labor for communal activities and as units for meeting social needs in health, education, communication, and water supply.

4. In fairness, it should be added that this is perceived as a fanciful story by some scholars of predynastic Egypt. For a useful review of evidence, see Bard (1994, pp. 265–88).

5. The New York fragments consist only of a small table and the early section of the problems (nos. 1–6), which helped to complete the whole text.

6. For the contents of these deciphered fragments, see Imhausen and Ritter (2004) and Imhausen (2006).

7. Friberg identfies a codex of six papyrus leaves and a small corpus of "non-Euclidean" Greek mathematical texts.

8. For further details, see Imhausen (2007, pp. 10–11).

9. One cubit is equal to approximately 52.5 cm.

10. See Fischer-Elfert (1986, pp. 118–57) for a discussion of the different interpretations.

11. There are different interpretations of the numbers shown on the mace head. Petrie, an early Egyptologist, suggested that the mace head depicted scenes of a political marriage of Narmer to a princess from the north at which he received tributes from different people. Others have interpreted it as recording the spoils of war of Narmer after his conquest of the north. A third interpretation suggests a census of the male population and their livestock taken during his reign.

12. For further details, see Resnikoff amd Wells (1984, p. 23).

13. A highly cursive form of hieratic known as "abnormal hieratic," derived from the script of Upper Egyptian administrative documents, was used primarily for legal texts, land leases, letters, and other texts. This type of writing was superseded by demotic—a Lower Egyptian scribal tradition—and became the standard administrative script throughout a reunified Egypt.

14. However, Bruckheimer and Salamon (1977) have argued that in a number of cases the selection criteria put forward by Gillings are inappropriate. A more recent critique of Gillings's procedure is found in Abdulaziz (2008).

15. Apart from table texts for computing unit fractions, other tables to convert different measuring units (including measures of volume) and tables used as aids to calculation have been discovered.

16. The figure given in the text is 2,301, which is incorrect.

17. The following were the units of measurement of length in ancient Egypt: 1 *chet* = 100 cubits; 1 cubit = 7 palms; 1 palm = 4 digits. In terms of modern measurement, 1

cubit ≈ 52.5 cm; 1 palm ≈ 7.5 cm, and 1 digit ≈ 19 mm. A "trigonometric" interpretation of the *seked* has been offered on the basis of its being a ratio of width to height, which is therefore equivalent to the cotangent of the relevant angle. Since there is no notion of measurement of an angle in Egyptian mathematics (see endnote 17, chapter 1), such an interpretation would seem somewhat far-fetched.

18. It may be argued that the explanations given are contrived, unconvincing, and lacking in any real evidence to back them. To the author, however, their merit lies in the fact that there is an underlying materialistic basis to these conjectures.

19. The Heronian mean of two positive numbers x and y is given by $(1/3)(x + y + \sqrt{xy})$. For further details, see Bullen (2003).

Chapter Four

The Beginnings of Written Mathematics: Mesopotamia

Fleshing Out the History

Studying ancient Mesopotamian history is rather like going on a long and unfamiliar journey: we are not sure whether we are on the right road until we reach our destination. The abridged chronology given in table 4.1 will be of some help in plotting our course across this difficult terrain; the places mentioned in the table are shown in the accompanying map (figure 4.1). The earliest protocuneiform written records are from around the last few centuries of the fourth millennium BC, and the last cuneiform records are from around the end of the first millennium BC. With the Persian conquest in 539 BC, Mesopotamia ceased to exist as an independent entity. The subsequent history of this region cannot be separated from the histories of other countries such as Persia, Greece, Arabia, and, more recently, Turkey.

Along the fertile crescent between the Tigris and Euphrates rivers emerged the first cities occupied by the people who had originally migrated from the present-day Armenian region of the Black and Caspian seas. By 3500 BC, the population pressure was such that the naturally irrigated floodplains could no longer sustain the basic needs of the inhabitants, especially since, unlike the case of Egypt, the flooding occurred somewhat erratically. The rivers were not navigable, making the city-states culturally and economically isolated from one another. The whole region was wide open and flat, lacking in natural defenses, making it vulnerable to external invasions. Thus the physical environment of Mesopotamia influenced both the economy and the habitat of its inhabitants.

It has been suggested by certain historians, notably Wittfogel (1957) that, just as in the case of Egypt, a society that had mastered the principles of hydraulics (irrigation) was well equipped to initiate the beginnings of

TABLE 4.1: CHRONOLOGY OF ANCIENT MESOPOTAMIA FROM 4000 BC TO 64 BC

Dates	Historical/socioeconomic background	Mathematical developments
4000–3500	Early urbanization in the south.	Early accounting practices based on tokens. Development of separate systems of notations for (1) counting numbers on base 60 (sexagesimal), (2) area numbers, (3) weight numbers, (4) grain capacity numbers. The earliest school texts from Uruk.
3500–2500	Early Bronze Age. Emergence of city-states of Sumeria with centers of power at Ur, Nippur, Eridu, and Lagash.	Discovery of earliest school texts from Fara (Shuruppak). Development of sexagesimal numerals and phonetic writing, more advanced accounting practices.
2500–2000	Establishment of the empires of Sumer and Akkad (centers of power: Ur, Agade). Notable rulers: Sargon I (c. 2350) and Shulgi (2100).	Old Akkadian school texts. About 2000: tables of reciprocals and use of sexagesimal place-value notation.
2000–1500	Conflicts and wars; rule by city-state; establishment of the Old Babylonian empire (center of power: Babylon). Notable ruler: Hammurabi (1792–1752).	Widespread evidence of early concrete algebra and geometry, quantity surveying, often found as adjuncts to scribal training. Sophisticated Babylonian mathematical texts.
1500–1000	Late Bronze Age. International contacts.	Spread of sexagesimal numeracy. Development of astronomy.
1000–600	Iron Age. Assyrian empire. Development of Aramaic language (center of power: Nineveh). Notable rulers: Sennacherib (705–681) and Ashurbanipal (668–627).	Computational and astronomical developments continue.
612–539	Second or New Babylonian empire (Chaldeans) (center of power: Babylon). Notable ruler: Nebuchadnezzer (605–562).	Astronomical observations.
539–311	Persian invasion (539): end of ancient Mesopotamia (centers of power: Babylon and Susa). Notable rulers: Cyrus the Great (c. 525) and Darius (521–485).	Revival of education in mathematics. Great advances in mathematical astronomy.
312–64	Seleucid dynasty, Late Babylonian period (center of power: Antioch).	Work on astronomy and algebra continues: construction of extensive mathematical and astronomical tables.

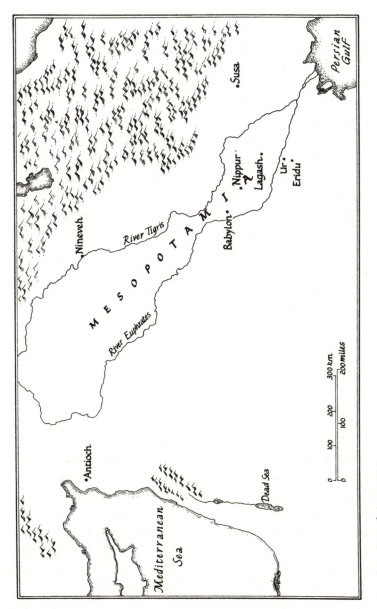

FIGURE 4.1: Map of Mesopotamia

two major and interrelated sciences: mathematics and astronomy. And, the argument continues, the pursuit of intensive agriculture and large-scale breeding of livestock, concentrated in the hands of a central power, necessitated a meticulous control of movements of the goods produced and exchanged. In an attempt to accomplish this task efficiently, writing first developed and was to remain for several centuries its only use. While this argument would seem to be somewhat simplistic today, there is the related point concerning the similarities between ancient Egypt and Mesopotamia in the emergence of priestly bureaucratic structures with both writing and mathematics developed to serve their ends. From the evidence we have so far, "mathematics" preceded writing, in that the earliest records that we have for both Egypt and Mesopotamia relate to inventories (or counting) of objects.[1] It would seem that the bureaucrats in both cultures needed accountants before writers or scribes.

The first city-states, such as Uruk Lagash, Ur, and Eridu, developed in Sumer, the most fertile region of Mesopotamia. They competed vigorously with one another for land, resorting to war at the slightest provocation. As a result, for the first time, there emerged empires, unions of city-states achieved through coercion or persuasion or both, often with a single city dominating the others. The history of Mesopotamia after 3000 BC is the history of one empire or dynasty succeeding another, with each developing its own "style" of dominance and survival.

The Akkadian empire (c. 2375–2225 BC) was the creation of Sargon (2371–2316 BC), who, initially taking advantage of internal dissension in Sumer, conquered most of the Mesopotamian river valley during his reign of fifty-six years. To retain his hold, Sargon arranged the marriage of his daughter, Enheduanna, the high priestess of the Akkadian religion, to the former king of Sumer (the high priest of the Sumerian religion). It was hoped that if two peoples worshipped the same gods, they were unlikely to go to war with one another. Enheduanna set herself the task of writing a text containing the liturgy and rituals from both religions. This text became the earliest-known writing by a woman anywhere in the world.

About 2000 BC, the Akkadian empire was overrun by the Amorites (or Babylonians), who swept down into Mesopotamia from the northern land of Nimrod (present-day upper Iraq). Hammurabi built the city of Babylon as his capital. He arranged for a written legal code, which was among the first in history. His approach to safeguarding his power was two-pronged: increase the prestige of the king and promote the use of organized law.

The royal prestige was safeguarded by centralizing the seat of power: locating the palace and temple within the same group of buildings and so enabling the ruler to perform similar rituals to underpin temporal and secular power. An audience with the king was no different from an audience with a god in a temple. To create an organized system of law, Hammurabi employed legal experts to collect, codify, and disseminate laws across his kingdom. His new Code of Law was carved onto pillars, situated in every city of his empire. All citizens could call upon the protection of Hammurabi's justice; the unity and stability of the empire would thereby be assured by popular support. Hammurabi's dynasty was a high point in Mesopotamian history and a period of flowering of mathematical achievements. Indeed, as indicated later in the chapter, much of the evidence on accomplishment within both disciplines, mathematics as well as law, belongs to this period. The Old Babylonian empire lasted about seven hundred years, finally breaking down as a result of internal disorder and weakness of later rulers.

The Assyrian rulers (c. 900–600 BC) that followed held their empire together through terror, proudly displaying the severed heads and flayed skin of conquered enemies. Like many tyrannies, they were adept in introducing new technology of war, such as iron weapons and horse-drawn chariots. Their power was therefore based on fear, terror, and superior military skill. But once the conquered peoples got over their fears and gained the new technology, they rebelled. The Assyrian capital, Nineveh, was finally destroyed in 612 BC. The Assyrian epoch is marked by relative stagnation in practically all scientific activities with the possible exception of astronomy.

The Chaldean empire (c. 600–550 BC) tried the "restoration" approach. Nebuchadnezzar (630–562 BC) conquered a little, but built a little and spent a lot. He conquered Israel and brought a substantial section of the population back to Babylon as slaves. He developed Babylon, which had fallen into bad times under the Assyrians, including restoring the famous Hanging Gardens. The New Babylonian empire came into existence, and there was a revival of what we would now call algebra. Nebuchadnezzar's approach was, on the whole, successful until the arrival of the Persians.

In his campaigns between 550 and 530 BC, Cyrus united the Persians and the Medes, then moved his show on the road to Mesopotamia. The Persian "style" was tolerance and benevolence to all. They respected local customs and traditions, thereby gaining many supporters. The Jews were allowed to go home; Cyrus even gave them money to rebuild their smashed temple. Many cities just opened their gates to Cyrus and asked to be part of his empire. The Persians gave the region uniform coinage, shared technology, trade, and roads, and asked only for allegiance and taxes in return. There was considerable work on astronomy during this period. Cyrus and his successors ruled in peace for over two hundred years, until they were conquered by the Macedonian Alexander and, in 311 BC, the Seleucid dynasty was established. This was a period when the temples dedicated to the god Marduk in Babylon and the sky god Anu served as the protectors of the Babylonian religion and culture. Babylonian mathematical activity continued, especially the mathematical astronomy relating to the timings of eclipses based on observational data collected in these temples over centuries, and with realization of the full potential for calculation using the sexagesimal (base 60) system. But this is another story that we will not pursue further, except to add that work on astronomy and geometrical algebra continued apace to merge into the swelling stream of Hellenistic mathematics and astronomy.

In discussing the mathematics of Mesopotamia, it is worth raising the same questions we did in the preface to this new (third) edition:

1. *What* was the content of the mathematics known to that culture?

2. *How* was that mathematics thought about and discussed?

3. *Who* was doing the mathematics?

The Material Basis of Mesopotamian Mathematical Culture

The empires that grew out of the city-states of Sumer required a large bureaucracy to carry out their wishes. From the middle of the third millennium this bureaucracy began recruiting scribes. A scribe was a member of a specialized profession, trained in special schools where, increasingly, the curriculum was dominated by applied mathematics.[2] Even when empires collapsed, the two lasting legacies that the scribes had helped to create over the centuries remained: a method of systematic accounting and the introduction of a place-value number system. Both innovations were a result of the ability of scribal schools to respond effectively to the increasingly sophisticated demands made by the administrative apparatus engaged in collecting taxes, conducting land surveys, supervising large-scale building programs, and ensuring the supply of young men for wars.

The centralized Sumerian states collapsed, in part at least, because of the weight of their top-heavy bureaucracy. This led to the emergence of

Hammurabi and the Old Babylonian period. A new economic, social, and ideological order asserted itself. Instead of large-scale agriculture or craft workshops, often owned by the ruler, the emphasis changed to small-scale enterprises owned by private individuals. This decentralization spread into many spheres, including the occupations of scribes. Scribes were no longer small cogs in the large wheel of government. They could be found writing letters for private individuals or tutoring children in private homes. The school for scribes probably continued with the old curriculum for a long time, teaching the student scribes accounting, surveying, and other administratively useful pursuits. The high-status jobs were still to be found in the state bureaucracy. However, one begins to discern from the clay tablets of the period that a new mathematics was developing: a mathematics that was no longer purely utilitarian. This was a period of great interest in what we would describe today as "second- and higher-degree equations" discussed later in this chapter. The Plimpton Tablet (also discussed later in this chapter) belongs to this period. Even the scribal schools were caught up in this wave of new thinking. There are signs that mathematics was developing as a separate discipline, loosened from the coattails of narrow utilitarian preoccupations.

This period of Mesopotamian history, however, came to an end. For about one thousand years, "pure" mathematics took a backseat, to be reestablished during the Chaldean period, when there was once again a resurgence of mathematics. But that is another story. However, what we seek to establish in a limited fashion is that mathematical development, howsoever defined, was shaped by institutions such as scribal schools, which in turn were products of the material and social forces driving the society.

To illustrate, consider the evidence in the form of clay tablets of the activities of a scribal school in Nippur (c. 1740 BC) run by a priest in the front courtyard of his house.[3] He had no more than one or two students (possibly his own sons), who began their education by learning how to make wedge-shaped marks in clay with a reed stylus, learning by copying and repetition a set of simple cuneiform signs. The education of the student scribes progressed to writing Sumerian words for different objects, followed by more complex exercises that involved writing and learning multiplication tables and lists of metrological terms. Only after this was the student introduced to writing sentences in Sumerian and learning Sumerian literature. The method of instruction was rote learning, so that where an opportunity to do mathematical calculation was offered, this may have come as a relief even if mistakes were not uncommon. Soon after 1739 BC, the fifteen hundred school tablets that had accumulated were used as bricks and building material to repair the priest's house. However, this mode of instruction continued for a long time, as shown by the discovery of clay tablets in large terra-cotta jars at the home of a family of healers and diviners in Uruk (420 BC), where younger males were taught by their elders to write and calculate in Sumerian and Akkadian. What these tablets and other evidence indicate is that mathematics was rarely pursued in ancient Mesopotamia as a leisure activity, nor was it generally supported by institutional patronage. It was part of the process of providing training in literacy and numeracy, necessary requirements for a future priest or healer or accountant or teacher.

The Persian and Hellenistic periods saw the dethroning of the bureaucratic scribal class, with administration now being carried out by another class in the language of Aramaic or Greek. The place of the scribes was taken by a class of mathematically trained priests known as the *kalu*, located mainly in the temples dedicated to the gods Marduk and Anu, whose ceremonial function was to weep and wail and beat drums during the solar and lunar eclipses. This was to placate the gods and drive away the evil that followed the eclipses. It was the search for accurate methods of predicting these ominous events that led to significant work in mathematical astronomy, which combined observations and calculations. Archaeological evidence (Rocherg 1993 and Robson 2005) indicates that the role of the scribe was taken over by the priest in promoting and preserving mathematical knowledge in general.

Sources of Mesopotamian Mathematics

Of the half a million inscribed clay tablets that have been excavated, fewer than five hundred are of direct mathematical interest. Apart from those in the hands of private collectors, collections of these mathematical tablets are scattered among the museums of Europe in Berlin, London, Paris, and Strasbourg and the universities of Yale, Columbia, Chicago, and Pennsylvania in the United States. Some of the more recent finds, notably from Tell Harmal, Tell Hadad, and Tell Dhibayi in Iraq, were kept in the Iraqi Museum in Baghdad, although the ravages of the recent war have resulted in a number being destroyed or stolen.[4] The tablets vary in size, from as small as a postage stamp to as large as a pillow. Some are inscribed only on one side, others on both sides, and a few even on their edges.

To make a tablet, clay that may have come from the banks of the Tigris or Euphrates was collected and kneaded into shape. It was then ready for recording. The scribe used a piece of reed about the size of a pencil, shaped at one end so that it made wedgelike impressions in the soft, damp clay. He had to work fast, for the clay dried out and hardened quickly, making corrections or additions difficult. Having completed one side, he might turn the tablet over and continue. When he had finished, the tablet was dried in the sun or baked in a kiln, leaving a permanent record for posterity.

The wedge-shaped cuneiform script of the Sumerians was deciphered as early as the middle of the nineteenth century through the pioneering efforts of George Frederick Grotefend (1775–1853) and Henry Creswicke Rawlinson (1810–1895), but only since the 1930s have the mathematical texts been studied seriously.[5] This delay may be partly explained by the different ways in which a mathematician and a philologist approach early literature. The average mathematician, unless presented with a text that falls within the limits of what "mathematics" is perceived to be, has little time for the past; rarely is historical curiosity aroused by mathematical teaching. The philologist seeks to revive the past in order to explore the growth and decline of ancient civilizations; but, probably because of a lack of mathematical training, the philologist rarely takes an interest in ancient mathematics. So the Mesopotamian mathematical texts lay undeciphered and ignored until the pioneering work by Otto Neugebauer, who published his *Mathematische Keilschrift-Texte* in three volumes from 1935 to 1937, and by Francois Thureau-Dangin, whose complete works, titled *Textes mathématiques Babyloniens*, were brought out in 1938. Since then new evidence and interpretations have continued to appear, even in recent years.

There are three main sources for Mesopotamian mathematics. Some of the oldest records, written in Sumerian cuneiform, date back to the last quarter of the fourth millennium BC. From that period, in the temple precincts of the city of Uruk, a single tablet has been discovered of the oldest recorded mathematics. This consists of two exercises on calculating the area of fields. However, much of the information on this period is of a commercial and legal nature, as would now be found on invoices, receipts, and mortgage statements, and details about weights and measures. There are some but not many mathematical records until we come to the Old Babylonian period, during the first half of the second millennium BC. It has been estimated that between two-thirds and three-quarters of all the Mesopotamian mathematical texts that have been found belong to this period; in our subsequent discussion of Mesopotamian mathematics we shall concentrate on the evidence from this period. A very large portion of the remaining texts belong to a period beginning with the establishment of the New Babylonian empire of the Chaldeans, around 600 BC, after the destruction of Nineveh, and continuing well into the Seleucid era. This was also a period of considerable accomplishments in astronomy.

The Origins of Mesopotamian Numeration

From about 8000 BC, a system of recording involving small clay tokens was prevalent in the Near and Middle East. Tokens were small geometric objects, usually in the shape of cylinders, cones, and spheres. They were first identified in societies evolved from a life based on hunting and gathering to one based on agriculture, like the Ishango in central Africa. The earliest tokens were simple in design: they stood for basic agricultural commodities such as grain and cattle. A specific shape of token always represented a specific quantity of a particular item. For example, "the cone . . . stood for a small measure of grain, the sphere represented a large measure of grain, the ovoid (a rough egg-shaped solid with one end being more pointed) stood for a jar of oil" (Schmandt-Besserat 1992, p. 161). Two jars of oil would be represented by two ovoids, three jars by three ovoids, and so on. Thus, the tokens became not only an abstraction for the things being counted but also constituted a system of great specificity and precision.

With the development of city-states and the emergence of empires came a more complex economic and social structure, reflected in both the diversity and the standardization of tokens. This increased the scope for record keeping and commercial contracts in a way that counting using pebbles or twigs could not do. A collection of tokens could represent a future promised transaction or, stored in a temple or palace, a record of a past transaction. Both contracts and archives required secure methods of preserving groups of tokens. The Sumerians devised two main systems of storage: stringing the tokens on a piece of cord and attaching the ends of the strings to a solid lump of clay marked with a security seal called a *bulla*; or storing the tokens inside a clay envelope bearing impressions of the enclosed tokens for identification purposes. "For reasons we do not know, plain tokens were most often secured by envelopes and complex tokens by *bullae*" (Schmandt-Besserat 1992, p. 110).

Of the two systems, the practice of storing tokens in clay envelopes was more significant for the development of mathematics. The last step in the evolution of tokens was a merging of the two systems of *bullae* and envelopes. Simple tokens were pressed to make marks on a solid lump or tablet of clay. Only the clay tablet was then kept. Within a couple of hundred years, this new system was also being used for the complex tokens, but here, because of their complicated shapes and designs, the image of the token did not transfer satisfactorily onto the clay. This new system, in place by about 3000 BC, afforded greater ease of use and storage, at the price of a certain loss of security. These pressed or drawn marks on the clay tablets were the beginnings of the Babylonian numeration system.

From about 3000 BC, among the Sumerians, tokens for different goods began appearing as impressions on clay tablets, represented by different symbols and multiple quantities represented by repetition. Thus three units of grain were denoted by three "grain marks," five jars of oil by five "oil marks," and so on. The limitations of such a system became evident with the increasing complexity of Sumerian economic life: a confusing proliferation of different-style tokens to be learned and the tedium of representing large magnitudes. Recording the sale of five jars of oil or of a limited range of commodities was a simple affair, but an increase in the quantity and range of commodities was a different matter. Temple complexes, such as the temple of the goddess Inanna at Eana in Uruk (3200 BC), were large-scale enterprises, dealing in considerable quantities of goods and labor. A new system of recording and accounting needed to be devised. The accountants at the temple adapted a long-used system of accounting with clay tokens by impressing stylized outlines of tokens to denote numbers, with pictograms and other symbols to denote the objects that were being counted. A number of different numeration and metrological systems were used depending on the objects counted.

The first great innovation, as we saw earlier in chapter 2, was the separation of the quantity of the goods from the symbol for the goods. That is, to represent three units of grain by a symbol for "three" followed by a symbol for "grain unit" in the same way that we would write three goats or three cows or, even more generally, three liters or three kilometers.

Whereas we use the same number signs regardless of their metrological meaning (the "three" for sheep is the same sign as the "three" for kilometers or jars of oil), the Sumerians resorted to a wide variety of different symbols. Nissen et al. (1993) have identified around 60 different number signs, which they group into a dozen or so systems of measurement. For example, the Sumerians used one system for counting discrete objects, such as people, animals, or jars, and other systems for measuring areas. Each system had a collection of signs denoting various quantities.[6]

In each Sumerian metrological system there were a number of different size-units with fixed conversion factors between them, similar to our system, for example, where there are 12 inches in a foot and 3 feet in a yard, and so on. And just as in our old weight and measure systems, Sumerian metrology featured all sorts of conversion factors, although it is notable that they were all simple fractions of 60.[7]

In the early stages, however, there were different systems of numerical representation in Mesopotamia, depending on what was being measured. For a short period, a "bisexagesimal" system (i.e., a system with the units in the ratios 1:10:60:120:1,200:7,200) was used to count products relating to grain and certain other commodities. It operated with conversion factors 10, 6, and 2, so that the symbol for the largest quantity, this time a large circle containing two small circles, represented 7,200 base units. Yet another system was used for measuring grain capacity: the conversion factors were 5, 10, and 3, so that the largest unit, a large cone containing a small circle, was worth 900 base units. To add to the confusion, a single sign could be used in several systems to denote different multiples of the base unit. In particular, the small circle could mean 6, 10, or 18 small cones, depending on context and the system in use.

Gradually, over the course of the third millennium, the round number-signs were replaced by cuneiform equivalents so that numbers could be written with the same sharp stylus that was being used for the words in the text. A detailed account of this innovative system follows in the next section.

The Mesopotamian Number System

Early clay tablets (c. 3000 BC) show that the Sumerians did not have a systematic positional system for all powers of 60 and their multiples. They used the following symbols:

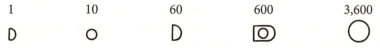

1	10	60	600	3,600
D	O	D	⊙	◯

The symbols for the first three numbers were written with the lower end of a cylindrical stylus, held obliquely for 1 and 60 and vertically for 10. The symbol for 600 was a combination of those for 10 and 60; the large circle for 3,600 was written with an extra large round stylus.

One of the most outstanding achievements of Mesopotamian mathematics, and one that helped to shape subsequent developments, was the invention of a place-value number system. From around 2000 BC there evolved a sexagesimal place-value system using only two symbols: Υ for 1 and $\blacktriangleleft$ for 10. In this system, the representation of numbers smaller than 60 was as straightforward as it was in the Egyptian notation. Thus

4:	28:	59:

$$\Upsilon\Upsilon\Upsilon\Upsilon \qquad \begin{array}{l}\blacktriangleleft\Upsilon\Upsilon\Upsilon\Upsilon\\\blacktriangleleft\Upsilon\Upsilon\Upsilon\Upsilon\end{array} \qquad \begin{array}{l}\blacktriangleleft\\\blacktriangleleft\Upsilon\Upsilon\Upsilon\Upsilon\\\blacktriangleleft\Upsilon\Upsilon\Upsilon\Upsilon\\\blacktriangleleft\end{array}\Upsilon$$

If the Mesopotamians had merely used these symbols on an additive basis (which they did not), their numeration and computations would probably have developed along Egyptian lines. But, from as early as 2500 BC, we find indications that they realized they could double, triple, quadruple (and so on) the two symbols for 1 and 10 by giving them values that depended on their relative positions. Thus the two symbols could be used to form numbers greater than 59:

$60 \ = \ 60(1)$: Υ

$95 \ = \ 60(1) + 35$: $\Upsilon \begin{array}{l}\blacktriangleleft\\\blacktriangleleft\end{array}\blacktriangleleft \Upsilon_\Upsilon\Upsilon_\Upsilon\Upsilon$

$120 \ = \ 60(2)$: $\Upsilon \Upsilon$

$4{,}002 \ = \ 60^2(1) + 60(6) + 42$: $\Upsilon \ ^\Upsilon_\Upsilon{}^\Upsilon_\Upsilon{}^\Upsilon \begin{array}{l}\blacktriangleleft\blacktriangleleft\\\blacktriangleleft\blacktriangleleft\end{array} \Upsilon \Upsilon$

It was a relatively simple matter, though one of momentous significance, to extend this principle of positional notation to allow fractions to be represented:

$1/2 \ = \ 60^{-1}(30) \ = \ 30/60$: $\blacktriangleleft\blacktriangleleft\blacktriangleleft$

$1/4 \ = \ 60^{-1}(15) \ = \ 15/60$: $\blacktriangleleft \Upsilon_\Upsilon\Upsilon_\Upsilon\Upsilon$

$1/8 = 60^{-1}(7) + 60^{-2}(30)$: 𒁹𒁹𒁹 𒌋𒌋

$532\frac{3}{4} = 60(8) + 52 + 60^{-1}(45)$: 𒁹𒁹𒁹𒁹 𒌍𒐈𒌍𒐈𒁹𒁹𒁹

Two important features of Mesopotamian positional notation are high-lighted by these examples: unlike our present-day system, there is no symbol for zero, and neither is there a symbol corresponding to our decimal point to distinguish between the integer and fractional parts of a number. There is also the more fundamental question of why the Mesopotamians should have constructed a number system on base 60 rather than the more "natural" base 10 (i.e., the decimal system). However, they used base 10 notation up to 59.

The absence of a symbol for a placeholder could lead to confusion over what number was being recorded. For example,

𒁹𒌋𒌋
𒁹𒌋𒌋

could be $60(2) + 40 = 160$, or $60^2(2) + 60(0) + 40 = 7{,}240$, or it could be $2 + 60^{-1}(40) = 2\frac{2}{3}$, or even $60^{-1}(2) + 60^{-2}(40) = 2/45$, since there is no "sexagesimal point" placeholder to indicate that the number is a fraction. In the absence of a special symbol for zero, the number might be identifiable from the context in which it appeared, or a space might be left to indicate a missing sexagesimal place. There again, it could have been that the lack of a zero symbol in ancient Mesopotamia was of little practical consequence, for the existence of a large base, 60, would ensure that most numbers of everyday concern could be represented unambiguously. For example, it is unlikely that the prices of commodities in ordinary use would have exceeded 59 "units" (discounting inflation of course!). Moreover, the relative positions of the two 𒁹 symbols and the four 𒌋 symbols in the example above would indicate that, if the number were an integer, it would not be less than 160. This was because the Mesopotamians, unlike the Egyptians, wrote their numerals the same way as we do, from left to right.

It was not until the Seleucid period that a separate placeholder symbol was introduced to indicate an empty space between two digits inside a number. Thus the number 7,240 would be written as where $\overset{\blacktriangle}{\blacktriangle}$ serves as the placeholder symbol. The problem still remained of how to represent the absence of any units at the end of a number. Nowadays we use the symbol for zero in the terminal position. Without something like that, it is difficult to know whether the number

𒐗𒐗𒐗𒌋

is $60(3) + 30 = 210$, or $60^2(3) + 60(30) = 12,600$, or even $3 + 60^{-1}(30)$ $= 3\frac{1}{2}$. It is therefore clear that while the Mesopotamians were consistent in their use of place-value notation, they never operated with an absolute positional system. When, in the second century AD, Claudius Ptolemy of Alexandria began to use the Greek letter o (omicron) to represent zero, even in the terminal position of a number, there was still no awareness that zero was as much a number as any other and so, just like any other, could enter into any computation. Recognition of this fact—"giving to airy nothing, not merely a local inhabitation and a name, a picture, a symbol but also a helpful power" (Halstead 1912)—was not to occur for another thousand years, in India and Central America.

If we are to make any further headway, we need a way of transliterating the Mesopotamian numerical representation into a notation more convenient for us. We shall adopt Neugebauer's convention of using a semicolon (;) to separate the integral part of a number from its fractional part, just as we use the decimal point today—the semicolon is in effect the "sexagesimal point." All other sexagesimal places are separated by a comma (,). Some examples, of numbers whose cuneiform representations have been given above, will make this convention clear:

60	=	60(1):	1,00
95	=	60(1) + 35:	1,35
120	=	60(2):	2,00
4,002	=	$60^2(1) + 60(6) + 42$:	1,06,42
1/2	=	$60^{-1}(30) = 30/60$:	0;30
1/8	=	$60^{-1}(7) + 60^{-2}(30)$:	0;07,30
$532\frac{3}{4}$	=	$60(8) + 52 + 60^{-1}(45)$:	8,52;45

With this scheme, the ambiguity in the representation of 7,240 in the Mesopotamian notation disappears: this number is now written as 2,00,40.

Different explanations have been offered for the origins of the sexagesimal system, which, unlike base 10 or even base 20, has no obviously anatomical basis. Theon of Alexandria, in the fourth century AD, pointed to the computational convenience of using the base 60. Since 60 is exactly divisible by 2, 3, 4, 5, 6, 10, 15, 20, and 30, it becomes possible to represent a number of common fractions by integers, thus simplifying calculations: the integers that correspond to the unit fractions 1/2, 1/3, 1/4, 1/5, 1/6, 1/10, 1/15, 1/20, and 1/30 are 30, 20, 15, 12, 10, 6, 4, 3, and 2, respectively. Of the unit fractions with denominators from 2 to 9, only 1/7 is not "regular" (i.e., 60/7 gives a nonterminating number). It is therefore quite a simple matter to work with fractions in base 60. In a decimal base, though, only three of the nine fractions above produce integers, and none of 1/3, 1/6, 1/7, and 1/9 is regular. Indeed, while base 10 may be more "natural," since we have ten fingers, it is computationally more inefficient than base 60, or even base 12.

However, this explanation for the use of base 60 is unconvincing because of its "hindsight" character. It is highly unlikely that such considerations were taken into account when the base was chosen. A second explanation emphasizes the relationship that exists between base 60 and numbers that occur in important astronomical quantities. The length of a lunar month is 30 days. The Mesopotamian estimate of the number of days in a year was 360, based on the zodiacal circle of 360°, divided into twelve signs of the zodiac of 30° each. The argument goes that either 30 or 360 was first chosen as the base, later to be modified to 60 when the advantages of such a change were recognized. Here again, there is a suggestion of deliberate, rational calculation in the choice of the base that is not totally convincing. A more plausible explanation is that the sexagesimal system evolved from metrological systems that used two alternating bases of 10 and 6, favored perhaps by two different groups, which gradually merged, and that the advantages of base 60 for astronomical and computational work then came to be recognized.[8]

The sexagesimal system was used in Mesopotamia in 1800 BC and continued to be used well into the fifteenth century AD. Sexagesimal fractions appeared in Ptolemy's *Almagest* in AD 150. The Alfonsine Tables, astronomical tables prepared from Islamic sources on the instruction of Alfonso X of Castile and written in Latin at the end of the thirteenth century AD, used a consistent sexagesimal place-value system. The Islamic astronomer al-Kashi (d. AD 1429) determined 2π sexagesimally as

6;16,59,28,01,34,51,46,15,50—the decimal equivalent of which is accurate to sixteen places. And Copernicus's influential work in mathematical astronomy during the sixteenth century contained sexagesimal fractions. The current use of the sexagesimal scale in measuring time and angles in minutes and seconds is part of the Mesopotamian legacy.

Before going on to look at operations with Mesopotamian numerals, let us pause to compare the Mesopotamian way of representing numbers with other systems. In assessing a notational system, the following questions are pertinent:

1. Is the system easy to learn and write?

2. Is the system unambiguous?

3. Does the system lend itself readily to computation?

The Mesopotamian system scores well on questions 1 and 3. It is easily learned, being one of the most economical systems in terms of the symbols used. The only other number system that operated with just two symbols (a dot and a dash) was the Mayan, though unlike the Mesopotamian system there was also a special sign for zero. If we compare the Mesopotamian with the Greek number system, which used twenty-seven symbols, the simplicity of the former notation is obvious. But one must contrast this simplicity with the awkwardness of representing a number such as 59, which in the unabridged Mesopotamian notation would require fourteen signs (though some would argue that they *together* represent a single cuneiform sign), as against just two in the Greek notation.[9] The Mesopotamian system is also remarkable for its computational ease, which arises from its place-value principle and its base of 60. Calculating in this base proved a distinct advantage when dealing with fractions. Until the emergence of decimal fraction representation, the Mesopotamian treatment of fractions remained the most powerful computational method available.

But the great disadvantage of Mesopotamian notation was its ambiguity, the consequence of having neither a symbol for zero nor a suitable device for separating the integral part of a number from its fractional part. It was not that the system of notation precluded the incorporation of these additional features, but that the Mesopotamians simply did not use them. (In the time of the New Babylonian empire, though, the placeholder symbol ▲▲ appeared.) All in all, compared with the Egyptian system, the Babylonian notation was computationally more "productive" and symbolically more economical (since the place-value principle made it unnecessary to invent new symbols for large numbers), but it had the disadvantage of being ambiguous. The Egyptian system had another advantage over the Babylonian: the order in which the symbols representing a number are written is of no consequence in Egyptian notation.

Operations with Mesopotamian Numerals

With a positional system of numeration available, ordinary arithmetical operations with Mesopotamian numerals would follow along the same lines as modern arithmetic. To relieve the tedium of long calculations, the Mesopotamians made extensive use of mathematical tables. These included tables for finding reciprocals, squares, cubes, and square and cube roots, as well as exponential tables and even tables of values of $n^3 + n^2$, for which there is no modern equivalent. These tables account for a substantial portion of the sources of Mesopotamian mathematics available to us.

Multiplication and division were carried out largely as we would today. Division was treated as multiplication of the dividend by the reciprocal of the divisor (obtained from a table of reciprocals). To take a simple example:

EXAMPLE 4.1 Divide 1,029 by 64.

Solution

In Neugebauer's notation, $1,029 = 60^1(17) + 60^0(9)$ is written as 17,09. Also, 1/64 becomes 0;00,56,15, since $1/64 = 60^{-2}(56) + 60^{-3}(15)$, found from a table of reciprocals.

Therefore

17,09 multiplied by 0;00,56,15 equals 16;04,41,15.

The long multiplication may have been carried out in the same way as we would today, apart from the sexagesimal base:

$$
\begin{array}{r}
0;00,56,15 \\
\times \qquad 17,09 \\
\hline
8,26,15 \\
15;56,15 \\
\hline
16;04,41,15
\end{array}
$$

Continued . . .

Continued . . .

The answer 16;04,41,15 can be converted to the decimal base:

$16 + 60^{-1}(4) + 60^{-2}(41) + 60^{-3}(15) \approx 16.0781.$

A complete set of sexagesimal multiplication tables was available not only for each number from 2 to 20, and for 30, 40, and 50, but also for many other numbers. This would be sufficient to carry out all possible sexagesimal multiplications, just as present-day multiplication tables for numbers from 2 to 10 are sufficient for all decimal products. Often, the tables of reciprocals were available only for those "regular" integers up to 81 that are multiples of 2, 3, or 5. The reciprocals of "irregular" numbers, or those containing prime numbers that are not factors of 60 (i.e., all prime numbers except 2, 3, and 5), would, in effect, have been nonterminating sexagesimal fractions. For example, the reciprocals of the "regular" numbers 15, 40, and 81 are

$$\frac{1}{15} = 0;04, \quad \frac{1}{40} = 0;01,30, \quad \frac{1}{81} = 0;00,44,26,40.$$

The reciprocals of the "irregular" numbers 7 and 11 are

$$\frac{1}{7} = 0;08,34,17,08,34,17,\ldots, \quad \frac{1}{11} = 0;05,27,16,21,49,\ldots.$$

The tables of reciprocals found on the older tablets are all for "regular" numbers. There is one tablet, from the period just before the Old Babylonian empire, which contains the following problem:

EXAMPLE 4.2   Divide 5,20,00,00 by 7.

Suggested Solution

Multiply 5,20,00,00 by the approximate reciprocal of 7 (i.e., 0;08,34,17,08) to get the answer: 45,42,51;22,40.

A later tablet from the Seleucid period gives the upper and lower limits on the magnitude of 1/7 as

$$0;08,34,16,59 < \frac{1}{7} < 0;08,34,18$$

Statements such as "approximation given since 7 does not divide" from the earlier periods, and the later estimates of bounds, give us a tantalizing glimpse of the Mesopotamians taking the first step (though it is not clear whether they were fully aware of the implications) in coming to grips with the incommensurability of certain numbers.

A Babylonian Masterpiece

Evidence that the Mesopotamians had no difficulty working with what we now know as irrational numbers is found on a small tablet, belonging to the Old Babylonian period, that forms part of the Yale collection.[10] It contains the diagram shown in figure 4.2a and "translated" in figure 4.2b. The number 30 indicates the length of the side of the square. Of the other two numbers, the upper one (if we assume that the "sexagesimal point" (;) occurs between 1 and 24) is 1;24,51,10, which in decimal notation is

$$1 + 60^{-1}(24) + 60^{-2}(51) + 60^{-3}(10) \approx 1 + 0.4 + 0.01416667 + 0.0000463$$
$$= 1.41421297.$$

To the same number of decimal places, the square root of 2 is 1.41421356, so the Babylonian estimate is correct to five places of decimals. The lower number is easily seen to be the product of 30 (the side of the square) and the estimate of the square root of 2.

The interpretation is now clear, particularly if one notes that on the back of this clay tablet there remains a partly erased solution to a problem involving the diagonal of a rectangle of length 4 and width 3. Let d be the diagonal of the square; applying the Pythagorean theorem then gives

$$d^2 = 30^2 + 30^2;$$
$$d = \sqrt{2}\,(30) \approx (1;24,51,10)(30;00) = 42;25,35.$$

The number below the diagonal is therefore the length of the diagonal of a square whose side is 30.

The solution to this problem highlights two important features of Mesopotamian mathematics. First, over a thousand years before Pythagoras, the Mesopotamians knew and used the result now known under his name.[11] (In a later section we discuss further applications of this result, as well as evidence that the Mesopotamians may have known the rules for generating Pythagorean triples a, b, c, where $a^2 + b^2 = c^2$.) Second, there is the intriguing question of how the Mesopotamians arrived at their remarkable estimate of the square root of 2, an estimate that would still be in use two thousand years later when Ptolemy constructed his table of chords.

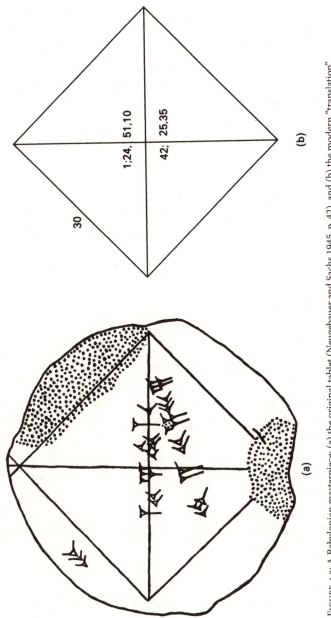

30

1;24, 51,10

42; 25,35

(a)

(b)

FIGURE 4.2: A Babylonian masterpiece: (a) the original tablet (Neugebauer and Sachs 1945, p. 42), and (b) the modern "translation"

One conjecture is that the method they used to extract square roots resembles the iterative procedure used by digital computers today. This procedure is as follows. Let x be the number whose square root you want to find, and the positive number a your rough guess of the answer. Then $x = a^2 + e$, where the difference (or "error") e can be positive or negative. We now try to find a better approximation for the square root of x, which we denote by $a + c$. It is obvious that the smaller the error e, the smaller is c relative to e. Thus we impose the following condition on c:

$$x = (a + c)^2 = a^2 + e, \tag{4.1}$$

from which

$$2ac + c^2 = e.$$

Now, if you make a sensible guess for a in the first place, c^2 will be much smaller than $2ac$ and can therefore be ignored. So equation (4.1) may be written as

$$c \approx \frac{e}{2a}. \tag{4.2}$$

Hence, from (4.1) and (4.2), an approximation for the square root of x is

$$a + c \approx a + \left(\frac{e}{2a}\right) = a_1.$$

Now, taking $a_1 = a + (e/2a)$ as the new "guesstimate," the process can be repeated over and over again to get $a_2, a_3, \ldots$ as better and better approximations.

Let us illustrate this approximation procedure with the example we started with—how did the Mesopotamians obtain their estimate for $\sqrt{2}$ as 1;24,51,10?

EXAMPLE 4.3 Find the square root of 2 up to the fourth approximation (i.e., a_3), starting with $a_0 = 1$.

Solution

The steps of the solution are summarized in table 4.2.

Continued . . .

i	a_i	e_i	c_i	a_{i+1}	
				Modern	Babylonian
0	1	1	1	1.5	1;30
1	1.5	−0.25	−0.0833	1.41667	1;28,0,7,12
2	1.41667	−0.00695	−0.00246	1.41421	1;24,51,10

Continued . . .

The value obtained after two steps is exactly the value in figure 4.2b!

This procedure for calculating square roots is widely known as Heron's method, after the Alexandrian mathematician who lived in the first century AD. A similar procedure could have been used to evaluate $\sqrt{2}$ in the earliest extant mathematical writings of the Indians, the *Sulbasutras*, which have been variously dated from 800 BC to 500 BC, although it is more likely that a protogeometric procedure involving "dissection and reassembly" may have been used. The Indian approximation procedure is discussed in chapter 7.

Other Tables and Their Uses

Like the Egyptians, the Mesopotamians were inveterate table makers. Apart from the multiplication and reciprocal tables already mentioned, there are two sets of tables that are worth examining. One contains the values of $n^3 + n^2$ for integers n from 1 to 20 and 30, 40, and 50. Stated in modern notation and terminology, the tables could be for solving mixed cubics of the form

$$ax^3 + bx^2 = c, \tag{4.3}$$

for if you multiply equation (4.3) through by a^2/b^3 you get

$$\left(\frac{ax}{b}\right)^3 + \left(\frac{ax}{b}\right)^2 = \frac{ca^2}{b^3},$$

from which

$$y^3 + y^2 = \left(\frac{a^2}{b^3}\right)c, \tag{4.4}$$

where $y = ax/b$. Equation (4.4) can be solved for y using the $n^3 + n^2$ table, and x obtained from y by using $x = by/a$. To illustrate:

EXAMPLE 4.4 Solve $144x^3 + 12x^2 = 21$.

Solution

Multiply both sides of the equation by 12 and substitute $y = 12x$:

$$(12x)(12x)(12x) + (12x)(12x) = (12)(21);$$

$$y^3 + y^2 = 252.$$

From the table $y = 6$, so $x = 0.5$ (or 0;30).

A solution involving cubics is a remarkable achievement in view of the high level of technical skill necessary to handle the algebraic concepts in the absence of a symbolic notation. With the benefit of modern symbolic algebra, it is easy to see that $(ax)^3 + (ax)^2 = c$ is equivalent to the equation $y^3 + y^2 = c$. Try to imagine the difficulties in recognizing this equivalence without the algebraic notation available to us, and you will appreciate the measure of the Mesopotamian achievement.

A number of tables in the collections at Berlin, Istanbul, the Louvre, and Yale contain values for exponents a^n, where n is an integer taking values 1, 2, ..., 10, and $a = 9, 16, 100,$ and 225—all perfect squares. What were these tables used for? The following problem, taken from a Louvre tablet of the Old Babylonian period, may provide the answer.

EXAMPLE 4.5 Calculate how long it would take for a certain amount of money to double if it has been loaned at a compound annual rate of 0;12 [20%].

Solution

Using modern symbolic notation, let P be the amount of the loan (the principal) and $r (= 0;12 = 1/5)$ the interest rate. The question may then be restated as follows: Find n, given $2P = P(1 + r)^n$. The problem is solved by first identifying from an exponential table that n must lie between 3 and 4 in order to satisfy the equation $(1;12)^n = 2$ [or in modern terms $(1.2)^n = 2$]. From the table,

Continued . . .

Continued . . .

$(1;12)^3 = 1;43,40,48$ (or 1.7280), and $(1;12)^4 = 2;04,25$ (or 2.0736).

Applying linear interpolation and working in modern notation would give

$$3 + \frac{2 - 1.7820}{2.0736 - 1.7280} = 3.7870$$

or, in Mesopotamian notation, 3;47,13,20 years—exactly the answer shown on the Louvre tablet!

Geometric Series

In the previous chapter on Egyptian mathematics, an example of a geometric series from an Old Babylonian text from Mari was mentioned. The problem was initially described as "an account of ants," and what it contained was the computation of five terms of a geometric progression, with the first term 99 and the common ratio 9. The standard transliteration of the text (on the obverse, or front, side of the tablet) by Friberg (2005, p. 4) gives the following data and computations (expressed both in sexagesimal and decimal bases). The computations are carried out twice, first in sexagesimal place-value numbers and then in "mixed decimal-sexagesimal" numbers, both being recorded on the obverse side of the tablet. On the reverse, the solution is offered in a "centesimal" number system peculiar to the Mari region.[12] The obverse side contains the terms of geometric progression increasing from 99 to $9^4 \times 99 = 649,539$ shown in table 4.3.

TABLE 4.3

Sexagesimal	Decimal	Items	Computation		
1,39	99	people	—		
14,51	891	birds	9×99	=	891
2,13,39	8019	ants	9×891	=	8019
20,02,51	72171	barley ears	9×8019	=	72171
3,00,25,39	649539	barley corns	9×72171	=	649539

The reverse side contains the same geometric series decreasing from 649,539 to 99 but also a sixth line that is the sum of the different items: $649,539 + 72,171 + 8,019 + 891 + 99 = 73,0719$.

This would lead naturally to the interpretation that Friberg (2005, p. 5) described as a "whimsical story" given in example 3.11 in chapter 3. A similar problem appears in Fibonacci's thirteenth-century text *Liber abaci*, where the problem starts with "7 women go to Rome and ends asking for the sum of old women, mules, sacks, breads, knives and sheaths." And the problem reappears in a different guise almost four thousand years after the Mari version in the modern *Mother Goose* riddle: "As I was going to St. Ives, I met a man with seven wives, each wife had seven sacks, each sack had seven cats, each cat had seven kittens. Kittens, cats, sacks, wives, how many were going to St Ives?" In the examples from the Ahmes Papyrus, Fibonacci's *Liber Abaci*, and the *Mother Goose* riddle, both the first term (a) and the common ratio (r) are seven, although the numbers of terms are different.

The Mari text seems to be the culmination of a number of Old Babylonian and later texts that are applications of doubling and halving algorithms. A table of doublings found in Mari (M 8631) has been interpreted as the growth of an initial capital (1 barleycorn) over a month of thirty days with a doubling every day. This has links with later stories (or legends) regarding the reward given to the inventor of chess in India, who asked the local ruler for one grain of rice on the first square of a chessboard and twice as much on each consecutive square, which would have wiped out the total stock of rice in the granary. A variant of this legend is later found in Europe, wherein a clever blacksmith who is offered the job of fitting a shoe for the king's horse asks a penny for the first nail and twice as much for each of the subsequent twenty-eight nails! These examples offer some intriguing possibilities of a chain of links that may have started with an Egyptian or Mesopotamian example thousands of years earlier.

Babylonian Algebra

Unlike the Egyptians, who had been constrained by the absence of an efficient number system, basic computations posed few difficulties for the Mesopotamians with their place-value number system and their imaginative use of various tables. As a result, even as early as the Old Babylonian period they had developed sophisticated numerical methods of solving equations and systems of equations, within the framework of a rhetorical algebra.

TABLE 4.4: SYMBOLIC NOTATION IN MESOPOTAMIAN ALGEBRA

Modern symbol	Geometric term	Mesopotamian quantity
x	length	ush
y	breadth	sag
x^2	square	lagab
z	height	sukud
xy	area	asha
xyz	volume	sahar

However, we should not be fooled by the general rhetorical nature of their algebra into overlooking traces of syncopated algebra present even then, and best exemplified by the use of certain geometric terms to denote unknown quantities. Analogous to the modern symbol x was the term "length" (of a square or rectangle); the square of this unknown quantity was referred to by the term "square." Where there was need to refer to two unknowns, they were called "length" and "breadth," their product being described as "area." Three unknowns became "length," "breadth," and "height," and their product "volume." Table 4.4 lists these terms and their modern equivalents.[13]

It is this peculiar form of reference to unknown quantities in Mesopotamian algebra that led some earlier commentators to be dismissive of statements such as "I have subtracted the side of the square from the area, and the result is 14,30," which we would now interpret as

$$x^2 - x = (14 \times 60) + (30 \times 1) = 870.$$

Here, if anywhere, appears for the first time the idea of an unknown quantity to be "found." And such problems were not necessarily ones that arose out of practical applications: they were, it seemed, exercises in "mental gymnastics." For example, consider the following example from the Old Babylonian period (c. 1800 BC), which is inscribed on a clay tablet belonging to the Yale collection:

EXAMPLE 4.6 I found a stone [but] did not weigh it; [after] I weighed [out] 8 times its weight, added 3 gin one-third of one-thirteenth I multiplied by 21, added [it], and then I weighed [it]: Result 1 mana. What was the [original] weight of the stone. [Answer] The (original) weight of the stone was $4\frac{1}{2}$ gin.

Continued . . .

Continued . . .

(Note: Words in brackets are not in the original text. Also, 1 *mana* = 60 *gin* ≈ 0.5 kg.)

The modern algebraic approach would start by assuming that the original weight of the stone is x *gin*. Multiply the weight x by 8 and add 3 to get $8x + 3$. Now do a separate calculation: "multiply one-third of one-thirteenth by 21" to get $(1/3)(1/13) = 21/39$. Multiply this result by $(8x + 3)$. Finally, add this product to the original $(8x + 3)$ and equate to 60. Thus

$$8x + 3 + \frac{21}{39}(8x + 3) = 60$$

Solving for x gives the answer of $4\frac{1}{2}$ *gin* as given in the text.

This is clearly not a practical problem of weighing stones! It resembles the rather contrived and artificial problems that one comes across in school arithmetic texts even today. And for that reason we would infer that it was a problem inflicted on scribe trainees as an exercise in mental gymnastics. What is more intriguing is how a Babylonian student would have proceeded to solve the problem without our knowledge of algebra. The tablet does not give us a clue, and so we are not any the wiser. But there are some inferences that we can make. It would be reasonable to suppose that the person who set the problem was aware of the fact that $39 + 21 = 60$, though the "language" of the problem setter would not be sufficient, in terms of our algebra, to conceive of the possibility of the original equation being reduced to $(8x + 3)(39 + 21)/39 = 60$. As mentioned in the previous chapter, the Egyptians and the others who followed them solved simple equations by guessing at a likely answer for x, finding it wrong, and scaling it to get the right one. This does not seem to work easily with the present problem. So we are left with the interesting question as to how the Babylonian scribe proceeded to solve this and similar problems. Fortunately, we have texts that throw some light on the procedures adopted.

The problems in Babylonian mathematics were mainly of three kinds: (1) problems that related to shape, area, and volume, which would be described today as geometry; (2) problems involving unknowns that were solved by methods that were a combination of algorithmic procedures and geometrical algebra, such as completing the squares; and (3) applied arithmetic involving problems in surveying, labor allocation, and construction.

Linear and Nonlinear Equations in One Unknown

The Mesopotamian solution to equations of the form $ax = c$ was no different from ours, which is $x = (1/a)c$. They would have taken $1/a$ from a table of reciprocals and obtained the product by referring to a multiplication table. If $1/a$ was not a regular sexagesimal fraction, they would have used a suitable approximation. An example from a mathematical text, found during excavations at Tell Harmal in 1949 and belonging to the Old Babylonian period, illustrates the approach. The statement of the problem and its solution are based on Taha Baqir's (1951) translation.

EXAMPLE 4.7 If somebody asks you thus: If I add to the two-thirds of my two-thirds a hundred qa of barley, the original quantity is summed up. How much is the original quantity?

Suggested Solution

1. First multiply two-thirds by two-thirds: result 0;26,40 (i.e., 4/9).

2. Subtract 0;26,40 from 1: result 0:33,20 (i.e., 5/9).

3. Take the reciprocal of 0;33,20: result 1;48 (i.e., 1 + 4/5).

4. Multiply 1;48 by 1,40 (i.e., 100): result 3,00 (i.e., 180).

5. 3,0 (*qa*) is the original quantity.

This procedure is identical to the one we would now use to solve a simple equation:

$$\left(\frac{2}{3} \text{ of } \frac{2}{3}\right)x + 100 = x \rightarrow \left(\frac{4}{9}\right)x + 100 = x \rightarrow \left(\frac{5}{9}\right)x = 100 \rightarrow x = 180.$$

The Mesopotamians were able to solve different types of quadratic equations. The two types that occurred most frequently have the forms

$$x^2 + bx = c, b > 0, c > 0; \tag{4.5}$$

$$x^2 - bx = c, b > 0, c > 0. \tag{4.6}$$

In solving (4.5), their approach was equivalent to the application of the formula

$$x = \sqrt{\left(\frac{b}{2}\right)^2 + c} - \left(\frac{b}{2}\right) \tag{4.7}$$

to get a positive solution. The corresponding formula for (4.6) gives the positive solution as

$$x = \sqrt{\left(\frac{b}{2}\right)^2 + c} + \left(\frac{b}{2}\right).$$

As an illustration of how the Babylonians solved these quadratics, consider a problem from a tablet of the Old Babylonian period, now in the Yale collection.

EXAMPLE 4.8 The length of a rectangle exceeds its width by 7. Its area is 1,00. Find its length and width.

Solution

The solution, shown below, establishes a close correspondence between the Babylonian approach and its modern symbolic variant.

SOLUTION GIVEN ON THE TABLET	SOLUTION EXPRESSED IN MODERN NOTATION
1. Halve 7, by which length exceeds width: result 3;30.	Let x be the width and y the length. Then $y = x + 7$, $xy = 60$. Or $x(x + 7) = 60$, from which $x^2 + 7x = 60$.
2. Multiply together 3;30 by 3;30: result 12;15.	Using equation (4.7) gives $x = \sqrt{(3.5)^2 + 60} - (3.5) = 5$.
3. To 12;15 add 1,00, the product: result 1,12;15.	The length is then obtained by adding 3.5 to the square root,
4. Find the square root of 1,12;15: result 8;30.	rather than adding 7 to the width.
5. Lay down 8;30 and 8,30. Subtract 3;30 from one (8;30) and add it to the other (8;30).	
6. 12 is the length, 5 the width.	Hence 12 is the length, and 5 is the width.

The Yale tablet also gives examples of solutions to quadratic problems of a more general type, such as

$$ax^2 + bx = c.$$

The technique here, expressed in our terms, was to multiply throughout by a to get

$$(ax)^2 + b(ax) = ac$$

and then to substitute $y = ax$ and $e = ac$ to obtain the standard Babylonian form of quadratic

$$y^2 + by = e, e > 0.$$

After solving for y, the value of y is divided by a to get the solution for x.

We have already discussed how the Babylonians handled cubics of the form $x^3 = c$ with the help of cube root tables, and problems of the form $x^2(x + 1) = c$ with the help of $n^3 + n^2$ tables. There is also a correct solution on the Yale tablet for a cubic of the form $x(10 - x)(x + 1) = c$, where $c = 2,48$. The correct solution of $x = 6$ is given. It is a tribute to the level of abstraction and manipulative skills of Mesopotamian mathematicians that they were solving higher-order equations such as $ax^4 + hx^2 = c$ and $ax^8 + hx^4 = c$ by treating them as if they were "hidden" quadratics in x^2 and x^4, respectively.

Linear and Nonlinear Problems in Two or Three Unknowns

The Babylonians approached these problems in two different ways. For a system of equations with two unknowns, they sometimes used the method of substitution, familiar to us, in which one of the equations is solved for one of the unknowns and the value found is substituted into the other equation. There was, however, another approach that remained uniquely Mesopotamian until it was adopted by Hellenistic and, probably, Indian mathematicians around the beginning of the Christian era. The method has been called Diophantine, after the Greek mathematician Diophantus, who lived in Alexandria during the third century AD. It has been remarked that his algebraic methods have much in common with the Babylonian procedures.

The Diophantine method is particularly suitable for solving a system of two equations where one is of the form $x + y = s$ and the other may be any type of equation (linear or nonlinear) in the two unknowns x and y.

The procedure is as follows: If x and y were equal, $x + y = s$ would imply that $x = y = (1/2)s$. Now, if we assume that x is greater than $(1/2)s$ by a quantity w, then

$$x = \frac{1}{2}s + w, \text{ and } y = \frac{1}{2}s - w.$$

If we substitute these expressions for x and y into another equation, which can be either linear or quadratic, we obtain an equation for w and can proceed to solve it. Next, we substitute the value for w into the above equations for x and y to obtain the solutions for the whole system of equations. An illustration is provided by a problem from a tablet found at Senkereh in the ancient city of Larsa, which dates back to the Hammurabi dynasty.

EXAMPLE 4.9 Length (*ush*), width (*sag*). I have multiplied *ush* and *sag*, thus obtaining the area (*asha*). Then I added to *asha*, the excess of the *ush* over the *sag*: result 3,03. I have added *ush* and *sag*: result 27. Required [to know] *ush*, *sag*, and *asha*.

Solution

EXPLANATION IN THE TEXT	EXPLANATION IN MODERN NOTATION
1. One follows this method: 27 + 3,03 = 3,30; 2 + 27 = 29.	Let x = length (*ush*), y = width (*sag*). Then the problem can be restated as $xy + x - y = 183;$ (4.8) $x + y = 27.$ (4.9) Then $xy + 2x = 210.$
2. Take one-half of 29 (14;30) and square it: 14;30 × 14;30 = 3,30;15.	Define $y' = y + 2$ (so $y = y' - 2$). Then $xy' = 210;$ (4.10) $x + y' = 29.$ (4.11)
3. Subtract 3,30 from the result: 3,30;15 − 3,30 = 0;15.	A general solution to the above set of equations may be expressed as $xy' = p$ $x + y' = s$

Continued...

Continued . . .

4. Take the square root of   or
   0;15: the square root of
   0;15 is 0;30.

$$x = \frac{1}{2}\,s + w$$

$$y' = \frac{1}{2}\,s - w$$

$$xy' = \left(\frac{1}{2}s\right)^2 - w^2 = p,$$

5. Then   where w is the square root of

$$\left(\frac{1}{2}s\right)^2 - p.$$

So if $s = 29$ and $p = 210$, then
$w^2 = 14.5^2 - 210 = 0.25$.

length (*ush*) =   So $w = 0.5$.
14;30 + 0;30 = 15;   Hence $x = 15$, $y = y' - 2 =$
14;30 − 0;30 = 14. $14 - 2 = 12$, and the area is 180.

6. Subtract 2 (which has been
   added to 27) from 14:
   width (*sag*) = 14 − 2 = 12.

7. I have multiplied 15 (*ush*)
   by 12 (*sag*) to get *asha*.
   Area (*asha*) = 15 × 12
   = 3,00.

Note that the transformations from equations (4.8) and (4.9) to equations (4.10) and (4.11) respectively are indicated in step 1 on the left-hand side. There is a close correspondence between the procedure as explained on the Babylonian tablet and the modern version given on the right.

There are a few examples in Babylonian algebra involving the solution of a set of equations in three unknowns. In a problem text kept at the British Museum, we find (in modern notation):

$$x^2 + y^2 + z^2 = 1,400 \text{ [or in base 60: 23, 20]}, \quad x - y = 10, \quad y - z = 10.$$

Its solution is unlikely to have caused many difficulties, for x and z can easily be expressed in terms of y as $y + 10$ and $y - 10$ respectively, so as to obtain a quadratic equation in y. The reader may wish to check the correctness of the solution set for (x, y, z), which is given as $(30,20,10)$; this solution set holds for positive y.

In recent years, there has been considerable discussion about how the Babylonians approached certain problems that we would today label as algebraic, but where the solutions offered have an explicit geometric rationale. What has been suggested by Høyrup (2002), Robson (2007), and others is that a substantial number of the Old Babylonian mathematical problems would fall under this category of geometrical algebra, that is, the manipulation of geometrical objects such as lines and areas to solve problems involving unknowns. A good example from the Yale collection (YBC 6967) would be as follows.

EXAMPLE 4.10 A reciprocal exceeds its [own] reciprocal by 7. What are the reciprocal and its reciprocal? [Given, by definition, the product of the two reciprocals is 60].

Modern Solution

The problem is one of finding the solution of a quadratic equation from the two equations

$$x - y = 7, xy = 60,$$

where x and y are the unknown reciprocals.

Solve the resulting quadratic equation $x^2 - 7x - 60 = 0$ using the algorithm given in (4.7) and (4.8):

$$x = \sqrt{\left(\frac{b}{2}\right)^2 + c} + \left(\frac{b}{2}\right) = \sqrt{\left(\frac{7}{2}\right)^2 + 60} + \left(\frac{7}{2}\right) = 12;$$

$$y = \sqrt{\left(\frac{b}{2}\right)^2 + c} - \left(\frac{b}{2}\right) = \sqrt{\left(\frac{7}{2}\right)^2 + 60} - \left(\frac{7}{2}\right) = 5.$$

The solution given in the text is essentially the same:

Break in two the 7 by which the reciprocal exceeds its reciprocal so that 3;30 ($3\frac{1}{2}$) results. Multiply 3;30 by 3;30: result 12;15 ($12\frac{1}{4}$). Add 1,00 (60) the area, to 12;15: result 1,12;15 ($72\frac{1}{4}$). Take the square root

Continued . . .

Continued . . .

of 1,12;15: result 8;30 ($8\frac{1}{2}$). Draw a square of side 8;30 and its counterpart of side 8;30. Take away 3;30, the holding square from one; add to one. One is 12 and the other is 5. The reciprocal is 12 and its reciprocal is 5.

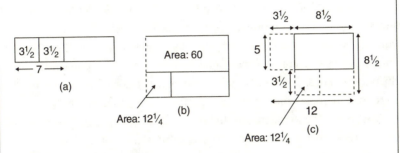

FIGURE 4.3: Algebra with a geometric rationale: an example from the Yale Collection

Figures 4.3a–c are self-explanatory. The numbers are given on the decimal base. What the solution assumes is that the two unknowns are the sides of a rectangle of area 1,00 (or 60) (shown in Figure 4.3b). Reassemble the rectangle as a gnomon (figure 4.3c) with the original lengths being obtained by completing the square. A detailed discussion of this problem and solution is found in Høyrup (2002, pp. 55–58).

This approach is similar to the ones used by the Chinese, who elevated it to a special place in their methodology, naming it the "out-in" principle. The approach may well have been used in the Indian *Sulbasutras*. There are traces of this procedure in the algebra of al-Khwarizmi. We will return to this subject when we consider the mathematics of these various traditions in later chapters.

Babylonian Geometry

Only a few years ago most historians of mathematics shared the view that although the Mesopotamians excelled in algebra, they were inferior to the Egyptians in geometry. One cannot deny the notable achievements of the Egyptians in the field of mensuration of spherical objects and pyramids. It was argued in the past that the Mesopotamian work in this area was relatively modest. The usual example taken to show the Mesopotamian lack of promise was the evidence pointing the way that they calculated the area of a circle, which was by taking three times the square of the radius. This is a cruder approximation than the square of 8/9 of the diameter found in the Ahmes Papyrus. However, in an Old Babylonian tablet excavated in 1950 appears the direction that 3 must be multiplied by the reciprocal of 0;57,36 to get a more accurate estimate of the area. This gives a value for π of 3.125.

We shall not be taking a detailed look at the Babylonian knowledge of simple rules of mensuration. Suffice it to say that there is evidence of their familiarity with general rules for the areas of rectangles, right-angled triangles, isosceles triangles, and trapeziums with one side perpendicular to the parallel sides. The most notable achievements of Babylonian geometry were in two areas where their calculation skills could be given full rein: their work on the Pythagorean theorem and on similar triangles foreshadowed Greek work in these areas by over a thousand years.

Plimpton 322: Pythagorean Triples or Prototrigonometry?

In 1945, Neugebauer and Sachs published their decipherment of a clay tablet (no. 322 in the Plimpton Collection at the University of Columbia) made sometime between 1800 and 1650 BC. This nearly four-thousand-year-old clay tablet is undoubtedly the most famous mathematical text from the Old Babylonian period. Since its decipherment it has been discussed in the "Babylonian" chapter of almost every general history of mathematics and in a number of specialists' works as well. The tablet, as it appears today, is shown in figure 4.4. Based on the curvature of the fragment that remains,

FIGURE 4.4: The Plimpton Tablet (Neugebauer and Sachs 1945, plate 25)

it has been suggested that a third or more of the tablet has been lost, which may have included an additional four narrow columns. The tablet is further marred by a deep chip in the middle of the right-hand edge and a flaked area at the top left-hand corner.

The tablet as we have it contains four columns of numbers arranged in fifteen rows, the last of these (column 4) giving the number of the row. Table 4.5 presents the four columns that can definitely be deciphered. (Column 5 is included for illustrative purposes.) There are errors in the original; asterisks indicate where correct values have been substituted. The break in column 1 throws some doubt on the accuracy of the first few terms in each sequence of numbers: it is not certain whether the entries set in italic type on the left side of column 1 were present on the original tablet. But what is beyond doubt is that a definite relationship exists between columns 2 and 3, which becomes clearer if we examine column 5.[14] In terms of the right triangle shown in figure 4.5, $b^2 + h^2 = d^2$. Hence, from row 1, $b = 1,59$ (119) and $d = 2,49$ (169). So

$$d^2 - b^2 = (169)^2 - (119)^2 = (120)^2,$$

and therefore

$$h = 120 \ (2,00).$$

TABLE 4.5: THE PLIMPTON TABLET DECIPHERED

Column 1 (?)	Column 2 (width, b)	Column 3 (diagonal, d)	Column 4 (row no.)	Column 5 (height, h)
1;59, 0,15	1,59	2,49	1	2,0
1;56,56, 58,14,50,6,15	56,7	1,20,25*	2	57,36
1;55,7,4, 1,15,33,45	1,16,41	1,50,49	3	1,20,0
1; 53,10,29,32,52,16	3,31,49	5,9,1	4	3,45,0
1; 48,54,1,40	1,5	1,37	5	1,12
1; 47,6,41,40	5,19	8,1	6	6,0
1; 43,11,56,28,26,40	38,11	59,1	7	45,0
1; 41,33,33,59,3,45	13,19	20,49	8	16,0
1; 38,33,36,36	8,1*	12,49	9	10,0
1; 35,10,2,28,27,24,26,40	1,22,41	2,16,1	10	1,48,0
1; 33,45	45	1,15	11	1,0
1; 29,24,54,2,15	27,59	48,49	12	40,0
1; 27,0,3,45	2,41*	4,49	13	4,0
1; 25,48,51,35,6,40	29,31	53,49	14	45,0
1; 23,13,46,40	56	1,46*	15	1,30

*Correct value substituted for incorrect one on the tablet

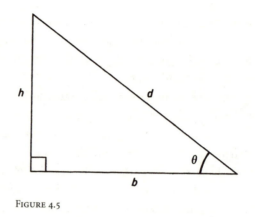

FIGURE 4.5

What was the purpose of this clay tablet? There have been a number of interesting suggestions, which involve three main approaches to interpreting the tablet:

1. It is a sophisticated listing of the so-called Pythagorean triples.

2. It is a remarkable trigonometric table (two thousand years before the apparent development of angle measurement in Alexandria).

3. It is a pedagogical tool intended to help a mathematics instructor of the period to generate a list of regular "reciprocal pairs" drawn up in a decreasing numerical order having known solutions and intermediate solution steps that are easily checked.

To understand the significance of Plimpton 322 it may be necessary to examine the "anthropology" of Babylonian mathematics, notably how the people of this area approached mathematical problems and what role these problems played in the wider society. It should be remembered that the tablet contains a table of fifteen rows and four columns. The usual presumption is that the first column contained all numbers beginning with 1. Then, it is easily seen that each number in the column is a perfect square. Also, subtracting 1 from each of these numbers gives a perfect square. Consider, for instance, the number in row 11 of table 4.5. The number 1;33,45 represents $1 + 33/60 + 45/3600 = 1 + 9/16 = 25/16$, which is the square of 5/4. One less than 25/16 is 9/16, the square of 3/4. The second and third entries in this row represent these fractions: 45 represents $45/60 = 3/4$, and 1;15 represents $1 + 15/60 = 5/4$.

The relationship would still hold if 45 represent 45 and 1:15 represents 75, in which case these two entries are proportional to the fractions. Considered as a table of Pythagorean triples, the second column (*b*) is one side of a right triangle or the width of a rectangle. In this interpretation, the third side of the triangle (column label *h*) does not appear in the fragment of the tablet that we have at present. However, the first column of the table has been interpreted as

$$\left(\frac{d}{h}\right)^2 = 1 + \left(\frac{b}{h}\right)^2.$$

This interpretation of the Plimpton Tablet views each first column entry, expressed in modern terms, as the square of the secant of an angle (i.e., the reciprocal of the cosine) between sides *d* and *h* of a right triangle with successive angles roughly one degree apart. In other words, the above equation may be written as

$$\sec^2\theta = 1 + \tan^2\theta.$$

So we have here the first table of a trigonometric function consisting of squares of secants for 45 degrees down to 30 degrees—the earliest case of degree measurement by about two thousand years. This is a claim not backed up by supporting evidence from other sources of Babylonian mathematics. In any case, there is no reason to believe that the Mesopotamians were familiar with the concept of a secant or, for that matter, any other trigonometric function; indeed, neither they nor the Egyptians had any concept of an angle in the modern sense, which first occurs in the work of Indian and Hellenistic mathematicians around the beginning of the first millennium AD.

It would seem, then, that column 1 may have served another function, that is, this column, and indeed the tablet itself, have to do with the derivation of Pythagorean triples (e.g., *b*, *h*, *d* in figure 4.5) for use in the construction of right triangles with rational sides. It is improbable that the values inscribed on the tablet were derived by using trial-and-error methods, for these would have given simpler triples. But if the Mesopotamians had a more systematic method of deriving *b*, *h*, and *d* to satisfy the equation $b^2 + h^2 = d^2$, what was it? We can only hazard a guess, helped by a possible clue in column 1.

Let us first assume that the height is normalized to 1 [i.e., that $(d/h)^2 - (b/h)^2 = 1$]. Now, if $\alpha = d/h$ and $\beta = b/h$, then

$$\alpha^2 - \beta^2 = (\alpha - \beta)(\alpha + \beta) = 1.$$

Let $\alpha + \beta = m/n$ and $\alpha - \beta = n/m$, where m and n are positive integers such that $m > n$. Then

$$\alpha = \frac{1}{2}\left[\left(\frac{m}{n}\right) + \left(\frac{n}{m}\right)\right] \quad \text{and} \quad \beta = \frac{1}{2}\left[\left(\frac{m}{n}\right) - \left(\frac{n}{m}\right)\right],$$

or

$$\alpha = \left[\frac{(m^2 + n^2)}{2mn}\right] \quad \text{and} \quad \beta = \left[\frac{(m^2 - n^2)}{2mn}\right].$$

But $b = \beta h$ and $d = \alpha h$. And if we put $h = 2mn$ so as to obtain a solution in integers, then

$$h = 2mn, b = m^2 - n^2, d = m^2 + n^2. \tag{4.12}$$

This method of generating integral Pythagorean triples is usually attributed to Diophantus (c. AD 250), who, as we have seen, may be thought of as working in the Babylonian "metric algebra"[15] tradition and introducing it into Greek mathematics. It is worth noting that to arrive at these formulae we need nothing more than the ability to add and subtract fractions, and a knowledge of the algebraic identity $\alpha^2 - \beta^2 = (\alpha - \beta)(\alpha + \beta)$.

We can use the formulas (4.12) to generate the first three triples on the Plimpton Tablet, as shown in table 4.5. With one exception, the integers chosen for m and n in the complete table are all products of prime factors of 60. For example, in the first row $m = (2)(2)(3)$ and $n = 5$.

One of the difficulties with this explanation of how the Babylonians generated Pythagorean triples is the lack of an underlying rationale for the choice of the particular values of (m, n), which seem to vary in an erratic fashion—there is certainly no discernible pattern to the first three sets of values in table 4.6. It is possible, though not very likely, that the rows may have been ordered so as to ensure an approximately linear increase in the

TABLE 4.6: GENERATING THE FIRST THREE PYTHAGOREAN TRIPLES FROM THE PLIMPTON TABLET

m	n	$h = 2mn$	$b = m^2 - n^2$	$d = m^2 + n^2$
12	5	120	119	169
64	27	3,456	3,367	4,825
75	32	4,800	4,601	6,649

values in column 1 of the tablet, which contains the square of the ratio of two sides of the triangle shown in figure 4.5 [i.e., $(d/h)^2$].

However, there is yet another explanation in terms of "reciprocal pairs." For the sake of simplicity, we present it here in modern algebra. First we put $\alpha + \beta = n$ and $\alpha - \beta = 1/n$, so that $(\alpha + \beta)(\alpha - \beta) = 1$. Then

$$a = \frac{1}{2}\left[n + \left(\frac{1}{n}\right)\right], \quad \beta = \frac{1}{2}\left[n - \left(\frac{1}{n}\right)\right], \quad h = 1,$$

which gives a fractional Pythagorean triple. This may be converted into a series of integer Pythagorean triples by multiplying each of the three numbers by $2n$. Bruins (1955), the originator of this explanation, shows how the entries as well as the scribal errors in table 4.5 can be explained by this method.

A choice between these explanations cannot be made on the basis of their mathematics. Robson (2001b) has introduced a set of criteria for judging the relative historical merit of each interpretation. She writes: "If we believe that Plimpton 322 was intended to be a list of parameters to aid the setting of school mathematics problems (and the typological evidence suggests that it was), the question 'how was the tablet calculated?' does not have to have the same answer as the question 'what problems does the tablet set?' The first can be answered most satisfactorily by reciprocal pairs, as first suggested half a century ago, and the second by some sort of right-triangle problems. That is perhaps as far as we can go on present evidence: without closer parallels we run the risk of crossing the fuzzy boundary from history to speculation. The mystery of the Cuneiform Tablet has not yet been fully solved" (p. 202).

Irrespective of which of these explanations, or any other, is valid, there can be little doubt that the Mesopotamians knew and used the Pythagorean theorem. This is confirmed by a problem from a tablet found at Susa, a couple of hundred miles from Babylon, belonging to the Old Babylonian period. It is one of the oldest examples of the use of the theorem in the history of mathematics:

EXAMPLE 4.11 Find the circum-radius of a triangle whose sides are 50, 50, and 1,0.

Solution

In terms of figure 4.6, the problem is to calculate the radius, r. The solution proceeds thus:

Continued...

Continued . . .

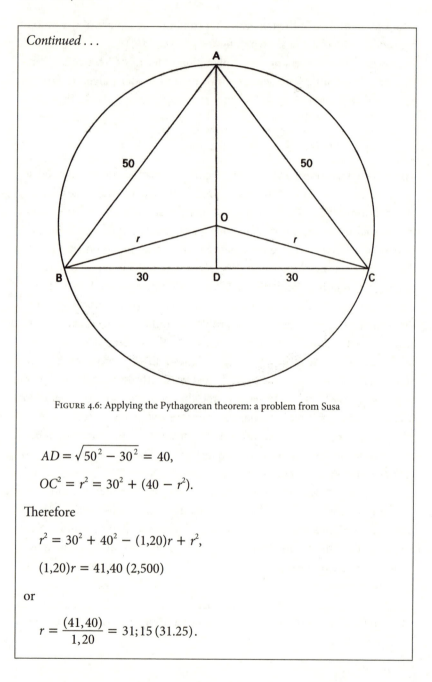

FIGURE 4.6: Applying the Pythagorean theorem: a problem from Susa

$$AD = \sqrt{50^2 - 30^2} = 40,$$

$$OC^2 = r^2 = 30^2 + (40 - r^2).$$

Therefore

$$r^2 = 30^2 + 40^2 - (1,20)r + r^2,$$

$$(1,20)r = 41,40 \ (2,500)$$

or

$$r = \frac{(41,40)}{1,20} = 31;15 \ (31.25).$$

There is another example of the application of the Pythagorean theorem that is notable in that it uses a rather long-winded method of solution, which has a stronger geometric rationale, instead of a shorter method based on algebraic techniques. The source of this evidence is interesting. In 1962, archaeologists working at Tell Dhibayi, near Baghdad, unearthed about five hundred clay tablets. Most of them deal with the commercial transactions and administrative matters of a city that flourished during the reign of Ibalpiel II of Eshunna (c. 1750 BC). One tablet presents a geometric problem in which the area and the length of diagonal of a rectangle are given, and what is apparently sought are its dimensions:

EXAMPLE 4.12 Find the length and width of [figure 4.7], given its area, 0;45 [0.75] and diagonal, 1;15 [1.25].

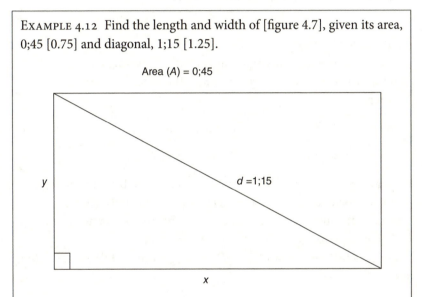

FIGURE 4.7

Suggested Solution

The tablet gives the following steps. The results at each step are given here both in sexagesimals and in decimals.

1. Multiply the area by 2: result 1;30 (1.5).

2. Square the diagonal: result 1;33,45 (1.5625).

3. Subtract result 1 from result 2: result 0;03,45 (0.0625).

4. Find the square root of result 3: result 0;15 (0.25).

5. Halve result 4: result 0;07,30 (0.125).

6. Find one-quarter of result 3: result 0;00,56,15 (0.015625).

Continued . . .

Continued . . .

7. Add the area to result 6: result 0;45,56,15 (0.765625).

8. Find the square root of result 7: result 0;52,30 (0.875).

9. Length = result 5 + result 8 = 1.

10. Width = result 8 − result 5 = 0;45 (0.75).

The procedure followed above is quite baffling at first sight. We might well have expected to see a solution along the following lines: Let x be the length, y the width, d the diagonal, and A the area.
Then

$$xy = A = 0.75 \qquad \text{(area of a rectangle);} \qquad (4.13)$$

$$x^2 + y^2 = d^2 = (1.25)^2 \qquad \text{(Pythagorean theorem).} \qquad (4.14)$$

We solve equation (4.13) for x (or y) and then substitute into equation (4.14) to solve for y (or x) after reducing the resulting biquadratic (i.e., quartic) equation to a quadratic one. Note that the general form of a quartic equation is: $ax^4 + bx^3 + cx^2 + dx + e = 0$, where $a \neq 0$.

Thus, substituting $y = (0.75)(1/x)$ into equation (4.14) and simplifying gives

$$16x^4 + 9 = 25x^2. \qquad (4.15)$$

Setting $x^2 = z$ in equation (4.15) yields

$$16z^2 - 25z + 9 = 0,$$

which has the solution $z = 1$ or 9/16, and so $x = 1$, $y = 3/4$ (or $x = 3/4$, $y = 1$).

The Babylonian solution is quite ingenious and follows from the recognition that

$$d^2 - 2A = x^2 + y^2 - 2xy = (x - y)^2 \qquad \text{(steps 1 to 3)}$$

$$d^2 + 2A = x^2 + y^2 + 2xy = (x + y)^2, \qquad \text{(the sum of steps 1 and 2)}$$

so that

$$\sqrt{d^2 - 2A} = \text{length} - \text{width};$$

$$\sqrt{d^2 + 2A} = \text{length} + \text{width}.$$

Hence the suggested solution for example 4.12 is largely a procedure for forming these two relationships: the result from step 5,

$$\frac{1}{2}\sqrt{d^2 - 2A} = \frac{1}{2}(\text{length} - \text{width}),$$

and the result from step 8 is

$$\sqrt{\frac{1}{4}(d^2 - 2A) + A} = \frac{1}{2}\sqrt{d^2 + 2A}$$

$$= \frac{1}{2}(\text{length} + \text{width}).$$

So the result from step 5 plus the result from step 8 gives the length, and the result from step 8 minus the result from step 5 gives the width.

This example epitomizes the versatility of Mesopotamian mathematics. Here was a group of people who for the first time combined what we would classify as arithmetic, algebra, and geometry in tackling problems—a remarkable feat.

Similar Triangles

One of the clay tablets (figure 4.8a) excavated at Tell Harmal in Iraq is thought to date back to about 2000 BC, making it one of the earliest problem texts we know of. The problem is stated thus (Robson 2007, p. 100):

EXAMPLE 4.13 A wedge. The length is 1, the long length 1;15, the upper width 0;45, the complete area 0;22,30. Within 0;22,30, the complete area, the upper area is 0;08,06, the next area 0;05,11,02,24, the third area 0;19,03,56,09,36, the lower area 0;05,53,53,39,50,24. What are the upper length, the middle length, the lower length, and the vertical?

Or restated in modern terminology and with reference to figure 4.8b:

Given the sides of $\triangle ABC$ and areas of $\triangle$s BAD, ADE, DEF, and EFC, find the lengths BD, DF, AE, and AD.

(It is possible to infer from the lengths of the sides of $\triangle ABC$ that it is a right-angled triangle. The original diagram and the procedures would indicate that AD and EF are perpendicular to BC, and that DE

Continued . . .

Continued . . .

is perpendicular to AC, as shown in figure 4.8b. So what we have is a series of similar right-angled triangles.)

Solution

The Babylonian procedure will be given in both its rhetorical and symbolic forms. In its rhetorical form, the steps are

1. Take the reciprocal of 1;00 and multiply it by 0;45: result 0;45.

2. Multiply the result by 2: result 1;30.

3. Multiply the result by the area of ΔABD: result (0;08,06)(1;30) = 0;12,09.

4. Find the square root of 0;12,09: result BD = 0;27.

Now that BD is known, the Pythagorean theorem can be used to show that the length of AD is 0;36. If the area and hypotenuse of ΔADE are known, the application of the above procedure would give the length of AE. This is followed by a further application of the Pythagorean theorem to evaluate ED. And this process may be continued ad infinitum to work out the required dimensions of an infinite series of similar right-angled triangles. (The reader is invited to check that AE and ED are 0;21;36 and 0;28;48 respectively.)

In symbolic terms, steps 1 to 3 of the Babylonian procedure are as follows.

1. (1/AC)AB.

2. (AB/AC).

3. (2AB/AC) (area of ABD) = BD^2.

To make any sense of these steps, it is necessary to introduce two results that must have been known to the Babylonians:

(a) If ΔABC is similar to ΔABD, then AB/AC = BD/AD.

(b) Area of ΔABD = $\frac{1}{2}$(BD × AD).

Applying (a) and (b) to step 3, we get (2BD/AD) × $\frac{1}{2}$(BD × AD) = BD^2.

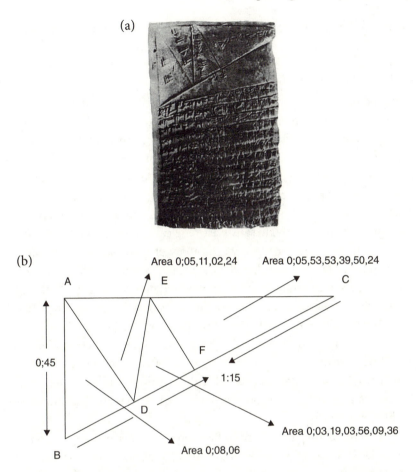

FIGURE 4.8: Similar triangles: (a) the original tablet (Baqir 1950, opp. p. 54), and (b) the modern translation

It is clear from this example that the Babylonians had some knowledge of the properties of similar triangles (though we know of no explicit contemporary statement). In particular they were familiar with one of the Euclidean theorems:

In a right-angled triangle, if a perpendicular is drawn from the right angle to the hypotenuse, the triangles on each side are similar to the whole triangle and to one another.

It is a plausible hypothesis that Euclid took the kernel of his ideas about similar triangles either directly or indirectly from the Mesopotamians and then imbued it with that peculiarly Greek contribution to mathematics, the method of axiomatic deductive logic, whose importance to the future development of the subject cannot be overestimated.

Astronomy: The Babylonian Beginnings

It was pointed out earlier in this chapter that angular measurement as we know it today only began with the Babylonians relatively late, around the eighth century BC.[16] The methods devised were primarily to accompany the much older interest in observing the skies and recording the celestial phenomena. As in the case of all early cultures of which we have information, the driving forces behind these activities were partly the need to construct calendars for both agricultural and religious purposes and partly (though very importantly) to satisfy astrological (or horoscopic) demands. As a result, the character of early astronomy of Babylonia and elsewhere was primarily *predictive* rather than *explanatory*. The emphasis was on predicting the location of heavenly bodies and determining the time and place at which celestial events (such as an eclipse) would occur rather than attempting to *explain* the behavior of celestial objects. As a result, computational schemes to determine when and where a periodic celestial event would next occur became the primary focus of early Babylonian astronomy. Only in the Hellenistic period, culminating in Ptolemy's *Almagest* (second century AD), was geometry put to use to explain the movement of the heavenly bodies.[17]

A seventh-century BC text from Babylonia contains an early version of the zodiac.[18] A zodiac is divided into 360 units (or later, degrees). Among the constellations on or near the zodiac are the twelve *zodiacal signs* such as Aries, Taurus, and so on. If these signs are equal in size and each is subdivided into 30 *ush* (or length), the sun is expected to travel 1 *ush* (or 1° per day). However, the speed at which the sun travels around the ecliptic is not constant. The Babylonians found that at that time it was slower in the spring but faster in the autumn. They devised two methods of measuring the sun's continuously changing speed: a zigzag algorithm in which the speed alternated between two fixed values, and a second algorithm in which the speed varied linearly between a maximum and a minimum. The arithmetical schemes implicit in these two representations allowed for direct predictions without the need for a geometric specification. Examples of the use of these procedures are found in the Babylonians' prediction of the monthly solar movement along the zodiac and lunar conjunctions for particular years.[19]

• • •

This is but a brief survey of Mesopotamian mathematics. Because of the wealth of sources and the tantalizing glimpses of some unusual mathematics from about four thousand years ago, this subject has attracted a number of specialist math historians, for whom the attractions of primary research on hitherto undeciphered tablets and the scope offered for constructing exciting conjectures remain unabated.

Before we attempt, in the next chapter, a final assessment of both Egyptian and Mesopotamian mathematics, it would be useful to summarize the overall character of Mesopotamian mathematics. The evidence presented in this chapter provides a compelling testimony to the quality and range of the mathematical achievements of this ancient civilization. It is clear that, with their numerical and algebraic skills, the Mesopotamian mathematicians produced work that compares favorably with what was being done in sixteenth-century Europe before the advent of modern mathematics. Yet in emphasizing the quality of their arithmetic and algebra, we should not ignore their achievements in other areas. Their work on Pythagorean triples and similar triangles provides fine examples of their interest in and contributions to geometry.

There is a tendency to label all mathematics before the Greeks as utilitarian and prescientific. This view should be critically reevaluated in the light of the content as well as the spirit of the work examined in this chapter. The connections discerned between Egyptian, Babylonian, and Greek mathematics will be discussed in the next chapter. Even after four thousand years, some of the Mesopotamian contributions to mathematics remain quite awe-inspiring.

Notes

1. The precedence of mathematical activities over a written language has already been observed in chapter 2, especially in the case of the Incas. It is interesting in this context to note that more recent attempts at deciphering the Harappan script of ancient India are based on the premise that the evidence available on seals contains numerical rather than literary records. For further discussion, see the relevant section in chapter 8.

2. It may be worth pointing out, as Robson (2007) does, that the scribe and others who were professionally literate had to be numerate and mathematically (in terms of the development of concepts, etc.) proficient.

3. This is based on the discussion contained in Robson (2007, pp. 64–65).

4. It is to be hoped that the gems from the looted cuneiform tablets in the Baghdad museum will be replaced by passable replicas in the display cabinets.

5. It should be noted that Hilprecht, who led the excavations at Nippur, published a book in 1906 in which he discussed multiplication tables and tables of reciprocals as well as metrological tables found in the sources that had been discovered up to that point.

6. Note that in chapter 2 we observed a similar practice in dealing with counting systems in Papua New Guinea and numeration systems in Mexico.

7. From the twenty-first to the sixteenth centuries BC, the standard Mesopotamian units of calculation with their modern equivalents were CAPACITY: 1 *sila* ($\approx$ 1 liter); LENGTH: 1 rod = 12 cubits ($\approx$ 6 meters); AREA: 1 area *sahar* = 1 rod square ($\approx$ 36 m^2); VOLUME: 1 volume *sahar* = 1 area *sahar* $\times$ cubit ($\approx$ 18 m^3); WEIGHT: 1 *mina* = 60 *shekels* ($\approx$ 0.5 kg). For further details, see Robson (2007, pp. 70–72).

8. The interpretation discovered independently by Friberg and by Proust (2002) of a Mari text containing three kinds of counting numbers should warn us to desist from "easy" explanations of the origins of the Mesopotamian sexagesimal system. See endnote 12 below for a summary of the three systems.

9. There is some fragmentary evidence that the Mesopotamians made use of a subtraction sign ($\mathbf{T}\vdash$) to relieve the tedium of their unabridged notation. In this scheme 39 would be represented as 40 − 1: $\mathbf{\ll\!\!\ll}\,\mathbf{T}\!\vdash\ \mathbf{T}$.

10. This is *not* to imply that the Mesopotamians had any concept of irrational numbers. This only came with the Greeks.

11. It is worth remembering that just as we think today, both the Mesopotamians and Egyptians thought of the Pythagorean theorem as a rule relating to numbers. But for the Greeks (of whom Euclid was a prime example), the theorem (Euclid's *Elements* 1.47) was a statement about actual squares. For the Greeks, Chinese, and possibly the Indians, if you cut up the squares on the two smaller sides and reassemble the pieces, it will make the square on the hypotenuse. For further details of the importance of the "dissection and reassembly" technique in mathematical proofs, see Joseph (2003).

12. Three different systems of recording numbers are mentioned here. There is the *Babylonian sexagesimal system*, already discussed in some detail earlier in this chapter. In this system, there are special number signs for the "units," from 1 to 9, and for the "tens" from 10 to 50. The *"mixed" decimal-sexagesimal system* (which was a nonpositional system

peculiar to Mari) consisted of signs for numbers "a hundred," "a thousand," and "a ten thousand." Numbers below 100 were written as sexagesimal numbers, with or without the word "sixty." The Mari *centesimal place-value system* operated the same way as the Babylonian sexagesimal place-value system but with the base 100 instead of 60, with signs not only for the "tens" from 10 to 50 but also for 60, 70, 80, and 90. What is very interesting here is the existence of three different number systems in a small urban center.

13. The metrology of the Old Babylonian period had a relatively standardized set of measurements, although there are plenty of exceptions. The key units were the *ninda* (rod) for length, *sahar* (garden plot) for area and volume, *sila* for capacity, and *mina* for weight. At the base of the system is the barleycorn, *she*, used for the smallest unit in length, area, volume, and weight. Note that 1 *ush* = 60 rods ≈ 360 meters; 1 *eshe* = 600 *sahar*, where 1 *sahar* ≈ 36 square meters.

14. The headings above the columns of the tablet also give us a clue. Column 2 is translated as "the square side of the front" (*b*); column 3 is the "square side of the crossover" (*d*); and column 4 is the name or row number. The mention of "crossovers" (diagonals), "fronts," and "squares" would indicate that the sides and diagonals of a series of rectangles are calculated though the application of the "diagonal rule" (the Babylonian name for the Pythagorean theorem).

15. Friberg (2007b, p. vi) uses the term "metric algebra" for a special kind of mathematics, an elaborate "combination of geometry, metrology, and linear or quadratic equations, first documented in proto-Sumerian texts from the end of the fourth millennium BC and which continued to be used in Mesopotamia until the Seleucid period, close to the end of the first millennium BC."

16. It is worth remembering that measurement of continuous quantities such as angles or lines requires an efficient system of numeration for fractional parts. The Mesopotamian system of numeration had devised as early as the third millennium BC a sexagesimal place-value system that became the standard system of numeration in astronomy and trigonometry. The hour is divided into 60 minutes and the minute into 60 seconds. Similarly, the angle is divided into minutes and seconds. Expression of fractional parts of angles thus becomes a relatively easy matter.

17. It is now well known that the remarkable work of a line of Greek astronomers and mathematicians of the stature of Eudoxus, Euclid, Aristarchus, Archimedes, Hipparchus, and Menelaus advanced mathematical astronomy to an unprecedented level. However, a discussion of their contributions is beyond the scope of this book. For a clear account of the Greek contribution, see Van Brummelen (2009, chapters 1 and 2).

18. The zodiac is a band around the celestial sphere known as the ecliptic, which in turn is the perfect circular path of the sun through the celestial sphere. Now the zodiac is split up among twelve constellations that would become the zodiacal signs. And as the

"sun moves on an inclined circle dividing it into four regions," spending three months in each, each of the twelve regions corresponds to a constellation on or near the zodiac. From this emerged the twelve names of the zodiacal signs, being Taurus, Gemini, Cancer, . . . , Aquarius, Pisces, and Aries.

19. For further details, see Neugebauer (1962, pp. 101–10).

Chapter Five

Egyptian and Mesopotamian Mathematics:
An Assessment

A particular view of Egyptian and Mesopotamian mathematics held not too long ago is crystallized in the writings of Morris Kline, a well-known American historian of mathematics. Dismissing all the evidence to the contrary marshaled by both ancient Greeks and modern scholars, he considers that the Egyptian and Mesopotamian contributions to mathematics were "almost insignificant." This is followed by his astonishing statement that, compared with what the Greeks achieved, "the mathematics of the Egyptians and Mesopotamians is the scrawling of children just learning to write, as opposed to great literature." In any case, Kline continues, these civilizations "barely recognized mathematics as a distinct discipline," so that "over a period of 4000 years hardly any progress was made in the subject."

I have quoted extensively from a single page (p. 14) of Kline's book *Mathematics: A Cultural Approach* because his views represent a concise summary of what we labeled in chapter 1 as "Eurocentric scholarship."[1] We identified the main characteristics of this Eurocentric outlook, the chief of which is a tendency to ignore new findings that go against deeply entrenched views about the origins of mathematics. Chapters 3 and 4 have spelled out in great detail some of the contributions to early mathematics made by these most ancient of civilizations. Evidence of their contributions is not all hidden away in obscure journals or expressed in languages that tend to be ignored by many Western scholars: much is published in English in "respectable" journals and books, brought out by major publishers on both sides of the Atlantic. The reason for the neglect was not that the relevant literature was inaccessible or "unrespectable" but something deeper—a serious flaw in Western attitudes to historical scholarship (one not confined to histories of mathematics or science). An excessive enthusiasm for everything Greek, arising from the belief that much that is desirable and worthy of emulation in Western civilization originated in ancient Greece, has led to a reluctance to allow other ancient civilizations any share in the historical heritage of mathematical discovery. The belief in a "Greek miracle" and the way of attributing any significant mathematical discoveries to Greek influences are part of this syndrome. And underlying this view is the belief that Egyptian, Mesopotamian, and Greek mathematics were drawn from three different mathematical traditions: never the three shall meet![2] Some of the most exciting work in recent years, which we will explore later in this chapter, has shown "unexpected links" between these traditions.[3]

Changing Perceptions

Mesopotamian Mathematics

For the most part Greek writers took Egypt to be the birthplace of mathematics but credited the Mesopotamians, especially the Chaldean priests, with astrological prowess in making predictions from the stars. Three Mesopotamian astronomers, two of whom have since been identified from cuneiform sources, were named by Strabo as notable for their time. The transmission of Mesopotamian observational data, values of periodicities,[4] and the use of the sexagesimal place-value system must have occurred during the Persian and Hellenistic periods (i.e., 550–150 BC) of Egyptian history. Around 300 BC Iamblichus claimed that Pythagoras had visited Babylon in the sixth century BC. No corroboration from earlier or contemporary sources for this claim has been found and hence the tendency on the part of contemporary historians to dismiss this as part of the fabrication of the Pythagorean tradition in late antiquity.[5] Much later and during the Middle Ages in Europe, the ancient Chaldeans gained the reputation of being skilled in mathematics and astronomy, according to Isidore of Seville, Bede, Bacon, Recorde, and, much later, Wallis. However, no written trace of Mesopotamian mathematical activity was available until the remarkable finds of the mid–nineteenth century.

In the 1840s rival British and French teams began to uncover and document the remains of vast stone palaces near Mosul, now in northern Iraq but then part of the Ottoman empire. These excavations led to the identification of the ruins of the ancient Assyrian city of Nineveh, already known through the stories in the Old Testament. These discoveries were instantly perceived as part of the early European heritage, and little attempt was made to locate them in the history of the area in which they were found. From this Eurocentric perception arose, however unwittingly, an approach to interpreting Mesopotamian remains, which were seen as part of a historical evolution toward European sophistication.[6] At the same time there occurred the birth of the idea of an exotic, decadent Orient—a view traced back to historians like Herodotus, who had written pejoratively of Mesopotamia in the fifth century BC at the height of the Greek wars against Persia. All this was occurring at a time when the emerging science of geology and the evolution of species posed real challenges to the literal truth of the events described in the Old Testament.

Decipherment of cuneiform from Mesopotamia continued apace throughout the latter part of the nineteenth century, with armies of archaeologists and adventurers engaged in uncovering Babylonian and Sumerian remains in the region between Baghdad and Basra in Iraq. The discovery of Sumerian language and culture created a new problem for cuneiform scholars: here was a major civilization, clearly older than Assyria or Babylonia, that had little biblical or Classical underpinnings. The result was that the study of ancient Mesopotamia freed itself from its biblical and Classical origins and took on an identity as an independent discipline.

Mathematical cuneiform tablets, as noted in the previous chapter, were first publicized in the late nineteenth century, although it took another fifty years before the works of Scheil, Thureau-Dangin, and Neugebauer were translated and interpreted. Neugebauer, who represents the culmination of this scholarship, was particularly thorough when it came to establishing a bridge between Mesopotamian and modern mathematics. But his "mathocentric" approach came under increasing critical scrutiny from 1970s onward for its neglect of the issues of context, function, and authorship of the Mesopotamian texts. Further, the excessive focus on the mathematics of the Old Babylonian period of the early second millennium BC and the relative neglect of the mathematics of other periods, particularly the Late Babylonian, presented a lopsided picture of Mesopotamian mathematics. Recent studies have attempted to restore a balance between the mathematical and social contexts of the Mesopotamian work.[7]

Egyptian Mathematics

The study of ancient Egyptian mathematics began with the discovery of the Ahmes Papyrus at the end of the nineteenth century. Since then, while the available primary sources of Egyptian mathematics have not increased significantly, surveys on the subject have continued unabated. In recent years, the communication barriers that have traditionally existed between historians of mathematics and Egyptologists have fragmented the math historians as a community. Math historians now disagree on issues regarding what are considered as "true" mathematical texts and what are the legitimate ways of presenting the mathematical content in these texts. In my view, Egyptian and Mesopotamian mathematical sources should include not only table texts and problem texts but also administrative texts involving calculations.[8] Further, while the "translation" of Egyptian procedures into modern mathematical language and symbols has no doubt come at the expense of losing some of their characteristic features, a refusal to do so makes it difficult to comprehend what lies behind these procedures. In the case of both Egyptian and Mesopotamian mathematics covered in the previous two chapters, we have not hesitated in "translating" texts into today's mathematical language and notation whenever it was felt that this would help our understanding of the procedures involved.[9]

A close examination of individual problems in Egyptian mathematics discussed in chapter 2 reveals a common structure. First, an introduction gives the title of the problem, the relevant data, and an indication of what result is being sought. This is followed by a series of instructions, expressed as a sequence of arithmetical operations that eventually give the solution to the problem.[10]

An important change in perspective relating to both Egyptian and Mesopotamian mathematics in recent years has been the recognition that, in order to achieve the correct interpretation of a particular mathematical text, there is need to understand the context in which the text was written in the first place. In the training of a scribe, mathematical texts and practices played an important part. And evidence of such training was to be found not only in the problem and table texts alone but also in the economic and administrative documents that were composed by the scribes. In the case of ancient Egypt, for example, today we have in hand around one hundred problems mainly from the two major problem texts, the Ahmes and Moscow papyri, covering a wide range of topics relating to daily life, including calculations of volumes of granaries, calculations of rations for workers, and calculations relating to baking and brewing. Supplementing these problems are tables as aids to calculations and administrative documents that show the results obtained through calculations.

Neglect of Egyptian and Mesopotamian Mathematics

A substantial reason for neglecting Egyptian and Mesopotamian mathematics is contained in Kline's view that these civilizations "barely recognized mathematics as a distinct discipline." Behind views such as this can be discerned a number of assertions. The mathematics of Egypt and Babylonia, it is argued, (1) had no general rules, (2) contained no "proofs," (3) lacked abstraction, (4) failed to distinguish clearly between exact and approximate results; and (5) generally there was no clearly discernible activity that we may label "mathematics" that was studied for its own sake. Let us examine each of these alleged shortcomings in detail.

1. No explicit general statements of algebraic rules and their appropriateness are found in the mathematical sources of either civilization. But this is hardly surprising, given both the nature of the mathematical evidence that has come down to us and the lack of symbolic notation. Also, it must not be forgotten that there was no general deductive algebra before the emergence of modern mathematics. Now, a rule can be general without being deductive. Consider two notable mathematical achievements, one from Egypt and the other from Babylonia, which we discussed earlier: the Egyptian discovery of the rule for finding the volume of a truncated pyramid,[11] as shown in the Moscow Papyrus, and the Mesopotamian calculation of Pythagorean triples, contained in the Plimpton Tablet. It cannot be argued that these were merely empirical rules arrived at through a painful process of trial and error for specific problems, without any awareness of their general application. In any case, the very fact that problems requiring specific algorithms for their solutions are grouped together in both the Ahmes Papyrus and the Mesopotamian tablets would indicate that there was some understanding of the generality of the underlying rules. It has also been pointed out that in a number of Greek geometric solutions each step is identical to the corresponding step in the algebraic solutions of the Mesopotamians. For example, the Babylonian "take square side a certain number A" would correspond to the Greek "take the side of a square whose area is A."

There is sometimes a tendency to devalue the role of algorithms in the development of mathematics. From the practical concerns of society have arisen a number of rules that should be judged for both their effectiveness and their intrinsic qualities. A "good" algorithm should have three properties:

a. It should be clear and simple, laying out step by step the procedures to be followed.

b. It should emphasize the general character of its applications by pointing out its appropriateness, not to a single problem but to a group of similar problems.

c. It should show clearly the answer obtained after the prescribed set of operations is completed.

It will be left to the reader to judge whether the mathematical sources that we have examined in the last two chapters contain algorithms that satisfy these requirements.

2. There is hardly a trace, according to the next argument, in any of the mathematical sources that we have examined, of what is commonly recognized as "proof"; this implies a nonscientific approach to the subject. However, what constitutes "proof" is a difficult question. Today, a rigorous mathematical proof that is not symbolic is inconceivable. A modern proof is a procedure, based on axiomatic deduction, that follows a chain of reasoning from the initial assumptions to the final conclusion. But is this not taking a highly restrictive view of what is proof? Could we not expand our definition to include, as suggested by Imre Lakatos (1976), explanations, justifications, and elaborations of a conjecture constantly subjected to counterexamples? Is it not possible for an argument or proof to be expressed in rhetoric rather than symbolic terms, and still be quite rigorous? As Gillings (1972, p. 233) states:

> A non-symbolic argument or proof can be quite rigorous when given for a particular value of the variable; the conditions for rigor are that the particular value of the variable should be typical, and that a further generalization to any value should be immediate.

It is possible to distinguish between logically deductive and axiomatically deductive algebraic reasoning. Once David Hilbert (1862–1943) and Bertrand Russell (1872–1970) had laid the foundations of mathematical logic, it became possible to construct an algebra from a limited set of axioms. Previously, what great mathematicians such as Euler, Gauss, and Lagrange had considered as proof was logically deductive proof.

These questions are relevant not just for Egyptian and Mesopotamian mathematics but for other mathematical traditions that we shall be examining in subsequent chapters. By posing the questions here, I am stressing how important it is not to be blinded by present-day preconceptions of what constitutes mathematical demonstration and proof when studying the mathematics of the past.

The mathematical papyri and tablets from Egypt and Mesopotamia show a considerable technical facility in computation, and also a recognition of the applicability of certain procedures to a similar set of problems, and of the importance of verifying the correctness of a procedure by checking, say, a division by multiplication or the solution of an equation by substitution of the calculated value of the unknown into the original equation. These procedures and checks in the mathematics of these early civilizations must be regarded as a form of "proof" in the broader sense.

3. It is the supposed absence of abstraction in the Egyptian and Mesopotamian sources that sways many critics in their judgment of whether these civilizations produced mathematics or merely some form of applied arithmetic. In a number of examples of Mesopotamian mathematics discussed in the previous chapter, we found close parallels between the steps of the ancient procedure and the steps of the corresponding modern analysis in algebraic symbols. The Mesopotamian symbols *ush* and *sag* for length and width, respectively, served the same purpose as our algebraic symbols x and y. The transition from specifics to abstract generalization was present. How else are we to interpret meaningfully the addition of length (*ush*) and area (*asha*)? It has also been pointed out that the addition of length and area can be interpreted differently. The "side to be added" may be thought of as the area of a rectangle with the length of that side and with width 1.

4. It is sometimes argued that ancient civilizations could not distinguish between exact and approximate results, and therefore that which they practiced was not mathematics, only something that resembled mathematics. We have seen that the Mesopotamians, in their evaluation of the reciprocal of 7 and their calculation of the square root of 2, were aware of the fact that their results were approximations and not the true values. Indeed, their omission of irregular sexagesimals from mathematical tables implied an uncertainty as to whether they would ever obtain accurate results. Balanced against this was the need for practical or computational precision in solving real-life problems. But were the Mesopotamians aware of the important distinction between mathematical precision and computational precision? On the existing evidence, it is impossible to tell.

5. A point often made about Egyptian and Mesopotamian mathematics is that each was more a practical tool than an intellectual pursuit. This implied criticism is symptomatic of the attitude, often attributed to the Greeks, that mathematics devoid of all utilitarian purpose is in some sense a nobler or better mathematics.

An important distinction running right through Greek thought is between *arithmetic*, the study of the properties of pure numbers, and *logistics*, the use of numbers in practical applications. Cultivation of the latter discipline was left mainly to the slaves. Legend has it that when Euclid was asked what was to be gained from studying geometry, he disdainfully told his slave to toss a coin at the inquirer. The notion that the Egyptian and Mesopotamian cultures were entirely utilitarian, with little or no interest in mathematics for its own sake, is not borne out by the nature of some of the problems we have examined, which appear to have no practical implications. In any case, the pursuit of mathematics as an aesthetic activity for its own sake presupposes the existence of a leisured class, freed from the concerns of survival, including the need to make a livelihood. Greek civilization, with its substantial slave population, allowed a small elite the freedom to pursue activities that had no practical significance. Both the character of the Greeks' mathematics and their conception of mathematics as a deductive science were to some extent influenced by this form of social stratification. In civilizations where such a luxury was not possible, mathematics would have had little chance of transcending its utilitarian origins.

Ultimately, whether one characterizes the activities of the Egyptians and Mesopotamians as mathematics will depend on how one perceives the long algorithmic phase that preceded the development of modern algebra. Was this phase an early stage in the emergence of "true" algebra, or did algebra begin only with the introduction of algebraic symbolism? The long period in which material was collected in the form of problems and valid methods were invented for their solution was algebra's gestation period; the rearrangement of old procedures into new deductive structures marks its birth.

The Babylonian-Egyptian-Greek Nexus: A Seamless Story or Three Separate Episodes?

An example of a geometric series, cited by Friberg, from an old Babylonian text found in Mari was discussed in chapter 4 on Mesopotamia. A parallel to this example found in the Ahmes Papyrus of Egypt (example 3.11 in chapter 3) was the antecedent of the modern English nursery rhyme that went through a number of incarnations in different countries over the centuries. From a study of the mathematical content and the structure of problems in the Ahmes and Moscow papyri, Friberg (2005) found significant "non-trivial" parallels in six out of eleven themes explored in the former and four of nine themes in the latter when they were compared to their Babylonian counterparts. We have already noted the connection in the case of the nursery rhyme discussed earlier. Yet another common theme treated by the two mathematical traditions (although the Babylonian one is based on more recently discovered texts) relates to the correct computation of the volumes of truncated pyramids.[12] The overall conclusion reached is that there existed significant interrelations between the mathematics of the Middle Kingdom of Egypt and the Old Babylonian mathematics of Mesopotamia. A study of the later texts, notably the Cairo demotic mathematical papyrus (third century BC), confirms Friberg's conjecture that a significant part of the themes and methods of the Late Babylonian mathematics was known to Ptolemaic Egyptians. Therefore, two parallel traditions may have existed around the time that Euclid lived in Alexandria: a practical (non-Euclidean) Greek-Egyptian mathematics (which Friberg calls "metric algebra"[13]) and a theoretical (Euclidean) Greek mathematics.

It is at this point that Friberg's related thesis comes into its own. In his book *Amazing Traces of of a Babylonian Origin in Greek Mathematics* (2007b), he argues that several of the well-known Greek mathematicians (for example, Euclid, Heron, Ptolemy, Diophantus, and Archimedes) were familiar with the Babylonian mathematical methodology (described as "metric algebra"), known from both Babylonian and pre-Babylonian mathematical clay tablets. He presents eighteen pieces of evidence in eighteen chapters to support his conjecture. For example, the connection between Babylonian methods and Euclid's *Elements* is the focus of a number of chapters. The connection of Babylonian methods with Diophantus's *Arithmetica* is the subject matter of chapter 13 of Friberg's book. Other chapters compare and contrast Babylonian geometry with corresponding topics from major and minor Greek mathematicians such as Theon of Smyrna, Theodorus of Cyrene, Heron, Ptolemy, and even the Indian mathematician Brahmagupta. Such comparisons not only lead to new questions and answers relating to important issues in the history of Greek mathematics but also highlight the interdependence between the three mathematical traditions, which have been considered more or less independent of one another for so long.

There is one important issue that Friberg's two studies do not address. Having pointed to the existence of mathematical links based on an examination of the structure and content of the texts, he has little to say about where, how, and when these connections were established in the first place. It is clear that these broader issues need to be addressed if Friberg's theses are to be fully sustained.

Notes

1. The author has been criticized for paying excessive attention to the views expressed by Kline, who is perceived as unrepresentative of the opinions of historians of mathematics these days. However, the assessment of Indian mathematics by a widely quoted historian of mathematics, Carl Boyer (1968), was not very different:

> (The Indian mathematicians) delighted more in the tricks that could be played with numbers than in the thoughts the mind could produce, so that neither Euclidean geometry nor Aristotelian logic made a strong impression upon them. The Pythagorean problem of the incommensurables, which was of intense interest to Greek geometers, was of little import to Hindu mathematicians, who treated rational and irrational quantities, curvilinear and rectilinear magnitudes indiscriminately. . . . Questions concerning incommensurability, the infinitesimal, infinity, the process of exhaustion, and other inquiries leading towards the conceptions and methods of calculus were neglected. (pp. 61–62)

2. As late as 2002, Høyrup wrote: "Apart from the family likeness between the filling problems in IM 53957 and Rhind Mathematical Papyrus # 37, no evidence suggests the slightest connection between Old Babylonian mathematics and 'classical' (Pharanoic) Egyptian mathematics as found in Middle and New Kingdom papyri—nor between the surveyors' tradition and classical Egyptian mathematics"—a view that is not uncommon even today among other scholars of Egyptian mathematics.

3. In the forefront of this activity of studying the links are the works of Friberg (2005, 2007b), who was referred to in an earlier chapter. It is Friberg who has highlighted these connections using terms such as "unexpected links" to describe the relationship between Egyptian and Babylonian mathematics and "amazing traces" to characterize the "Babylonian origin in Greek mathematics."

4. Periodicity is the recurrence at regular intervals of an astronomical phenomenon relating to a celestial body. For example, the lunar synodic period is the period of time from one full or new moon to another, that is, the time between consecutive align-

ments of the sun, earth, and the moon on a plane perpendicular to the plane of solar revolution.

5. The relationship between Babylonian and Greek astronomy has been studied in recent years, a good example being Jones (1996). Burkert (1972) contains a useful discussion of the late antique construction of Pythagoreanism.

6. The political basis of Western scholarship's appropriation of the Middle East's past needs to be noted. Both postcolonial historians and anthropologists (e.g., Said 1978 and Fahim 1982) have written influential critiques of the strategies adopted in the West to "domesticate" historical interpretations toward the West and not to the area studied.

7. The innovations include Powell's study (1976) of the evolution of the Mesopotamian numeration system in the third millennium BC, Høyrup's analysis (2001) of the language of Old Babylonian algebra without forcing it into a modern symbolic package, Friberg's study (1999, 2000) of the mathematics of the New Babylonian period, and Robson's (2008) emphasis on the social history of Mesopotamian mathematics.

8. As discussed in chapter 2, the differentiation is usually made between table texts, such as tables of decomposition into unit fractions, and problem texts, which state a problem and the method of solution. The administrative texts containing calculations involve a wide range of sources that include accounts, parceling out food, work, and land.

9. The use of modern mathematical notations to decode and "transform" original obscure texts written in dead languages remains controversial among the small band of professional historians of mathematics. As far as this book is concerned, the intended audience is general readers, and for such readers a translation of these texts into present-day mathematical notation and language, together with the occasional insertion of excerpts from the original texts, will hopefully suffice.

10. For further discussion of this characterization, see Imhausen (2007, pp. 25–28).

11. It should be noted that the Babylonians could also compute the volume of truncated pyramids. For further details, see chapter 9 in Friberg (2007b).

12. Computations relating to pyramids and cones are also found in other mathematical traditions, notably Greek, Chinese, and Indian. For a useful survey, see Friberg (1996).

13. As indicated in the previous chapter, the term "metric algebra" was coined by Friberg (2007b, p. vi) to describe a type of mathematics that resulted from an "elaborate combination of geometry, metrology and linear or quadratic equations."

Chapter Six

Ancient Chinese Mathematics

Background and Sources

To understand the history of Chinese mathematics requires some familiarity with Chinese history. The history of China is a vast subject, as be-
fits a country that can trace the continuity of its civilization through 4,500
years. The period we are concerned with runs from prehistoric times to
the end of the Ming dynasty (AD 1386–1644) and the beginnings of European contact. It will help if we divide this long time span into five shorter
periods. The reader may find it useful to refer to the map of China, figure
6.1, for places mentioned in the following account.

The first period began with the civilization that developed along the
banks of the Yangzi and Huang He rivers during the legendary Xia kingdom in the third millennium BC. It continued through the Shang rule,
which began around 1500 BC and lasted for five hundred years. The earliest evidence of numeration in China is from this period and consists of
oracle bones, so named because inscriptions on them indicate that they
were used for divination or fortune-telling. The social organization during
the Shang dynasty was a primitive form of feudalism, with extensive use of
bronze for weaponry and armor as well as the practical arts. Cowrie shells
were widely used as money.

The Zhou invaders completed their conquest of the Shang around 1030
BC. They extended their territorial control and instituted a more developed form of feudalism, reminiscent in its structure of the feudal system
that would emerge in parts of Europe some two thousand years later. The
development of a written language that had come into use during the
Shang period was well under way. But from around 700 BC the Zhou dynasty came under increasing attack from groups of insurgents. The empire
disintegrated, and for a period of two hundred years from about 400 BC

FIGURE 6.1: Map of eastern China

there existed a number of independent states virtually at war with one another for most of the time.

This period, usually known as the period of the Warring States, is notable from a mathematical viewpoint because from it comes one of the oldest sources of Chinese mathematics, the *Zhou Bi Suan Jing* (The Mathematical Classic of the Zhou Gnomon).

It is believed that the version that survived is a compilation of earlier texts, some of which may even date back to the Shang dynasty.[1] The *Zhou Bi* is written in two parts. The first part contains a dialogue between a historical figure from the eleventh century BC, the Duke of Zhou, and Shang Gao, who is described as a skilled mathematician. Their conversation ranges widely and includes the first statement of what is known as the *gou gu* (or the Pythagorean) theorem and also what some believe is a "proof" of that theorem.[2] While it is no longer believed that this treatise predates

Pythagoras by five centuries, it is still thought that it was likely composed before the time of the Greek mathematician. Apart from the consideration of the right-angled triangle and a brief discussion of simple arithmetic operations, the *Zhou Bi* is primarily an archaic astronomical text. (This and other Chinese mathematical sources are listed in table 6.1, together with their authorship, date, and important subjects covered.)

The time of the Warring States was also a period when "a hundred schools of philosophers" flourished. Feudal lords, faced with uncertain times of popular unrest and technological innovation (iron was probably introduced into China at this time), employed itinerant philosophers as advisers. One of these philosophers was Confucius, but his deep and abiding loyalty to the Zhou dynasty permitted him only a short spell as an adviser to the local ruler of Lu. The emphasis Confucius placed on unity and stability in his philosophy may have been a reaction to the troubled times he lived in. As a pillar of social orthodoxy, Confucianism seemed singularly uninterested in science. On the other hand, the reverse was true of the other philosophical stream, Taoism, which was founded on the teachings of Lao Zi, an older contemporary of Confucius. Indeed, Lao Zi's *Dao De Jing* provides one of the earliest references to the practice of mathematics: "Good mathematicians do not use counting rods." With hindsight, one may well disagree with Lao Zi, but the comment provides an indication of both how old the practice of computing with counting rods was as well as the existence of views of what constituted a "good" mathematician.

It would be useful to put this period in a global context. It is one of the more remarkable coincidences of history that the middle of the first millennium BC saw the emergence of many of the great religious and ethical leaders whose influence is felt even today. Between 650 and 450 BC lived Confucius, Gautama Buddha, Mahavira, and, probably, Zoroaster. The same period saw Babylon fall to the Persians (538 BC), India invaded by the Persian emperor Darius (512 BC), and the Persian advance to the west halted by the Greeks (480 BC). And the last two centuries of the Warring States saw the conquests of Alexander (c. 330 BC), the foundation of the Mauryan empire in India, of which Asoka (273–232 BC) was the most illustrious ruler, and the protracted Punic Wars in the Mediterranean (c. 250–150 BC).

The Second Punic War was contemporaneous with the successful reunification of China by the Qin emperor Shi Huang Di, who was master of all China between 221 and 207 BC. He rebuilt the Great Walls; a more

TABLE 6.1: MAJOR CHINESE MATHEMATICAL SOURCES UP TO THE SEVEN-
TEENTH CENTURY

Title	Author	Date	Notable subjects covered
Zhou Bi Suan Jing (The Mathematical Classic of the Gnomon and the Circular Paths of Heaven)	Unknown	c. 500–200 BC	Pythagorean theorem; simple rules of fractions and arithmetic operations
Suanshu Shu (A Book on Arithmetic)	Unknown	300–150 BC	Operations with fractions; areas of rectangular fields; fair taxes
Jiu Zhang Suan Shu (Nine Chapters on Mathematical Arts)	Unknown	300 BC–AD 200	Root extraction; ratios (including the rule of three and the rule of false position); solution of simultaneous equations; areas and volumes of various geometrical figures and solids; right-angled triangles
Ta Tai Li Chi (Records of Rites Compiled by Tai the Elder)	Unknown	AD 80	Magic square order of 3
Commentary on *Jiu Zhang*	Chang Heng	130	π = square root of 10
Shu Shu Chi Yi (Manual on the Traditions of the Mathematical Arts)	Xu Yue	c. 200	Theory of large numbers; magic squares; first mention of the abacus
Commentary on *Zhou Bi*	Zhao Zhujing	c. 200–300	Solution of quadradic equations of the type $x^2 + ax = b^2$
Hai Dao Suan Jing (Sea Island Mathematical Manual)	Liu Hui	263	Extentions of problems in geometry and algebra from the *Nine Chapters*
Sun Zu Suan Jing (Master Sun's Mathematical Manual)	Sun Zu	400	A problem in indeterminate analysis; square root extraction; operations with rod numerals

continued

TABLE 6.1: CONTINUED

Title	Author	Date	Notable subjects covered
Sui Shu (Official History of the Sui Dynasty)	Zu Chongzhi	450	Evaluation of π; method of finite differences
Ji Gu Suan Jing (Continuation of Ancient Mathematics)	Wang Xiaotong	625	Solution of third-degree equations; practical problems for engineers, architects, and surveyors
Suan Jing Shi Shu (The Ten Mathematical Manuals)	Li Chungfeng	656	An encyclopedia of mathematical classics of the past
Meng Xi Bi Tan (Dream Pool Essays)	Shen Kuo	1086	Summation of series by piling up a number of kegs in a space shaped like a dissected pyramid
Shu Shu Jiu Zhang (Nine Sections of Mathematics)	Qin Jiushao	1247	Numerical solutions of equations of high degree; indeterminate analysis
Ce Yuan Hai Jing (The Sea Mirror of the Circle Measurements)	Li Ye	1248	Solutions of high-degree equations; applications of the Pythagorean theorem to practical problems; use of a diagonal line across a digit to indicate a minus quantity
Xiang Jie Jiu Zhang Suan Fa Zuan Lei (Detailed Analysis of the Nine Chapters)	Yang Hui	1261	Arithmetic progressions; decimal fractions; quadratic equations with negative coefficients of x
Yuan Shi (Official History of the Yuan Dynasty)	Guo Shoujing	1280	Foundations of spherical trigonometry; cubic interpolation formula; biquadratic equations
Si Yuan Yu Jian (The Precious Mirror of the Four Elements)	Zhu Shijie	1303	Pascal's triangle; solutions of simultaneous equations with five unknowns by matrix methods
Suan Fa Dong Zong (A Systematic Treatise on Arithmetic)	Cheng Tai We	1593	Magic squares; introduction to abacus
Ji He Yuan Pen (Elements of Geometry)	Ricci and Xu	1607	Six books of Euclid's *Elements* translated into Chinese

*Works most influential in the development of Chinese mathematics dubious claim to fame is his order to burn all books. Since its recent excavation, the famous army of life-size ceramic soldiers buried with him has attracted considerable attention outside China, and may be seen as one of the more lasting manifestations of his egomania.

During the subsequent period, which saw the emergence of the Han dynasty (200 BC to AD 220), scholars devoted a considerable amount of their time to transcribing from memory literary and scientific texts and seeking out manuscripts that had escaped destruction. This was the period when the earliest Chinese mathematical text was composed. Known as *Suanshu Shu* (A Book on Arithmetic), it consisted of around 200 bamboo strips, of which about 180 are reasonably well preserved. It was found in a tomb, and documentary evidence indicates that this tomb had been closed in 186 BC. The occupant of the tomb may have been a minor local government official. The *Suanshu Shu* is anonymous, in the sense that we do not know the name of the person who assembled this material.[3] While it is not a systematic introduction to mathematics, it contains a collection of problems involving arithmetic operations, including operations with fractions, the determination of proportional payments, the calculation of areas of fields and areas and volumes of different figures and shapes.[4]

A little later but in the same dynasty appears the most influential of all Chinese mathematical texts: the *Jiu Zhang Suan Shu* (Nine Chapters on the Mathematical Arts), which occupies a similar position in Chinese mathematics to that of Euclid's *Elements* in Western mathematics. Possibly an original product of the late Qin and early Han dynasties, it was written at a time when the Roman empire was at its height and Buddhism was starting to have an impact in China. It was reputed to have been arranged and commented upon by Zhang Shang (c. 150 BC) and later Geng Shouzhang (c. 50 BC), from some earlier texts that have not survived. However, the version put together and commented on by Liu Hui (c. AD 250) remains one of the more authoritative early texts. The book has now been translated into a number of European languages.[5]

The Han period is noted for significant developments in Chinese science and technology. Much was achieved in astronomy and calendar construction. Xu Yue (c. AD 200) wrote a treatise titled *Shu Shu Chi Yi* (Manual on the Traditions of the Mathematical Arts), which discusses calendar construction and gives an early description of the classical 3×3 magic square.[6] Foundations were laid for the systematic study and classification of plants and animals. One of the greatest technological inventions of mankind—paper—was a product of this period. An extensive bibliography was compiled by experts in medicine, history, military science, philosophy, astronomy, magic, and divination; some of the volumes, recorded on wooden or bamboo tablets or on silk, have survived. This was also a period when the main features of the Chinese society started to take shape, with the particular characteristic of a burgeoning bureaucracy recruited from scholars through the medium of examinations. Palace revolutions, peasant rebellions, and religious uprisings weakened the Han dynasty until it was overthrown in AD 220, when a united China split into the Three Kingdoms. It is interesting that Liu Hui, an inhabitant of the Wei kingdom (one of the Three Kingdoms) and the influential commentator on the *Jiu Zhang*, advocated that mathematics be one of the six subjects taught to recruits for the state bureaucracy.

The next period saw considerable divisions and upheavals, lasting for about a hundred years, which were brought to an end by a second unification of China under the Qin dynasty. However, the troubled times apparently did not disrupt mathematical activity, for during this period lived Liu Hui, mentioned earlier, on whose work we are so dependent for information on early Chinese mathematics. Foreign contacts through the spread of Buddhism, which began during the last decades of the Han dynasty, continued—in art, sculpture, medicine, and the sciences as well as religion. Fa Xian, the great Buddhist pilgrim, set out for India in 399 and traveled the length and breadth of northern India, and over central Asia, for fifteen years.

This period produced two great mathematicians: Sun Zu (c. 300), in whose work we find the beginnings of indeterminate analysis, and Zu Chongzhi (c. 450), who accurately approximated π to be equal to 355/113 and whose life span takes us into the succeeding Liu Song dynasty (420–479). The Chinese work on both these topics will be discussed in the next chapter. From the beginning of the Tang dynasty, an encyclopedia of Chinese mathematical classics titled *Suan Jing Shi Shu* (The Ten Mathematical Manuals) began to appear. At first, the encyclopedia consisted of more than ten volumes, but only ten were published and hence the title. This was to remain an influential text for several centuries.

A number of northern and southern dynasties followed in relatively quick succession after the second partitioning of China, until the short-lived Sui dynasty (589–618) reunified the country once more. It was a period of construction of large-scale waterworks, of which the most notable was the Grand Canal linking the Huang and the Yangzi rivers. The labor requirement was enormous: at times over five million workers were needed, which in certain districts meant all adult males between the ages of fifteen and fifty The benefit to posterity was immense, for a transport system had been developed that linked the productive agriculture of southern China with the politically and demographically more influential north.

The immense cost borne by ordinary people under the Sui dynasty bore fruit during the rule of the Tang dynasty, which many would consider one of the most productive periods of Chinese history. For about three hundred years (618–906) the Tang emperors ruled over a China whose territorial boundaries and cultural dominance had never been so extensive. This was a period characterized by a remarkable openness to foreign influences. Just as Baghdad, the center of intellectual activity to the west, welcomed scholars from all lands, the great Tang capital Chang'an numbered Arabs, Koreans, Japanese, and Indians among a population estimated to have reached a million. This was a period of literary and artistic renaissance, while major technological innovations included printing and gunpowder. Surprisingly, no important mathematical work from this period has been discovered, but it could well have been the fertilization through foreign contacts and the scrutiny of past Chinese texts which took place in this period that led to the upsurge in Chinese mathematics of a few centuries later.

The Tang dynasty was followed in relatively quick succession by the years of the Five Dynasties and the Ten Independent States (907–960). Although these were chaotic times, there were great advances in block printing, which began initially as a means of disseminating religious texts but then slowly spread into secular fields. A great desire for unity elevated Zhao Kuangyin to the throne in 960, marking the beginning of one of longest dynasties in Chinese history, the Song (900–1279), in which may be seen the culmination of the developments of the previous two centuries. In reality, there were two dynasties: the Northern Song (960–1127) with its capital at modern Kaifeng, and the Southern Song (1127–1279), whose capital was situated in modern Hangzhou.

The scientific and technological achievements of the Song period are too numerous to list here; we shall concentrate on the mathematics. The Song period produced some of the greatest mathematicians of China, especially during the thirteenth century. In 1247 Qin Jiushao wrote *Shu Shu Jiu Zhang* (Nine Sections of Mathematics, not to be confused with the *Jiu Zhang* of a thousand years before). In this work Qin explained the numerical solution of equations of all degrees, and extended the work on indeterminate analysis begun by Sun Zu. A year later appeared Li Ye's *Ce Yuan Hai Jing* (The Sea Mirror of the Circle Measurements), which explained how to construct equations of various degrees from a given set of data, thus complementing Qin's methods of solving these equations. Between 1261 and 1275 appeared a series of works by Yang Hui, the most influential of which was *Xiang Jie Jiu Zhang Suan Fa Zuan Lei* (Detailed Analysis of the Mathematical Methods in the Jiu Zhang). It starts as a commentary on the *Jiu Zhang* and goes on to present some remarkable extensions of the original work in a number of areas including mathematical series, quadratic equations with negative coefficients of x, and higher-order numerical equations. The fourth name in this grand quartet of mathematicians is Zhu Shijie, who lifted Chinese algebra to the highest level it was ever to attain. In the two treatises he wrote, *Suan Xu Ji Meng* (Introduction to Mathematical Studies, 1299) and *Si Yuan Yu Jian* (The Precious Mirror of the Four Elements, 1303), are to be found a treatment of "Pascal's" triangle 350 years before Pascal, the solution of simultaneous equations with five unknowns using the "method of rectangular arrays" (or what we would now call matrix methods), and detailed applications of what was known as the "celestial element method" for solving equations of higher degree. This list of notable mathematicians would be incomplete without the mention of Guo Shoujing (1231–1316). While there is no extant treatise on mathematics that can be traced to Guo, there are records from the Ming period (1368–1648) that show his influence on astronomy and calendar construction. The first Chinese work on what we would now consider as spherical trigonometry based on approximation formulas is directly attributed to Guo. By the closing stages of the Song dynasty, the Chinese algebraists had forged so far ahead that it was only during the eighteenth century that the gap between Chinese and European algebra, particularly with respect to the solution of equations, was finally closed.

The last period of this short historical survey of China and Chinese mathematics stretches over four hundred years, and takes in the Yuan (Mongol) and Ming dynasties, which ruled over the whole of China. The Yuan dynasty began with Shi Zu (better known in Europe as Kublai Khan, the grandson of Genghis Khan) in 1280 and ended with Shun Di (Togan Timur) in 1367. This was not a period of great creative activity in mathematics, but it was a time when contacts between China and Europe were at their height. Mongol control extended right across the Asian heartland, from the Yellow Sea in the east to the Black Sea in the west. Appointment of non-Chinese to positions of authority was state policy, as an important means of preserving control. Ideas and technology were transmitted via the trade routes across central Asia. Marco Polo was one of many travelers who came to China and wondered at its marvels; he later served at the court of the khan. It was also during this period and the subsequent two centuries that four technological innovations from China, which, in the words of Francis Bacon (1561–1626), "changed the whole face and state of things throughout the world," started their way slowly westward to Europe: block printing, gunpowder, papermaking, and the magnetic compass. It is an interesting reflection that while there was a time lag of ten centuries between use of paper in China and in western Europe, the time lags for gunpowder and the magnetic compass were four and two centuries respectively. The last two would later be used by Europeans, to dramatic and devastating effect, on the rest of the world.

The Ming period, which began in 1368, saw the restoration of indigenous culture and values after a century of foreign rule, as well as expanding Chinese influence abroad. Great maritime expeditions were sent to the southern parts of India, Sri Lanka, the eastern coast of Africa, and the Persian Gulf. Exotic items, including African animals such as giraffes, zebras, and ostriches, were brought back from these areas.[7] However, this interest in things foreign was not matched by a desire to learn from foreigners. Chauvinism and intolerance of other people prevailed and led to a stagnation of science that would be briefly relieved with the arrival of the Jesuits during the later part of the sixteenth century. It is not surprising, given the spirit of the age, that the first translation of Euclid's *Elements* into Chinese had little impact in China. It was felt more in post-Meiji Japan, which had until then been very much under the influence of its larger neighbor. At the beginning of the seventeenth century, the Manchus from Manchuria mounted an invasion that led to the establishment in 1644 of a dynasty that would survive until its overthrow in 1911. Foreign interventions became increasingly common, particularly in the nineteenth century; and Chinese mathematics became influenced by the West once more, this time in a way that was to change its character completely. But this takes us far beyond the period covered by this book.

The following discussion of Chinese mathematics takes a thematic approach rather than examining in detail the various sources of Chinese mathematics listed in table 6.1. However, because of the great importance of the *Jiu Zhang Suan Shu* (hereafter referred to simply as the *Jiu Zhang*) in the development of Chinese mathematics, its contents will be subjected to a detailed analysis.

In this chapter we consider developments in Chinese mathematics up to the end of the first millennium AD; subsequent progress, particularly during the thirteenth century, will be examined in the next chapter.

The Development of Chinese Numerals

Types of Chinese Numerals

It is possible to distinguish four main types of Chinese numerals, all based on the decimal system:

1. The "standard" or "modern" numerals may have originated from common number-words in use from about the third century BC.

2. The "official" numerals are merely a highly decorative (and rather more "complex") version of the standard numerals, used on legal documents and banknotes that need to be protected from forgery or unauthorized alterations.

3. The "commercial" numerals, again based on the standard ones, were devised for writing quickly and were widely used in trade and commerce. They are of more recent origin, dating from the sixteenth century, and are found most commonly today on price tags or bills in Chinese shops and restaurants.

4. The "stick" or "rod" numerals served until recently as the principal instrument for mathematical and scientific work, and had been in use from at least the second century BC. The name derives from their origins in arrangements of sticks or rods on Chinese counting boards. We shall be concentrating on these numerals.

However, the earliest-known Chinese numerals appear in the form of Shang "oracle bones," dating from sometime between 1500 and 1200 BC. A group of farmers, tilling their fields near Anyang in Henan at the end of the nineteenth century, came across a collection of tortoise shells and animal bones with inscriptions on them. They were eventually sold to an apothecary, who believed them to be the bones of a dragon, endowed with

medicinal properties. Fortunately they were rescued before being con-
sumed, and attracted the interest of several Chinese scholars, who were
instrumental in deciphering the inscriptions they bore. It turned out that
the bones had belonged to Shang nobles who were in the habit of appealing to the spirits of their ancestors for advice on the best times for traveling, harvesting, celebrating feasts, and other activities. Such questions,
together with answers recorded after the prophecies had been fulfilled,
were inscribed on the bones. New finds over the last century have provided thousands of examples of inscriptions on bones, giving us a better
understanding not only of the numeral system in use and its role as a divinatory instrument but also of the socioeconomic climate of the times. For
example, on some of the oracle bones are records of numbers of prisoners
captured, numbers of animals and birds captured on hunting expeditions,
numbers and types of animals sacrificed in ritual ceremonies.[8]

The oracle bones recorded integer numbers from 1 to 30,000. The numbers 1 to 10 as found on these bones were represented by the following
symbols:

This numeral system was more advanced than all contemporary systems
except the Mesopotamian, since it enabled any number, however large, to
be expressed by the use of ten basic symbols (1 to 10) and a selected number of additional symbols to represent twenties, hundreds, thousands, and
ten thousands. Thus, the numbers 537 and 1,348 were written as

(5 hundreds, 3 tens, 7)

(1 thousand, 3 hundreds, 4 tens, 8)

where the symbol ⟩ represents thousands, ⊖ hundreds, and ∪ twenties.

In the centuries that followed, different variants of the above characters were used, as can be seen on artifacts such as coins and bronze vessels,
mostly from the Zhou period (eleventh to the third century BC). All were
used in a decimal system of notation, with extra symbols to denote different
orders of magnitude. The standard number system is a direct descendant of
the ancient Shang system: its symbols for the numbers from one to ten are

1	2	3	4	5	6	7	8	9	10
一	二	三	四	五	六	七	八	九	十

The number 842 may be written from left to right as

八百四十二

where 百 and 十 represent hundreds and tens respectively.

It is important to recognize that just as in the case of the Egyptian numerals, the Chinese had no need for a zero to distinguish between positional values, since each positional value was paired with a symbol representing its power of ten. For example, 2,005 would be represented as "2 thousands and 5" while 250 would be represented as "2 hundreds and 5 tens." This way of writing numbers undoubtedly influenced the way that computational methods developed in China. And indeed, as we shall see later, it had considerable bearing on how the Chinese would work with fractions, extract roots, and solve equations.

However, over the years the demands of commerce, administration, and science led to the development of a distinctive Chinese place-value number system that involved the use of counting rods. These rods, originally of ivory or bamboo, were arranged in columns from right to left representing increasing powers of ten. Positive and negative numbers were represented by red and black rods, respectively. Our information on their use goes back only to the Qin and early Han dynasties, though the system was probably invented earlier.

By the third century AD these rod numerals were being described as *heng* (horizontal) and *zong* (vertical) numerals. A later notation for these two variants of rod numerals took the form shown in table 6.2. The *zongs* represent units, hundreds, ten thousands, etc., and the *hengs* tens, thousands, hundred thousands, etc.[9]

Thus the number 3,614 would be written as

Note that the columns of numbers need not be demarcated, since the type of numeral used defines the column value. Thus, in common with

TABLE 6.2: CHINESE *HENG* AND *ZONG* ROD NUMERALS

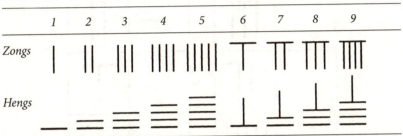

| | 1 | 2 | 3 | 4 | 5 | 6 | 7 | 8 | 9 |

Note: The *zongs* represent units, hundreds, tens of thousands, etc., and the *hengs* tens, thousands, hundreds of thousands, etc.

our present-day number system, to show 3,614, a reckoner places rods to represent 4 in the extreme right column, 1 in the second, 6 in the third, and 3 in the fourth column. The alternating *heng* and *zong* representation would help to distinguish units of different orders in odd-numbered and even-numbered columns of the counting board. This provided a built-in means of checking whether the digits were correctly represented before undertaking arithmetical operations. And as computation could be carried out by placing the rods on any flat surface, Chinese reckoners needed no materials other than their bundle of counting rods.[10]

During the Han period, the counting rods were round bamboo sticks about 2.5 mm in diameter and 140 mm long, tied together in a hexagonal bundle that could be conveniently carried by hand. But from the sixth century AD they became shorter and rectangular in shape. Besides bamboo, counting rods were also made from wood, cast iron, jade, or ivory. Counting rods of other shapes and sizes were also used in Korea and Japan, which came within the Chinese sphere of influence in mathematics.

Operations with Rod Numerals

Elementary operations with counting rods were carried out as one would calculate on an abacus. If numbers were to be added or subtracted, then the rods were repositioned column by column. For example, the addition of 8 and 7 would be shown as in figure 6.2a, where the first and second rows represent 8 and 7, respectively, while the third row shows the sum, 15. The subtraction of 6 from 12 would proceed as shown in figure 6.2b, where 12 and 6 are first laid down in the first and second rows. The horizontal rod in the first row is then converted into an additional set of ten vertical rods before subtraction. The third row shows the answer, 6.

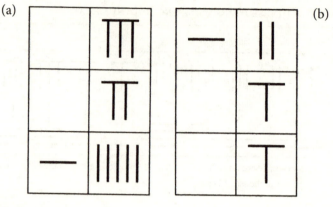

FIGURE 6.2

For multiplication the top row is the multiplier, the middle row is left blank for the intermediate steps to be entered as multiplication proceeds, and the bottom row shows the multiplicand. As an illustration we take the multiplication of 387 by 147, using modern numerals for clarity. The product of the two numbers is found via the five sets of arrangements of counting rods shown in figure 6.3. As a first step, the rods are arranged as in (a). In (b), 441 is obtained by multiplying 147 by 3. In the next step, shown in (c), 147 is multiplied by 8 and the result added to 4,410 to give 5,586. Then, in (d), 147 is multiplied by 7 and the result added to 55,860 to give the final product, 56,889. The multiplication process is complete in (e). At each stage the rods are rearranged so that the numerals fall in the appropriate columns, 147 being moved one place to the right at each step. Both canceling and correcting mistakes are easier to do with counting rods than with paper and pencil. With sufficient practice, the rods could be manipulated with such speed that a writer in the eleventh century comments on the rods "flying so quickly that the eye loses track of their movements."

The method of division begins with the divisor (*fa*) in the bottom row and the dividend (*shi*) in the middle row, with the top row left blank for the quotient (*shang*). As an illustration, we shall divide 56,889 by 147; the

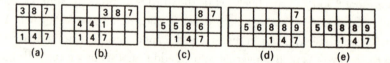

FIGURE 6.3: Multiplication with counting rods

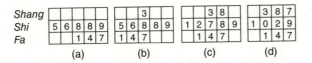

FIGURE 6.4: Division with counting rods

process takes four steps, as shown in figure 6.4. In (a) the rods are set up for division. The number to be divided, 56,889, is inserted in the middle row and the divisor, 147, in the bottom row. The divisor is moved left to the point where it is exactly below 568. Dividing 568 by 147 gives 3, which is inserted in the top row in (b). The divisor 147 is multiplied by the quotient and subtracted from 568 to give 127, which is placed in the middle row in (c) followed by the last two digits, 8 and 9, of 56,889. The divisor is moved one place to the right and divided into the number above it, 1,278. The result, 8, is inserted in the top row in (c). Then 147 is multiplied by 8 and the result subtracted from 1,278, giving 102, which is inserted in the middle row followed by the last digit, 9, of 56,889, as in (d). The division of 1,029 by 147 gives 7, with no remainder, which is inserted in the top row in (d). The answer is 387.

The representation of zero in this system poses no problem: a space is simply left blank. But there was a difficulty, as with the Mesopotamian notation, when the rod numerals were written down. A circular sign for zero makes its first appearance in Qin Jiushao's work in 1247, probably an influence from India. However, the blank space standing for what we now call zero in Chinese numerals was conceptually different from the blank space in Mesopotamian numerals. In the Chinese system the blank space is itself a numeral, whereas in the Mesopotamian system it represents the absence of a digit—that is, a placeholder.[11] This is immediately evident if we consider what is implied by the absence of a blank space as a right-hand terminal indicator in each of the two systems. In the Mesopotamian sexagesimal place-value notation, the symbols T ⟨ could have stood for 70, or 3,660, or some other number. Such ambiguities did not occur with the Chinese rod numerals, since the reckoner would always have been aware of the positional values of the digits in the numbers operated with. Also, the ingenious device of alternating the orientation of the rods in successive place values meant that it was easy to check that the numerals were correctly positioned relative to one another. However, there is an underlying ambiguity that the device of alternating the orientation of the rods in successive place values does not resolve. In the absence of a sign to represent zero, how does one distinguish between 12 and 1002 in the Chinese system? Thus

12: | || 1002: — ||

An important advantage of a place-value number system, as we saw earlier, is that the fractional and integral parts of a mixed number can be represented as economically as possible. In the rod numeral system the integral and fractional parts are separated by a line. Thus, the number 48,125 is represented as

The counting rods were not merely a calculating device: they eventually became a vehicle for expressing certain mathematical concepts. Apart from its notational facility, the rod numeral system was helpful in suggesting new approaches to algebraic problems, and also ways of operating with negative numbers. It may even be argued that geometric algebra, which began with the Babylonians and was extended by the Greeks, was given an arithmetic dimension by the Chinese counting rods. It was merely a matter of time before the positions of the counting rods came to stand for algebraic symbols, and operations with the rods for algebraic operations. Let us now look at how the rods were used as a representational device. (Their use for algebraic operations will be examined in a later section.)

To represent a system of equations, the counting rods were arranged in such a way that one column was assigned to each equation of the system and one row to the coefficients of each unknown in the equations. The elements of the last row consisted of the entries on the right-hand side of each equation. Red rods were used to represent positive (*zheng*) coefficients and black rods negative (*fu*) coefficients.

Figure 6.5 shows the representation of the following system of equations in three unknowns:

$$2x - 3y + 8z = 32,$$
$$-6x - 2y - z = 62,$$
$$3x + 21y - 3z = 0.$$

It has been argued that because equations were represented in this form, it was only a matter of time before matrix methods of solving systems of

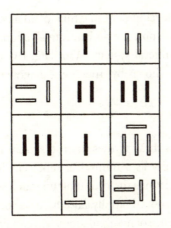

FIGURE 6.5: Representing simultaneous equations with counting rods. In this drawing, solid (black) rods represent negative numbers and open rods positive numbers.

linear simultaneous equations or numerical methods of solving higher-order equations would logically suggest themselves. Whether or not this happened, it is clear that these methods became an integral part of Chinese mathematics, probably as early as the third century BC (they made their first appearance in the *Jiu Zhang*), and remained unique to Chinese mathematics until the eighteenth century AD. We shall examine these methods more thoroughly later.

It is easy to be deceived by the simplicity of our present-day decimal number system into overlooking the advantages of other systems. We have seen how the Mesopotamians developed a sexagesimal place-value system whose lack of a symbol for zero hindered both number representation and computation. The distinctive feature of the modern decimal system, which it shares with the Chinese rod numeral system, is an economy both in the number of symbols used and in the space occupied by a written number. In the Chinese system, each of the first nine numbers is represented by no more than five rods. By letting a single rod represent the number 5, the numbers from 6 to 9 became much simpler to represent than in, say, the Egyptian system, and arithmetical operations were easier to carry out. Alternating the orientation of the rods provided a clear indication of the place positions of different digits of a number, and all that was needed to carry out computations was a bundle of rods, together with a flat surface to place them on. This led to the development of a set of quick and easy algorithms, not only for multiplication and division and extraction of square and cube roots, but also for the solution of simultaneous linear and higher-order equations. The early development in China of an algebra of negative numbers (*zheng fu shu*, or "positive and negative" operations) is another by-product of the rod numeral system. With later (written) rod numerals, a negative number was shown by drawing a diagonal line through its last column, for example:

-12:   ⏤ 卅

However, as well as these virtues the rod system had its shortcomings, apart from its ambiguity as a numerical representational device. In long and complicated calculations the rods took up much space. Errors might be made when rods were moved rapidly, and there was no possibility of detecting errors once a sequence of manipulations was complete. It may be that the remarkable success of these devices for calculation in China inhibited the development of alternative mechanical devices such as the abacus, which did not become widespread until the time of the Ming dynasty, and delayed the adoption of the modern decimal number system until recent times.

There is, however, a view that the *idea* of decimal place-value computation embodied in the counting-rod system was transmitted from China, first to India and then to the Islamic world. This is because the manner in which computations were carried out in the Arabic texts of al-Khwarizmi and other Islamic mathematicians was almost identical to the Chinese procedure. No tangible evidence exists to support this conjecture despite its plausibility. However, one difficulty with substantiating this conjecture is the need to determine at what point a computational device that did not depend on any writing was transformed into a procedure that was wholly dependent on written numerals.[12]

Chinese Magic Squares (and Other Designs)

A magic square is a square array of numbers arranged in such a way that the numbers along any row, column, or principal diagonal add up to the same total. In most magic squares of n rows and n columns the n^2 "cells" are occupied by the natural numbers from 1 to n^2. For example, a magic square of four rows and four columns (i.e., of order 4) would contain all the integers from 1 to 16.

Magic squares have some interesting mathematical properties. If s is the constant sum of the numbers in each row, column, or principal diagonal, and S is the grand total of the numbers in the n^2 cells, then

4	9	2
3	5	7
8	1	6

FIGURE 6.6: A regular magic square of order three

$$S = \frac{1}{2}n^2(n^2 + 1), \text{ and } \quad s = S/n.$$

If n is odd, the number of the central cell is given by S/n^2, which is also the mean of the series $1 + 2 + \ldots + n^2$. (For magic squares of even order, for which there is no single central cell, S/n^2 is not a whole number.) This is a key number for all odd-order magic squares, since from this number and the value of n it is possible to work out the partial sums s and total sum S. For example, if $n = 3$ and the middle number is 5,

$$S = 9 \times 5 = 45, \text{ and } \quad s = \frac{45}{3} = 15.$$

A magic square of odd order in which every pair of numbers on opposite sides of the central cell add up to twice the middle number is known as a regular magic square. For example, in figure 6.6, in each of the pairs (4, 6), (2, 8), (3, 7) and (1, 9), the two numbers lie on opposite sides of the middle number (5) and add up to 10, which is twice 5. Figure 6.6 represents the most famous of the regular Chinese magic squares, known as the *Luo shu*. It was seen as a symbol of the universe itself.

Magic squares are only of marginal interest today, forming part of a peripheral area known as "recreational mathematics." Yet until four hundred years ago, in almost all mathematical traditions, magic squares engaged the interest of notable mathematicians as a challenging object of study. Their attractions were heightened not only by their aesthetic appeal but also by their association with divination and the occult; they were engraved on ornaments worn as talismans.

Magic Squares in China

In China, where the earliest recorded magic squares have been found, there was a long-established fascination with number patterns and the associated combinatorial analysis. The Chinese shared with the Greeks an

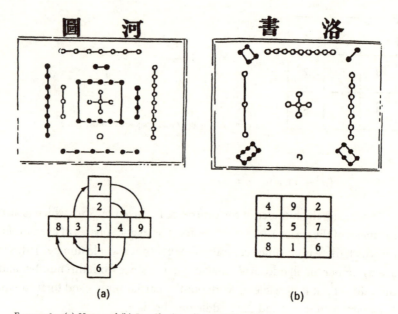

FIGURE 6.7: (a) *He tu* and (b) *Luo Shu* (Needham 1959, p. 57)

interest in numerology and number mysticism (e.g., odd numbers were thought to be lucky, even numbers unlucky), but there is nothing in Chinese mathematics that resembles the Pythagorean fascination for figurate numbers (i.e., triangular, square, pentagonal numbers) or for types of numbers such as perfect or amicable numbers.[13] Neither is there anything in Greek mathematics that suggests even the slightest interest in magic squares.

The first record of a magic square in China goes back to the time of the semimythical emperor Yu, who was reputed to have lived in the third millennium BC. There is a legend that Yu acquired two diagrams, the first one (*He tu*, meaning the River Chart) from a magical dragon-horse that rose from the waters of the Huang He (Yellow River) and left its footprints along the river in the form of an imprint of the *He tu*, and the second (*Luo Shu*, meaning Luo River Writing) copied from the design on the back of a sacred turtle found in the Luo, a tributary of the Huang He. Figure 6.7 shows these gifts: *He tu*, a cruciform array of numbers from 1 to 10, and *Luo Shu*, a regular magic square of order 3. (The number 10 is shown in the *He tu* as a square of black beads surrounding the five central white beads.) The *He tu* is arranged so that, disregarding the central 5 and 10, both the odd and even sequences of numbers add up to 20. The *Luo Shu*, which we discussed earlier, is a magic square in which the figures in any diagonal, row, or column add up to 15, a remarkable balance being maintained between the odd (white beads) and even (black beads) numbers around the central number, which is again 5. Both diagrams represent an important principle of Chinese philosophy—balancing the two complementary forces of *yin* (female) and *yang* (male) in nature, represented here by odd and even numbers respectively.

There can be no doubt about the antiquity of this story, and aspects of it suggest that it cannot have originated any later than the second century BC. Certain passages in the ever-popular manual of divination, *Yi Jing* (The Book of Changes), written at about that time, emphasize not only the magical/divinatory nature of these diagrams but also their numerical properties. Gradually an extensive folklore grew up around the magical properties of the *Luo Shu*, and more generally of magic squares of order 3. It came to be described as the nine rooms or halls of the cosmic temple, the Ming Tang, and later writers would refer to the construction of this square as the "nine halls calculation" (*Jiu Gong Suan*). The belief in the magical powers of this square spread into neighboring areas, among the Tibetans, Koreans, and Mongolians, who depicted it as an arrangement of black and white knots or beads on short lengths of cord, as shown in figure 6.7b.

Given the early start the ancient Chinese had in the development of magic squares, one would expect them to have progressed to squares of order higher than 3. However, there is no evidence that they did, though brief quotations from existing texts show that commentaries were written on the *Luo Shu*, especially during the periods of disorder and unrest that occurred between the fourth and sixth centuries AD. It can only be assumed that this great interest in magic and divination was but a sign of the times—a desperate search for a better tomorrow.

The long hiatus ended with the emergence of Yang Hui, who in 1275 published his *Xu Gu Zhai Suan Fa* (Continuation of Ancient Mathematical Methods for Elucidating the Strange Properties of Numbers). In the preface to his book, Yang Hui pointed out that he was merely passing on the works of earlier scholars and would make no claim to originality. Some of the magic squares he constructed were very complicated, and instructions for building them were either absent or cryptic to the point of obscurity. Let us consider briefly the methods he outlined for constructing magic squares of orders 3, 4, and 5.

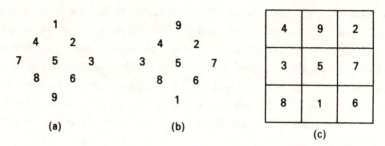

FIGURE 6.8: Construction of the *Luo Shu*

Construction of the *Luo Shu*

Yang Hui's instructions may be expressed as follows:

1. Arrange the numbers 1 to 3, 4 to 6, and 7 to 9 (as shown in figure 6.8a) so that they slant downward to the right.

2. Replace 1 on the "head" with 9 on the "shoe," and vice versa.

3. Interchange 7 and 3, other numbers remaining in their old positions. (The new positions are shown in figure 6.8b.)

4. Lower 9 to fill the slot between 4 and 2, and raise 1 to fill the slot between 8 and 6. (Figure 6.8c shows the *Luo Shu* that results, after similarly moving 7 and 3 inward.)

Construction of a Magic Square of Order 4

To construct a magic square of order 4, place the numbers 1 to 16 in an array of four columns and four rows, as shown in figure 6.9a. Then interchange the numbers at the corners of the outer square (16 and 1, 4 and 13), to give the arrangement in figure 6.9b. Finally, interchange the numbers at the corners of the inner square (6 and 11, 7 and 10). This will produce a magic square in which all the columns, rows, and diagonals add up to 34, as shown in figure 6.9c.

As Yang Hui points out, different variants of this "method of interchange" can produce different magic squares. For example, figure 6.10 was obtained by arranging the sixteen numbers in four columns beginning at the top left-hand corner. Figure 6.11 was obtained by first listing the sixteen numbers beginning at the top right-hand corner and going down the four columns from right to left. This arrangement for a magic square was

1	2	3	4
5	6	7	8
9	10	11	12
13	14	15	16

(a)

16	2	3	13
5	6	7	8
9	10	11	12
4	14	15	1

(b)

16	2	3	13
5	11	10	8
9	7	6	12
4	14	15	1

(c)

FIGURE 6.9: Constructing a magic square of order 4

16	5	9	4
2	11	7	14
3	10	6	15
13	8	12	1

FIGURE 6.10

4	9	5	16
14	7	11	2
15	6	10	3
1	12	8	13

FIGURE 6.11

2	16	13	3
11	5	8	10
7	9	12	6
14	4	1	15

FIGURE 6.12

referred to as the yin (female) square. "The diagram of the sixteen flowers," shown in figure 6.12, was constructed from an initial arrangement of the numbers 1 to 16 in four rows running from right to left.

Construction of a Magic Square of Order 5

Yang Hui provides no explanations of how he constructed magic squares of order 5 onward. However, by examining the two squares of order 5 that he included in his book, we can get an idea of this method. The first (figure 6.13c) is a magic square within a magic square, the inner one of order 3 having a constant of 39. The central number is 13, which is also the middle number of the sequence 1 to 25. The numbers in this inner square are (in ascending order) 7, 8, 9; 12, 13, 14; and 17, 18, 19; this suggests that to construct the complete magic square, we should proceed as follows.

First write the numbers from 1 to 25 in columns starting from the top right (figure 6.13a). Then proceed to make the inner square into a magic square of order 3 by following the method of forming a diamond-shaped pattern, as for the construction of the *Luo Shu*, discussed earlier, interchanging the numbers 8 and 18 and rearranging the sequence 12, 13, 14 to

21	16	11	6	1
22	17	12	7	2
23	18	13	8	3
24	19	14	9	4
25	20	15	10	5

(a)

```
           7
    14          18
17       13         9
    8           12
          19
```

(b)

1	23	16	4	21
15	14	7	18	11
24	17	13	9	2
20	8	19	12	6
5	3	10	22	25

(c)

FIGURE 6.13: Constructing a magic square of order 5

14, 13, 12, to produce figure 6.13b. Around the inner magic square arrange the other numbers in complementary pairs such as (1, 25), (2, 24) and (3, 23), such that the rows, columns, and diagonals of the outer square sum to the constant of the outer square, 65 (figure 6.13c). This arrangement of the numbers around the inner magic square follows no particular pattern, and is done by trial and error.

Yang Hui constructs magic squares up to order 10, although the squares of order 10 are "incomplete." Both the ingenuity and the "number sense" represented by these constructions are remarkable, and provide yet further evidence of the Chinese knack in computation, surely a product of their facility with rod numeral operations. The other striking feature of their work on magic squares is the crucial importance of the *Luo Shu* in all their constructions.

Figure 6.14 shows one of Yang Hui's magic squares of order 7. It encompasses three magic squares, all consistent with the *Luo Shu* principles of associated pairs and the yin-yang balance, but Yang Hui does not explain how he derived it. The name given to this magic square is *yen shu tu*, which means the "diagram of the abundant number." This must be a reference to 50, which is twice the central number 25 and also the sum of each associated pair: 46 + 4 = 50, 3 + 47 = 50, 14 + 36 = 50, and so on.

Yang Hui also provides six magic circles of varying complexity. The simplest of them, shown in figure 6.15, consists of a total of seven circles arranged in such a manner that each has four numbers on its circumference, with one other number at the center of the diagram. Each of the numbers on the central circle lies on one of the four circles that touch it. Thus, associated with each of the seven circles is a group of five numbers, and each of these groups adds up to 65. If we compare this magic circle

46	8	16	20	29	7	49
3	40	12	14	18	41	47
44	37	33	23	19	13	6
28	15	11	25	39	35	22
5	24	31	27	17	26	45
48	9	38	36	32	10	2
1	42	34	30	21	43	4

FIGURE 6.14: A magic square of order 7

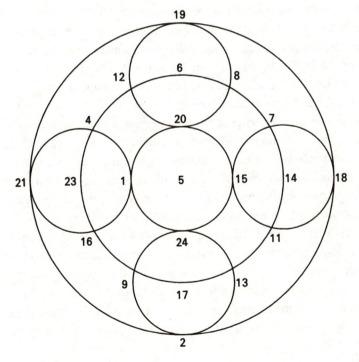

FIGURE 6.15: Magic circles

with the pattern of numbers in one of Yang Hui's magic squares of order 5 (figure 6.13c), we can see that there is a correspondence between some circles and rows.

Yang Hui's work continued to arouse interest among later mathematicians. But apart from Zhang Zhao (c. 1650), who produced the first complete magic square of order 10, and Bao Jishou (c. 1880), who constructed three-dimensional magic cubes, spheres, and tetrahedrons, there were hardly any innovations after Yang Hui. Ever more elaborate magic circles continued to be constructed, though the decline in the hold of the *Luo shu* principle may have contributed to the emergence of incomplete designs. But China's influence spread abroad to produce interesting developments in Japan and India. In Japan, Isomura's (1660) interest in magic shapes was directly derived from his work in geometry; and Seki Kowa (or more correctly Seki Takakazu) (1683), who devoted one of his seven books to the theory of magic squares and magic circles, used his algebraic talents to produce one of the first scientific treatises on the subject. These developments in Japan are well covered by Mikami (1913).

West of China, the subject of magic squares was first discussed in the Islamic world, toward the end of the ninth century AD. Possible influences from China through trade links cannot be ruled out. In the works of Thabit ibn Qurra (c. 850), al-Ghazzali (c. 1075), and al-Buni (c. 1225), and also of the Indian mathematician Narayana Pandit (c. 1350), one finds more or less the same ingredients of the occult, numerology, and combinatorial analysis as in the Chinese sources. One of the later exponents of the numerical properties of magic squares was a Fulani from northern Nigeria, Ibn Muhammad. In an Arabic manuscript written in 1732, he discusses procedures for constructing magic squares up to order 11. Al-Buni's book *Kitab al-Khawass* (The Book of Magic Properties) provided the inspiration for the first systematic treatment of magic squares in the West. This was by a Byzantine Greek, Manuel Moschopoulos (c. 1300), who described methods of arranging the numbers 1 to n^2 in a square of dimension n such that the sum of the elements in each row, column, or diagonal equals $\frac{1}{2}n(n^2 + 1)$. Like Yang Hui, who was his contemporary, Moschopoulos was interested in the mathematical rather than the magical properties of the square. His work was responsible for introducing and then popularizing magic squares in Europe—an interest that remained for many centuries.[14]

Mathematics from the *Jiu Zhang* (*Suan Shu*)

The *Jiu Zhang* is one of the oldest and certainly the most important of the ancient Chinese mathematical texts. We have no reliable knowledge of its authorship or of the exact date of its composition, but as far as can be judged it is a product of the late Qin or early Han dynasty, which places it near the beginning of the first century AD. It presents a detailed summary of contemporary Chinese mathematical knowledge, and in subsequent generations it attracted a line of distinguished commentators including Liu Hui (third century) and Yang Hui (thirteenth century), who elaborated and extended its contents and served to stimulate the creation of new topics of study.

The *Jiu Zhang* consists of nine sections or chapters, with a total of 246 problems. Some of these problems are similar to those found in the *Suanshu Shu*, mentioned earlier, although there are sufficient innovations to make it the premier text of its time. Each chapter deals with a topic in mathematics relevant to the Chinese society of the time. Information is provided through the statement of a specific problem and the answer, followed by the rule for solution, which is often terse and occasionally obscure—hence the invaluable role of the later commentators. While there is nothing in the way of algebraic notation or proofs as we understand them today, an examination of the general context in which the problems were solved firmly places the book within an algebraic/arithmetic tradition similar to that of the Mesopotamian mathematics we examined in chapter 4.

Field Measurement and Operation with Fractions

The first chapter, titled *Fang Tian* (Field Measurement or, more literally, Square Fields), begins where the *Suanshu Shu* ends. Its central theme is the calculation of areas of fields (*tian*) of different shapes, and the basic unit of measurement is the *fang*, "square unit." Correct rules are given for finding the areas of rectangles, triangles, trapezoids, and circles (π having an implicit value of 3). Liu Hui[15] was at pains to point out in his commentary that 3 is the ratio, not of the circumference, but of the perimeter of an inscribed regular hexagon to the diameter, and that the more sides the inscribed regular polygon had, the closer its perimeter would approach the circumference of the circle. However, an approximate value of 3 was sufficient for most practical purposes for which such calculations were required. Liu

Hui's remarks were taken up by Li Chunfeng and his group in composing their encyclopedia of mathematical classics in AD 686.

The chapter also contains a discussion of methods for adding, subtracting, and multiplying fractions. And here appears a rule for simplifying that is identical to that of the "repeated subtraction" algorithm found in the work of the Hellenistic mathematician Nicomachus of Gerasa, who lived around the first century AD. Consider a reducible fraction of the form m/n. The rule for the "reduction of fractions" is given as follows in the text:

> If the [numerator and denominator, i.e., m and n in our notation] can be halved, then halve them. Otherwise set down the denominator below the numerator, and subtract the smaller number from the greater number. Repeat this process to obtain the greatest common divisor (*teng*). Simplify the original fraction by dividing both numbers by *teng*.

In his commentary, Liu explains and illustrates in some detail.[16] Consider the following example.

EXAMPLE 6.1 Simplify the fraction 49/91.

Solution

Following the text, lay out the solution as follows:

49	49	7	7	7	7	7	7
91	42	42	35	28	21	14	7

So the common divisor (or *teng*) is 7, and the simplified fraction is 7/13.

The *Jiu Zhang* contains rules for adding and subtracting fractions identical to the ones we would use today if we were operating without a lowest common multiple—multiply the numerator of each fraction by the denominator of the other, and add or subtract the product before dividing the result by the product of the denominators; in modern notation,

$$\frac{a}{b} \pm \frac{c}{d} = \frac{ad \pm cb}{bd}.$$

If necessary, the resulting fraction can be simplified by using the algorithm explained above. Multiplication would proceed in the same way as today, but without cancellations. The algorithm may again be used to simplify the fraction that results. Division is not different from the way that we proceed today. It is worth noting that this algorithm has a close affinity to the Euclidean algorithm.

EXAMPLE 6.2 Given $3\frac{1}{3}$ persons share $6\frac{1}{3} + \frac{3}{4}$ coins. Tell how much does each person get? Answer: Each gets $2\frac{1}{8}$ coins.

LIU'S EXPLANATION

Multiply the numerator and denominator of the dividend [i.e., the number of coins] by the denominator of the divisor [i.e., the number of persons], and multiply those of the divisor by the denominator of the dividend. If both the dividend and the divisors are [mixed] fractions, we first convert them to improper fractions, and then multiply both the dividend and the divisor by the denominator of the other to get a uniform denominator so that the numerators can be added. Divide the divisor by the dividend and simplify and convert the answer to a mixed fraction.

IN MODERN NOTATION

Given that all fractions have been expressed as improper fractions (i.e., the number of coins is 85/12 and the number of persons is 10/3), the following steps are suggested:

Step 1.

$$\text{Divisor} = \frac{10}{3} \times \frac{12}{12} = \frac{120}{36};$$

$$\text{dividend} = \frac{85}{12} \times \frac{3}{3} = \frac{255}{36}.$$

Step 2.

$$\frac{\text{Divisor}}{\text{Dividend}} = \frac{120}{36} \div \frac{255}{36} = \frac{255}{120}.$$

Step 3. Simplifying the fraction by the algorithm as explained in example 6.1,

$$\frac{255}{120} = \frac{17}{8} = 2\frac{1}{8}.$$

Proportions and Rule of False Position

The second chapter, titled *Su mi* (Different Grains), deals with questions of simple percentages and proportions relating to these commodities.[17] The third chapter, "Distributions by Proportions" (*Shuai fen*), is concerned with the distribution of property and money according to prescribed rules, which lead, in some cases, to arithmetic and geometric progressions. Solutions often use the "rule of three" for determining proportions.[18] This rule, according to our present knowledge, was first applied in China. Here is an example from the text.

EXAMPLE 6.3 Two and one half piculs [of paddy] are purchased for 3/7 of a tael of silver. How many [piculs of paddy] can be bought for 9 taels? [A picul is a measure of weight carried by a man on his back, approximately 65 kg.]

Solution

The suggested solution, expressed in modern terms, is to let x be number of piculs of paddy bought for 9 taels of silver; applying the "rule of three" then gives

$$\frac{2\frac{1}{2}}{\frac{3}{7}} = \frac{x}{9}, \quad \text{or} \quad x = 52\frac{1}{2} \; piculs.$$

Our next example illustrates how the Chinese tackled simple problems involving series by using the rule of false position (or assumption), which we came across in the section on Egyptian algebra in chapter 3.

EXAMPLE 6.4 A weaver, improving her skills daily, continues to double her previous day's output. In five days she produces five chi of cloth [1 *chi* = 10 *cun* corresponds to a Chinese "foot" and is about 23 cm]. How much does she weave in each successive day?

Answer: On the first day she weaves 50/31; on the second day 100/31; on the third day 200/31; on fourth day 400/31; and fifth day 800/31. [This totals 1,550/31 = 50 *cun* or 5 *chi*.]

Method

Lay down the rates for distribution: 1, 2, 4, 8, and 16. Take their sum as divisor. Multiply 5 *chi* by each rate as dividend. Divide, giving the number of *chi*.

[Implied in this method is the argument of false position, which would proceed as follows: If the total output is 1, then the weaver would have produced only 1/31 of it on the first day. But since the total output is 5, the output on the first day would be 5/31. The outputs of successive days until the fifth would therefore be 10/31, 20/31, 40/31, and 80/31, which when added to 5/31 give 155/31, or 5 *chi*.]

Extraction of Square and Cube Roots

The fourth chapter, *Shao guang*, contains twenty-four problems on land mensuration (*shao* means "short," and *guang* means "width"). An important objective was to parcel out squares of land, given the area and one of the sides. This chapter is notable for the first occurrence of an important topic in the development of Chinese mathematics—how to find square and cube roots. Although the original text is very vague, later commentators on the *Jiu Zhang*, notably Liu Hui and Yang Hui, have no doubts about the geometric basis of the method as used for both square and cube roots. To illustrate the calculation of square roots, here is a problem from the chapter:

EXAMPLE 6.5 There is a [square] field of area 71,824 [square] *bu* [or paces]. What is the side of the square? Answer: 268 *bu*.

Solution

In the text only the answer is given. Fortunately, a detailed description of how the problem was solved is given in a fifteenth-century encyclopedia, *Yong Luo Da Dian*, reproduced from Yang Hui's commentary on the *Jiu Zhang*. (The Encyclopedia of Yong Le's Reign, as its title translates, originally consisted of 11,095 volumes, of which only about 370 survive. It covered almost every field of human knowledge and was compiled by over three thousand scholars under the supervision of Xie Jin.) Here we give both the algorithmic and the geometric approach so that the correspondence between the two can be easily established.

Figure 6.16 illustrates the steps of the algorithm for finding the square root of 71,824. On the left-hand side of each diagram is the algebraic rationale for the numerical calculations. In the diagrams N is a number whose square root is a three-digit integer, and α, β, and γ are the digits standing in the "hundreds," "tens," and "units" places, respectively. Thus, if the square root of N is the three-digit number *abc*, then $\alpha = 100a, \beta = 10/b$, and $\gamma = c$. Therefore

$$N = (100a + 10b + c)^2 = (\alpha + \beta + \gamma)^2$$
$$= \alpha^2 + (2\alpha + \beta)\beta + [2(\alpha + \beta) + \gamma]\gamma. \tag{6.1}$$

Continued...

Continued . . .

It is a simple matter to extend this formula to numbers with more than three digits by expanding $(\alpha + \beta + \gamma + \delta + \dots)^2$.

The Chinese method uses the above relationship but reverses the procedure and hence the ensuing calculations. The procedure is begun by finding an appropriate value for α by "inspection." It is, for example, easily deduced that $\alpha = 200$ (i.e., $\alpha = 100a$, where $a = 2$) if we are seeking the square root of $N = 71{,}824$. The procedure continues with a calculation of α^2, and this quantity is subtracted from N. We next estimate β (= $10b$), the second place of the square root, and form $(2\alpha + \beta)\beta$. We can now work out

$$N - \alpha^2 - (2\alpha + \beta)\beta = N - (\alpha + \beta)^2, \qquad (6.2)$$

and the procedure continues along similar lines until the third component on the right-hand side of equation (6.1) is calculated. If N is a perfect square, the final subtraction of this component from equation (6.2) would leave a remainder of 0.[19]

How did the Chinese apply this method in calculating the square root of 71,824? They began by laying out counting rods in four rows, as shown in figure 6.16a. The top row (*shang*) shows the result obtained at each stage of the rod operations. The second row (*shi*) contains the number on which further operations are carried out. The third row, known as the "square element" (or *fang fa*), shows the adjustments made to the element in the previous row in the process of extracting the square root. The final row has two different interpretations, depending on the context. In this context it is called a "carrying rod" row and is used to fix the positions of the digits as calculation proceeds.

This method of extracting square roots was eventually extended to the solution of quadratic equations. Indeed, a clear connection was established between the extraction of roots of any degree and the solution of equations of the same degree—at the time a unique feature of Chinese mathematics. It was eventually adopted with interesting variations by both the Koreans and the Japanese. The interested reader may wish to consult Smith and Mikami (1914) for further details.

Continued . . .

Continued . . .

The procedure used by the Chinese may be summarized in the following steps.

Step 1. Lay out the counting rods (shown as modern decimal numbers for clarity) as shown in figure 6.16a. This is the initial configuration before calculations begin. Empty cells represent zero.

Step 2. Find the value of the "hundreds" component of the square root (i.e., the value $\alpha = 100a$). In this example $\alpha = 200$ is entered in the first, "result" row. Move the "carrying rod" in the fourth row to the "ten thousands" position. Multiply the value of the "carrying rod" by $a = 2$, and place the result, 2, in the "square element" row (*fang fa*). Multiply the "square element" row by $a = 2$ and subtract the result from the "given number": $71,824 - 2(20,000) = 31,824$. The new entries are shown in figure 6.16b.

Step 3. Double the value α to get 2α (400) and enter it as the new "square element" in the third row of figure 6.16c after moving the entry one step backward. Make the necessary adjustment to the "carrying rod" by moving its entry two spaces forward. The "tens" value is then obtained by estimating the value of b so that, in this case, $4,000b + 100b^2$ is less than or equal to 31,824. The resulting entries are shown in figure 6.16c.

Step 4. The product of $b = 6$ and the "carrying rod," when added to the "square element" in the third row of figure 6.15c, gives $2\alpha + \beta$. Then β is multiplied by the new "square element" $(2\alpha + \beta)$ and subtracted from the given number $(N - \alpha^2)$. The new entries are shown in figure 6.16d.

By continuing this line of reasoning, passing through the stages shown in figures 6.16e and 6.16f, the digit in the "units" position ($\gamma = c$) is found to be 8, and thus the square root of 71,824 is obtained as 268, as shown in the top row of figure 6.16g.

Continued . . .

Continued . . .

(a)

Result (*shang*)						
Given number (*shi*)	N	7	1	8	2	4
Square element (*fang fa*)						
Carrying rod (*jie suan*)	1					1

(b)

Result (*shang*)	a			2		
Given number (*shi*)	$N - a^2$	3	1	8	2	4
Square element (*fang fa*)	a	2				
Carrying rod (*jie suan*)	1	1				

(c)

Result (*shang*)	$a + \beta$			2	6	
Given number (*shi*)	$N - a^2$	3	1	8	2	4
Square element (*fang fa*)	$2a$		4			
Carrying rod (*jie suan*)	1			1		

(d)

Result (*shang*)	$a + \beta$		2	6	
Given number (*shi*)	$N - a^2 - \beta(2a + \beta)$	4	2	2	4
Square element (*fang fa*)	$2a + \beta$	4	6		
Carrying rod (*jie suan*)	1		1		

Continued . . .

Continued . . .

(e)

Result (shang)	$a+\beta$		2	6	
Given number (shi)	$N-(a+\beta)^2$	4	2	2	4
Square element (fang fa)	$2(a+\beta)$	5	2		
Carrying rod (jie suan)	1		1		

(f)

Result (shang)	$a+\beta+\gamma$		2	6	8
Given number (shi)	$N-(a+\beta)^2$	4	2	2	4
Square element (fang fa)	$2(a+\beta)$		5	2	
Carrying rod (jie suan)	1			1	

(g)

Result (shang)	$a+\beta+\gamma$	2	6	8
Given number (shi)	$N-(a+\beta)^2-\gamma[2(a+\beta)+\gamma]$			
Square element (fang fa)	$2(a+\beta)+\gamma$	5	2	8
Carrying rod (jie suan)	1			1

FIGURE 6.16: The Chinese method of finding square roots: an algorithmic approach

A correspondence is easily established between this algorithmic approach and the geometric approach shown in figure 6.17. We begin by constructing a square with side $\alpha = 200$ *bu*, and therefore of area $A = 40,000$ square *pu*. Two additional rectangular sections, each of dimensions 200 by 60 ($B = \alpha\beta$, $C = \alpha\beta$), have a combined area of 24,000. To complete a square figure, we add a smaller square, of side 60 and

Continued . . .

Continued . . .

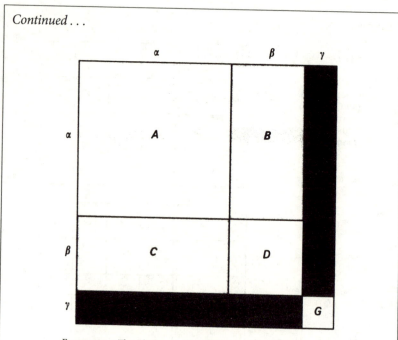

FIGURE 6.17: The Chinese method of finding square roots: a geometric interpretation

area 3,600 ($D = \beta^2$). The area of the larger square, $A + B + C + D$, is 40,000 + 24,000 + 3,600= 67,600. This falls short of 71,824, the number whose square root we are seeking, by 4,224. It is seen that this is equal to the area of two rectangular (black) strips of dimensions 260 by 8 ($E = F = (\alpha + \beta)\gamma$), and a small square of side 8 ($G = \gamma^2$): 2(260 × 8) + 82,4224. Thus the geometric representation of the procedure for extracting the square root of 71,824 is equivalent to finding the length of the side of a square of area 71,824 square *bu*. Figure 6.17 indicates that the side has length $\alpha + \beta + \gamma = 200 + 60 + 8 = 268$ *bu*.

The method of extracting cube roots is based on the following expansion:

$$(\alpha + \beta + \gamma)^3 = \alpha^3 + (3\alpha^2 + 3\alpha\beta + \beta^2)\beta + \{[(3\alpha^2 + 3\alpha\beta + \beta^2)$$

$$+ (3\alpha\beta + 2\beta^2)] + [3(\alpha + \beta)\gamma + \gamma^2)]\}\gamma$$

$$= (\alpha + \beta)^3 + [3(\alpha + \beta)^2 + 3(\alpha + \beta)\gamma + \gamma^2)]\gamma. \quad (6.3)$$

The following example from the *Jiu Zhang* illustrates the method.

EXAMPLE 6.6 Find the cube root of 1,860,867 [or, solve the cubic equation $x^3 = 1,860,867$].

Solution

The solution algorithm is begun by laying out the counting rods in five rows, as shown in figure 6.18a. The arrangement is similar to the one used for extracting square roots, except that the third and fourth rows, now known as the "upper" and "lower" elements, respectively, are for numbers obtained during the operations. For simplicity, we ignore operations with the "carrying rods."

It is easily deduced from the problem that the cube root of N is a three-digit number, abc, where the "hundreds" value is $\alpha = 100a$, the "tens" value is $\beta = 10b$, and the "units" value is $\gamma = c$. Thus the cube root of N is $\alpha + \beta + \gamma$. When we start the calculation, the first thing to note is that $a = 1$, or that $\alpha = 100$. Then we reverse the calculation implied in the identity (6.3) and arrange the rods as in figure 6.18b. The procedure continues with the calculation of $\beta = 20$, giving the configurations shown in figures 6.18c and 6.18d. Then we identify $\gamma = 3$ and obtain the final rod arrangement shown in figure 6.18e. Here the "given number" is zero, so the calculation is complete: the cube root of 1,860,867 is 123.

(a)

Result (shang)								
Given number (shi)	N	1	8	6		8	6	7
Upper element (shang fa)								
Lower element (jia fa)								
Carrying rod (jie suan)	1							1

FIGURE 6.18: The Chinese method of finding cube roots: an algorithmic approach

Continued . . .

Continued . . .

(b)

Result (*shang*)	a				1		
Given number (*shi*)	$N - a^3$	8	6		8	6	7
Upper element (*shang fa*)	$3a^2$		3				
Lower element (*jia fa*)	$3a$				3		
Carrying rod (*jie suan*)	1	1					1

(c)

Result (*shang*)	$a + \beta$					1	2
Given number (*shi*)	$N - a^3 - [3a^2 + (3a+\beta)\beta]\beta$	1	3	2	8	6	7
Upper element (*shang fa*)	$3a^2 + (3a+\beta)\beta$		3	6	4		
Lower element (*jia fa*)	$3a + \beta$				3	2	
Carrying rod (*jie suan*)	1						1

(d)

Result (*shang*)	$a + \beta$					1	2
Given number (*shi*)	$N - (a+\beta)^3$	1	3	2	8	6	7
Upper element (*shang fa*)	$3(a+\beta)^2$			4	3	2	
Lower element (*jia fa*)	$3(a+\beta)$				3	6	
Carrying rod (*jie suan*)	1						1

Continued . . .

Continued . . .

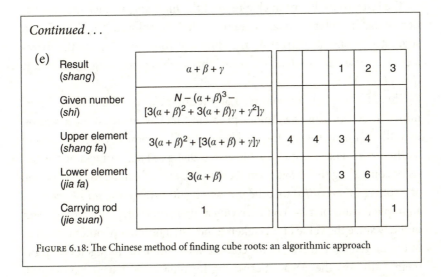

(e)

Result (*shang*)	$a + \beta + \gamma$					1	2	3
Given number (*shi*)	$N - (a + \beta)^3 -$ $[3(a + \beta)^2 + 3(a + \beta)\gamma + \gamma^2]\gamma$							
Upper element (*shang fa*)	$3(a + \beta)^2 + [3(a + \beta) + \gamma]\gamma$	4	4	3	4			
Lower element (*jia fa*)	$3(a + \beta)$					3	6	
Carrying rod (*jie suan*)	1							1

FIGURE 6.18: The Chinese method of finding cube roots: an algorithmic approach

The difficulties of presenting all the stages of the solution here should not lead us to suppose that the Chinese method was particularly laborious or time-consuming. Indeed, contemporary records suggest that solving problems of this nature was a matter of a few minutes and made use of mechanical routines (and also perhaps an auxiliary set of counting rods where necessary) that were second nature to the reckoners.

As with the procedure for square roots, it is possible to provide a geometric interpretation of this method of extracting cube roots. The resulting geometric figure (figure 6.19) is three-dimensional, and its analysis, although more complex than that for the plane region shown in figure 6.17, is along similar lines. Note that figure 6.19, from Dauben (2007, p. 248), uses a different notation so that our $a = x_1 = 1$ represents the "hundreds" digits, $b = y_1 = 2$ represents the "tens" digits, and $c = z_1$ represents the "units" digits.

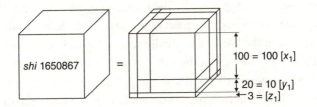

shi 1650867 $=$

$100 = 100 [x_1]$
$20 = 10 [y_1]$
$3 = [z_1]$

FIGURE 6.19: The Chinese method of finding cube roots: a geometric interpretation

The last two problems in chapter 4 of the *Jiu Zhang* involve the calculation of the diameter (d) of a sphere with a known volume (*V*). The solution suggested is one of "extracting the spherical root" to calculate the diameter. Expressed in modern notation, this involves applying the formula $d = \sqrt[3]{(16/9)\,V}$.

Thus, for a sphere of volume 4,500 (cubic) *chi*, the diameter is determined as 20 *chi*. This is a famous problem that has attracted the attention of both commentators on the *Jiu Zhang* as well as some of the historians of mathematics. For example, Yushkevich (1964, p. 61) has argued on the basis of comparing the above formula with the exact formula for the volume of a sphere ($V = \pi d^3/6$) that the unknown author(s) of the *Jiu Zhang* used an implicit value of $\pi = 3\frac{3}{8}$ in arriving at their value of *d*. This would seem unlikely since the value of π is taken to be 3 in a number of other instances in the *Jiu Zhang*. Liu Hui, in his commentary, presents an ingenious argument as to how the elegant but wrong formula could have been derived, but admits that the correct solution to the problem is beyond him. It was left to Zu Chongzhi and his son Zu Gengzhi (whose evaluation of π will be discussed in the next chapter) to arrive at the correct formula for finding the volume of the sphere. Their method bears a strong resemblance to that of the Islamic mathematician Abu Sahl al-Kuhi and the Italian mathematician Cavalieri (1598–1647) a thousand years later.[20] The Zus calculated the volume of the sphere of diameter *d* for a "precise rate" for π (i.e., $\pi = 22/7$), using the formula $V = (11/21)\,d^3$.

Engineering Mathematics

The fifth chapter of the *Jiu Zhang*, titled *Shang Gong* (or, literally, Construction Consultations), contains twenty-eight problems involving computing the volumes of a variety of three-dimensional shapes that would be familiar to the builders of castles, houses, and canals. They include the correct formulas for several solids, which were referred to by the names of common objects (see figure 6.20). The formula for the truncated triangular prism was later to appear in Adrien Marie Legendre's *Éléments de géométrie* (1794), and in the West he is usually credited with its discovery.

Particularly noteworthy is the rule for the volume of a tetrahedron (or *bienuan*) whose edges, of lengths *w* and *g*, share a common perpendicular, a third edge of length *h*. This problem is seen in modern mathematics as forming part of the theory relating to volumes of polyhedrons. The *Jiu Zhang*'s rule, in modern notation, is

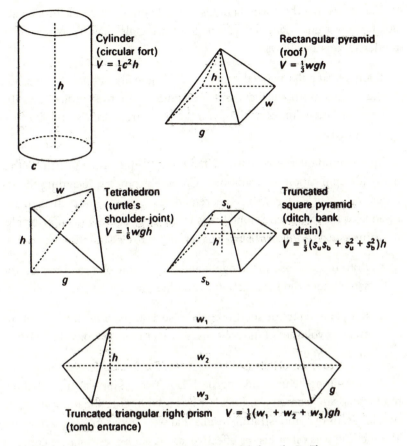

Cylinder
(circular fort)
$V = \frac{1}{4}c^2h$

Rectangular pyramid
(roof)
$V = \frac{1}{3}wgh$

Tetrahedron
(turtle's
shoulder-joint)
$V = \frac{1}{6}wgh$

Truncated
square pyramid
(ditch, bank
or drain)
$V = \frac{1}{3}(s_u s_b + s_u^2 + s_b^2)h$

Truncated triangular right prism $V = \frac{1}{6}(w_1 + w_2 + w_3)gh$
(tomb entrance)

FIGURE 6.20: Formulas for the volume V of various solids from the *Jiu Zhang*

$$V = \tfrac{1}{6}wgh.$$

In his commentary on the *Jiu Zhang*, Liu Hui explains why the volume so measured is half that of a rectangular pyramid (a *yangma*) with base wg and height h. A cuboid of sides w, g, and h can, he says, be divided into three congruent *yangma* so that the volume of each of them is one-third that of the cube ($\frac{1}{3}wgh$). He proceeds to show that a *yangma* and a *bienuan* can be slotted together to produce a right triangular prism (a *qiandu*) whose volume is $\frac{1}{2}wgh$. So the volume V of the *bienuan* is

$$V = \tfrac{1}{2}wgh - \tfrac{1}{3}wgh = \tfrac{1}{6}wgh.$$

Liu shows that this formula holds not only for a prism cut from a cube but for any right triangular prism. Liu's principle may be expressed in modern terms as follows:

> If any rectangular parallelepiped is cut diagonally into two prisms, and the prisms are further cut into pyramids and tetrahedrons, the ratio between the volumes of the pyramid and tetrahedron so produced is always 2:1.

The method of "proof" used by Liu reminds one of the principles that had wide applications in traditional Chinese geometry, including the famous *gou gu* (or Pythagorean) theorem, discussed in the next chapter. The essence of this procedure, sometimes referred to as the "out-in principle" in Chinese texts,[21] follows from two commonsense assumptions:

1. Both the area of a plane figure and the volume of a solid remain the same under rigid translation to another place.

2. If a plane figure or solid is cut into several sections, the sum of the areas or volumes of the sections is equal to the area or volume of the original figure.

The reasoning behind this approach is not very different from that behind Euclidean geometry, although it was used in Greek mathematics as a simple means of demonstrating results that were not immediately obvious. The method was often just as effective, as we shall see when we come to look at how it was applied in obtaining a proof of the Pythagorean theorem. However, it is wrong to dismiss it as a trial-and-error method, wherein the rule is merely a result of experimenting with concrete structures: the rules are stated far too precisely for that. (We have discussed this point in relation to Egyptian geometry in chapter 3.)

Li, in his commentary, uses four types of components to calculate the volume of solids: namely the cube (*lifang*), the *qiandu* (a right triangular prism), the *yangma* (a rectangular pyramid), and the *bienuan* (tetrahedron). Consider the illustration from Martzloff (1997a, p. 282) given here as figure 6.21. It shows how a cube (*lifang*) cut along its diagonals produces two solid *qiandu*, which when cut along its diagonal produces a *yangma* and a *bienuan*. A *yangma* can be cut in half to yield two *bienuan*. Reassembling these sections (or evaluating the sum of the volumes of the sections) gives the volume of the original cube. Martzloff (1997a, pp. 283–293) has an extensive discussion of various heuristic ("rule of the thumb") methods

FIGURE 6.21: Volume of a cube (Martzloff 1997a, p. 282)

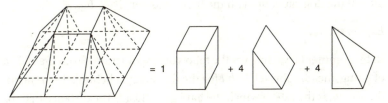

FIGURE 6.22: Volume of a square pavilion (Martzloff 1997a, p. 290)

which builds on the "out-in" principle. For example, as shown in figure 6.22, Li attempts to calculate the volume of the *fang ting* (square pavilion) by decomposing it into nine basic components: a cube, four *qiandu*, and four *yangma*. But since these components do not correspond to those which constitute the square pavilion (pyramid),[22] he proceeds to evaluate further and eventually arrives at the formula given for the volume (V) of the truncated square pyramid (or pavilion) in figure 6.22:

$$V = \tfrac{1}{3}\left(s_u s_b + s_u^2 + s_b^2\right)h,$$

where s_u and s_b are the sides of the upper and lower squares respectively and h is the height.

What the calculation of the volume of the square pavilion encapsulates is the merging of algebraic and geometric operations. And where this was not sufficient, Li would introduce the notion of passage to the limit as the case of calculating the volume of a *yangma*. Further applications and extensions of the "out-in" principle will be found in later sections of this book.

Fair Taxes

The sixth chapter, *Zhun shu* (literally, Fair Taxes), contains a medley of twenty-eight problems—some relating to the distribution of grain, some to the number of conscripts to be supplied proportionately from different sections of the population, and others to the time required to transport taxes (paid in grain) from outlying towns to the capital. Among the third group are "pursuit" problems, which were introduced into Europe by the Arabs and enjoyed considerable popularity between the twelfth and fifteenth centuries. Many of them have a hound chasing a hare, as in the following example from the chapter.

EXAMPLE 6.7 A hare runs 100 *bu* [paces] ahead of a dog. The dog pursues the hare for 250 *bu*, but the hare is still 30 *bu* ahead. In how many *bu* will the dog catch up with the hare? Answer: $107\frac{1}{7}$ *bu*.

Explanation

In his commentary, Liu gives the following explanation. From the 100 *bu* of going ahead, subtract 30 *bu* for the dog lagging behind. The remainder, 70 *bu*, is the rate at which the hare goes ahead. The distance that the dog pursues the hare is 250 *bu*. So for every 25 *bu* that the dog pursues the hare, it gains 7 *bu*. Applying the "rule of three" to 30 *bu* of lagging behind as the given number, 25 is the sought rate and 7 as the given rate, we get the answer. In other words: Answer $= (25 \times 30)/7 = 107\frac{1}{7}$ *bu*.

The last few problems in this chapter would be familiar to past generations of schoolchildren studying arithmetic.

EXAMPLE 6.8 Water flows into a cistern at different rates through five canals. Open the first canal and the cistern would be filled in 1/3 of a day; open the second, the cistern fills in a day; open the third, the cistern fills in $2\frac{1}{2}$ days; open the fourth, the cistern fills in 3 days; and open the fifth, in 5 days. If all of them are opened at the same time, how long will it take to fill the cistern? Answer: 15/74 days.

Explanation

Two methods are outlined in the *Jiu Zhang* and elaborated on by Liu. The first involves finding the sum of the number of times that the cistern is filled by each of the five canals in one day and taking the reciprocal of that sum. The second method involves arranging the information given in the example in columns, as shown in table 6.3, so the first row

Continued . . .

Continued . . .

TABLE 6.3: INFORMATION USED FOR SOLVING EXAMPLE 6.8

Canals (A)	No. of days (B_i)	Times cistern is filled (C_i)	Times cistern is filled in one day (D_i) = C_i/B_i	"Homogenization" factor $E_i = (75 \times D_i)$
1	1	3	3	225
2	1	1	1	75
3	5	2	2/5	30
4	3	1	1/3	25
5	5	1	1/5	15
	Column product $\prod_{i=1}^{5} B_i = 75$		$\sum_{i=1}^{5} D_i = 4 + \dfrac{14}{15}$	$\sum_{i=1}^{5} E_i = 370$

indicates the fact that with the first canal, it takes 1 day to fill the cistern three times; the second row represents the fact that with the second canal, it takes 1 day to fill the cistern once, and so on. The second and third columns represent the number of days and the times taken by each canal to fill the cistern respectively. The solution is arrived at by multiplying the corresponding elements of the columns and then "normalizing" to a single day and one full cistern.

Method 1: Answer $= \dfrac{1}{\sum_{i=1}^{5} B_i} = \dfrac{1}{4 + 14/15} = \dfrac{15}{74}$ part of a day.

Method 2: Answer $= \dfrac{\prod_{i=1}^{5} B_i}{\sum_{i=1}^{5} E_i} = \dfrac{75}{370} = \dfrac{15}{74}$ part of a day.

Excess and Deficit

The seventh chapter is titled *Ying Bu Zu*, which may be translated as Too Much and Not Enough. The origin of the phrase may lie in the Chinese description of the phases of the moon, from full (too much) to new (too little). There are twenty problems in this chapter in which numbers are placed on a counting board in the form of an array resembling a matrix. Operations are then performed to eliminate unknown variables such that only one will be left on the board at the end, for which a solution is obtained. Later commentators, notably Liu Hui and Yang Hui, have shown how many of these problems can be tackled by alternative methods that are less cumbersome than the ones proposed in the text. An illustration is problem 2, which may be restated as follows:

EXAMPLE 6.9 A group of people buy hens together. If each person gave 9 wen, there will be 11 wen of money left after the purchase. If, however, each person contributed only 6 wen, there would be a shortfall of 16 wen. How many persons are there in the group, and what is the total cost of hens?

Suggested Solution

[What follows is the procedure as given in the text, with some modifications for clarity.] Arrange the two types of contribution made by members of the group toward the purchase of the hens along the first row. The excess and deficit that result are arranged as a row below the first row, which contains the members' contributions. Cross-multiply diagonally. Add the products together and label the sum as *shi*. Add the excess and deficit and label the sum as *fa*. If a fraction ocurs in either *shi* or *fa*, make them have the same denominator. Divide *shi* by the difference between the two contributions to get the total cost of hens. Divide *fa* by the difference between the contributions to get the number of persons in the group. Expressed in algebraic terms, the application of the rule is simple: Let the two contributions be a and a', and let the excess and deficit be b and b', respectively;[23] then

$$\begin{pmatrix} a & a' \\ b & b' \end{pmatrix} = \begin{pmatrix} 9 & 6 \\ 11 & 16 \end{pmatrix},$$

$$\begin{pmatrix} ab' & a'b \\ b & b' \end{pmatrix} = \begin{pmatrix} 144 & 66 \\ 11 & 16 \end{pmatrix},$$

$$\begin{pmatrix} ab' + a'b \\ b + b' \end{pmatrix} = \begin{pmatrix} 210 \\ 27 \end{pmatrix}.$$

Therefore, the total cost of the hens is

Continued . . .

Continued . . .

$$\frac{shi}{a - a'} = \frac{ab' + a'b}{a - a'} = \frac{210}{3} = 70,$$

and the number of persons in the group is

$$\frac{fa}{a - a'} = \frac{b + b'}{a - a'} = \frac{27}{3} = 9$$

The above problem may be restated in terms of a system of equations in two unknowns x and y, where x is the number of persons and y is the cost:

$$ax - cy = b, \quad a'x - c'y = -b'. \tag{6.4}$$

If $a = 9$, $a' = 6$, $b = 11$, $b' = 16$, $c = 1$, and $c' = 1$, equations (6.4) become

$$9x - y = 11, \quad 6x - y = -16. \tag{6.5}$$

Equations (6.4) and (6.5) may be experrssed in matrix form as

$$\begin{pmatrix} a & -c \\ a' & -c' \end{pmatrix}\begin{pmatrix} x \\ y \end{pmatrix} = \begin{pmatrix} b \\ -b' \end{pmatrix}, \quad \begin{pmatrix} 9 & -1 \\ 6 & -1 \end{pmatrix}\begin{pmatrix} x \\ y \end{pmatrix} = \begin{pmatrix} 11 \\ -16 \end{pmatrix}.$$

Applying the rule attributed to the Swiss mathematician Gabriel Cramer (1750) for solving simultaneous equations by determinants gives

$$x = \frac{bc' + b'c}{ac' - a'c}, \quad y = \frac{ab' + a'b}{ac' - a'c},$$

so that

$$x = \frac{11 + 16}{9 - 6} = 9, \quad y = \frac{144 + 66}{9 - 6} = 70.$$

For $c = c' = 1$, the result is identical to the solution given in the *Jiu Zhang*.

So, as early as the beginning of the Christian era, a variant of Cramer's rule for solving two equations in two unknowns was known in China, though there is nothing in either this treatise or any of the subsequent commentaries that hints at an awareness of the rule for three equations in three unknowns, or of the general rule for p equations in p unknowns.[24] However, we can detect in this method an early hint of the concept of a determinant. Apparently this escaped the notice of Chinese mathematicians, but it was taken up by the Japanese mathematician Seki Takakazu (alias Seki Kowa) in 1683—ten years before Leibniz, to whom historians of mathematics usually attribute the discovery of determinants.

There was another variant of the method for tackling problems involving "excess and deficit." The "rule of double false position" was particularly popular during the period when lack of a suitable symbolic notation made the solution of even simple linear equations a difficult undertaking. This rule was brought to Europe by the Arabs and is found in the works of the ninth-century Islamic mathematician al-Khwarizmi under the name *hisab al-khataayn*. Whether this method was of Chinese origin or not is difficult to say with the existing evidence. However, what we know is that the method that was first introduced in the *Jiu Zhang* was commented on and refined by commentators from Liu Hui (c. third century) to Yang Hui (thirteenth century).

The method is best explained in present-day notation.[25] We want to find an unknown quantity x in a linear equation of the form

$$ax + b = 0.$$

Let g_1 and g_2 be two preliminary (incorrect) guesses for the value of x, and let f_1 and f_2 be the errors arising from these guesses; then

$$ag_1 + b = f_1, \qquad ag_2 + b = f_2. \tag{6.6}$$

Hence

$$a(g_1 - g_2) = f_1 - f_2. \tag{6.7}$$

The first of equations (6.6) is multiplied by g_2 and the second by g_1; subtracting the two resulting equations one from the other gives

$$b(g_2 - g_1) = f_1 g_2 - f_2 g_1. \tag{6.8}$$

Equation (6.8) is divided by equation (6.7) to give

$$x = \frac{f_1 g_2 - f_2 g_1}{f_1 - f_2}. \tag{6.9}$$

This rule is illustrated by problem 7 from the seventh chapter of the *Jiu Zhang*. The "excess and deficit" is not explicitly stated but has to be inferred from the question.

EXAMPLE 6.10 A tub of full capacity 10 *dou* contains a certain quantity of coarse [i.e., husked] rice. Grains [i.e., unhusked rice] are added to fill up the tub. When the grains are husked, it is found that the tub contains 7 *dou* of coarse rice altogether. Find the original amount of rice in the tub. [Assume that 1 *dou* of grains yields 6 *sheng* of coarse rice, where 1 *dou* is equal to 10 *sheng*.]

Suggested Solution

If the original amount of rice in the tub is 2 *dou*, a shortage of 2 *sheng* occurs; if the original amount of rice is 3 *dou*, there is an excess of 2 *sheng*. Cross-multiply 2 *dou* by the surplus 2 *sheng*, and then 3 *dou* by the deficiency of 2 *sheng*, and add the two products to give 10 *dou*. Divide this sum [i.e., 10] by the sum of the surplus and deficiency [i.e., 4] to obtain the answer: 2 *dou* and 5 *sheng*.

This method gives the same answer as the rule of double false position: equation (6.9) is used, with $g_1 = 2$, $g_2 = 3$, $f_1 = -2$, and $f_2 = 2$, to give

$$x = \frac{f_1 g_2 - f_2 g_1}{f_1 - f_2} = \frac{-10}{-4} = 2\frac{1}{2} \text{ dou.}$$

Method of Rectangular Arrays (Solution of Simultaneous Equations)

Chapter 8 in the *Jiu Zhang*, called *Fang cheng* (Method of Rectangular Arrays) contains one of the more innovative parts of Chinese mathematics. It deals with the solution of simultaneous equations with two to five unknowns by placing them in a table and then operating on the columns in a way that is identical to the row transformations of the modern matrix method of solution. A notable feature of this method is that it is just as easy to use with negative as with positive numbers. The best way to explain it is by examples, and we begin with problem 14 from chapter 7. (There are strong affinities between the solutions we discussed from chapter 7 of the text and those in chapter 8, and for that reason a problem similar to one discussed from chapter 7 was chosen to illustrate this chapter's method of "rectangular array.")

EXAMPLE 6.11 5 large containers and 1 small container have a total capacity of 3 *hu*; 1 large container and 5 small containers have a capacity of 2 *hu*. Find the capacities of 1 large container and 1 small container. [1 *hu* = 10 *dou*.]

Solution

The method of tables starts by setting up the information contained in the problem in the form of a matrix:

Large containers $\begin{pmatrix} 1 & 5 \\ 5 & 1 \\ 2 & 3 \end{pmatrix}.$
Small containers
Total capacity

Step 1. Multiply the first column by 5 and then subtract the second column from the result. Put this down as the first column of the next matrix:

Large containers $\begin{pmatrix} 0 & 5 \\ 24 & 1 \\ 7 & 3 \end{pmatrix}.$
Small containers
Total capacity

Step 2. Multiply the second column by 24 and then subtract the first column from the result. Put this down as the second column of the next matrix:

Large containers $\begin{pmatrix} 0 & 120 \\ 24 & 0 \\ 7 & 65 \end{pmatrix}.$
Small containers
Total capacity

Thus a small container has a capacity of 7/24 *hu*, and a large container has a capacity of 65/120 or 13/24 *hu*.

There are in all eighteen problems requiring the method of rectangular arrays to be used to solve systems of simultaneous linear equations with up to five unknowns. The text explains how counting rods can be set up for column operations that would finally yield the solutions. Negative numbers that appear in the course of such operations do not pose any problem, since they could be represented (as we saw earlier in this chapter) by black rods, red rods, being used for positive numbers. Here is an example where there are three unknowns and negative quantities appear in the course of solution:

EXAMPLE 6.12 The yield of 2 sheaves of superior grain, 3 sheaves of medium grain, and 4 sheaves of inferior grain is each less than 1 *dou*. But if 1 sheaf of medium grain is added to the superior grain or if 1 sheaf of inferior grain is added to the medium, or if 1 sheaf of superior grain is added to the inferior, then in each case the yield is exactly 1 *dou*. What is the yield of one sheaf of each grade of grain?

Solution

In modern notation, the solution begins by letting x, y, and z be the yields from 1 sheaf of superior, medium, and inferior grain respectively. The question may then be posed as follows: given $2x \leq 1$, $3y \leq 1$, and $4z \leq 1$, solve

$$2x + y = 1, 3y + z = 1, 4z + x = 1.$$

The information contained in the problem is first arranged on a counting board so that it resembles a matrix:

$$\begin{pmatrix} 1 & 0 & 2 \\ 0 & 3 & 1 \\ 4 & 1 & 0 \\ 1 & 1 & 1 \end{pmatrix}.$$

The column on the extreme right contains the coefficients and constant of the first equation $2x + 1y + 0z = 1$, and similarly for the other columns. The extreme left column is multiplied by 2, and the extreme right subtracted from the result; the extreme left column is replaced by the new column. We then have

$$\begin{pmatrix} 0 & 0 & 2 \\ -1 & 3 & 1 \\ 8 & 1 & 0 \\ 1 & 1 & 1 \end{pmatrix}.$$

The extreme left column is multiplied by 3, and the middle column is added to the result:

$$\begin{pmatrix} 0 & 0 & 2 \\ 0 & 3 & 1 \\ 25 & 1 & 0 \\ 4 & 1 & 1 \end{pmatrix}.$$

Continued . . .

Continued . . .

Back substitution is now needed to obtain the full solution of the problem. Thus, given 1 sheaf of inferior grain yields 4/25 *dou*, then 3 sheaves of medium grain yield $1 - 4/25 = 21/25$ *dou*, or 1 sheaf of medium grain yields 7/25 *dou*. Also, given 1 sheaf of medium grain yields 7/25 *dou*, then 2 sheaves of superior grain yield $1 - 7/25 = 18/25$, or 1 sheaf of superior grain yields 9/25 *dou*.

Hence:

25 sheaves of inferior grain yield 4 *dou*, so 1 sheaf of inferior grain yields 4/25 *dou*.

3 sheaves of medium grain and 1 sheaf of inferior grain yield 1 *dou*, so 1 sheaf of medium grain yields 7/25 *dou*.

2 sheaves of superior grain and 1 sheaf of medium grains yield 1 *dou*, so 1 sheaf of superior grain yields 9/25 *dou*.

This method of solving simultaneous linear equations is essentially the same as the one we use today, the development of which is attributed in the West to the famous German mathematician Karl Friedrich Gauss (1777–1855). But, over fifteen hundred years before Gauss, Chinese mathematicians were using a variant of one of Gauss's methods. It is interesting to note that a variant of Gauss's elimination procedure was proposed by Tobias Mayer (1723–1762), an earlier German contemporary of Gauss. The three procedures, including the Chinese, are algebraically similar but computationally distinct. The principal defect of the Chinese method is that it uses only whole numbers in the cells, and these can become very large in relatively simple problems.[26] This innovation, together with the use of a special case of Cramer's rule for tackling certain problems in the previous chapter of the *Jiu Zhang*, raises some interesting questions about the subsequent treatment of these subjects in Chinese mathematics. First, despite the early promise, there was no work on determinants until the Japanese mathematician Seki Takakazu, working very much within the Chinese mathematical tradition, developed the concept of a determinant in his book *Kai Fukudai no Ho* (1683). There is a brief discussion of Seki Takakazu's work in the next chapter, which contains a survey of Japanese mathematics.

Second, this way of tackling simultaneous equations is not found in any other mathematical tradition until the advent of modern mathematics. We are therefore driven to the conclusion that the method may have been a logical outcome of rod numeral computational techniques. However, it has been argued that the reliance on these very techniques inhibited the development of abstract algebra—a prerequisite for more advanced work on matrices and determinants.

The ninth and last chapter of the *Jiu Zhang* is called *Gou gu* (Base-Height). In the Chinese mathematical literature the shortest side of a right-angled triangle was called the *gou*, the longer side the *gu*, and the hypotenuse the *xian*. Twenty-four problems on right-angled triangles[27] are presented in this chapter. The demands of accurate surveying and astronomical observation must have required an understanding and application of the Pythagorean theorem well before the *Jiu Zhang* was written, and this would explain its presence in the earlier text, the *Zhou Bi Suan Jing*. We shall examine this subject in chapter 7.

We have devoted many pages to the *Jiu Zhang*. The range of topics it covers is impressive, and indicates the level of sophistication reached by Chinese mathematicians at the beginning of the Christian era. It is one of the oldest mathematical texts in the world, with problems more varied and richer than in any Egyptian or Mesopotamian text. But although it was to have a powerful influence on the course of Chinese mathematics, with a number of notable mathematicians writing commentaries on it, by becoming a classic it also acted as an impediment to progress. Its influence on the Song mathematicians of the thirteenth century was perhaps even counterproductive, since they were obliged to refer to it, just as some of today's academics routinely cite standard authorities to make their work "respectable." (There is a parallel too with generations of students in the West being taught from Euclid's *Elements*, a practice abandoned only in the twentieth century.) When the *Jiu Zhang* was written, the status of mathematicians was high. This status was gradually eroded as mathematics came to be perceived as a diligent and unquestioning application of ancient wisdom rather than a process of building on the solid foundations that the early texts had laid.

Notes

1. However, Cullen (1996, p. 1) argues that the *Zhou Bi* "can best be understood as a product of the Han dynasty," which would place it about fifteen hundred years later!

2. It may be remembered from the discussion of the Plimpton Tablet in chapter 4 that the Babylonians were aware of the existence of the Pythagorean result much earlier than both the Greeks and the Chinese.

3. A few sections of the text are marked with the common surnames Wang and Yang. However, these are most likely names of scribes or reviewers rather than the names of authors because their location in the text of *Suanshu Shu* corresponds to what are found in many other rod manuscripts in the Han period. Generally speaking, names of authors are mentioned before texts and not inside.

4. For further details, see Dauben (2007, 2008).

5. These translations include English (Shen et al., 1999), Russian (Berezkina 1957), German (Vogel 1968) and French (Chemla and Guo 2004).

6. Swetz (2008) contains an interesting discussion of the mathematics and metaphysics of the classical 3 × 3 magic square.

7. One example of China briefly looking "outwards" is the naval expedition of Zheng He. Nearly ninety years before Christopher Columbus, in 1405, he led the first expedition (there were six more to follow) of almost four hundred ships with a crew of over thirty-seven thousand and a flagship (the *Treasure Ship*) measuring 450 feet long with nine masts. These expeditions would explore regions of the southern seas (as far as Java), the coasts of eastern Africa, and the long shorelines of Africa, Asia, and Eurasia. These expeditions were not undertaken with any "commercial" or "evangelical" goal in mind. Their purpose was to show the outside world the glories of Ming China. Zheng He's treasure ship contained large quantities of gold and silver pieces, quantities of silk, and perfume jars and other gifts that were distributed among those whom the explorers came across during the journeys. However, this short-lived interest in foreign climes disappeared as a result of changes of personnel in the imperial court with the pendulum swinging the other way, so that imperial edicts appeared banning any maritime travel outside the immediate Chinese coasts at the pain of being put to death. For a fuller account of this fascinating episode in Chinese history, see Dreyer (2006).

8. Some of the oracle bones also contain early records of celestial events such as lunar and solar eclipses; and one bone in particular may record the earliest-known observation of a supernova, which occurred in the thirteenth century BC. For further details, see Keightley (1985).

9. In Sun Zu's *Mathematical Manual* (AD 400), the following verse explains this method of recording numbers:

> Units are vertical, tens are horizontal,
> Hundreds stand, thousands lie down,
> Thus thousands and tens look the same,
> Ten thousands and hundreds look alike.

It is interesting in this context to note that a widespread knowledge of how to use the counting rods must have been assumed, since in only two of the ancient texts is there any explanation of computation with counting rods. One of the texts is that of Master Sun referred above.

10. There is a view that counting rods were manipulated on a specially devised surface such as a counting board or other similar frames. However, as Martzloff (1997a, p. 209) points out, there is no evidence of the existence of such boards.

11. The comment of a scholar relating to this comparison of a blank space in Mesopotamian numerals with that of the Chinese system is illuminating. "Chinese or not, a blank space cannot be isolated in itself. We cannot list it alone in a dictionary for example. Moreover, it necessarily exists absolutely everywhere, for example in front of, after, and inside any number between successive digits, between words too, and inside words, as well, regardless of the intention of the person who has unavoidably 'written' it (and not only in the case of writing but also with rod digits as well). In other words, a void space is compulsory in a large number of situations where no zero is involved at all and it is not liable to be interpreted correctly without an associated counting board or any other such exterior help, while this is not the case for non-zero digits."

12. For further details of this conjecture see Lam (1987), Lam and Ang (1992), and Hodgkin (2005, pp. 86–88).

13. Figurate numbers can be represented by a geometrical array of dots; the first four triangular numbers, for example, are 1, 3, 6, and 10. The meanings of perfect and amicable numbers will be made clear in chapter 11, on Islamic mathematics.

14. For further details, see "The Magic Squares of Manuel Moschopoulos" at http://convergence.mathdl.org.

15. The importance of Liu Hui cannot be overstated. As Shen et al. (1999) states: "The *Chiu Chang* [*Jiu Zhang*] would have remained a mere recipe book and not a complete classical mathematical textbook without Liu's work" (p. 5).

16. This is the same as Euclid's "antiphairesis": starting with two homogeneous quantities, one subtracts successively the smaller of two quantities from the larger one, so that with each subtraction, the excess takes the place of the larger quantity while the smaller one stays unchanged. The final result consists of a couple of rational numbers expressing the relation between the quantities. This procedure provides a systematic way for finding the highest common factor (HCF) of two given positive integers.

17. The translation used in the earlier editions of this book was "Millet and Rice." However, rice does not figure in the list of grains given in the conversion table at the beginning of this chapter in the *Jiu Zhang*. Hence, the translation that is now favored is of different types of grain such as shelled or hulled or dried. A table of exchange rates for bartering different kinds of millet, beans, wheat, sesame seeds, and malt is offered in

Liu's commentary. This table contains an item known as paddy or rough rice, which has its husk and bran. For further details, see Martzloff (1997a, p. 132).

18. The "rule of three" is a method for solving proportions using algebra. The rule states that if you know three numbers a, b, and c, and want to find d such that $a/b = c/d$ (or $a{:}b = c{:}d$), then $d = cb/a$.

19. Although all numbers in the relevant problems in the *Jiu Zhang* turn out to be perfect squares, the commentators were aware that they would at times need to extract square roots of other numbers. For example, Liu Hui in discussing the extraction of the square root of 10 suggests that "the more the digits, the finer the fractions, till the number omitted from the area of the red areas is negligible." The color referred to appears in Liu Hui's attempt at explaining geometrically the iteration procedure implied in this method of root extraction. In relation to our example, Liu represents the first approximation of the square root as the "yellow" area, the second approximation as the "red" area, and the final approximation as the "blue" area. An interesting question that arises is whether Liu Hui was aware that the square root of 10 could not be exactly determined (or in modern parlance it was an irrational number). It would seem that the jury is still out!

20. Al-Kuhi's incorrect derivation of the spherical volume is contained in his correspondence with al-Sabi and discussed by Berggren (1983) in his translation of that correspondence. For a detailed discussion of the Chinese work on the volume of the sphere, see Wagner (1978).

21. The term "out-in" comes from labeling the portion of a rectangle below the diagonal as "out" and that above the diagonal as "in." Depending on the application of this principle, the portions of the areas considered "out" have their equivalence in areas that are considered "in." This is referred to today as the "dissection and reassembly" principle.

22. Note that the central figure in figure 6.22 is not a cube unless the height is equal to the length of the upper side.

23. Note that b' is positive even though it is a deficit.

24. Cramer's rule should really be regarded as an orthogonalization procedure rather than as a procedure for solving systems of linear equations. Bill Farebrother worked out that attempting to solve a system of 240 equations in 240 unknowns by Cramer's rule using the University of Manchester's CDC 7600 mainframe computer would take about 150,000 years, since the computer was able to perform only one million multiplications per second!

25. It is worth noting that the method described is useful in problems where a and b are unknown (but determined), as in the problem that follows. Otherwise we would just use $x = -b/a$. This method makes it necessary to find a and b first.

26. For further discussion of these and other procedures, see Farebrother (1999).

27. There was no specific term for a "triangle" in the Chinese mathematical literature. So by inference, since the "base-height" relationship is expressed in connection with a hypotenuse, the subject of this chapter is the right-angled triangle.

Chapter Seven

Special Topics in Chinese Mathematics

The last half of the thirteenth century and the early fourteenth marked the culmination of over a thousand years of development of Chinese mathematics, built on the solid foundation of the *Jiu Zhang*. In the historical introduction to the previous chapter we saw that the Song period produced outstanding scientific and technological achievements. Four of the greatest Chinese mathematicians—Qin Jiushao, Li Ye, Yang Hui, and Zhu Shijie—lived during this period, and there were more than thirty mathematical schools scattered across the country. As in the Tang dynasty, when a number of ancient texts were collected and then designated as classics under the leadership of Li Chunfeng (AD 656), the same compendium of classics, *Suan Jing Shi Shu* (The Ten Mathematical Manuals), became for a short period the recommended text to set the standards for teaching at the Imperial Academy and for the evaluation of major examinations. However, mathematics as a subject tended to fall in and out of favor depending on the whims of the Imperial Court. And in the case of the Southern Song dynasty, the subject was finally removed from the curriculum of the civil service examinations, not to be reintroduced again. At the same time, mathematics was finding more practical applications in a widening number of disciplines—calendar making, surveying, chronology, architecture, and meteorology, as well as areas relating to trade and barter, the payment of wages and taxes, and simple mensuration.

The most striking feature of the Chinese mathematics of this period is its essentially algebraic character; there was little in contemporary geometry, particularly mensuration, that was not to be found in the *Jiu Zhang* or its early commentaries. Many of the algebraic innovations of the period were extensions of previous work. They fall into three main categories:

1. *Numerical equations of higher order.* Although the procedure for solving higher-order equations had its origins in the method of extracting square and cube roots found in the *Jiu Zhang*, discussed in the

previous chapter, its fullest development is contained in the works of Qin Jiushao (c. 1202–1261). The breakthrough came with the use of what we know as Pascal's triangle for the extraction of roots. Not until the beginning of the nineteenth century would European mathematicians, notably Horner and Ruffini, make any substantial progress in this area.

2. *Pascal's triangle.* Although this triangular array of numbers is named after the seventeenth-century French mathematician Blaise Pascal, it received detailed treatment in the hands of Yang Hui and Zhu Shijie, some 350 years previously. And Yang reported that his discussion of Pascal's triangle was derived from an earlier work by Jia Xian (c. 1050), which has not survived. Jia had differentiated between two methods of extracting square and cube roots. The first, known as *zeng cheng fang fa*, the method of "extraction by adding and multiplying," is similar to the method examined in the previous chapter. The second was known as *li cheng shi shuo*, "unlocking the coefficients by means of a chart," which uses binomial coefficients taken from Pascal's triangle to solve numerical equations of higher order.

3. *Indeterminate analysis.* Interest in this subject arose in both China and India in connection with calculations in calendar making and astronomy. The basic problem was one of finding a procedure for solving a system of n equations with more than n unknowns. At the simplest level, how does one go about solving an equation in two unknowns,

$$3x + 8y = 100,$$

where the solutions for x and y are either real numbers or positive integers? Such problems were successfully tackled in Europe only in the eighteenth century, by Euler and Gauss.

There were other specific developments worthy of note, though they will not be examined in any detail in this book. They include

1. The derivation of a cubic interpolation formula popularized by Guo Shoujing (c. 1275), which later came to be known in the West as the Newton-Stirling formula

2. The *tien yüan* notation for nonlinear equations, first used by Li Ye and Zhu Shijie, which made possible the development of algorithms for solving different types of equation

3. The use of geometric methods to study mathematical series by a number of mathematicians of the period (We shall also examine Indian work in this area in chapter 9.)

There were also two subjects of a geometric character to which the Chinese made significant contributions: they investigated the properties of right-angled triangles, and they continued the age-old search for more accurate estimates of π. Chinese interest in both these subjects goes back to well before AD 1000, in the former case to one of the earliest sources of Chinese mathematics, the *Zhou Bi*.

The "Piling-Up of Rectangles": The Pythagorean Theorem in China

The Pythagorean theorem is generally held to be one of the most important results in the early history of mathematics. From it came important discoveries in theoretical geometry as well as practical mensuration. We saw in chapter 4 how the Mesopotamians' understanding of geometry, based on similar triangles and circles, was enhanced by the discovery of the Pythagorean result, and how their algorithmic procedure for extracting square roots of "irregular" (irrational) numbers was also based on this result. In China too, a study of the properties of the right-angled triangle had a considerable impact on mathematics.

The earliest extant Chinese text on astronomy and mathematics, the *Zhou Bi*, is notable for a diagrammatic demonstration of the Pythagorean (or *gou gu*) theorem. Needham's translation of the relevant passage is illustrated by figure 7.1a, drawn from the original text. The passage reads:

> Let us cut a rectangle (diagonally), and make the width 3 (units) wide, and the length 4 (units) long. The diagonal between the (two) corners will then be 5 (units) long. Now, after drawing a square on this diagonal, circumscribe it by half-rectangles like that which has been left outside, so as to form a (square) plate. Thus the (four) outer half-rectangles, of width 3, length 4 and diagonal 5, together make two rectangles (of area 24); then (when this is subtracted from the square plate of area 49) the remainder is of area 25. This (process) is called "piling up the rectangles." (Needham 1959, pp. 22–23)

In terms of figure 7.1b, the larger square ABCD has side 3 + 4 = 7 and thus area 49. If, from this large square, four triangles (AHE, BEF, CFG,

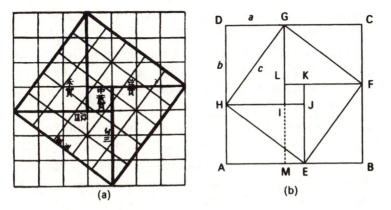

FIGURE 7.1: The *gou gu* (Pythagorean) theorem: (a) the original illustration from the *Zhou Bi* (Needham 1959, p. 22), and (b) the modern translation

and DGH), making together two rectangles each of area $3 \times 4 = 12$, are removed, this leaves the smaller square HEFG. And implicitly,

$$(3 + 4)^2 - 2(3 \times 4) = 3^2 + 4^2 = 5^2.$$

The extension of this "proof" to a general case was achieved in different ways by Zhao Zhujing and Liu Hui, two commentators living in the third century AD. In modern notation, Zhao's extension may be stated thus: if the shorter (*gou*) and longer (*gu*) sides of one of the rectangles are a and b respectively, and its diagonal (*xian*) is c, then the above reasoning would produce

$$c^2 = (b - a)^2 + 2ab = \text{square IJKL} + \text{rect DGIH} + \text{rect CFLG}$$
$$= \text{square AMIH} + \text{square MBFL}$$
$$= a^2 + b^2.$$

An alternative explanation is based on the identity

$$(a + b)^2 = a^2 + 2ab + b^2;$$
$$c^2 = (a + b)^2 - 2ab = a^2 + b^2$$
$$= \text{square ABCD} - 4\triangle\text{DGH}.$$

A geometric interpretation of this identity is fairly easily established. The result was certainly known to the authors of the *Sulbasutras* (c. 500 BC) of Vedic India, which we shall examine in the next chapter. There is also the possibility that it was known to the Babylonians of the

Hammurabi dynasty. Later, geometric proofs of this identity are found in Euclid's *Elements* (c. 250 BC) and al-Khwarizmi's *Algebra* (c. AD 800). The Chinese may have deduced the identity from the drawing reproduced in figure 7.1a itself, where two squares of areas $a^2 = 4^2$ and $b^2 = 3^2$, together with two rectangles of area ab, together make up the large square of area $(a + b)^2$.

However, there is a third explanation, found in Liu's commentary, which does not refer to the diagram in the *Zhou Bi* (i.e., figure 7.1a) but is based on the "out-in" technique taken as an axiom by Chinese mathematicians and discussed in the previous chapter. We saw that this technique is based on two commonsense assumptions:

1. Both the area of a plane figure and the volume of a solid remain the same under rigid translation to another place.

2. If a plane figure or solid is cut into several sections, the sum of the areas or volumes of the sections is equal to the area or volume of the original figure.

If these conditions hold, it is possible to infer simple arithmetic relations between the areas or volumes of various sections of the plane or solid figures resulting from dissection or reassembly. It was this principle that Liu used to "prove" the Pythagorean theorem. Liu also refers to a diagram that is now lost. Lam and Shen's (1984, p. 95) translation of the relevant passage from Liu reads:

> Let the square on *gou* (*a*) be red and the square on *gu* (*b*) be blue. Use the [principle] of mutual subtraction and addition of like kinds to fit into the remainders, so that there is no change in [area on the completion of] a square on the hypotenuse (*c*).

The principle referred to is the "out-in" principle. There have been a number of attempts to construct the missing diagram that accompanied this statement. One of the more plausible is shown in figure 7.2. Here ABC is a space for a right-angled triangle. BCIS is the red square on the *gou* (or *a*), while AQDC (= FERI) is the blue square on the *gu* (or *b*). From SB-CFER cut off the triangle GBS and put it in the space ABC. Next, cut off the triangle EGR and place it over the triangle EAF. What we have now is the square AEGB, which is the square on the hypotenuse (or *c*). This completes the "proof" of the *gou gu* theorem.

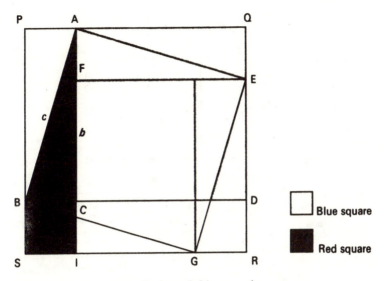

FIGURE 7.2: A reconstruction of Liu's proof of the *gou gu* theorem

Yet another reconstruction of this diagram, shown in figure 7.3, would involve covering the square on the hypotenuse with pieces cut from the red and blue squares.

Figure 7.3 shows the original right-angled triangle (in bold) together with two squares on the sides *gou* and *gu* of the right angle. Following Liu, these squares are color-coded blue and red on the diagram. By construction, initially the square on the hypotenuse (*xian*) is already partly covered by sections of the red and blue squares. To show that these squares together cover the same space as that of the square on the hypotenuse, we need to remove the pieces of the jigsaw (identified by the word "remove") and replace them by congruent pieces (identified by the word "insert"). Once the process of substitution is complete, we have demonstrated the validity of the Pythagorean result using the Chinese "out-in" technique.

It is important to recognize a basic difference between this Chinese proof and the Euclidean proof of the Pythagorean theorem. Considerable geometric knowledge of properties relating to congruent triangles and areas is required to understand the Euclidean proof, which probably explains why the theorem does not appear in Euclid's *Elements* until the end of Book I. The Chinese proof is a matter of common sense, enabling the theorem to be applied to many practical problems with relative ease.[1] We now examine a few of these applications.

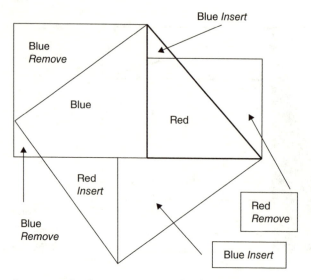

FIGURE 7.3: Another reconstruction of Liu's proof of the *gou gu* theorem

Applications of the Gou Gu Theorem

The *gou gu* theorem is applied via various permutations built on the relation *gou*² + *gu*² = *xian*², or $a^2 + b^2 = c^2$. The reader is invited to solve the following problems, which are mainly from the ninth chapter of the *Jiu Zhang*, before consulting the solutions given below them.

EXAMPLE 7.1   Under a tree 20 *chi* high and 3 *chi* in circumference, there grows an arrowroot vine that winds seven times round the stem of the tree and just reaches its top. How long is the vine? [1 *chi* is about 23 cm.]

The problem is of the form: given *a* and *b*, find *c* (see figure 7.4).

Suggested Solution

Take 7 × 3 = 21 as one of the sides (*a*) of a right-angled triangle. Take the height of the tree as a second side (*b*). Find the hypotenuse (*c*), which is then the length of the vine:

$$\text{length of vine} = \sqrt{(21^2 + 20^2)} = 29 \ chi.$$

Continued . . .

Continued . . .

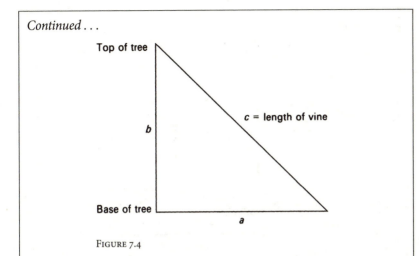

Top of tree

b

c = length of vine

Base of tree

a

FIGURE 7.4

EXAMPLE 7.2 There is a rope hanging from the top of a tree with 3 *chi* of it lying on the ground. When it is tightly stretched, so that its end just touches the ground, it reaches a point 8 *chi* from the base of the tree. How long is the rope?

The problem is of the form: given *b* and *c* − *a*, find *c*.

Suggested Solution

$$\text{Length of rope} = \frac{1}{2}\left(\frac{64}{3} + 3\right) = \frac{73}{6} = 12\frac{1}{6} \ chi.$$

This is an interesting solution, displaying a considerable degree of sophistication. With modern notation and figure 7.5, we can attempt to reconstruct the algebraic route underlying the solution offered above. Let

c = length of rope,

d = *c* − *a* = length of rope on the ground,

b = distance of end of rope, if tightly stretched, from base of tree.

The following relationship may be deduced from figure 7.5:

$$c^2 = (c − d)^2 + b^2.$$

Continued . . .

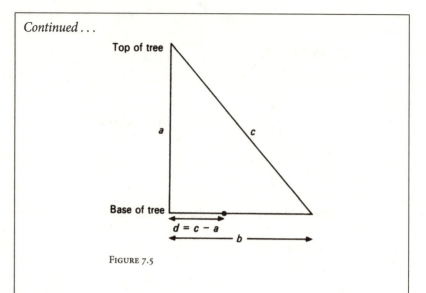

FIGURE 7.5

Therefore

$$0 = -2cd + d^2 + b^2;$$
$$c = \tfrac{1}{2}\left(b^2/d + d\right).$$

And it was this expression for c that could have been used to solve this problem.

However, Martzloff (1997a, p. 95) draws our attention to two problems similar to this example, one of which appears in Mesopotamian and the other in Chinese mathematics. A comparison between the two could be instructive.

EXAMPLE 7.3 Mesopotamian problem (c. 1800–1600 BC): A reed is placed vertically against a wall. If it comes down by 3 cubits, it moves away by 9 cubits. What is the [length] of the reed, what is the [height] of the wall? Answer: 15 cubits.

EXAMPLE 7.4 Chinese problem (from *Jiu Zhang*, chapter 9, 200 BC–AD 200): Suppose a wall is 10 *chi* high. A wooden pole (tree) is rested against

Continued . . .

Continued...

it so that its end coincides with the top of the wall. If one steps backward a distance of one *chi* pulling the pole, the pole falls to the ground. What is the length of the pole?

The similarity between the statement of the two problems is immediately clear. But what is interesting is that the rules for their solution are identical. This is easily established if the rules are expressed in modern symbolic notation:

$$\text{Length of the reed} = c = \frac{a^2 + d^2}{2d} = 15 \text{ cubits,}$$

where $a = 9$ cubits and $d = c - a = 3$ cubits.

$$\text{Length of the wooden pole} = c = \frac{1}{2}\left(\frac{a^2}{d} + d\right) = 50\frac{1}{2} \text{ chi,}$$

where $a = 10$ *chi* and $d = 1$ *chi*.

This similarity is found in a number of other comparisons made by van der Waerden (1983) and others, which should make us cautious about overemphasizing the algebraic/arithmetic nature of early Chinese mathematics. It is more likely than not that the similarities between the Mesopotamian and the Chinese approaches to mathematics show a strong geometric bent in both traditions, not only in the way that problems were stated but also in the way that the solutions were arrived at.

EXAMPLE 7.5 The height of a door is 6 *chi* 8 *cun* larger than its width. The diagonal is 10 *chi*. What are the dimensions of the door? [1 *chi* = 10 *cun*.]

The problem is of the form: given $b - a$ and c, find a and b (see figure 7.6).

Suggested Solution

From the square of 10 *chi*, subtract twice the square of half the given difference (6 *chi* 8 *cun*). Halve this result, and find its square root. The

Continued...

Continued . . .

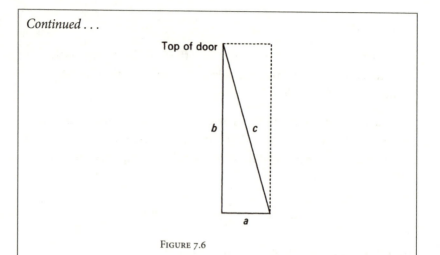

FIGURE 7.6

width of the door is equal to the difference of this root and half of 6 *chi* 8 *cun*. The height of the door is equal to the sum of the root and half of 6 *chi* 8 *cun*. Thus

width = 2 *chi* 8 *cun*,
height = 9 *chi* 6 *cun*.

The algebraic basis of this solution is derived from the equation (in modern notation)

$$(a + b)^2 = 2c^2 - (b - a)^2,$$ (7.1)

where a is the width of the door, b the height of the door, and c the diagonal. This is easily seen, since

$$(a + b)^2 + (b - a)^2 = 2a^2 + 2b^2 = 2c^2.$$

However, Qin Jiushao gave a geometric demonstration (discussed by Lam and Shen [1984]).

From the numerical values given in the problem, it is easily established that $b - a = 6.8$ *chi* and $c = 10$ *chi*. Substituting these values into equation (7.1) and taking the square root gives

$$a + b = \sqrt{2(10)^2 - (6.8)^2}.$$ (7.2)

Adding $b - a = 6.8$ to equation (7.2) and halving gives the value of b; subtracting the same quantity and halving gives the value of a.

Continued . . .

Continued . . .

EXAMPLE 7.6 There is a bamboo 10 *chi* high, the upper end of which, being broken, touches the ground 3 *chi* from the foot of the stem. What is the height of the break?

This is a famous problem in the history of mathematics. Figure 7.7a shows the problem as illustrated in Yang Hui's *Xiang Jie Jiu Zhang Suan Fa Zuan Lei* (1261). It kept reappearing in the works of Indian mathematicians, from Mahavira in the ninth century to Bhaskaracharya in the twelfth century, and eventually in European works, probably thus charting a westward migration of Chinese mathematics via India and the Islamic world.

The problem is of the form: given a and $b + c$, find b.

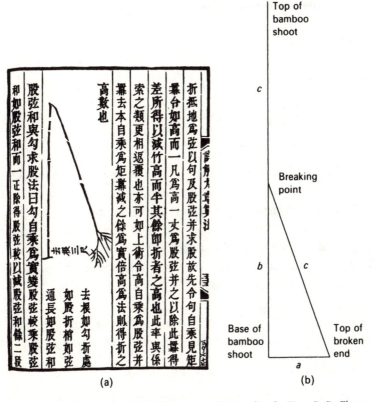

(a)　　　(b)

FIGURE 7.7: The "broken bamboo" problem: (a) as illustrated in the *Xiang Jie Jiu Zhang Suan Fa Zuan Lei* (Needham 1959, p. 28), and (b) the modern translation

Continued . . .

Continued...

Suggested Solution

Take the square of the distance from the foot of the bamboo to the point at which its top touches the ground, and divide this quantity by the length of the bamboo. Subtract the result from the length of the bamboo, and halve the resulting difference. This gives the height of the break.

These instructions may be expressed in modern notation, with reference to figure 7.7b, as follows. Let

a = distance from foot of bamboo,

$b + c$ = length of bamboo,

b = height of erect section of bamboo.

Then the above rule is equivalent to

$$b = \frac{1}{2}\left(b + c - \frac{a^2}{b + c}\right),$$

which yields

$$b = \frac{1}{2}\left(10 - \frac{9}{10}\right) = \frac{91}{20} \ chi.$$

Problem 16 of the ninth chapter of *Jiu Zhang* contains a circle inscribed in a triangle between the *gou, gu,* and *xian* sides of the familiar figure. The problem is one of calculating the diameter of the inscribed circle. Although the dimensions given here are *gou* = 8 and *gu* = 15, Liu in his commentary generalizes the solution for any *gou gu* figure. We will consider his explanation briefly. Further details are found in Dauben (2007, pp. 287–88) and Martzloff (1997a, pp. 300–301).

EXAMPLE 7.7 Given a [right-angled triangle whose] *gou* and *gu* are 8 and 15 respectively. Tell: what is the diameter of the inscribed circle?

Suggested Solution

The method of solution involves finding the area of a rectangle whose width is equal to the diameter of the circle and length is the sum of the

Continued...

Continued . . .

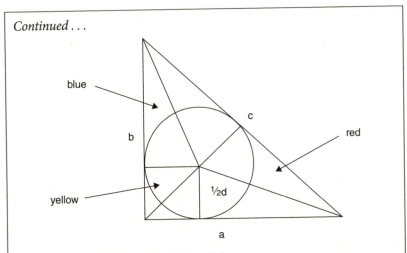

FIGURE 7.8: (Source: Dauben 2007, p. 285)

three sides of the triangle. Liu's demonstration of the result involves color-coding the diagram shown in figure 7.8. If *gou* = *a*, *gu* = *b*, *xian* = *c*, and *d* = diameter of the circle, then it can be shown that by adding the areas of different colored triangles in figure 7.8,[2]

$$d = \frac{2ab}{a+b+c}.$$

Also, Liu states the following equivalent ways of obtaining *d*:

$$d = a - (c - b);$$

$$d = \sqrt{2(c-a)(c-b)}.$$

The reader is invited to show the equivalence of these three results and check that the answer to this example is 6.

The first extant commentary on *Zhou Bi* by Zhao Zhujing, a name we came across earlier, contains examples that involve fifteen different applications of the *gou gu* triangle. They and other subsequent applications are listed in table 7.1, following the lead given by Martzloff (1997a, pp. 294–95). It may be remembered that *a*, *b*, and c denote *gou* (smaller side), *gu* (larger side), and *xian* (hypotenuse) respectively. The diameter of a circle that is inscribed in a *gou gu* triangle is denoted as *d*.

TABLE 7.1: APPLICATIONS OF THE *GOU-GU* TRIANGLE

Type	Given Data for	Answer Sought for	Examples Discussed
1	a, b	c	7.1
2	b, c	a	
3	$b, c - a$	a, c	7.2
	or	or	
	$a, c - a$	b, c	7.4
4	$c, b - a$	a, b	7.5
5	$c - a, c - b$	a, b, c	
6	$a, b + c$	b	7.6
7	$a, a + c = \varphi b$, where φ is a given number	b, c	
9	a, b	d	7.7
10	$ab, c - a$		
	or	a, b, c	
	$ab, c - b$		
11	$ac, c - b$ or		
	$bc, c - a$	a, b, c	
12	$ab/2, c$	a, b, c	
13	$ab/2, a + b$	a, b	
14	$ab/2, c - (b - a)$	c	
15	$ab/2, c + b - a$	a, b, c	

Table 7.1 shows the different permutations of problems that could arise. We could continue this discussion of applications of the *gou gu* theorem in Chinese mathematics almost indefinitely. The applications are remarkable for their range and ingenuity. The reader who wishes to know more will find, apart from Dauben (2007) and Martzloff (1997a) referred to frequently in this chapter, the works Ang (1978), Gillon (1977), Lam and Shen (1984), and Swetz and Kao (1977) informative. The title of Swetz and Kao's book is particularly pertinent: *Was Pythagoras Chinese? An Examination of Right Triangle Theory in Ancient China.*

Our detailed treatment of the properties and problems of right-angled triangles is justifiable for a number of reasons. Of all the ancient mathematical traditions, the Chinese contained the most extensive, sustained, and ingenious treatments. In contrast to the mathematics of ancient Greece, the corpus of knowledge built up was primarily for the purpose of practical applications in height and distance mensuration. But the discovery of the *gou gu* theorem gave rise to work on the proportionality of sides in similar triangles, the extraction of square and cubic roots, methods of solving different types of quadratic equations, and the numerical solution of higher-order equations (to be discussed in a later section).

The importance of the *gou gu* theorem in establishing geometric-algebraic solution schemes and in its contribution to the broader development of Chinese algebra cannot be overestimated. It founded a tradition in geometric reasoning which belies the notion that all mathematical traditions not influenced by the Greeks were essentially algebraic and empirical. The commentaries by Liu and Zhao clearly show a deductive geometry that was moving beyond the mere numerical relations connecting the sides of a right-angled triangle in an active search for general proofs of the relationship. The reasoning in the proofs was based on geometry, with the basic concept being that figures of dissimilar shape can have the same area, and the basic procedure being the "out-in" principle. Terms such as *ji ju* (piling up rectangles) were widely used, underlining the importance of pictorial representation in Chinese mathematics. But the work on the right-angled triangle also highlighted one of the negative aspects of the development of Chinese mathematics—the excessive reverence accorded the *Jiu Zhang*. The above examples, chosen to show the wide range of applications, were all either taken directly from this text or derived from it. So why were mathematicians of the caliber of Liu and Zhao, and later the brilliant thirteenth-century quartet, unable to free themselves from the constricting influence of the *Jiu Zhang*? Was the astonishing continuity and stability of Chinese civilization responsible, coupled perhaps with a great reverence for the past? Most likely we shall never know.

Estimation of π

The Greek symbol π was first used to denote the ratio of the circumference of a circle to its diameter in 1706, by the Welshman William Jones. The same symbol had previously been used for just the circumference, at a time when the idea that a ratio could be a number was quite a novel one. In his major work *Introductio in analysin infinitorum* (1748), Leonhard Euler gave his personal approval to this use of it, thereby popularizing it. It is now perhaps the most widely known mathematical symbol.

If π represented merely the ratio of the circumference of a circle to its diameter, the determination of its numerical value would have been of limited mathematical interest. However, there are other reasons why the evaluation of it has been a continuing quest for some four thousand years, from 1800 BC to the present day:

1. A practical requirement for increasingly accurate determinations of π in fields as diverse as building and electronics

2. The perennial fascination with the problem of "squaring the circle"

3. A growing interest in the nature of the constant represented by π

Stated simply, the problem of squaring the circle (or, more formally, the quadrature of the circle) is: can one construct a square whose area is exactly equal to that of a circle of given diameter using only a straightedge and a compass? Only in the nineteenth century was it demonstrated that, since squaring the circle is equivalent to constructing a line segment whose length is equal to the product of the square root of π (which is not a constructable quantity) and the radius of the given circle, it cannot be done. However, the search for more accurate estimates of π, which began with the Egyptians, continues even today as powerful computers are used to calculate its value to millions of decimal places. Table 7.2 contains some of the historical highlights of this search, before the advent of modern mathematics. Two ways of calculating π are represented here, though only one example (Madhava's calculation) of the second method was in use before AD 1600.

1. *The "classical" (or geometric/empirical) method.* This consists of calculating the perimeters of regular polygons inscribed in and/or circumscribed about a circle of given radius whose circumference lies between these perimeters. This method or a variant of it was used in all but one of the calculations listed in table 7.2.

2. *The "modern" (or analytical) method.* This consists of evaluating the circumference of a circle for a given diameter by applying finite series approximations to a slowly converging infinite series. It was first used in Kerala, India, in about the fifteenth century and attributed to Madhava (c. 1350). Details of this method are found in chapter 10 in the section on medieval Indian mathematics.

For a long period the Chinese were content with taking the ratio of circumference to diameter as 3 in their calculations. One of the earliest attempts to get a better estimate was by an astronomer and calendar maker called Liu Xin, who lived around the beginning of the Christian

TABLE 7.2: ESTIMATES OF π BEFORE AD 1600

Date	Source/mathematician	Method and value
c. 1650 BC	Ahmes Papyrus (Egypt)	Equating a circular field of 9 units to a square of side 8 units implies $\pi \simeq 3.16$.
c. 1600 BC	Susa Tablet (Babylonia)	Equating a regular hexagon to a circle; the ratio of the perimeter of the hexagon to the circumference of the circle, given as 0;57,36 implies $\pi \simeq 3;7,30$ (3.125).
c. 800–500 BC	*Sulbasutras* (India)	Baudhayana gave the following rule: (s = side of a square, d = diameter of a circle): $s = d[1 - 28/(8 \times 29) - 1/(6 \times 8 \times 29) + 1/(6 \times 8 \times 29 \times 8)]$, which implies, if the areas are equal, that $\pi \simeq 3.09$.
c. 250 BC	Archimedes (Greece)	Calculating the perimeters of inscribed and circumscribed regular polygons with 12, 24, 48, and 96 sides within and around a circle, to obtain $223/71 < \pi < 22/7$, where both limits give $\pi \simeq 3.14$, correct to 2 decimal places.
c. 150 BC	Umasvati (India)	Inscribing a regular hexagon and then a 12-sided polygon, and applying the Pythagorean result gives a value equal to the square root of 10: $\pi \simeq 3.16$.
c. AD 260	Liu Hui (China)	Inscribing a regular hexagon within a circle and calculating by successive applications of the Pythagorean theorem the perimeters of polygons with 12, 24, . . . , 96 sides. By considering the last of these polygons, he arrived at $\pi \simeq 3.1416$. (The method is examined later in this chapter.)
c. 480	Zu Chongzhi (China)	Similar method to Liu's, except for the successive applications of the Pythagorean theorem to polygons with up to 24,576 sides! The result was $3.1415926 < \pi < 3.1415927$.

continued

Table 7.2: Continued

Date	Source/mathematician	Method and value
c. 500	Aryabhata (India)	Probably by calculating the perimeter of a regular inscribed polygon of 384 sides: $\pi \simeq 3.1416$.
c. 1400	Madhava (India)	Using an infinite-series expansion for π (to be discussed in chapter 10): $\pi \simeq 3.14159265359$, correct to 11 decimal places.
1429	Al-Kashi (Persia)	By calculating the perimeter of a regular polygon with 3×2^{28} sides. Expressed in decimals: $\pi \simeq 3.1415926535897932$, correct to 16 decimal places.
1579	François Viète (France)	Calculating the perimeter of a regular polygon with 393,216 sides gave $\pi \simeq 3.141592654$, correct to 9 decimal places.

era. Instructed by his ruler, Wang Mang, to construct a standard measure of capacity, Liu Xin produced a cylindrical vessel made of bronze, which became the prototype for hundreds of such vessels produced and distributed throughout China. From an examination of the dimensions of one of these vessels (now kept in a museum in Beijing), some commentators have inferred that Liu Xin used the value $\pi \approx 3.1547$. This view is supported by an entry in *Sui Shu* (Official History of the Sui Dynasty) stating that Liu Xin found a new value for π to replace the old one of 3. Another piece of conjectural evidence comes from a stray remark by Liu Hui that Zhang Heng, a court astrologer who lived in the first century AD, made an implicit estimate of π as the square root of 10, a value also found in the Jaina mathematics of India and reported by Umasvati a few centuries earlier.

The first systematic treatment of this topic is contained in Liu Hui's notable commentary on the *Jiu Zhang*, written in the third century AD. He began by examining the underlying assumption in problems 31 and 32 of the first chapter of the *Jiu Zhang*, in which the area of a circle is calculated by taking the product of half the circumference and half the diameter.[3] After explaining the rationale for this rule and the use of the inaccurate value of 3 for the ratio of circumference to diameter, Liu proceeded to work out a

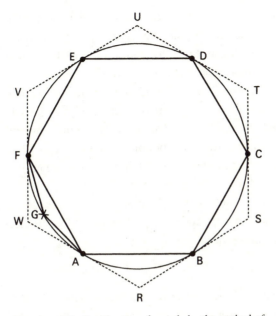

FIGURE 7.9: Finding the area of a circle by the method of exhaustion

method for obtaining a more accurate value for it. Since the method bears some similarity to Archimedes's innovative approach four hundred years earlier, it is interesting to compare them.

Archimedes based his method on the simple observation that if a circle is enclosed between two polygons of *n* sides, then as *n* increases, the gap between the circumference of the circle and the perimeters of the inscribed and circumscribed polygons keeps diminishing, with the perimeters of the polygons and the circle becoming nearly, although never exactly, identical. Or in other words, as *n* increases, the difference in area between the polygons and the circle will be gradually exhausted without ever becoming zero. (This approach is an example of a Greek method known as "exhaustion.")

Figure 7.9 shows a circle enclosed between an inscribed regular hexagon ABCDEF and a circumscribed regular hexagon RSTUVW. It is convenient to begin with a regular hexagon, since its construction with straightedge and compass is a fairly simple matter. First draw the circle. Then, with the compass still set to the radius of the circle, start at any point on the circumference and mark off the vertices A, B, C, D, E, and F. Draw tangents to the circle at each of these points, and join them to give the circumscribed hexagon RSTUVW. Then given the inscribed polygon ABCDEF, it is possible next to construct a 12-sided polygon by bisecting the arc subtended on each of the sides of ABCDEF and joining each of the additional six points to the two original vertices that are adjacent.

As an illustration, if G is the midpoint of the arc AF, then joining F to G, and G to A, will give two of the sides of the 12-sided inscribed polygon. And if, at each of the 12 vertices of the new polygon, tangents are drawn to the circle, we obtain the corresponding circumscribed 12-sided polygon. This process can be carried on to obtain regular polygons of 24, 48, 96, . . . sides. Archimedes stopped at 96-sided polygons.

Now, if the perimeters of the inscribed and circumscribed polygons of n sides are denoted by p_n and P_n respectively, and C is the circumference of the circle, it follows that

$$p_6 < p_{12} < p_{24} < p_{48} < p_{96} < \ldots p_{n/2} < p_n < C < P_n < P_{n/2} \ldots < P_{96} < P_{48} < P_{24} < P_{12} < P_6.$$

It can be shown with modern mathematics that p_n and P_n converge to C as n tends to infinity. Next, it is easily established that the perimeter of the inscribed regular hexagon p_6 is $3d$, where d is the diameter, and also that the perimeter of the circumscribed regular hexagon, P_6, is $2\sqrt{3}d$. Starting with p_6, Archimedes found close approximations to p_{12}, p_{24}, p_{48}, and p_{96}. Then from P_6 he approximated P_{12} to P_{96}. Since $p_{96} < C < P_{96}$, he concluded that

$$3 + \frac{10}{71} < \pi < 3 + \frac{1}{7}.$$

In his computations of the square root of 3 and other calculations, Archimedes appears to have followed the Mesopotamian methods we discussed in chapter 4. But the idea of finding approximate numerical values of π by establishing narrower and narrower limits, between which the value must lie, turned out to be a peculiarly Greek innovation.

Liu Hui's method of evaluating π required only inscribed regular polygons (see figure 7.10). Starting with the known perimeter of a regular polygon of n sides inscribed in a circle, the perimeter of the inscribed regular polygon of $2n$ sides was calculated by applying the Pythagorean theorem twice. The circle in figure 7.10 has its center at O and radius r. Let PQ = s be the side of a regular inscribed polygon of n sides having a known perimeter. Then, given s and n, the Pythagorean theorem is used to calculate, in turn, that

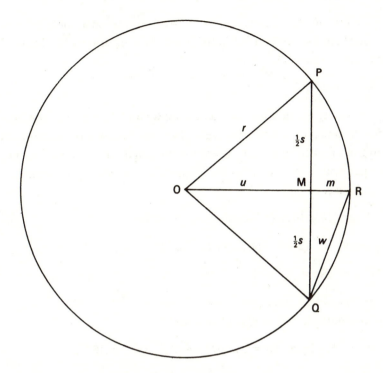

FIGURE 7.10: Liu's method of finding π

$$OM = u = \sqrt{r^2 - \left(\tfrac{1}{2}s\right)^2},$$

$$MR = m = r - u,$$

$$RQ = w = \sqrt{m^2 + \left(\tfrac{1}{2}s\right)^2},$$

where w is one of the sides of the regular polygon with $2n$ sides. Repetition of this procedure will produce closer and closer approximations to the circumference of the circle, in terms of which π may then be defined. A similar procedure can be used with circumscribed polygons.

In Liu's own example, $OR = r = 1$ *chi* $= 10$ *zuen*, $n = 6$, and $PQ = s = 10$ *zuen*, and so

$$u = \sqrt{100 - 25} \approx \sqrt{r^2 - \left(\tfrac{1}{2}s\right)^2} = 8.660254,$$

$$m = 10 - 8.660254 = 1.339746,$$

$$w_1 \approx 5.176381.$$

The first iteration produces $w_1 \approx 5.176381$ as the length of the side of a 12-sided regular polygon. Repeating the same process for a 12-sided polygon with $s = 5.176381$ gives

$w_2 \approx 2.610524.$

The iterative process is continued until we find the length of the side of a 96-sided regular inscribed polygon. Now, the area of a regular inscribed polygon of $2n$ sides is $\frac{1}{2}nsr$, where s is the length of the side of a regular polygon with n sides and r is the radius of the circle. For the dimensions given above,

$A_{12} = \frac{1}{2}(6 \times 10 \times 10) = 300$

is the area of a 12-sided polygon. Similarly,

$A_{24} = \frac{1}{2}nsr = \frac{1}{2}(12 \times 5.176381 \times 10) = 310.5829$

is the area of a 24-sided polygon.

Continuing in this way, Liu Hui found

$A_{48} = 313.2629$, $A_{96} = 313.9344$, and $A_{192} = 314.1024.$

Thus

$A_{192} < A < 2A_{192} - A_{96},$

where A is the area of the circle. This leads to

$314.4024 < A < 314.2704,$

and by interpolation Liu Hui then arrived at an estimate of the circumference of the circle to its diameter as 3,927 to 1,250, that is,

$\pi = 3.1416.$

Liu provides the rationale underlying the following rules for calculating the area of circular fields given in the *Jiu Zhang*:[4]

1. Multiply half the circumference by half the diameter.

2. Multiply one-fourth the of the circumference and the diameter.

3. Evaluate three times one-fourth of the diameter squared.

4. Evaluate the square of the circumference divided by 12.

It is worth noting that while (1) and (2) are perfectly correct, (3) and (4) would give "inaccurate" results since they make the implicit assumption that $\pi = 3$.

The Chinese fascination with the ratio of circumference to diameter reached its climax in the work of Zu Chongzhi and his son Zu Gengzhi, who established new boundaries for π. In his text *Su Shu* (Method of Interpolation), written in AD 479 but unavailable today, the elder Zu may have explained his method for approximating it. All we have are passages in the *Sui Shu* (Official History of the Sui Dynasty) recording the efforts made by Zu (who is described as a historian) to improve the accuracy of the value of π. Using a circle of diameter 10 *chi*, subdivided into 108 units, he obtained an (implicit) estimate of an upper limit of 3.1415927 and a lower limit of 3.1415926 for π.[5] The same official history states that Zu gave the Archimedean value of 22/7 as an inaccurate approximation, and 355/113 as an accurate one. Zu's book was adopted by the Tang government as a text for its civil service examinations.

We can only offer conjectures about how Zu achieved his highly accurate estimate of π. Most probably, Zu extended the method of polygons well beyond the bounds achieved by Liu, to $A_{24,576}$, which yields an approximation for π correct to the seventh decimal place. An accurate (implicit) approximation for π, correct up to the eleventh decimal place, appears, as we shall see later, in the work of medieval Indian mathematicians of the fifteenth century using the "analytical" (infinite series) method. Around the same time, the Persian mathematician Al-Kashi obtained an implicit estimate of π correct to sixteen decimal places using the method of polygons with 3×2^{28}. In 1585 a Dutch mathematician, Valentin Otho, discovered the value 355/113 by subtracting the numerators and denominators of the Ptolemaic and Archimedean values, 377/120 and 22/7! A plausible explanation (although unsupported by any positive evidence at present) is that the Indian and European knowledge of this highly accurate value for π came from China as Zu's discovery spread southward and westward over a number of centuries. In China itself the work of Liu and Zu was soon forgotten, and various inaccurate approximations—including 3—continued to be used. In 1247 Qin Jiushao stated that 3, 22/7, and $\sqrt{10}$ were all in use! In 1275, Yang Hui gave five formulas for finding the area of a circle. In one of them it was 3, two had it as 22/7, and in the other two it was 3.14. There was no guidance on choosing which value to use. One is left with the impression that the innovative work of Liu and Zu was ignored by the mathematicians who came after them.

There are similarities between the work of Archimedes and that of Liu and Zu. But the epistemological differences between the methods of Archimedes and those of the two Chinese commentators are very considerable, the most fundamental being the recourse to the use of "double reduction to the absurd" in proof demonstration by the former compared with the latter.[6] All three used procedures akin to the "method of exhaustion," whereby increasing the number of sides of a polygon inscribed in a circle made the sides so short that it eventually became possible to closely approximate the circle by the polygon.[7] Both Archimedes and Liu used the method of exhaustion to show that the area of the circle is half the product of its circumference and its radius, and both proceeded to establish an equality between the ratio of the area of a circle to the square of its radius, and the ratio of the circumference to the diameter. However, Archimedes' achievement seems the more remarkable when weighed against the limited scope offered by the numerals and computational techniques available to him, and the strongly ingrained aversion to experimentation and computation that marked Greek mathematics until the Egyptian and Mesopotamian empirical influences subtly altered its character. However, Liu and Zu's highly elaborate calculations were products of a mathematical culture sympathetic to computation and offering methods that made its mathematicians far better equipped to carry out complicated calculations. We shall return to this point, about how a penchant for numerical work determined the character of Chinese mathematical achievements, particularly in the solution of higher-order and indeterminate equations.

Solution of Higher-Order Equations and Pascal's Triangle

The Chinese Development of Pascal's Triangle

The invention of the "arithmetical" triangle named after Blaise Pascal cannot be traced or credited to any individual. Its earliest use was among the Indians and the Chinese. In India it was an outgrowth of interest in a subject called *vikalpa*, or what is known today as permutations and combinations. It is first mentioned in the *Chandasutra* by Pingala (c. 200 BC) as a method of determining the number of combinations of n syllables taking *p* at a time. And this was further elaborated on by the commentator Hala-yudha, who lived in the tenth century AD. A discussion of the Indian work on this subject will be found in chapter 8. The triangle also appears up to the twelfth power in a work by Nasir al-Din al-Tusi in 1265.

During the first half of the eleventh century there appeared in China the earliest reference to a basic diagram for solving equations. In a book unfortunately no longer extant, Jia Xian is believed to have constructed a table of binomial coefficients up to the sixth power, the exponents being positive integers.[8] In modern notation, Jia was selecting the coefficients of the binomial expansion

$$(a + b)^n = {}_nC_0 a^n b^0 + {}_nC_1 a^{n-1} b^1 + {}_nC_2 a^{n-2} b^2 + \ldots + {}_nC_n a^0 b^n, \text{ for } r = 1, 2, \ldots, n.$$

Here

$$ {}_nC_r = \frac{n!}{r!(n-r)!}, $$

where $n!$ is the usual notation for "n factorial,"

$$ n! = n \times (n-1) \times (n-2) \times \ldots \times 2 \times 1, $$

with 0! taken to be 1.

For $n = 0$, the coefficient is ${}_nC_0 = 1$.

For $n = 1$, the coefficients of $(a + b)^1$ are ${}_1C_0$ and ${}_1C_1$, or 1, 1.

For $n = 2$, the coefficients of $(a + b)^2$ are ${}_2C_0$, ${}_2C_1$, and ${}_2C_2$, or 1, 2, 1.

For $n = 3$, the coefficients of $(a + b)^3$ are ${}_3C_0$, ${}_3C_1$, ${}_3C_2$, and ${}_3C_3$, or 1, 3, 3, 1.

For $n = 4$, the coefficients of $(a + b)^4$ are ${}_4C_0$, ${}_4C_1$, ${}_4C_2$, ${}_4C_3$, and ${}_4C_4$, or 1, 4, 6, 4, 1.

In a similar fashion, the coefficients can be determined for any value of n. The Pascal's triangle for $n = 0, 1, 2, \ldots, 8$ depicted at the beginning of Zhu Shijie's book *Si Yuan Yu Jian* (The Precious Mirror of the Four Elements) is shown in figure 7.11a, with a transliteration in our numerals given in figure 7.11b. Below Zhu's representation of the triangle is a comment that provides both an explanation of how it was constructed and an indication of the uses to which it might be put. Recast in present-day terminology, it reads:

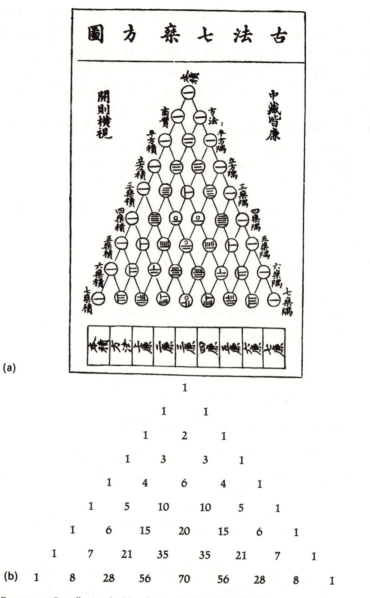

(a)

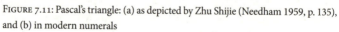

```
                        1
                   1         1
                1       2       1
             1       3       3       1
          1       4       6       4       1
       1       5      10      10      5       1
    1      6      15      20      15      6       1
 1      7      21      35      35      21      7      1
(b)  1     8     28     56     70     56     28     8     1
```

FIGURE 7.11: Pascal's triangle: (a) as depicted by Zhu Shijie (Needham 1959, p. 135), and (b) in modern numerals

The numbers in the $(n + 1)$st row show the coefficients of the binomial expansion of $(a + b)^n$, n being a positive integer. The unit coefficients along the extreme left slanting line (the *chi shu*) and the extreme right slanting line [the *yu suan*] are the coefficients of the first and last term respectively in each expansion. The inner numbers, "2," "3, 3," "4, 6, 4," ... on the third, fourth, and fifth, etc. rows [the *lien*], are the inner terms of the binomial equations of the second, third, fourth, etc. degree.

Zhu continues by indicating the close relationship between the construction of the triangle and the solution of numerical equations of higher order:

Multiply the coefficients of the $(n + 1)$st row by a suggested value for the root; then subtract the nth power of the suggested row from *shi* [i.e., the constant whose root is to be extracted]; and divide the difference by the product of the suggested value and the coefficient to obtain a new value for the root.

The Chinese Origin of the Horner-Ruffini Method

The Horner-Ruffini method is named after an English schoolteacher and mathematician, William George Horner, who in 1819 published a numerical method of finding approximate values of the root of equations of the type

$$f(x) = a_0 x^n + a_1 x^{n-1} + \ldots + a_{n-1}x + b_n = 0,$$

and an Italian, Paolo Ruffini (1765–1822), who was awarded a gold medal for his independent discovery of the same. The procedure that Horner and Ruffini rediscovered is identical to the computational scheme used by the Chinese over five hundred years earlier. And unlike other computational methods for finding such roots, this method is more efficient in that it requires fewer iterations to achieve the same result.

In our earlier discussion of the extraction of square and cube roots in the *Jiu Zhang*, we found that the basic procedure had a strong geometric rationale: the method consisted of adding and subtracting sections from a given geometric figure. But the limitation of such a geometric approach is obvious: it cannot be used to solve equations beyond the third degree. Indeed, the geometric approach, even when used in the context of cubic equations, is highly cumbersome, which was why we avoided it in solving the cubic equation. What Pascal's triangle did for Chinese mathematics was to help break the geometric mold by establishing a clear link between the patterns of the coefficients of the triangle and the derivation of transformed equations.

A few examples will help. Let us begin with a simple one, expressed in symbolic notation. Suppose we want to find the square root of N, that is, to solve the quadratic

$$x^2 = N. \tag{7.3}$$

We take the following steps:

1. Let $x = h + y$. Equation (7.3) is then transformed as

$$N = x^2 = (h + y)^2 = h^2 + 2hy + y^2 = h^2 + (2h + y)y, \tag{7.4}$$

where h is a guesstimate of the root x, and the equation is formed with the coefficients from the third row of Pascal's triangle shown in figure 7.11.

2. Obtain y by dividing $N - h^2$ in equation (7.4) by $2h$ to get

$$\frac{N - h^2}{2h} = \frac{2hy}{2h} + \frac{y^2}{2h}, \text{ or } \frac{N - h^2}{2h} = y + \frac{1}{2h}y^2.$$

Since the error term y is small relative to h, we can ignore y^2 to get

$$y \approx \frac{N - h^2}{2h}.$$

3. Therefore, it would follow from the initial definition in step 1 that

$$x \approx h + \frac{N - h^2}{2h}.$$

If the square of this quantity is N, then you are done. Otherwise, repeat steps 1 and 2 with the new estimate

$$h^* = h + \frac{N - h^2}{2h}.$$

Repeat this process until you get the desired result.

A more difficult example is from the fourth chapter of Qin Jiushao's *Shu Shu Jiu Zhang*, which requires the solution of the equation

$$-40,642,560,000 + 763,200x^2 - x^4 = 0. \tag{7.5}$$

The first step in the solution is to guesstimate the number of digits in the answer and also the value of the first digit; this was done by trial and error. Here it is easily established that the final answer will be a three-digit number starting with 8. Note also that $x = 240$ is another solution to the equation. It is not clear why Qin chose 8 rather than 2 for the first digit. The approach used by Qin is analogous to the present-day method of synthetic division to factor out $y = x - 800$ from equation (7.5). The first division gives

$$-40,642,560,000 + 763,200x^2 - x^4$$
$$= (x - 800)(98,560,000 + 123,200x - 800x^2 - x^3) + 38,205,440,000,$$

or

$$-40,642,560,000 + 763,200x^2 - x^4$$
$$= y(98,560,000 + 123,200x - 800x^2 - x^3) + 38,205,440,000.$$

Four repetitions of this procedure, each giving a new remainder, finally produce

$$38,205,440,000 - 826,800,000y - 3,076,800y^2 - 3,200y^3 - y^4 = 0, \tag{7.6}$$

from which we can estimate that the first digit of y, a two-digit number, is 4.

The process of synthetic division is repeated, with equation (7.6) divided by $y - 40$. The first division leaves no remainder, so the solution to equation (7.5) is $x = 840$.

Clearly, Qin had neither the notation nor the technique to follow the synthetic division used above. Instead, he proceeded by setting up a row of numbers corresponding to the coefficients in equation (7.5) and then applying a procedure resembling synthetic division. We shall examine Qin's procedure for solving equation (7.5) by using the counting-rods format, but with decimal rather than rod numerals. Each row in figures 7.12a–c represents a certain quantity. The first is reserved for the root, the eventual result, and is labeled R. The next five rows represent the coefficients of the zeroth, first, second, third, and fourth powers of the unknown x in equation (7.5) and are labeled accordingly. If counting rods were used, the negative constant and the coefficient of the fourth power of x would be shown by black rods, while the positive coefficient of the second power of x would be shown by red rods. Further, instead of leaving blank spaces for zeros, as the Chinese did, here we insert the symbol 0 for clarity.

FIGURE 7.12: Solving higher order equations: Qin's procedure

After arranging the numbers as in figure 7.12a, the calculation begins by advancing the elements in the rows labeled 1st to 4th by one to four places respectively, into the positions shown in figure 7.12b. We can now deduce that the solution will be a three-digit number, and that the "hundreds" digit is 8. Next the "hundreds" digit is multiplied by −1 from the 4th row, and added to 0 from the 3rd row. This gives the new entry for the 3rd row in figure 7.12c, which is −800. This −800 is multiplied by the root 800 and added to 763,200, which is the entry in the 2nd row of figure 7.12b. This gives the new entry in the 2nd row of figure 7.12c as

$$(-800)(800) + 763,200 = 123,200.$$

Next 123,200 is multiplied by the root 800 and added to the quantity in the 1st row in Figure 7.12b, which is 0. This gives the new entry for the 1st row of figure 7.12c, 9,856,000. Finally, this quantity is multiplied by the root 800 and added to −40,642,560 to give

$$(9,856,000)(800) + (-40,642,560) = 38,205,440,000,$$

which is the new entry in the 0th row of figure 7.12c.

These calculations are equivalent to taking $h = 800$ as the first approximation, so that $x = h + x$, and then carrying out the steps 1 to 3 outlined to obtain the new equation

$$38,205,440,000 + 9,856,000x + 123,200x^2 - 800x^3 - x^4 = 0. \quad (7.7)$$

Subsequent iterations follow the same procedure as above. It is possible to skip some iterations by making use of the property that the coefficient of each of the transformed equations will involve the numbers of Pascal's triangle lying on a line slanting across the triangle.

Thus for $n = 4$ we use the first five rows of the triangle, and the coefficients of the general equation of the fourth degree:

<table>
<tr><td>1</td><td>$R_0 = a_4 = -1$</td></tr>
<tr><td>1 1</td><td>$R_1 = 4a_4h + a_3 = -3,200$</td></tr>
<tr><td>1 2 1</td><td>$R_2 = 6a_4h^2 + 3a_3h + a_2 = -3,076,800$</td></tr>
<tr><td>1 3 3 1</td><td>$R_3 = 4a_4h^3 + 3a_3h^2 + 2a_2h + a_1 = -826,880,000$</td></tr>
<tr><td>1 4 6 4 1</td><td>$R_4 = a_4h^4 + a_3h^3 + a_2h^2 + a_1h + a_0 = 38,205,440,000,$</td></tr>
</table>

where R_0 to R_4 are the last five rows of figures 7.12a–c written from top to bottom. They also show how these quantities are calculated. The transformed equation in y (at the end of the fourth iteration, which is the same as equation (7.6) above) can then be written, using the coefficients of Pascal's triangle, as

$$-y^4 - 3{,}200y^3 - 3{,}076{,}800y^2 - 826{,}880{,}000y + 38{,}205{,}440{,}000 = 0. \quad (7.8)$$

Now, taking equation (7.8) and setting $y = h' + z$ or $x = h + h' + z$, where $h' = 40$ is taken as the next approximation, proceed exactly in the same manner as before. The next transformed equation, in z, is

$$z(-z^3 - 3{,}240z^2 - 3{,}206{,}400z - 955{,}136{,}000) = 0, \quad \text{so } z = 0.$$

Since $x = h + h' + z$ and since $h = 800$, $h' = 40$, and $z = 0$, it would follow that the solution is $x = 840$. We are advised by Qin Jiushao to check the correctness of this solution by substituting $x = 840$ in equation (7.5).

We conclude with two similar problems from contemporary texts, one from Qin Jiushao's book *Shu Shu Jiu Zhang* and the other from Li Ye's *Ce Yuan Hai Jing*.[9]

EXAMPLE 7.8 From *Shu Shu Jiu Zhang*: There is a round, walled town of which the circumference and diameter are unknown. Entrance into the town is through four gates in the wall. Three *li* outside the northern gate is a high tree. When we go outside the southern gate and turn east, we have to walk 9 *li* before we see the tree. Find the diameter and the circumference of the town [1 *li* = 500 meters].

The suggested solution for example 7.8 in modern terminology may be expressed thus (see figure 7.13). Let the distance from the northern gate to the tree be a and the distance from the southern gate to where we can get sight of the tree be b. Let x^2 be the diameter of the town. Then in general terms the equation to be solved for x is

$$x^{10} + 5ax^8 + 8a^2x^6 - 4a(b^2 - a^2)x^4 - 16a^2b^2x^2 - 16a^3b^2 = 0.$$

If $a = 3$ and $b = 9$, then the equation becomes

$$x^{10} + 15x^8 + 72x^6 - 864x^4 - 11{,}664x^2 - 34{,}992 = 0.$$

The solution given is $x = 3$, or the diameter of the town is 9 *li*. The reader is invited to find out how the equation was derived in the first

Continued . . .

Continued . . .

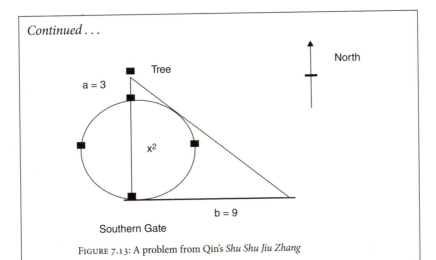

FIGURE 7.13: A problem from Qin's *Shu Shu Jiu Zhang*

place. The derivation of these equations will take us beyond the scope of this book. For details, see Libbrecht (1973).

EXAMPLE 7.9 From *Ce Yuan Hai Jing*: 135 *bu* out of the south gate of a circular town is a tree. If one walks 15 *bu* out of the north gate and then turns east for a distance of 208 *bu*, the tree can be seen. Find the diameter of the town [100 *bu* ≈ 180 meters].

Figure 7.14 is the diagram representing the problem given in example 7.9. The solution involves more than basic geometry; the properties of the *gou gu* triangle, similar triangles, and of tangents are needed.

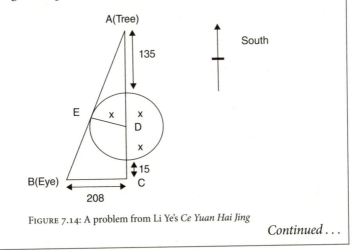

FIGURE 7.14: A problem from Li Ye's *Ce Yuan Hai Jing*

Continued . . .

Continued . . .

Since the triangles ABC and AED are similar, it follows that

AD/AB = ED/BC = x/BC,

where x is the radius of the circle, so that

AB = 208 (x + 135)/x.

Applying the *gou gu* theorem gives

$(AB)^2 = (AC)^2 + (BC)^2,$

or

$(AB)^2 - (AC - BC)^2 = 2(AC \times BC),$

from which the following equation may be obtained:[10]

$-4x^4 - 600x^3 - 22{,}500x^2 + 11{,}681{,}280x + 788{,}486{,}400 = 0.$

Solving the equation for the root of the quartic equation gives the value of x = 120 *bu*, which is the radius of the circular town.

There are certain general features of this method that need further elaboration:

1. The representation of an equation on the counting board conformed to what has been described as the *tien yuan* notation. For example, the equation

$$x^5 + 4x^3 - 5x^2 + 8x - 65 = 0$$

would be represented as in figure 7.15.

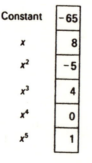

Constant	-65
x	8
x^2	-5
x^3	4
x^4	0
x^5	1

FIGURE 7.15: The *tien yuan* notation

It is characteristic of this representation that the constant term was always expressed as either positive or negative so as to put the equation into the form f(x) = 0, allowing the method described above to be used. It is interesting that the idea of rearranging an equation to give a sum of terms equal to zero did not arise in European mathematics until the seventeenth century, with René Descartes.

2. The absence of symbols for "equals" and "minus" was unimportant in Chinese mathematics, since the color of the counting rods indicated sign and their arrangement indicated the relations between terms.

3. While handling negative quantities posed few problems in Chinese mathematics, negative solutions to equations were ignored. This was a reflection of the practical nature of their problems, which rendered such solutions meaningless.

4. The origins of the method described may go back to the time of the *Jiu Zhang*, when techniques for extracting square and cube roots were first used to solve numerical equations. The method remained unique to China until the eleventh century AD, when it appears in the works of Islamic mathematicians such as al-Nasawi (c. 1025) and al-Samaw'al (c. 1172), who used it to extract cube roots, and later al-Kashi (c. 1450), who used it to extract roots of any degree. In chapter 11 we shall examine al-Kashi's numerical solution of a cubic equation for evaluating the sine of 1°, which bears some resemblance to the method discussed above. This has raised the possibility that the Chinese mathematicians of the thirteenth century may have borrowed a version of the Horner-Ruffini method from the Islamic world. But this raises difficulties regarding the facts that negative numbers were used only by the Chinese and that mathematicians in the Islamic world were more adept at using sexagesimal rather than decimal numbers in their solutions of higher-order equations. There have been suggestions of a reverse transmission, whereby a simple version of what is now known as the Horner-Ruffini method may have slowly evolved from the *Jiu Zhang* and made its way from China to the Islamic world, and may even have been known to Fibonacci through the Islamic mathematicians. If this is so, then this method entered the mainstream of European mathematics, via the Islamic world, as early as the thirteenth century.[11]

The Chinese Method in the West

The possibility of later Chinese mathematical influence on the West cannot be ruled out altogether. It is true that there is as yet no direct evidence of either William Horner or his Italian contemporary and rival, Paolo Ruffini, being aware of the Chinese solution. However, there is a tendency to dismiss too easily any circumstantial evidence of Chinese influence in this area. This is at least partly a consequence of ignoring the possibility that the Jesuit link established during the closing decades of the sixteenth century may have led to a two-way exchange of ideas. The emphasis has always been on how European mathematics reached China through the Jesuits.

In 1582 an Italian Jesuit, Matteo Ricci (1552–1610), was sent on a mission to China. He was one of the most remarkable men of his time, not only an accomplished linguist, with an extraordinary mastery of the Chinese language, but also a scientist and mathematician of note. A particular feature of his mission that is often ignored was that he had been given specific instructions by his superiors to gather information on scientific matters from the East. He was received warmly by the Imperial Court, where he, together with a talented group of fellow Jesuits and Chinese, set out to acquaint the Chinese with scientific works from the West. To that end, they translated the first six books of Euclid. They also made detailed reports of aspects of Chinese science to their parent organization, the Society of Jesus. It is not unreasonable, given Ricci's and his younger contemporary Johann Schreck's (1576–1630) knowledge of mathematics (they were acquaintaned with the works of algebraists of the caliber of Girolamo Cardano [1501–1576] and Francois Viète [1540–1603]), to suggest that Ricci studied Chinese mathematics with some care and reported back his findings.[12] But this must remain a conjecture until further research is undertaken on the mathematical content of the communications from the Jesuits in China, and on the extent of their contacts with the Italian mathematicians of the sixteenth and seventeenth centuries who were mainly responsible for the revival of algebra in Europe. However, as will be pointed out in a subsequent chapter of this book, there is some circumstantial evidence to support the thesis that the Jesuit conduit may have played a part in transmitting the mathematics of India to the West.

Indeterminate Analysis in China

Indeterminate analysis arose in China primarily as a method of calculating calendars. In calculating the starting point of the calendar, Chinese astronomers had to solve systems of relationships with data so vast that it was impossible to get unique solutions without some special algorithms. The Chinese remainder theorem was one such algorithm. Similar problems of calendar construction were faced in India and the Islamic world. It was with the work of European mathematicians such as Lagrange, Euler, and Gauss during the seventeenth and eighteenth centuries that the subject was detached from the coattails of astronomy and attached to the realm of pure mathematics.[13]

The following problem occurs in the third chapter of the fourth-century mathematical text *Sun Zu Suan Jing* (Master Sun's Mathematical Manual):[14]

> There is an unknown number of objects. When counted in "threes," the remainder is 2; when counted in "fives," the remainder is 3; and when counted in "sevens," the remainder is 2. How many objects are there? Answer: 23.

In modern notation, what we have here is the following set of simultaneous equations of the first degree:

$$N = 3x + 2, N = 5y + 3, N = 7z + 2,$$

where N is the total number of objects, x the number of "threes," y the number of "fives," and z the number of "sevens." This information can be expressed even more concisely in the notation of linear congruences as

$$N \equiv 2 \,(\mathrm{mod}\ 3) \equiv 3 \,(\mathrm{mod}\ 5) \equiv 2 \,(\mathrm{mod}\ 7),$$

where an integral value for N is required.[15]

Now, we know that in order that a solution to a given set of equations may be obtained, there must be as many equations as unknowns. Here there are three equations but four unknowns (N, x, y, z), so there are an infinite (or indeterminate) number of solutions. However, there is a further constraint implied by the answer given to the question: what we are seeking is the least (or minimum) integer value for N.

There are four main approaches to the solution of indeterminate equations. The most obvious is an arithmetic one in which, for the example given above, a solution set for (x, y, z) that satisfies the three equations is obtained by trial and error. By a long and laborious process it is possible to work out that the solution set ($x = 7, y = 4, z = 3$) will give a least-integer value of $N = 23$. The scope of such a method, apart from its tedium, is very restricted.

A second approach is the one that probably originated with the Indian mathematician Aryabhata I (c. AD 500) and was refined and extended by later mathematicians, notably Brabmagupta (c. 625), Mahavira (c. 800), and Bhaskaracharya (c. 1100). The method, referred to as *kuttaka*, consisted of continuous divisions and substitutions. This approach is discussed in chapter 9.

The third method of solving indeterminate equations of the first degree bears some resemblance to the Indian approach. It is the method favored in more recent times. To illustrate it, consider the following problem:

EXAMPLE 7.10 Solve $5x + 8y = 100$, in integers ≥ 0.

Solution We have

$$8 = 1 \times 5 + 3 \tag{7.9}$$

$$5 = 1 \times 3 + 2 \tag{7.10}$$

$$3 = 1 \times 2 + 1 \tag{7.11}$$

$$2 = 2 \times 1 \tag{7.12}$$

The greatest common denominator (GCD) of 8 and 5 is 1, since any number that divides 8 and 5 must divide 3 by equation (7.9), then 2 by equation (7.10), and then must divide 1 by equation (7.11). Backsubstituting, starting with equation (7.11), gives

$$
\begin{aligned}
1 &= 3 - (1 \times 2) \\
&= 3 - 1[5 - (1 \times 3)] \quad \text{from equation (7.10),} \\
&= (2 \times 3) - (1 \times 5) \\
&= 2 \times [8 - (1 \times 5)] - (1 \times 5) \quad \text{from equation (7.9).}
\end{aligned}
$$

Therefore

$$1 = (2 \times 8) - (3 \times 5). \tag{7.13}$$

The method is perfectly general, and with it we can always obtain from any pair of integers x and y their GCD h, and a relation

$$h = Mx + Ny,$$

where M and N are positive or negative integers.[16] Returning to the solution, multiplying equation (7.13) by 100 gives

Continued . . .

Continued . . .

$$100 = (200)(8) + (-300)(5),$$

and

$$100 = (200 - 5t)8 + (8t - 300)5,$$

for any t chosen, covers all solutions. Hence the solution sets are obtained from the equations

$$x = 8t - 300, \ y = 200 - 8t.$$

The solutions in integers ≥ 0 correspond to $t = 38, 39, 40$:

$$x = 4, \qquad y = 10;$$
$$x = 12, \qquad y = 5;$$
$$x = 20, \qquad y = 0.$$

Finally, there is the Chinese procedure (or *da yan*). Before we examine it, though, we must consider why there was such a long and sustained interest in this subject in both India and China. The answer for both countries lies in problems to do with time that arose in calendar making and astronomical calculations. One problem in calendar making attracted the attention of both Sun Zu (c. AD 300), the originator of indeterminate analysis in China, and Qin Jiushao (c. 1250), whose statement of the *da yan* rule for the general solution of indeterminate equations of the first degree predated the work of Euler and Gauss by five hundred years. We shall consider the role of astronomy in motivating Indian work in this area in a later chapter.

All calendars need a beginning. A calendar constructed during the Wei dynasty (220–265) took as its starting point the last time that winter solstice coincided with the beginning of a lunar month and was also the first day of an artificial sexagenary (60-day) cycle known as *Jia Zu*. The objective was to locate exactly the number of years (measured in days) since the beginning of the calendar. To restate the problem in modern symbolic notation, let y be the number of days in a tropical year, N the number of years since the beginning of the calendar, d the number of days in a synodic month, r_1 the number of days in the 60-day cycle between the winter solstice and the last day of the preceding *Jia Zu*, and r_2 the number of days between the winter solstice and the beginning of the lunar month. The number of years since the beginning of the calendar can then be calculated from

$$yN \equiv r_1 \pmod{60} \equiv r_2 \pmod{60}.$$

More complex alignments, including planetary conjunctions, were built into models for estimating the beginnings of calendars, and as early as the fifth century AD the mathematician-astronomer Zu Chongzhi solved a set of ten linear congruences. However, the first general mathematical formulation for solving problems in indeterminate analysis of the first degree is found in the work of Qin Jiushao (c. 1250).

The Early Approach

Let us return to the problem from the *Sun Zu Suan Jing*. The solution offered by *Sun Zu* reads:

If you count in "threes" and have the remainder 2, then put 140.

If you count in "fives" and have the remainder 3, then put 63.

If you count in "sevens" and have the remainder 2, then put 30.

Add these numbers and you get 233; from this subtract 210, and you have the answer (23).

A popular folk song of the time, "The Song of Master Sun," offered the following mnemonic for the problem:

Not in every third person is there one aged three score and ten,
On five plum trees only twenty-one boughs remain,
The seven learned men meet every fifteen days,
We get our answer by subtracting one hundred and five over and over
    again.

In modern algebraic notation we would say that, given

$$N = 3x + 2 = 5y + 3 = 7z + 2,$$

or

$$N \equiv 2 \pmod{3} \equiv 3 \pmod{5} \equiv 2 \pmod{7},$$

then the solution is obtained as

$$70 \equiv 1(\text{mod } 3) \equiv 0(\text{mod } 5) \equiv 0(\text{mod } 7),$$
$$21 \equiv 1(\text{mod } 5) \equiv 0(\text{mod } 3) \equiv 0(\text{mod } 7),$$
$$15 \equiv 1(\text{mod } 7) \equiv 0(\text{mod } 3) \equiv 0(\text{mod } 5).$$

Hence

$$N = [(2 \times 70) + (3 \times 21) + (2 \times 15)] = 233 - (2 \times 105) = 23.$$

The key to understanding this method is to find out where the numbers 105, 70, 21, and 15 come from. First, we find the smallest integers a_1, a_2, a_3 such that

$$a_1 \equiv 1(\text{mod } 3) \equiv 0(\text{mod } 5) \equiv 0(\text{mod } 7),$$
$$a_2 \equiv 1(\text{mod } 5) \equiv 0(\text{mod } 3) \equiv 0(\text{mod } 7),$$
$$a_3 \equiv 1(\text{mod } 7) \equiv 0(\text{mod } 3) \equiv 0(\text{mod } 5).$$

Since $a_1 \equiv 1(\text{mod } 3) \equiv 0(\text{mod } 5) \equiv 0(\text{mod } 7)$, it would follow that $a_1 \equiv 0(\text{mod } 35)$ and thus a_1 must be a multiple of 35. Now the smallest multiple that is congruent with $1(\text{mod } 3)$ is 70. So we let $a_1 = 70$. Similarly, we can show that the appropriate values for $a_2 = 21$ and for $a_3 = 15$.

Now let

$$N = 2a_1 + 3a_2 + 2a_3.$$

It can be seen from the congruences shown above for a_1, a_2, a_3 that

$$N \equiv 2(\text{mod } 3) \equiv 3(\text{mod } 5) \equiv 2(\text{mod } 7).$$

Specifically,

$$N = (2 \times 70) + 5(3 \times 21) + (2 \times 15) = 233.$$

To obtain the smallest possible solution, we subtract multiples of 105, the least common multiple of 3, 5, and 7, since doing so will preserve the congruences modulo 3, 5, and 7:[17]

$$N = 233 - (2 \times 105) = 23.$$

We shall not attempt to generalize this procedure; the interested reader may wish to consult Libbrecht's splendid monograph (1973), which contains an extensive discussion of this subject.

However, what is interesting is the application of this procedure in Chinese mathematics. In his *Methods of Computation*, Yang Hui identifies an ancient problem of indeterminate analysis that has come to be known as the "hundred fowls problem":

EXAMPLE 7.11 If cockerels cost 5 *qians* each, hens cost 3 *qians* each, and 3 chickens cost 1 *qian*, and if 100 fowls are bought for 100 *qians*, how many cockerels, hens, and chickens are there [the *qian* is a copper coin]?

Various alternative answers are offered:

- Answer: 4 cockerels, 18 hens, and 78 chickens costing 20, 54, and 26 *qians* respectively

- Answer: 8 cockerels, 11 hens, and 81 chickens costing 40, 33, and 27 *qians* respectively

- Answer: 12 cockerels, 4 hens, and 84 chickens costing 60, 12, and 28 *qians* respectively

Given our powerful tool of symbolic algebra, this problem is easy to solve by specifying the following set of equations, given that x, y, and z are the number of cockerels, hens, and chickens respectively:

$$5x + 3y + \tfrac{1}{3}z = 100,$$
$$x + y + z = 100.$$

Substitute $z = 100 - x - y$ in the first equation to obtain $7x + 4y = 100$ or $y = 25 - (7/4)x$. Since y must be an integer, it would follow that x must be a multiple of 4. In other words, the solutions are obtained from the equations

$$x = 4t, \; y = 25 - \left(\frac{7}{4}\right)4t, \; z = 100 - 4t - \left[25 - \left(\frac{7}{4}\right)4t\right] = 75 + 3t.$$

For $t = 1, 2, 3$, we get solution sets (x, y, z): $(4, 18, 78)$, $(8, 11, 81)$, and $(12, 4, 84)$, which are the solution sets given above. For $t = 4$ or greater, y become negative, which is plainly absurd. However, the solution set corresponding to $t = 0$ (or no cockerels) was not offered. Neither is a plausible explanation or justification of how the answers were obtained.[18]

The interest in this problem lies in a different context: the appearance of very similar problems in other mathematical traditions. The examples may vary in terms of the groups involved: men, women, and children (Alcuin,

735–804); or pigeons, cranes, swans, and peacocks (Sridhara, fl. 850–950); or birds, ducks, hens, and sparrows (Abu Kamil, c. 900). However, the totals amount to one hundred in each case. It is more than likely that there was diffusion in this case, but the direction and its chronology remain unclear.

The Grand Quartet

In the course of this chapter the names of four notable mathematicians occur with regularity; their innovative work in the areas discussed must rank as some of the greatest contributions of Chinese mathematics. We know little about Yang Hui, except that he lived around 1250 and came from Hangzhou, now a small city in the Yangzi River Delta, and was probably a minor civil servant. Unlike the other three, he was essentially a prolific arithmetician, as indicated by the content of his output, which was mostly concerned with bringing to light earlier computational methods found in the *Jiu Zhang* or in everyday arithmetic, including different methods of multiplication and division. His work on magic squares, discussed in the previous chapter, vouches for his interest in recreational mathematics.

All we know regarding Zhu Shijie, who lived around the end of the thirteenth century near present-day Beijing, is contained in a short passage that reads:

> Master Songting [the literary name of Zhu Shijie] of Yanshan became famous as a mathematician. He travelled over seas and lakes for more than 20 years and the number of those who came to be taught by him increased each day. (Quoted in Martzloff 1997a, 153)

We therefore infer that he was a wandering teacher. His major work, *Si Yuan Yu Jian* (The Precious Mirror of the Four Elements[19]), shows that he was not merely an uncritical admirer of the past reflected in the long tradition of the *Jiu Zhang*. In his mathematics he showed considerable daring: like the Mesopotamians, whom we discussed in chapter 4, he had no scruples adding areas to volumes and prices to lengths. He must have been an irritant to the military establishment, for when asked to help with their recruitment drive, he suggested that it should be based on an arithmetical progression. The originality of his work and his manner of presentation could have militated against his work being noticed by those who came after him in China. In more recent years, this neglect has also masked the extent to which Chinese algebra had outpaced its European counterpart.

We know more about the life of Li Ye. Born in 1192 in what is the present-day Beijing, he was appointed initially an assistant magistrate, a post that he was unable to take up because of war. He was then appointed a governor in Hunan Province, an appointment that was short-lived because of the Mongol invasion. Fleeing from the invaders, he took refuge in Shansu Province, where lived a reclusive life. It was during this period that he composed his main work, *Ce Yuan Hai Jing* (The Sea Mirror of the Circle Measurements). This is a book in twelve sections involving 170 problems all telling the same story (relating to the same diagram, a circle inscribed in a right-angled triangle) of people wandering along certain roads around a circular town. Each person tries to catch sight of one another or of a given object such as a tree, which is hidden from them by the town walls. The question is invariably to find the diameter of the town, given the distances they have walked, and the answer is often the same: 120 *bu* (≈ 200 meters). As we have seen in an earlier section, there is a striking similarity between problems posed by Li and Qin (see examples 7.9 and 7.10), and the solutions suggested involve higher-order equations. It is interesting that the two, although contemporaries, lived in mutually hostile parts of China and were therefore unlikely to have met or communicated. Yet, here they were, working independent of each other, producing work that was both original and similar.

Qin Jiushao is generally regarded as one of the most accomplished mathematicians to come out of China. Indeed, in any list of great mathematicians who lived before the emergence of modern mathematics, Qin should figure prominently. He was one of the four brilliant Song mathematicians of the first half of the thirteenth century who were responsible for developing algebra to a level that was far in advance of anything that would be achieved elsewhere until the middle of the seventeenth century. However, it should not be thought that in China at that time there was anyone like today's professional mathematicians. Of the four notable mathematicians of the period, Zhu was a wandering teacher, Yang a minor civil servant, Li a scholarly recluse, and Qin a man of many parts, the most important of which were his work for the military and civil service.

Born in Anyue in what is now Sichuan Province, Qin's father occupied various posts in the local administration, which meant that the family moved around. In his youth, according to his own report, Qin moved to Hangzhou, which enabled him to study both mathematics and astronomy. A central part of his study was the *Jiu Zhang*. He served in the army when the Mongol armies invaded Sichuan and held posts in various administrations. He died in 1261.

The little we know of Qin's character is not very flattering. He was considered by some of his contemporaries as unprincipled, extravagant, and boastful, and his penchant for sexual imbroglio made him the equal of Casanova himself. But nobody denies his remarkable versatility. He was well versed in astronomy, harmonics, mathematics, and even architecture. In sports, there were few to match him in polo, archery, or swordplay. He made notable contributions in two areas of mathematics: in the solution of numerical equations of higher degree (as we have discussed) and, more importantly, the derivation of the *da yan* rule for solving indeterminate equations of the first degree.

His treatment of indeterminate analysis is found in his best-known book, *Shu Shu Jiu Zhang*, written in 1247. The book contains eighty-one problems divided into nine sections, but there the resemblance to the *Jiu Zhang* stops. Neither the examination of the subjects covered nor the illustrative problems included owes much to the revered text. In his preface, Qin notes that he intends to introduce a method of indeterminate analysis (*da yan shu*) which, though known to calendar makers and astronomers such as the famous Buddhist sage Yi Xin (c. AD 700), is not found in the *Jiu Zhang*.

Qin's approach is best illustrated by an example from his book; we do not, however, follow him in the details of the solution he offered. His procedure for solving this problem is summarized by Libbrecht (1973, p. 408).

EXAMPLE 7.12 Three thieves, A, B, and C, entered a rice shop and stole three vessels filled to the brim with rice but whose exact capacity was not known. When the thieves were caught and the vessels recovered, it was found that all that was left in Vessels X, Y, and Z were 1 *ge*, 14 *ge*, and 1 *ge* respectively. The captured thieves confessed that they did not know the exact quantities that they had stolen. But A said that he had used a "horse ladle" (capacity 19 *ge*) and taken the rice from vessel X. B confessed to using his wooden shoe (capacity 17 *ge*) to take rice from vessel Y. C admitted that he had used a bowl (capacity 12 *ge*) to help himself to the rice from Vessel Z. What was the total amount of rice stolen?

Continued . . .

Continued . . .

Solution

The problem, restated concisely in modern notation, is to find N given that

$$N \equiv 1 \,(\text{mod}\, 19) \equiv 14 \,(\text{mod}\, 17) \equiv 1 \,(\text{mod}\, 12).$$

Qin's answer is that the total amount of rice stolen (N) is

$$22{,}573 - (5 \times 3{,}876) = 3{,}193 \, ge.$$

The reader who wishes to follow the computation method adopted by Qin should consult Libbrecht.

The Shu Shu Jiu Zhang contains a number of practical problems of indeterminate analysis. They include problems in calendar calculations, engineering and military applications, and architecture. Qin worked not only with integers but also with fractions, for which he devised special procedures. Libbrecht (1973) gives a detailed technical discussion of Qin's work and fills in the social and economic background to the various problems. For its rigor, clarity, and originality, Qin's work must rank as one of the outstanding pieces of mathematical literature. It is a measure of its quality that later Chinese mathematicians found it difficult to comprehend. It thus remained neglected until the eighteenth century, when it began to arouse some interest. By this time, though, work on linear congruence had already begun in Europe with seminal contributions from Euler (1743) and Gauss (1801), initially using techniques similar to the ones pioneered by Qin. European developments soon eclipsed the Chinese work that had begun with Sun Zu some fifteen hundred years before.

Mathematics and Music in China

It is a widely held view that mathematics in China suffered a general decline after the heights attained during the Song and Yuan dynasties between the tenth and the fourteenth centuries. The succeeding Ming dynasty (1368–1644) was believed to be a period during which crucial mathematical texts and techniques were lost and the mathematical creativity of the earlier period was stultified by the rigidities inherent in civil service and the court bureaucracy.[20]

Such an assertion is not based on any exhaustive studies of mathematics during the Ming period. One would suspect, with Hart (1997), that claims of mathematical decline during the Ming period emanated in part from Jesuit propaganda, notably the comments of Matteo Ricci, which was then uncritically accepted by later historians of Chinese science.[21]

A neglected contribution from the Ming period has been the work of the musician Zhu Zaiyu (1536–1611), the first person to solve the mathematical problem of "equal temperament."[22] A ninth-generation grandson of the founder of the Ming dynasty, his father enjoyed the status and wealth of a provincial king until he was stripped of his title and placed under house arrest as a result of some trumped-up charge accusing him of treason against the emperor. Zhu, as a loyal son, built himself a thatched hut with mud floor outside his father's palace and decided to dedicate himself to scholarly pursuits. These included, in the main, mathematics, calendar reform, music theory, performance of music instruments, and ritual dance.

To understand the context of Zhu's work on equal temperament, it is important to recognize that as early as 2700 BC, Chinese theorists had been preoccupied in establishing the gong pitch (also known as the "Yellow Bell"), which can be translated as the "fundamental pitch." They had struggled to work out the mathematical complexities involved in calculating the eleven tones that should rise above it. To understand this preoccupation, it is important to recognize that, for the ancient Chinese, music was not a matter of entertainment and amusement alone but also had an exceptionally important component in the rites and rituals of the court. There was a strong belief that the downfall of a dynasty was, therefore, caused by a flaw in the ritual music of that court. So every new dynasty was dictated by the imperative to establish the correct ritual music to prolong its survival. Also, correct ritual music must ensure that the scale and tonic pitch it used agreed with the numerical ratios that reflected the relative positions of the planets in the zodiac.[23] Zhu was particularly conscious of the corruption and decadence of the court that was responsible for his father's plight. He believed that by introducing a new system of ritual music, the fortunes of the court could be restored and the decline of the Ming dynasty arrested. Equal temperament was part of his armory.

An equal temperament in music is a system of tuning in which every pair of adjacent notes has an identical frequency ratio.[24] Equal temperaments are often intended to approximate some form of "just intonation" (i.e., any tuning system in which the frequencies of notes are related by ratios of integers). In such cases, an interval—generally an octave—is divided into a series of equal steps (or equal frequency ratios). In the case of modern Western music, the most common tuning system is to divide the octave into twelve equal parts (semitones), usually starting with a standard pitch of 440 Hz. This system is normally referred to as the "twelve-tone equal temperament system." The concept of equal temperament begins with the recognition that the ratio of tones between octaves is 1:2 (or 4:8, 8:16, ...). This ratio has been known among various historical cultures for a long time.

Vincenzo Galilei (father of Galileo), in a 1581 treatise, may have been the first person in the West to advocate equal temperament.[25] But Zhu Zaiyu was the first to obtain the correct solution to the problem of how to achieve equal temperament. He came up with the solution in 1584, and about thirty years later the same ideas were published in Europe by Marin Mersenne (1588–1648) and Simon Stevin (1548–1620).[26] An interesting question, to be discussed later, is whether the solution may have spread from China to Europe through either the Jesuits or some other agency.

In an equal temperament, the distance between each step of the scale is the same length. In a twelve-tone equal temperament system, which divides the octave into twelve equal parts, the question arises as to the ratio of frequencies (r) between two adjacent semitones. Zhu concludes that this ratio, expressed in modern mathematical language, is

$$r = \sqrt[12]{2} \approx 1.05946309.$$

Or, in other words, the division is achieved by calculating the thirteen notes of the scale (including the fundamental note, assumed here to have the value 1) in the following ratios:

$$1:1;\ 1:2^{1/12};\ 1:2^{2/12};\ 1:2^{3/12};\ \ldots;\ 1:2^{11/12};\ 1:2^{12/12}\ [\text{i.e., } 2].$$

Every interval between each pair of the adjacent notes within the octave is $\sqrt[12]{2}$; that is, the scale is equidistant (or equally tempered).

The manner in which Zhu arrives at this answer is interesting. He begins with the familiar Pythagorean result for a right-angled isosceles triangle of length 10 *cun*. The length of its hypotenuse is

$$\sqrt{10^2 + 10^2} = 10\sqrt{2} = 10 \times 1.41421356\ldots.$$

Ignoring the length of the fundamental, given as 10 *cun*, the square root of 2 (which is the length of the hypotenuse) represents the note that is the midpoint (denoted by C) between the octave of the fundamental 1 (denoted by A) and the terminal 2 (denoted by B). Zhu used an abacus to calculate the square root of 2 correctly to twenty-four decimal places, which is probably the first time in the long history of Chinese mathematics that the square root, and later the cube root, were calculated using the abacus!

Now the note that is the midpoint between the calculated value (C) and the terminal value B can be obtained by the same procedure, as follows:

$$\sqrt{10 \times 10\sqrt{2}} = 10\sqrt[4]{2} = 10 \times 1.18920711....$$

This number represents the ratio for the interval of a minor third, or the interval separating three semitones.

The final step involves obtaining the ratio for a semitone that is equivalent to calculating the cube root:

$$\sqrt[3]{10 \times 10 \times 10\sqrt[4]{2}} = 10\sqrt[12]{2} = 10 \times 1.05946309....$$

The number $r = 1.059463094359295264561825$ is the twelfth root of 2 (correct to the twenty-fourth decimal place). This is what is required for generating an "equal temperament" series. That is, if the fundamental is 1, repeated multiplication by r, starting with multiplying 1 by r, would generate all the terms of the series.

Zhu's actual computational procedure has an elegance and simplicity worthy of note. He needs only three steps to arrive at the value of twelfth root of 2. These steps are given below. No further explanation is needed, except to add that he uses the following result in step 3 to simplify his calculations:

$$\sqrt[12]{2} = \left\{ \left[(2)^{1/2} \right]^{1/2} \right\}^{1/3}$$

Step 1. Calculate $\sqrt{2} = 1.41421356....$

Step 2. Calculate $\sqrt{1.4142135623} = 1.18920711....$

Step 3. Calculate $\sqrt{1.189207115} = 1.05946309....$

In other words, in a twelve-tone equal temperament scale, which divides the octave into twelve equal parts, the ratio of frequencies between two adjacent semitones is the twelfth root of 2. Or, more generally, the smallest interval in an equal-tempered scale is the ratio

$$r^n = p, \text{ so } r = \sqrt[n]{p},$$

where the ratio r divides the ratio p (= 2/1 in an octave) into n equal parts.[27]

The possibility of Zhu's work being transmitted to Europe should not be dismissed outright. The evidence needs to be collected and evaluated further. However, from Matteo Ricci's *Journals*, it is clear he was aware that the way of gaining entry into the royal court was to introduce its members to the latest in astronomy, calendar reforms, and Western music. He had asked to be sent a harpsichord, which he taught court officials to play. He was on good terms with the court officials and the literati, although he never met Zhu, nor is there any documentary record of him being aware of Chu's work on equal temperament. Nevertheless, Ricci must have been aware of the close relationship in China between calendrical science and the musical rites, and thus the revolutionary results of Zhu's musical theorizing, even if he was not aware of Zhu himself.[28] After all, two of Ricci's most influential converts and disciples had strong connections with music: one became the director of the Bureau of Rites, and the other was an accomplished instrumentalist.

The Influence of Chinese Mathematics

Three main cultural areas came under the influence of Chinese mathematics: Korea, Tibet, and Japan. The circumstances under which the influence manifested itself, the response to it, and the assimilation of Chinese mathematics differ considerably between each of these areas.

Mathematics in Korea

In 682 AD, under the Shilla dynasty, a system of mathematics education was established in Korea. Based on the *Tang liu tian*, a codified mathematics program established during the Tang dynasty in China (AD 618–906), its purpose was to train professional mathematicians. A closely controlled program, it prescribed the number of mathematicians to be trained, the length of study, and the content of the curricula. Students could be enrolled at any age between fifteen and thirty, and their studies lasted for nine years or longer. The full curriculum, according to Kim (1994, p. 112), consisted of elementary ("six-chapter arithmetic"), intermediate ("nine-chapter arithmetic"), and advanced ("continuation techniques") components. The original Tang model, on which the Korean system was based, recruited students at a younger age, with the duration of study being seven years, and a more elaborate curriculum was offered, based on the *Jiu Zhang* and its commentaries. The Korean program remained in place as the official program until the end of the Choson dynasty in 1910. It influenced the system set up in Japan around AD 701, known as the *Taihorei*. China and Japan soon discontinued their program of studies.

During the Choson dynasty in Korea (1392–1910), the program was revised and strengthened, with the requirement that all bureaucrats had to undergo mathematical training. A class of bureaucrats, called the *chungin* ("middle men"), were put in charge of the technical civil service examinations. Mathematicians were recruited to become an increasingly important section of this class.

The *chungin* soon became an exclusive class, intermarrying among themselves and establishing an almost castelike line of descent, whereby a father who was a mathematician would be succeeded by a son who was also trained in the same discipline. The discipline was further regimented during the reign of Sejong (AD 1419–1450), when a Bureau of Mathematics and an Agency for Calendars were established at the same time as a system of awarding titles such as *sanhak paksa* (doctor of mathematics), *sanhak kyosu* (professor of mathematics), and *sansa* (mathematician).

The growing professionalism in the discipline was not, however, matched by greater creativity on the part of its members. We have an account of one Hong Chong-ha, a professor of mathematics and a *chungin*, who wrote a text titled *Kuilchip* (Nine Chapters of Arithmetic in One). He was born in 1684 into a family where his father, both grandfathers, and a great-grandfather, as well as his wife's father, were all mathematicians! (Kim 1994, p. 113). His book could well have been written three hundred years earlier. Its content was no different from the older Chinese methods of the original program, even to the extent of retaining calculation rods when, in China, they had already been replaced by the abacus.

The stranglehold that Chinese mathematics had on Korean mathematics was never significantly loosened. Even when new ideas came to China from Europe, they had virtually no impact on Korean mathematics. The intellectual climate and the institutionalization of mathematics as a discipline in Korea seem to have hindered the understanding and assimilation of modern mathematics.

Astronomy in Tibet

Tibet has a living astronomical heritage influenced by both China and India. As a product of the cultural mixing of religions, languages, and astronomical knowledge, Tibetan astronomy provides another model of how outside influences are fused in the development of science.

Tibetan astronomy (or *rtsis)* may be broadly divided into four branches: *nag-rtsis* (black calculation), *rgya-rtsis* (Chinese calculation), *skar-rtsis* (star calculation), and *dbayabis-char* (divination). The former two branches are of Chinese origin while the latter two are of Indian origin. The first and the third may be loosely interpreted in modern terms as astronomy, while the other two are closer to astrology. We will concentrate on the Chinese components.

A popular work of the seventeenth century on *nag-rtsis* lists nine integral components of the subject, being a blend of Chinese natural philosophy such as the theory of the five basic elements and the *yin-yang* principle and of Chinese astrological practices that include the association of each year in a twelve-year cycle, and each month in a year, with a particular animal (e.g., rat, tiger, horse, sheep). There is also a description of the Chinese lunar zodiac, which contains 28 lunar mansions or houses.

The *rgya-rtsis* is based on a Tibetan translation of an original Chinese text on the construction of a Shixian calendar, the last lunisolar calendar to be constructed before being replaced by the Gregorian solar calendar in China in 1912. A lunisolar calendar consists of "short" months of 29 days and "long" months of 30 days. The idea is to arrange short and long months so that the new moon will occur on the first day of each month and the full moon on the fifteenth day of that month. A solar calendar, on the other hand, requires an independent system of solar intervals, which, in the case of a tropical year, consists of twelve equal intervals of time, with the interval center being the middle point of each equal interval. Since the synodic month (which is based on the moon and so varies between 29 and 30.1 days) is always slightly shorter than an equal interval in a solar calendar, an interval center will not occur in certain months. The month of nonoccurrence is known as an intercalary month; the year in which an intercalary month occurs has 13 synodic months. This occurs roughly once in every three years.

The Tibetans have had a long tradition of constructing a lunisolar calendar and making adjustments for intercalary months; this has gone hand in hand with astronomical and calendar construction practices that originated in India. The coexistence of these different calendars used for different purposes, reminiscent of the Mayan practice discussed in chapter 2, is one of the more interesting aspects of Tibetan astronomy.

The culmination of traditional Tibetan astronomy, however, is found in the works of Bu-ston Rin-chen grub (1290–1344). Bu-ston wrote an astronomical text called *Kas-pa dga'byed*, which summarizes the subject matter of Tibetan astronomy in seven chapters. Apart from the final chapter of the text, which contains a discussion of word-numerals based on Indian texts (to be discussed in the next chapter), mathematics (as opposed to "calculation") is treated here as primarily a tool for astronomical calculations. Indeed, there is no known Tibetan text that deals with mathematical work outside astronomy.

Mathematics in Japan

The introduction of Chinese mathematics into Japan took place during two periods: in the eighth and ninth centuries AD, and in the sixteenth and seventeenth centuries AD.[29] In the first period, when the *Jiu Zhang* was brought to Japan, its impact was limited, probably because of the lack of understanding of the text on the part of the Japanese. However, the text was preserved and a Japanese translation made, to be studied by a few scholars.

The introduction of Chinese mathematics during the second period was a more creative encounter. Two important thirteenth-century texts found their way to Japan via Korea. An encounter occurred between two cultures, which, despite other differences, shared the same mathematical language. This experience would be different from the Japanese encounter with Western mathematics that occurred a few centuries later. The language of one had to be translated and interpreted for the other.

We will approach the Sino-Japanese encounter through a case study of Seki Takakazu (alias Seki Kowa, 1642–1709). He is a central figure in Japanese mathematics in a number of ways and was mentioned briefly in chapter 6 for his work on magic squares and the discovery of determinants. He is generally accepted as the greatest Japanese mathematician of his time. His role was crucial in giving final shape to an original and individualistic creation of Japanese mathematics, an approach known as the *Wasan*, the importance of which is reflected in the fact that the word *Wasan* is sometimes used to describe the whole indigenous mathematical tradition of Japan.

Although Seki Takakazu was strongly influenced by Chinese mathematics, it is difficult to determine precisely which Chinese works he studied. However, he was a prolific writer, and a number of his publications are either transcriptions of mathematics from Chinese into Japanese or commentaries on certain works of well-known Chinese mathematicians.

There is little known about the personal life of Seki Takakazu. He was born in Fujioka, the second son in a samurai warrior family, and began writing mathematical texts when he was employed as an auditor by

Shogun Tokugawa Ienobu. In later life he became the landlord of a "300 person" village.

Seki Takakazu's interests ranged widely over a number of fields of mathematics (fifteen, according to certain Japanese historians of mathematics), including constructions of calendars, recreational mathematics, magic squares and magic circles, and solutions of higher-order and indeterminate equations. He had two able disciples, the brothers Katahiro Takebe (1664–1739) and Kataaki (1661–1739). Between them they brought out an encyclopedia of the mathematics of their forerunners, titled *Taisei Sankyo* (Large Account of Mathematics), which ran to twenty volumes. Seki Takakazu died in Edo (now Tokyo) on December 5, 1708.

According to Shigeru and Rosenfeld (1997), Seki Takakazu made contributions in advance of, or contemporaneous with, the works of European mathematicians of the same period: he discovered determinants ten years before Leibniz, extended the Chinese work on solving numerical equations of higher order using the Horner-Ruffini method, discovered the conditions for the existence of positive and negative roots of polynomials, did innovative work on continued fractions, and discovered the Bernoulli numbers a year before Bernoulli. In geometry, he calculated the value of π correct to nine decimal places by applying an ingenious extrapolation to a polygon with 2^{17} sides. This work was extended by his disciple Takebe, who obtained an accuracy, remarkable for the time, of thirty-one decimal places, from a formula equivalent to Taylor's expansion of $\sin^{-1}x$, fifteen years before Euler. Other notable geometrical contributions include the calculation of the volume of a sphere using an original integral method called *enri*, calculations on conic sections, and on Archimedes' spirals. The last few years of his life were spent mainly on astronomical works, which are found in *Shiyo Sampo* (Mathematical Methods of Computing Four Points on the Lunar Orbit) and *Tenmon Sūgaku Zatcho* (Notes on Astronomy and Mathematics).

Many of Seki Takakazu's works are remarkable considering the standard of Japanese work of that time. The subjects he took up were mainly those that interested Chinese mathematicians of the Song and Yuan dynasties. These included Yang Hui's *Suan Fa* (Method of Computation), Shen Kuo's *Meng Ji Xi Tan* (Dream Pool Essays), Zhu Shijie's *Si Yuan Yu Jian* (Precious Mirror of the Four Elements), Li Ye's *Ce Yuan Hai Jing* (Sea Mirror of Circle Measurements), and Qin Jiushao's *Shu Shu Jiu Zhang* (Nine Sections of Mathematics).

As an illustration of the way that Seki Takakazu put his own stamp on the Chinese work, consider the first and the last works in the above list. Yang Hui's work on magic squares indicates that the Chinese interest in the subject had both mathematical and philosophical significance. As mentioned in the previous chapter, Chinese philosophers believed that the smallest magic square (*Luo Shu*) had mysterious power. Chinese mathematicians were constrained because of this "magical" element to confine themselves only to squares of certain sizes. Seki Takakazu was not under any such constraint: his interest in the subject was purely mathematical. As a result, the work on this topic found among Seki Takakazu and other Japanese mathematicians after him ranged widely to include larger squares as well as other shapes such as circles, cubes, and spheres.

One may detect a similar disposition in Seki Takakazu's approach to the problem of solving indeterminate equations. In China, indeterminate equations arose from astronomical studies in particular to compute "accumulated years from an initial epoch." Shigeru and Rosenfeld (1997) maintain that it is likely, as in the case of magic squares, that Seki Takakazu ignored the metaphysics contained in Qin Jiushao's works and concentrated instead on the mathematics. However, Seki Takakazu's work on indeterminate equations does not represent any significance advance on *Shu Shu Jiu Zhang*.

There is an area in which Seki Takakazu had a far-reaching impact on Japanese mathematics: in his book *Jinko-ki*, he invented a form of notational algebra that helped him to understand and decipher the solutions of higher-order numerical equations that Qin Jiushao had obtained using the *tian yuan* notation. Seki Takakazu proceeded to develop this approach, known as the *Wasan*, to study parabolas, hyperbolas, and the spirals of Archimedes. He applied this approach to calculate a determinant of order 5 in his work *Kaifukudai-no-ho*, published in 1683. Seki Takakazu's disciples took up his work and developed the Japanese mathematical tradition of *Wasan* to a significant degree, climaxing in the work of Yoshiro Kurushima (d. 1757) and Ryohitsu Matsunaga (d. 1744). For further details of the strengths and weaknesses of the *Wasan* tradition, the publications of Murata (1975, 1980, and 1994) are particularly valuable.

Chinese Mathematics: A Final Assessment

From our survey in this chapter and the last, we can identify the areas where Chinese contributions were notable:

1. As early as the Shang dynasty (c. fourteenth century BC), there emerged a system of notation consisting of fourteen symbols, with

thirteen representing numbers (1 to 10 and signs for hundreds, thousands, ten thousands) and an additive sign corresponding to "and." Numbers were written using a decimal system. For example, 4,876 was written as $(4 \times 1,000) + (8 \times 100) + (7 \times 10) + 6$ using specific symbols for units and powers of ten.

2. The development of an algorithm for extracting square and cube roots was first explained in the *Jiu Zhang*, and elaborated and refined by Sun Zu (c. AD 300) and other commentators. The thirteenth-century mathematicians extended the algorithm to the extraction of roots of any order using a Chinese version of the Horner-Ruffini method. It is possible that the Islamic mathematician Al-Kashi (c. AD 1400) and later Europeans were influenced by the Chinese method.

3. It is in China, near the beginning of the Christian era, that the concept of negative numbers and of operations with them appears for the first time. These numbers appeared in India over five hundred years later.

4. In a third-century AD commentary on the *Zhou Bi Jing* there appears one of the earliest visual "proofs" of the *gou gu* (Pythagorean) theorem. A detailed discussion of the applications of this theorem to practical problems is found in the *Jiu Zhang* and its commentaries.

5. Early applications of the "rule of three" and the "rule of false position" are found in Chinese mathematics during the first few centuries AD. The methods also appear in India soon after that and then in the Islamic world, from which they passed on to the West.

6. A notable contribution of Chinese mathematics was the development of numerical methods of solving higher-order equations in the thirteenth century—methods bearing an uncanny resemblance to the Horner-Ruffini method, which was discovered in Europe at the beginning of the nineteenth century.

7. From Sun Zu (c. AD 300) onward, the Chinese forged ahead with the solution of indeterminate equations of the first degree, using an approach further developed on the basis of continued fractions by Lagrange, Euler, and Gauss many years later.[30]

8. Pascal's triangle of binomial coefficients was known in China as early as AD 1100. Chinese mathematicians used it as an aid to root

extractions. Later appearances of the triangle occur in the Islamic world (in Samarkand) and then in Europe. Pascal's interest in the triangle that was named after him was more as an aid toward establishing his theory of probability. Work on what looks like Pascal's triangle is found in India, but the triangles were used more as representational devices to show combination than as analytical tools for numerical solutions of equations. We will discuss the Indian work in the next chapter.

9. In the work of some Chinese mathematicians can be seen attempts to blend geometric and algebraic approaches that are reminiscent of the work of al-Khwarizmi (c. AD 800), discussed in chapter 11.

10. The values of π estimated by Liu Hui (c. AD 200) and Zu Chong-zhi (c. AD 400) remained the most accurate values for a thousand years.

11. Practical geometrical problems from the *Jiu Zhang* (early years of the common era), such as the broken bamboo problem, are found in the work of the ninth-century Indian mathematician Mahavira. In his work there also appears an erroneous rule for calculating the area of a segment of a circle, which one also finds in the *Jiu Zhang*.

12. The earliest appearance of the Chinese "remainder" theorem is in the work of Sun Zu (fourth century AD). It appears later in India in the work of Brahmagupta and then later in Europe. This is also true of the problem of the hundred fowls, which appears in China in the fifth century AD and then reappears later in India in the work of Bhaskara-charya (twelfth century) and in Europe in the works of the Italian Leonardo of Pisa (also known as Fibonacci) in the thirteenth century.

13. The rule of "double false position" has its origins in the *Jiu Zhang*. It reappears in Europe via the Islamic world.

14. Numerical solutions of equations of order 3 are found in the *Ji Gu Suan Jing* (Contination of Ancient Mathematics) of Wang Xiaotong in the seventh century AD. They then appear in the work of Leonardo of Pisa via the Islamic world.

Correspondences similar to these can be established in other cases. However, more research needs to be done before we can be more certain about the nature, direction, and mode of the interchange of mathematical ideas that took place between China and other cultural centers.

Throughout history, China has been relatively isolated from other cultures, partly by sheer geographical distance. Archaeological evidence suggests that at the time of the river valley civilizations there were contacts between Egypt, Mesopotamia, and the Harappan cultures; the civilization that developed along the Yellow River, however, was remote and separated from areas to the south and west by natural barriers such as the Himalayas and the central Asian plains. But these geographical barriers were not sufficient to exclude all contacts throughout history. The major impact of Chinese culture and mathematics (particularly the *Jiu Zhang*) on Japan, Korea, and other neighboring countries is clear, but Chinese contacts with areas to the south and west are more difficult to establish.

By the second century AD trade over the silk routes from China to the West was at its height, and along with the goods went ideas and techniques. In the centuries to come, the Classical civilizations of both the East and the West would suffer invasions small and large, culminating in Mongol hegemony over vast stretches of the Eurasian plains, which both served as an instrument for diffusion and led to the convergence of ideas and technological practices. In examining the dissemination of Chinese mathematics, one needs to look at the Indian and Islamic connections.

There is only fragmentary evidence of Chinese-Indian cultural and scientific contacts before the rise of Buddhism around the fourth century AD. A number of Chinese Buddhist scholars (notable among the early travelers were Fa Xian, c. AD 400, and Xuan Zang, c. AD 650) made their pilgrimage to holy places in India, bringing back many texts for translation. Among the places they visited were monasteries such as Nalanda and Taxila, which were Indian centers of scholarship not only in religion but in medicine, astronomy, and mathematics too. Few of the writings or commentaries of these Buddhist pilgrims from China have been examined for what they reveal about Indian science; the main interest has been in their religious and sociological content.

Then there is the evidence of Chinese diplomats posted at the court of the Guptas in India around the fifth century AD. And from the seventh century, there is evidence that translations were made of Indian astronomical and mathematical texts, such as the *Bo-luo-men Suan Fa* (Brahman Arithmetical Rules) and *Bo-luo-men Suan Jing* (Brahman Arithmetical Classic) mentioned in the records of the Sui dynasty. These works are no longer extant, and so it is difficult to assess how influential they were on Chinese science. However, there is clearer evidence of Indian influence on

Chinese astronomy and calendar making during the Tang dynasty. Indian astronomers were employed in the Imperial Bureau of Astronomy and charged with the tasks of preparing accurate calendars, some of which contain the names of Indian astronomers. One of the Indians, whose Chinese name was *Xi Da* (Siddharta), was reputed to have constructed in AD 718 a calendar, based on the Indian *Siddhanta* of Varamahira (c. AD 550), on the orders of the first emperor of the Tang dynasty. The text contains sections on Indian numerals and operations, and sine tables. There are also sine tables at intervals of 3° 45′ for a radius of 3,438 units, which are the values given in the Indian astronomical texts *Aryabhatiya* and *Suryasiddhanta*.[31] This is the earliest record of a sine table in any Chinese text. Unfortunately, there is little evidence of Chinese science in any of the extant Indian texts.

The Islamic connection is probably better documented. One of the better-known *hadiths* (i.e., utterances of the prophet Muhammad that have religious sanction) is: "Seek learning, though it be as far away as China." There are a number of reports of political and diplomatic links between the Islamic world and China to supplement trade relations. Arab travelers, including ibn Battuta (c. AD 1350), reputedly the greatest traveler of medieval times, gave detailed accounts of Chinese society and science, including shipbuilding, the manufacture of porcelain, the use of paper money, and even a comprehensive system of old-age pensions. Chinese mathematics may have made specific borrowings from Islamic sources: it is possible that trigonometric methods used in astronomy may have been transmitted through Arab and Indian contacts. In constructing a calendar in the fourteenth century, Gou Shoujing used spherical trigonometric methods reminiscent of Islamic work. It is possible that Euclid's geometry may have reached China at the end of the thirteenth century via the Islamic scholars, though the lack of interest in the Euclidean method—or, more probably, a lack of sympathy—meant that this knowledge was forgotten until it was reintroduced by the Jesuits. Finally, the lattice method of multiplication, which will be examined in chapter 10, appears in Chinese mathematical texts at the end of the sixteenth century. It is not clear whether the agents of transmission were the Arabs or the Portuguese.

Whether there was any direct transmission of mathematical knowledge from China to the West remains a matter of conjecture. However, the possibility should not be dismissed out of hand, as many historians of mathematics are inclined to do—either because they find the idea unpalatable or because there is insufficient documentary evidence. The fact remains that, as early as the third century BC, Chinese silk and fine ironware were to be found in the markets of imperial Rome. And a few centuries later a whole range of technological innovations found their way slowly to Europe. It is not unreasonable to argue that some of China's intellectual products, including mathematical knowledge, were also carried westward to Europe, there perhaps to remain dormant during Europe's intellectual Dark Ages but coming to life once more with the cultural awakening of the Renaissance.

During the late seventeenth and the eighteenth centuries, Europe became aware of the Chinese intellectual heritage. The Jesuits were responsible through their translations for awakening the interest of people like Voltaire, Gottfried Leibniz, and François Quesnay in Chinese thought and science. Leibniz (1646–1716), one of the founders of modern mathematics, was in the forefront of promoting a universal system of natural philosophy based on Confucian writings. He founded the Berlin Society of Science with the express purpose of "opening up China and the interchange of civilizations between China and Europe." While there is already some recognition of Europe's debt to China in the realms of philosophy and the arts (Edwardes 1971), the possibility of an "east to west" passage of scientific ideas during this period through the Jesuit connection has hardly been explored.

If these conjectures are implausible, then so too must be the attribution of Greek or European origins to so many developments in mathematics and astronomy in other cultures. However, it is my belief that if the idea of a westward transmission of mathematics were to be taken more seriously, and research were to be channeled in this direction, perhaps it would be only a matter of time before further evidence of east-west links comes to light.[32]

Notes

1. It should be recognized here that the value of each of the two proofs mentioned varies, depending on the cultural context. Thus, trying to axiomatize a Chinese proof involving what we would today describe as a "cut-and-paste" method would be a cumbersome exercise, just as any attempt to remove Euclid's proof from an *Elements*-type approach would run counter to common sense. For an attempt by the Islamic mathematician al-Kuhi to refocus proofs contained in Euclid's *Elements* from an axiomatic basis to one of analysis and synthesis (a project close to the spirit of the "cut-and-paste" method of the Chinese), see Berggren and Van Brummelen (2005).

2. A reconstruction of Liu's explanation is given in Dauben (2007, p. 285).

3. Problem 32 in the *Jiu Zhang* reads: *Given a circular field, the circumference is 181 bu and the diameter $60\frac{1}{3}$ bu. Tell what is the area.* Liu points out that this corresponds to

an inaccurate ratio of circumference to diameter of 3. A more accurate diameter of $57\frac{13}{22}$ *bu* would result from the ratio of 22/7.

4. For a detailed discussion of Liu's demonstration of these results, see Dauben (2007, pp. 235–39) and Martzloff (1997a, pp. 277–80).

5. It is worth reiterating that none of the ancient societies discussed in this book had the modern concept of π, and hence the qualification suggested by the term "implicit."

6. Rogers (2009) has argued that a closer study of the works of Euclid and of Archimedes indicates that anything you can do with circumscribed polygons can be done just as well with inscribed ones. His argument is quite complex and best left to the interested reader to follow up.

7. It may be argued that the concept of the "method of exhaustion" strictly involves a process of limiting the number captured between two converging bounds. In that case, it cannot be claimed that the "method of exhaustion" existed in China.

8. This was reported by Yang Hui in *Xiang Jie Jiu Zhang Suan Fa* (Detailed Analysis of the Mathematical Mehods in the Nine Chapters), who describes a table of binomial coefficients up to the sixth power that he attributes to Jia Xian.

9. These examples have been adapted from Dauben (2007, pp. 323–27) and Hodgkin (2005, pp. 92–93).

10. For details of the derivation, see Dauben (2007, pp. 325–27).

11. This is a view that originated with an article by Wang Ling and Joseph Needham (1955), who argued that the method was already present in the *Jiu Zhang*. This is not a view that is universally accepted. For a brief discussion see Martzloff (1997a, pp. 247–49).

12. Johann Schreck (1576–1630) was a fellow student of Viète. He arrived in China in 1619 and, after learning Chinese, wrote and translated several Chinese textbooks on mathematics, engineering, medicine, and astronomy, alongside other Jesuit and Chinese scholars. He was in contact with important scientists of his time, including Johannes Kepler, who sent Schreck his newest astronomical opus, the *Rudolphine Tables*, which arrived only after Schreck's death. For further details, see Iannaccone (1998).

13. For an interesting technical discussion of the historical development of the subject, with special emphasis on China, see Shen Kangsheng (1988).

14. It is this problem of Sun Zu from which the name "Chinese remainder theorem" originates, a name that was given after the problem came to be known in the West.

15. If two integers A and B are divided by a common integer m (called the modulus) and leave the same remainder r, then A is said to be "congruent to B modulus m." This is written as $A \equiv r \pmod{m} \equiv B$.

16. Underlying this procedure is the use of the Euclidean algorithm, an algorithm to determine the greatest common denominator (GCD) of two integers. Its major significance is that it does not require finding the factors of the two integers. It is one of the oldest algorithms known, appearing in Euclid's *Elements* around 300 BC (Book VII, proposition 2). Given two natural numbers a and b, not both equal to zero: check if b is zero; if yes, a is the GCD. If not, repeat the process using, respectively, b, and the remainder after dividing a by b. The remainder after dividing a by b is usually written as a mod b. The Euclidean algorithm can be used in any context where division with remainder is possible.

17. Let n be a positive integer. Integers a and b are said to be congruent modulo n if they have the same remainder when divided by n. This is shown by writing $a \equiv b \pmod{n}$.

18. In another work, Yang Hui offers a similar problem that involves the purchase of three types of oranges: the total cost of 100 oranges is 100 coins, and the answer sought is how many oranges of the three types are bought. Yang Hui attempts to explain the method by which he obtains the solution, but the logic of his solution is not clear. See Dauben (2007, p. 333).

19. This English translation, suggested by Mikami (1974, p. 89), has been questioned by Hoe (1977, p. 41), who offers his own as "Mirror [trustworthy as] jade [relative to the] four origins [unknown]," which is incomprehensible unless explained!

20. For expressions of this widely held view among the historians of Chinese mathematics and science, see Liu Dun (1994, pp. 103–4), Martzloff (1994, p. 99), and Sivin (1995, p. 172). Needham (1959, p. 50) talks about "decay," and Mikami (1974, p. 112) labels the Ming scholars "degenerate."

21. Ricci was initially full of praise for Chinese science and technology. In a journal that he kept, he wrote soon after his arrival in China: "In their sciences, the Chinese are very learned: in medicine, moral [sciences], mathematics and astronomy, arithmetic, and finally all the liberal or mechanical arts. It is admirable that a nation, which has never had any relations with Europe, should have reached by its means almost the same results as we with the collaboration of the whole universe." This opinion changed radically over time.

22. In the book titled *Big Bangs: The Story of Five Discoveries That Changed Musical History*, Goodall (2000, pp. 111–12) writes: "Equal Temperament is probably the single most important development in Western European music in the last 400 years and yet most people haven't heard of it. Even musicians don't really understand it, but an enormous amount of the world's most beautiful music wouldn't exist without it. Equal Temperament is . . . to music what the calendar is to the days and nights or what the 24-hour clock is to the minutes and seconds."

23. It is interesting to note that, as early as the second century BC, the Han emperor Wu Di had set up a Bureau of Rites with the task of regulating the performance of ritual

music to ensure harmony between Man and Heaven. The twelve tones represented the cyclic return of a twelve-month year. Each month had its own fundamental pitch—and thus symbolic meaning and cosmological significance were bestowed on the relationship between the musical pitch and the calendar month.

24. For those unaware of musical terminology, a few explanations may be useful. Musical *notes* are the periodic oscillations in air pressure felt though the eardrums. If we plot air pressure against time, we get wavelike shapes. *Frequencies*, measured in hertz (Hz), are the number of cycles a wave completes in a second. The human ear can pick up frequencies from 20 Hz to 20,000 Hz, although aging lowers the upper limit to below 10,000 Hz. Now some *note combinations* (i.e., two or more notes played together) are more pleasing than others, and they obey certain mathematical laws. For example, it has been found in a number of musical traditions that the most pleasing combinations are those in which the numerical ratios between the frequencies are in the ratio of 1:2 (the "octave") or in the ratio of 2:3 (the "perfect fifth"). Thus 440 Hz (considered as the fundamental "tuning" note in Western music) is one *octave* above 880 Hz, and the interval between two notes whose frequencies are 440 Hz and 660 Hz respectively constitutes a "perfect fifth." Now if a *melody* (i.e., a sequence of notes played together) with frequencies 440, 660, and 733.3 Hz is played together with another melody exactly one octave apart, whose note frequencies are 880, 1,320, and 1,466.6 Hz, the result will be pleasing to the ear. However, if the second melody has frequencies 550, 825, and 916.6 Hz, the result will sound discordant and not pleasing to the ear. Pythagoras and his disciples developed a theory to connect numbers, musical notes, and the movement of the planets. Some of the esoteric and metaphysical elements of the theory have been discarded, but the credit for recognizing the mathematics behind musical notes rests with the Pythagoreans.

25. Galilei's solution in his *Dialogo* was to use the ratio of 18:17 (i.e., 1.058823529 . . .) between the semitones, which was very close to equal temperament tuning although not as accurate as Chu's based on taking the twelfth root of 2.

26. These dates are highly conjectural and the "thirty years" stated may well be an underestimate. We know that the preface to Chu's first publication on the subject was written in January 1581, which implies that he must have arrived at his theory at least three years before, or earlier. Stevin's undated work was first published after his death in 1620.

27. Traditions conforming to five- and seven-tone equal temperament are quite common. The Indonesian gamelan is tuned to a five-tone equal temperament, while a Thai or Ugandan *Chopi* xylophone approximates to a seven-tone equal temperament. Further details of different tone temperaments found in ethnomusicology are discussed in Tenzer (2006).

28. Cho (2003, p. 200) writes: "The justification that Zhu [i.e., Chu] gave in first fixing this *juibin* [i.e., a particular note] by equalizing the distance between all the adjacent

tones within the octave is of considerable interest: it is based on calendrical science. That is, the correct length of (or the number of days within) a solar year can be accurately determined by measuring the length of the periods between the summer and winter solstices or the vernal and autumnal equinoxes." In other words, what Chu was recommending was the validity of a solar calendar measurement to replace what had been in use in China since time immemorial, a lunar calendar in which equinoxes fall on different days and even in different months.

29. This section on Japanese mathematics concentrates on the work of one man, Seki Takakazu, arguably the greatest mathematician produced by Japan before the eighteenth century. However, there was a period between the seventh and nineteenth centuries when Japan was cut off from the outside by imperial decree. During that period an unusual form of indigenous mathematics flourished independent of what was happening in the rest of the world. A wide selection of geometry problems were inscribed on wooden tablets called *sangaku* and posted in Buddhist temples and Shinto shrines all across the country by a variety of people including farmers, merchants, and warriors (or samurai). For further details of this unique mathematical tradition, see the book by Fukagawa and Rothman (2008). Their book also provides a useful survey of Japanese mathematics before the introduction of Western mathematics.

30. For further details, see Shen Kangsheng (1988).

31. The choice of a radius of value 3,438 was determined by the practice of dividing the circumference (C) of a circle into $360 \times 60 = 21,600$ equal parts. If the length of the arc of each of these equal parts is one unit, and the value of π is taken as 3.1416 (Aryabhatiya's value), then the radius (r) of the circle can easily be established from the formula $C = 2\pi r$ to be $r = 21,600/62,832 = 3,438$ (to the nearest integer). Furthermore, in the construction of a sine table for angles between 0 and 90°, Aryabhata divided the quadrant into 24 equal parts so that the table would give the sines of the multiples of the basic angle (90/24), or $3°45'$.

32. Since the first edition of this book in 1991, there has been considerable work on transmissions to and from India, the Islamic world, and China. Chapter 10 contains a discussion of the methodologies of establishing transmissions across cultures and their application to a possible migration of Kerala mathematics to Europe through the Jesuit conduit. Further bibliographical information is available in that chapter. Another useful reference is Y. Dold-Samplonius et al. (2002), which contains the proceedings of a conference on mathematical transmissions held in 2000.

Chapter Eight
Ancient Indian Mathematics

A Restatement of Intent and a Brief Historical Sketch

Ancient Indian history raises many problems. The period before the Christian era takes on a haziness that seems to have prompted opposing reactions. There are those who make excessive claims for the antiquity of Indian mathematics, and others who go to the opposite extreme and deny the existence of any "real" Indian mathematics before about AD 500. The principal motive of the former is to emphasize the uniqueness of Indian mathematical achievements. In this view, if there was any influence, it was always a one-way traffic from India to the rest of the world. The motives of the latter are more mixed. For some their Eurocentrism (or Graeco-centrism) is so deeply entrenched that they cannot bring themselves to face the idea of independent developments in early Indian mathematics, even as a remote possibility.[1]

A good illustration of this blinkered vision is provided by a widely respected historian of mathematics at the turn of the twentieth century, Paul Tannery. Confronted with the evidence from Islamic sources that the Indians were the first to use the sine function as we know it today, Tannery devoted himself to seeking ways in which the Indians could have acquired the concept from the Greeks. For Tannery, the very fact that the Indians knew and used sines in their astronomical calculations was sufficient evidence that they must have had it from the Greeks.[2] But why this tunnel vision? The following quotation from G. R. Kaye (1915) is illuminating:

> The achievements of the Greeks in mathematics and art form the most wonderful chapters in the history of civilisation, and these achievements are the admiration of western scholars. It is therefore natural that western investigators in the history of knowledge should seek for traces of Greek influence in later manifestations of art, and mathematics in particular.

It is particularly unfortunate that Kaye is still quoted as an authority on Indian mathematics. Not only did he devote much attention to showing the derivative nature of Indian mathematics, usually on dubious linguistic grounds (his knowledge of Sanskrit was such that he depended largely on indigenous *pandits* for translations of primary sources), but he was prepared to neglect the weight of contemporary evidence and scholarship to promote his own viewpoint. So, while everyone else claimed that the Bakhshali Manuscript (discussed at the end of this chapter) was written or copied from an earlier text dating back to the first few centuries of the Christian era, Kaye insisted that it was no older than the twelfth century AD. Again, while the Islamic sources unanimously attributed the origin of our present-day numerals to the Indians, Kaye was of a different opinion. And the distortions that resulted from Kaye's work have to be taken seriously because of his influence on Western historians of mathematics, many of whom remained immune to findings that refuted Kaye's inferences and established the strength of the alternative position much more effectively than is generally recognized.

This tunnel vision is not confined to mathematics alone. Surprised at the accuracy of information on the preparation of alkalis contained in an early Indian textbook on medicine (*Sushruta Samhita*)[3] dating back to a few centuries BC, an eminent chemist and historian of the subject, M. Berthelot (1827–1909), suggested that this was a later insertion, after the Indians had come into contact with European chemistry!

While non-European chauvinism (on the part of, for example, the Arabs, Chinese, and Indians) does persist, "arrogant ignorance"—as J. D. Bernal (1969) described the character of Eurocentric scholarship in the history of science—is the other side of the same coin. But the latter tendency has done more harm than the former because it rode upon the political domination imposed by the West, which imprinted its own version of knowledge on the rest of the world.

Table 8.1 offers a brief summary of the main events in the long history of India as a backdrop to the development of mathematics; it divides Indian history up to the beginning of the sixteenth century into six periods. The map of India in figure 8.1 shows places mentioned in the text. The earliest evidence of mathematics is found among the ruins of the Indus Valley civilization, which goes back to 3000 BC. (It is perhaps more appropriately referred to as the Harappan civilization, since at its peak it spread far beyond the Indus Valley itself.) Around 1500 BC, according to the traditional —though increasingly contentious—view among historians, a group of

TABLE 8.1: CHRONOLOGY OF INDIAN HISTORY AND MATHEMATICS

Period	Main historical events	Mathematics	Notable mathematicians
3000–1500 BC	The Indus Valley civilization (script undeciphered) covering 1–2 million square km; main urban centers Harappa, Lothal, and Mohenjo-Daro	Weights, artistic designs, "Indus scale"; brick technology probably influenced the construction of Vedic altars in the next period	
1500–500 BC	The coming of the Aryans; the formation of Hindu civilization; the emergence of the *Code of Manu*; the recording of the Vedas and *Upanishads*	*Vedangas* and *Sulbasutras*; problems in astronomy, arithmetical operations, Vedic geometry	Baudhayana, Apastamba, Katyayana
500–200	The establishment of Indian states; the rise of Buddhism and Jainism; contacts with Persia maintained; the Mauryan empire, culminating in the reign of Asoka, who spread Buddhism abroad	Vedic mathematics continues during the earlier years but declines with ending of ritual sacrifices; beginnings of Jaina mathematics: number theory, permutations and combinations, the binomial theorem; astronomy	
200 BC–AD 400	Triple division: Kushan dynasty (North), Pandyas (South), Bactrian-Persian (Punjab); pervading influence of Buddhism in art and sculpture	Jaina mathematics: rules of mathematical operations, decimal-place notation, first use of 0; algebra including simple, simultaneous, and quadratic equations; square roots; details of how to represent unknown quantities and negative signs	

continued

Table 8.1: Continued

Period	Main historical events	Mathematics	Notable mathematicians
400–1200	Imperial Guptas reaching their height in the reign of Harsha (606–647); flowering of Indian civilization as shown in science, philosophy, medicine, logic, grammar, and literature	The Classical period of Indian mathematics; important works: the Bakshali Manuscript, *Aryabhatiya, Pancasiddhantika, Aryabhatiya Bhasya, Maha Bhaskariya, Brahma Shputasiddhanta, Patiganita, Ganita Sara Samgraha, Ganitilaka, Lilavati, Bijaganita*	Aryabhata I, Varahamihira, Bhaskara I, Brahmagupta, Sridhara, Mahavira, Bhaskara II (also known as Bhaskaracharya)
1200–1600	Early Muslim dynasties; birth of Sikhism; the Hindu kingdom of Vijaynagar in the South	Decline of mathematics and learning in the North; the rise of the Kerala school of astronomy and mathematics; work on infinite series and analysis	Narayana, Madhava, Nilakantha

people descended from the north and destroyed the Harappan culture, but not before they had absorbed some of its features. These invaders are often referred to as "Aryans"—a term that has acquired an unfortunate connotation in modern times through its association with the Nazis.

The Aryans were a pastoral people, speaking a language that belonged to the Indo-European family. It remained for a long time a spoken rather than a written language, with writing initially restricted to the vernaculars. Over the years this language, Sanskrit, developed sufficiently to become a suitable medium for religious, scientific, and philosophical discourse. Its potential for scientific use was greatly enhanced as a result of the thorough systematization of its grammar by Panini, about 2,600 years ago. In a book titled *Astadhyayi* (Eight Chapters), Panini offered what must be the first attempt at a structural analysis of a language. On the basis of just under four thousand sutras (i.e., rules expressed as aphorisms), he built virtually

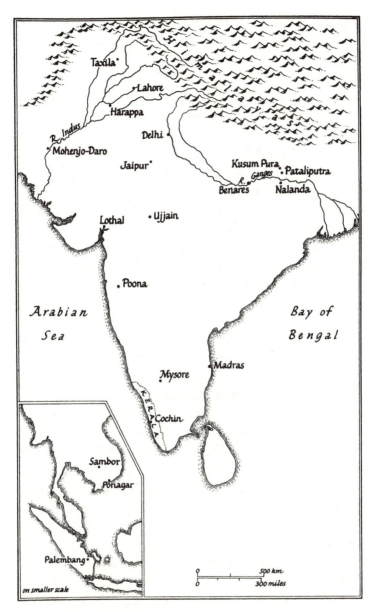

FIGURE 8.1: Map of India and (inset) Southeast Asia

the whole structure of "Classical" Sanskrit language, whose general "shape" hardly changed for the next two thousand years. Sanskrit served as a useful medium for recording early scriptural texts such as the Vedas and *Upanishads*, early scientific literature such as the *Vedangas* (or Limbs of the Vedas), and early rules of social conduct such as the *Code of Manu*.

An indirect consequence of Panini's efforts to increase the linguistic facility of Sanskrit soon became apparent in the character of scientific and mathematical literature. This may be brought out by comparing the grammar of Sanskrit with the geometry of Euclid—a particularly apposite comparison since, whereas mathematics grew out of philosophy in ancient Greece, it was, as we shall see, partly an outcome of linguistic developments in India.

The geometry of Euclid's *Elements* starts with a few definitions, axioms, and postulates and then proceeds to build up an imposing structure of closely interlinked theorems, each of which is in itself logically coherent and complete. In a similar fashion, Panini began his study of Sanskrit by taking about seventeen hundred basic building blocks—some general concepts, vowels and consonants, nouns, pronouns and verbs, and so on—and proceeded to group them into various classes. With these roots and some appropriate suffixes and prefixes, he constructed compound words by a process not dissimilar to the way in which one specifies a function in modern mathematics. Consequently, the linguistic facility of the language came to be reflected in the character of mathematical literature and reasoning in India. Indeed, it may even be argued that the algebraic character of ancient Indian mathematics is but a by-product of the well-established linguistic tradition of representing numbers by words.

The third period of Indian history began around 800 BC. It saw not only the establishment of two of the great religions originating in India, Buddhism and Jainism, but also the growth of independent states, a number of which were later merged to form the first of the great empires of India, the Mauryan empire. This period marked the decline of Vedic mathematics and the gradual emergence of the Jaina school, which was to do notable work in number theory, permutations and combinations, as well as other abstract areas of mathematics.

The fourth period, from about 200 BC, was a period of instability and fragmentation brought about by waves of foreign invasions. But it was also a time of useful cross-cultural contacts with neighbors and with the Hellenistic world, bringing fresh ideas into Indian science and laying the foundation for great advances in the next period. The Kushan empire became an important vehicle for spreading not only Buddhist religion and art but also Indian science, particularly astronomy, into western Asia. Probably the only piece of existing mathematical evidence from this period is the Bakhshali Manuscript. However, the earlier dating of this manuscript to the third century is based on an estimate made by Hoernle, who was the first to study it. On the basis of recent evidence, notably that of Hayashi (1995), the manuscript cannot be dated earlier than the eighth century.

The fifth period, from the third to the twelfth centuries, is often referred to as the Classical period of Indian civilization. The earlier part of this period saw much of India ruled by the imperial Guptas, who encouraged the study of science, philosophy, medicine, and other arts. Mathematical activities reached a climax with the appearance of the famous quartet: Aryabhata, Brabmagupta, Mahavira, and Bhaskaracharya. Their lives and works will be examined in the next chapter. Indian work on astronomy and mathematics spread westward, reaching the Islamic world, where it was absorbed, refined, and augmented before being transmitted to Europe.

The last period, which we may describe as the "medieval" period of Indian history, saw the rise of great states in southern India and a migration of mathematics and astronomy from the North to the South, probably as a result of political upheavals. It was believed for a long time that mathematical development came virtually to a stop in India after Bhaskaracharya in the twelfth century. There may be some element of truth in this as far as the North was concerned, but in the South—and particularly in the Southwest, in the area corresponding to the present-day state of Kerala—this was a period marked by remarkable studies of infinite series and mathematical analysis that predated similar work in Europe by about three hundred years.

The mathematics of Kerala will be presented in a separate chapter. In this chapter we examine Indian mathematics from its early beginnings to just before the Classical period; in the next chapter we consider mainly Classical Indian mathematics. The development of Indian numerals is dealt with in this chapter, though there is some historical overlap, particularly when one considers the spread of the numerals into countries such as Cambodia and Java to the east, and into the Islamic world to the west. The reader may wish to refer to table 8.1 and figure 8.1 whenever necessary to sketch in the historical and geographical background to this and the next chapter.

Math from Bricks: Evidence from the Harappan Culture

Between 1921 and 1923 a series of archaeological excavations along the banks of the Indus uncovered the remains of two urban centers, at Harappa and Mohenjo-Daro, dating back to about 3000 BC. Subsequent searches over the last four decades have revealed further remains spread across an area of about 1.2 million square kilometers, including not only the Indus Valley but parts of East Punjab and Uttar Pradesh, northern Rajasthan, the coastal areas of Gujerat (which contained the major port and the third of the large urban settlements of this civilization, Lothal), and northern areas near the Persian border. Over seventy sites, large and small, have been excavated to uncover this most dispersed of the early civilizations, hereafter referred to as the Harappan culture.

It was a highly organized society, with the towns supplied by surrounding agricultural communities, which cultivated wheat and barley and raised livestock. Urban development was regulated by planning and characterized by a highly standardized architecture. There is every possibility that the town dwellers were skilled in mensuration and practical arithmetic of a kind similar to what was practiced in Egypt and Mesopotamia. Alas, the Harappan script remains undeciphered, so our evaluation of the mathematical proficiency of this civilization must be based on excavated artifacts. The archaeological finds described below do provide some indication, however meager, of the nature of the numerate culture that this civilization possessed.

1. A number of different plumb bobs of uniform size and weight, showing little change over the five hundred years for which evidence is available, have been found throughout the vast area of the Harappan culture. This uniformity of weights over such a wide area and time is quite unusual in the history of metrology. Rao (1973), who examined the considerable finds at Lothal, showed that the weights could be classified as "decimal": if we take the plumb bob weighing approximately 27.584 grams as a standard, representing 1, the other weights form a series with values of 0.05, 0.1, 0.2, 0.5, 2, 5, 10, 20, 50, 100, 200, and 500. Such standardization and durability is a strong indication of a numerate culture with a well-established, centralized system of weights and measures.

2. Scales and instruments for measuring length have been discovered at Mohenjo-Daro, Harappa, and Lothal. The Mohenjo-Daro scale is a fragment of shell 66.2 mm long, with nine carefully sawn, equally spaced parallel lines, on average 6.7056 mm apart. The accuracy of the graduation is remarkably high, with a mean error of only 0.075 mm. One of the lines is marked by a hollow circle, and the sixth line from the circle is indicated by a large circular dot. The distance between the two markers is 1.32 inches (335 mm) and has been named the "Indus inch."

There are a number of interesting links between this unit of measurement (if indeed this is what it was) and others found elsewhere. A Sumerian *shushi* is exactly half an Indus inch, which would support other archaeological evidence of a possible link between the two urban civilizations. In northwestern India a traditional yard, known as the *gaz*, was in use from very early times; in the sixteenth century the Mughal emperor Akbar even attempted (unsuccessfully) to have the *gaz* adopted as a standard measure in his kingdom. The *gaz*, which is 33 inches (840 mm) by our measurement, equals 25 Indus inches. Furthermore, the *gaz* is only a fraction (0.36 inches) longer than the megalithic yard, a measure that seems to have been in use in northwestern Europe around the second millennium BC. This has led to the conjecture that a decimal scale of measurement may have originated somewhere in western Asia and spread widely—as far as Britain, and to ancient Egypt, Mesopotamia, and the Indus Valley (Mackie 1977).

A notable feature of the Harappan culture was its extensive use of kiln-fired bricks and the advanced level of its brick-making technology. Chattopadhyaya (1986) has argued for a closer examination of this activity, which should give us some vital clues about the direction and character of later mathematical developments in India. So let us examine the socioeconomic origins of making and using kiln-fired brick in the Harappan culture.

There is general agreement among archaeologists of the Harappan culture that, during its formative stages, farmers had to produce a substantial agricultural surplus to support a rapidly growing urban population. Presumably this occurred, not as a result of the introduction of any revolutionary agricultural technology, but because of improved knowledge about how to exploit the annual flooding and thus raise agricultural productivity. When the floods receded, principal food crops such as wheat and barley would be sown on the land that had been submerged, to be harvested in March or April. The land would need no plowing, no manure, and no additional irrigation. With a minimum of labor and equipment, a substantial yield could be achieved. The return of the floods would mark the period when the autumn crops, such as cotton and sesame, would be sown for harvesting at the end of the autumn. By recognizing how the floods could be utilized to prepare the land for cultivation, and by following a strict sequence of sowing and harvesting spring and autumn crops, it became possible to build up a large agricultural surplus. However, before such a system of cultivation could be considered, an effective system of flood control was necessary. In areas where stone was not readily available (and this included most of the Harappan sites), there was a need for something more solid than mud brick, which was easily destroyed by rain or floodwater. The technology for firing bricks was thus a momentous discovery. There is evidence, especially from Kalibanga, a pre-Harappan site, that kiln-fired bricks were already in use, but by the time the Harappan culture had matured there had been a veritable explosion in the production and use of such bricks.

The story is told of how William Brunton, a nineteenth-century railway builder, dug up the ruins of Harappa for bricks to use as ballast for a railway line between Multan and Lahore, a distance of over a hundred miles! Despite this and other new uses for antique bricks, a massive quantity of them remain at Harappa. They are exceptionally well baked and of excellent quality, and may still be used over and over again provided some care is taken in removing them in the first place. They contain no straw or other binding material. While fifteen different sizes of Harappan bricks have been identified, the standard ratio of the three dimensions—the length, breadth, and thickness—is always 4:2:1. Even today this is considered the optimal ratio for efficient bonding.

A correspondence between the Indus scales (from Harappa, Mohenjo-Daro, and Lothal) and brick sizes has been noted by Mainikar (1984). Bricks of different sizes from these three urban centers were found to have dimensions that were integral multiples of the graduations of their respective scales. This apparent relationship between brick-making technology and metrology was to reappear about fifteen hundred years later during the Vedic period, in the construction of sacrificial altars of brick. We take up the story again later in this chapter.

The argument for Indian mathematics beginning in Harappa is not based on any direct evidence: nothing like the Mesopotamian clay tablets or the Egyptian papyri exist to testify to its origins. However, the elaborate constructions excavated there cannot be understood without attributing knowledge of a number of geometrical propositions: propositions relating to the shapes and mensuration of rectilinear figures and circles.

Notwithstanding these conjectures, an important source of evidence, the written Harappan script, has so far thrown no light on this subject. It remains unread, despite years of ingenious attempts to do so. The script poses certain problems that were not present in the case of other ancient scripts. The Harappan writing is available only through objects of a very restricted medium: typically in the form of seals made of steatite, each seal, on average, containing a text of only five graphemes (or signs). No bilingual or multilingual text, such as the Rosetta stone in the Egyptian case, is available. Also the language, or the language family, of the Indus script is unknown, although the common assumption made in the past was that it is some form of proto-Dravidian language and had "disappeared" sometime before the middle of the second millennium BC. There is, apparently, a long hiatus between this disappearance and the emergence of the so-called historical period of the Indian subcontinent (the Vedic period), thereby causing a big "hole" in the chronology of Indian mathematics right from its inception. Whether this "hole" has been partly filled by recent research evidence is a moot point to be discussed in a later section.

Deciphering the Indus Script

There have been a number of attempts to read the inscriptions on the steatite seals ever since a substantial collection of them became available around the 1920s. Many of the early attempts were phonetic interpretations based on unverifiable a priori linguistic attributes and other speculations sometimes involving mythological elements of Hindu traditions. A notable "objective" attempt was that of Hunter (1934), who carried out a positional and functional analysis of the signs of the Indus script and suggested methods for splitting the texts into certain sign combinations that constituted "words," irrespective of their linguistic attributes. Following Hunter's work, more recent investigations have involved detailed structural analysis of the texts with the aim of classifying the signs or sign combinations into linguistic units, such as root morphemes, attributes, and other grammatical suffixes, and then reading the texts phonetically, adapting a form of Dravidian as the underlying language.

Any fresh approach to the deciphering of the Indus script needs to take account of three distinctive features that have been identified in earlier studies. First, rich structural regularities exist in the texts, which makes them distinct from other ancient writings. Second, the texts occur in almost all cases on seals, so that the purposes of these seals become a matter of some importance. Finally, a closer examination should be made of the nature and significance of a number of animal and other motifs, named "field symbols" by archaeologists, that occur on many of the seals together with the writing. And the examination of all three features should be made within the historical context of the emergence of the Harappan culture and its aftermath. Some more recent studies (Jeganathan 1993; Kak 1989, 1990;

Subbarayappa 1993) are noteworthy attempts to incorporate these features and look at the seals with fresh eyes.

It is generally accepted that the Indus seals are records of administration and of internal as well as external trade. Indus seals have been found in sites of West Asia, commercial contact between the Harappans and the neighboring areas. There is also general agreement among archaeologists of the existence of an efficient and centralized administration, governing the vast area that constituted the Harappan culture, ensuring a degree of uniformity, whether it was in the construction of houses and public amenities or the promotion of arts and commerce or other activities.

What do the inscriptions on seals mean? To attempt an answer, some features of the Harappa culture need to be highlighted. The geographical spread of this culture makes it highly unlikely that the language was the same throughout the length and breadth of that culture. Even the Mesopotamian civilization, which was contained in a much smaller area than that of the Harappan culture, had regional languages. India has had a multilingual culture from ancient times. There have been marked variations not only in spoken but also in written languages. As to a common script, whatever may have been the situation in other ancient civilizations, the picture could well be different on the Indian subcontinent. Even the earliest extant written records from the first half of the first millennium BC used two scripts—Kharosti and Brahmi—as vehicles of the same language.

Are we therefore justified in assuming that there was one language throughout the Harappan culture—in the urban centers such as Mohenjo-Daro, Harappa, Chanhu-Daro, Lothal, and Kalibangan as well as in the rural settlements—spanning a total area of 1.2 million square kilometers? Yet the assumption of a similar literary script is necessary since the seals, and other inscribed objects found in different parts of the Harappan culture, have more or less identical forms with a noticeable uniformity of their own. Such continuity is more plausible in the case of a well-established system of numerical notation. Subbarayappa (1993) compared the Indus signs to Brahmi, Kharosti, and the Chinese "oracle-bone" numerals and found a number of similarities between these numeral forms. This finding has important implications for examining the origins of our decimal numerals. Future work based on longer written records found from more recent excavations is needed to verify other conjectures linking the Indus script with the Brahmi script, in the case of Kak (1989), or with the Dravidian counting etymology in the case of Jaganathan (1993).

Mathematics from the Vedas

The Sources

Vedic literature[4] went through four stages of development: the *Samhitas* (c. 1000 BC), the *Brahmanas* (c. 800 BC), the *Aranyakas* (c. 700 BC), and the *Upanishads* (c. 600–500 BC). The *Samhitas* were lyrical collections of hymns, prayers, incantations, and sacrificial and magical formulas. From what must have been a vast corpus of such writings, four great collections have come down to us—some of the oldest surviving literary efforts of mankind. In order of their age[5] they are

1. *Rig-veda* (Praise-Knowledge), which contains hymns and prayers to be recited during the performance of rituals and sacrifices

2. *Sama-veda* (Song-Knowledge), which contains melodies to be sung on suitable occasions

3. *Yajur-veda* (Sacrifice-Knowledge), which contains sacrificial formulas for ceremonial occasions

4. *Athara-veda* (Knowledge of the legendary sage Atharavan), a collection of magical formulas and spells

The *Brahmanas*, the second great division of Vedic literature, have been described as practical handbooks for those conducting sacrifices. Of these, an important source of early mathematics, which will be referred to later, is the *Satapatha Brahmana* (Brahmana of a Hundred Paths). As the Brahmana communities gradually dispersed from the north to the eastern and southern parts of the country, there arose a need for a record of ritual procedures and duties for a traveling class of priests, and a means of allocating special tasks among different priests.

For mathematics, a more important source is provided by the "limbs" (or appendixes) to the main *Vedas*, known as the *Vedangas*. These were classified into six branches of knowledge: (1) phonetics, the science of articulation and pronunciation, (2) grammar, (3) etymology, (4) the art of prosody (*chandah*), (5) astronomy, and (6) rules for rituals and ceremonials (*kalpa*). In the last two *Vedangas* are found the most important sources of mathematics from the Vedic period. The evidence is usually in the form of sutras, a peculiar form of writing that aims at the utmost brevity and often uses a poetic style to capture the essence of an argument or result.

By avoiding the use of verbs as far as possible and compounding nouns at great length, a vast body of knowledge was made easier to memorize. Condensation into sutras was also a way of eking out scarce writing materials. This was the form in which the contents of the *Brahmanas* were preserved, and it was adopted later not only by various philosophical and scientific schools but also by writers of books on statecraft (*arthasastra*) and sex manuals (*kamasastra*).

We have referred to the *Kalpasutras* as an important source of Vedic mathematics. This ritual literature included *Srautasutras*, which gave directions for constructing sacrificial fires at different times of the year. Part of this literature dealt with the measurement and construction of sacrificial altars, and came to be known as the *Sulbasutras*. The term originally meant rules governing "sacrificial rites," though later the word *sulba* came to refer to the rope used to lay out altars. Most of what we know of Vedic geometry comes from these sutras.

The Early Antecedents of Vedic Geometry

The *Satapatha Brahmana,* which is about three thousand years old,[6] contains one of the earliest mentions of the technical aspects of altar construction. A section of this text deals with constructing altars to carry out a twelve-day *Agnicayana* (Fire Altars) ceremony to mark the passage of time. The ceremony is carried out on an open ground divided into two sections:

1. The *Mahavedi* (Great Altar) was laid out as an isosceles trapezium with the bases 24 *prakrama* and 30 *prakrama* and width 36 *prakrama*. A *prakrama* is about 0.8 meters. The choice of these dimensions may have been dictated by a calendrical consideration: to reconcile the discrepancy between the two calendars in use at the time—the lunar and the solar—by choosing a nominal year of 360 days, which is four times the sum of the above dimensions. The *Mahavedi* contained within it a *vakrapaksa-syena* (i.e., an altar shaped like a falcon with curved wings), shown in figure 8.2. There were other constructions in the *Mahavedi* of functional and ritual significance, but of little mathematical interest.

2. The *Pracinavamsa* was a smaller rectangular section that lay to the west of the *Mahavedi* and contained three fire altars consisting of *Garhapatya* (of circular shape symbolizing the earth), *Dakshinagni* (of semicircular shape representing space), and *Ahvaniya* (of a square shape representing the sky). These three altars had to be of equal area

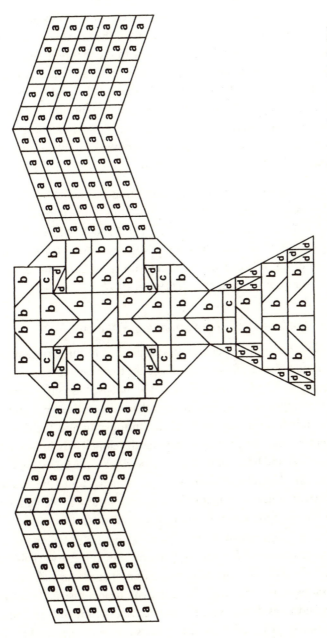

FIGURE 8.2: The first layer of a Vedic sacrificial altar in the shape of a falcon. The wings are each made from 60 bricks of type a, and the body from 46 of type b, 6 of type c, and 24 of type d. (After Thibaut 1875)

of one square *purusha* (approximately five square meters). Seidenberg (1962, 1983) has an interesting discussion of the ambiguities in the Vedic texts relating to equivalence of area as well as the philosophical underpinnings of such a requirement. The last of the fire altars mentioned (the sky altar) was laid out in five layers, with the first representing the earth, the second being the joining of earth and space, the third space, the fourth representing the joining of space and sky, and the fifth the sky. The need to maintain equivalence of areas among altars of various shapes was a preoccupation that continued for a long time. As discussed in the next section, it raised a number of geometric problems, the solutions of which led to early Indian geometry.

Another problem that led to some interesting mathematics related to ensuring the precise distance and relative positions of the three fire altars. The general requirement was that *Dakshinagni* should lie south of the line joining the other two fire altars and at a distance from *Garhapatya* of one-third the distance between the other two fire altars. A discussion of how this was solved is given in Joseph (1996a, 1996b).

The *Sulbas*: Mathematics in the Service of Religion?

There is a view that Indian mathematics originated in the service of religion. The proponents of this view have sought their main support in the complex motives behind the recording of the *Sulbasutras*. Since time immemorial, they argue, the needs of religion have determined not only the character of Indian social and political institutions but also the development of scientific knowledge. Astronomy was developed to help determine the auspicious day and hour for performing sacrifices. The thirty-six verses attributed to one Lagadha known as the *Vedanga-Jyotisa* (the *Vedanga* containing astronomical information) gave procedures for calculating the time and position of the sun and moon in various *naksatras* (signs of the zodiac).[7] Also, a strong reason in Vedic India for the study of phonetics and grammar was to ensure perfect accuracy in pronouncing every syllable in a prayer or sacrificial chant. And the construction of altars (or *vedi*) and the location of sacred fires (or *agni*) had to conform to clear instructions about their shapes and areas if they were to be effective instruments of sacrifice.

The *Sulbasutras* provided such instructions for two types of ritual, one for worship at home and the other for communal worship. Square and circular altars were sufficient for household rituals, while more elaborate altars whose shapes were combinations of rectangles, triangles, and trapeziums were required for public worship. One of the most elaborate of the public altars was shaped like a falcon just about to take flight, as shown in figure 8.2. It was believed that offering a sacrifice on such an altar would enable the soul of the supplicant to be conveyed by a falcon straight to heaven. There were other shapes of fire altars such as one in the form of a tortoise to be constructed by one "desiring to win the world of Brahman" and another in the shape of a rhombus to enable one to "destroy existing and future enemies" (Sen and Bag 1983, pp. 86, 98).

Early researchers on the *Sulbasutras*, notably Thibaut in the second half of the nineteenth century, were at pains to stress the religious element of these texts but ignored their secular side. It is worth merely mentioning at this stage an argument to be elaborated later that the *Sulbasutras* may well provide a connecting thread between the Harappan culture, which came to an end around 1750 BC, and the emergence of a literate Vedic culture around the beginning of the first millennium BC. The highly developed brick-making technology of the Harappan culture was replicated in the construction of sacrificial altars during the Vedic period. According to this view, then, the instructions given in the *Sulbasutras* were mainly for the benefit of craftsmen laying out and building altars. To overemphasize the religious and ritual features of altar design and construction at the expense of the technological aspects is to diminish the role of craftsmen in ancient Indian society, at the same time buttressing the stereotypical view of a society dominated by priests and overwhelmed by ritual.

Three of the more mathematically important *Sulbasutras* were the ones recorded by Baudhayana, Apastamba, and Katyayana. Little is known about these *sulbakaras* (i.e., authors of *Sulbasutras*), except that they were not just scribes but probably also priest-craftsmen performing a multitude of tasks including constructing *vedi* (sacrificial altars), maintaining *agni* (sacred fires), and instructing worshippers on the appropriate choice of both sacrifices and altars. It is difficult to assign firm dates to these three texts. All we can say is that the earliest of them, the one composed by Baudhayana, was probably first recorded between 800 and 500 BC, and that the other two were recorded one or two centuries later. (From their style, they predate the Sanskrit grammarian Panini, who lived in the fourth century BC).

Baudhayana's *Sulbasutra* is complete in three chapters and offers instructions to those conducting the sacrifices as to how to construct altars of various shapes using stakes and marked cords. With our present hindsight, we can discern from these instructions a general statement of the Pythagorean theorem, an approximation procedure for obtaining the square root of 2 correct to five decimal places, and a number of area-preserving transformations for "squaring the circle" (approximately) and constructing rectilinear shapes whose area was equal to the sum or difference of areas of other shapes. The next-oldest text, by Apastamba, contains six chapters and treats in more detail the topics examined by Baudhayana. Katyayana's *Sulbasutra* adds little to the work of his predecessors.

Sulba Geometry

The geometry of the *Sulbasutras* grew out of the need to ensure strict conformation of the orientation, shape, and area of altars to the prescriptions laid down in the Vedic scriptures. Such accuracy was just as important for the efficacy of the ritual as was the meticulous pronunciation of Vedic chants (or *mantras*). However, while accurate geometric methods were used, the principles underlying the constructions were often not discussed. It will therefore be useful, while going through the illustrative examples given below, to bear in mind the distinction between three aspects of the geometry found in the *Sulbasutras*:

1. Geometric results and theorems explicitly stated

2. Procedures for constructing different shapes of altars

3. Algorithmic devices contained in (1) and (2)

In the first category the most notable is the Pythagorean theorem for a right-angled triangle. As we have seen, knowledge of this result has been found in a number of early mathematical traditions. Here we shall look only at a specific application of it to the design of a particular type of altar. It is the second aspect of *sulba* geometry that forms a substantial part of the mathematical evidence of the period. Again, the emphasis is on designing altars with the minimum of tools. We shall briefly examine just three of the fifteen such constructions discussed in the texts. Finally, as an illustration of an algorithmic device born out of constructional needs, we consider the approximation procedure for evaluating the square root of 2.

The actual statement of the Pythagorean theorem, expressed in terms of the sides and diagonals of squares and rectangles, is found in both the Baudhayana and Apastamba *Sulbasutras*. Almost the same statement with exactly the same content occurs in Katyayana's *Sulbasutra*. Baudhayana states:

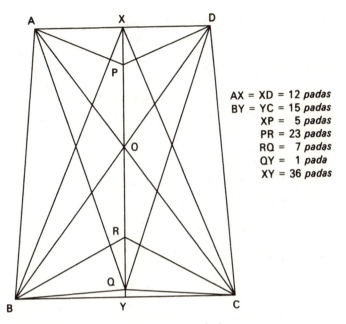

AX = XD = 12 *padas*
BY = YC = 15 *padas*
XP =   5 *padas*
PR = 23 *padas*
RQ =   7 *padas*
QY =   1 *pada*
XY = 36 *padas*

FIGURE 8.3: The layout of the *Mahavedi* (Great Altar)

The rope which is stretched across the diagonal of a square produces an area double the size of the original square.

All three *Sulbasutras* give a more general proposition:[8]

The rope [stretched along the length] of the diagonal of a rectangle makes an [area] which the vertical and horizontal sides make together.

Figure 8.3 illustrates how this proposition was applied in the construction of altars. It shows a drawing of the base of the *Mahavedi* (Great Altar) for the *Soma* ritual (at which an intoxicating drink called *soma* was offered as a sacrifice to the gods).[9] Its base had to be constructed to precise dimensions if the sacrifice was to bear fruit. It had to be an isosceles trapezium like ABCD, with AD and BC being 24 and 30 *padas* (literally feet). The altitude of the trapezium (i.e., the distance between the midpoints X and Y of AD and BC) had to be precisely 36 *padas*.

The instructions given for the construction of this altar in Apastamba's *Sulbasutra* are, in modern notation, as follows:

1. With the help of a rope mark out XY, which is precisely 36 *padas*.

2. Along this line, locate points P, R, and Q such that XP, XR, and XQ equal 5, 28, and 35 *padas* respectively.

3. Construct perpendiculars at X and Y.

4. Use the fact that the triangles APX, DPX, BRY, and CRY are right-angled triangles with integral-valued sides to locate points A, B, C, and D. In other words, make AXD 24 *padas* and BYC 30 *padas*. Join AB, BC, CD, and DA.

Implied in these directions for construction are the following right-angled triangles with integral sides:

$\triangle$APX and $\triangle$DPX with sides 5, 12, 13

$\triangle$AOX and $\triangle$DOX with sides 12, 16, 20

$\triangle$BRY and $\triangle$CRY with sides 8, 15, 17

$\triangle$BOY and $\triangle$COY with sides 15, 20, 25

$\triangle$AQX and $\triangle$DQX with sides 12, 35, 37

$\triangle$BXY and $\triangle$CXY with sides 15, 36, 39

Besides these integral "Pythagorean triples," two involving fractions ($2\frac{1}{2}$, 6, $6\frac{1}{2}$; and $7\frac{1}{2}$, 10, $12\frac{1}{2}$) were also used to construct right-angled triangles. Furthermore, the construction of some altars required the use of triples such as 1, 1, $\sqrt{2}$; $5\sqrt{3}$, $12\sqrt{3}$, $13\sqrt{3}$; and $15\sqrt{2}$, $36\sqrt{2}$, $39\sqrt{2}$. These numbers probably arose from ritual requirements that dictated constructions of altars whose areas were either integral multiples or fractions of the areas of other altars of the same shape. For example, the dimensions of the *Mahavedi* for the *Sautramani* ritual (one with a triangular base of sides $5\sqrt{3}$, $12\sqrt{3}$, and $13\sqrt{3}$) were arrived at by starting with a 5, 12, 13 triangle, the unit of measurement being the *purusha* (nearly 2.5 meters, or the height of a man with his arms stretched above him).

The *Sulbasutras* were, however, primarily instruction manuals for geometric constructions: squares, rectangles, trapeziums, and circles that had to conform to specified dimensions or areas. Any inaccuracy would make the consequent rituals and sacrifices ineffective. Here are three examples:

1. To *merge two equal or unequal squares to obtain a third square.* The method is reported in all three *Sulbasutras*. Figure 8.4 shows

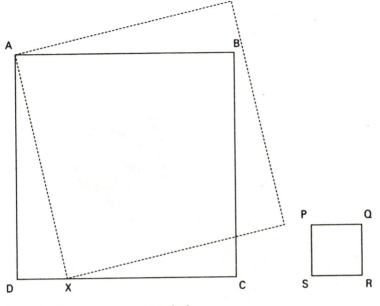

FIGURE 8.4: Turning two squares into a third

the construction. In modern notation, let ABCD and PQRS be the two squares to be combined, and let DX be equal to SR. Draw a line to join A and X. The square on AX is equal to the sum of the squares ABCD and PQRS. The original explanation then points out that $DX^2 + AD^2 = AX^2 = SR^2 + AD^2$, which shows the use of the Pythagorean theorem.

2. *To transform a rectangle into a square of equal area.* The result, from Baudhayana's *Sulbasutra*, is shown in figure 8.5. ABCD is a rectangle of length AD and breadth AB. The procedure begins by completing the square ABKH. Let E and M be the midpoints of HD and KC respectively, so that EM bisects the rectangle HKCD. Move the rectangle EMCD so that its new position is KBJG. Complete the square KGFM. Draw an arc of radius JF to cut BC at W, and draw a line through W, parallel to MF, to cut JF at S. The required square is then the square on JS, JSTR. A demonstration that the square JSTR is equal in area to the rectangle ABCD follows easily from the Pythagorean theorem:

$$JS^2 = JW^2 - WS^2 = AJ^2 - BJ^2 = (AJ + BJ)(AJ - BJ)$$
$$= AD \times AB = \text{area of ABCD}.$$

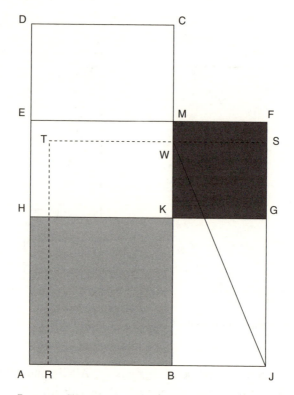

FIGURE 8.5: Turning a rectangle into a square

3. *Squaring a circle and circling a square.* No geometric method can achieve this exactly; the *Sulbasutras* provide only approximate constructions. In figure 8.6 ABCD is a given square of side *a*, centred at O. Join OD, and construct an arc from D to the point P, which lies on the line passing through O and the midpoint E of the side CD. Thus OD = OP. To construct a circle centered at O and of radius *r* equal in area to the square, we are advised to take the radius of the circle as the sum of half the length of the side of the square (i.e., $\frac{1}{2}a$) and one-third of the length of OP that remains outside the square.

In other words,

$$r = ON = OE + EN = OE + \frac{1}{3}EP = OE + \frac{1}{3}(OP - OE).$$

Now,

$$\sin\theta = \sin 45° = OE/OD = OE/OP = 1/\sqrt{2}.$$

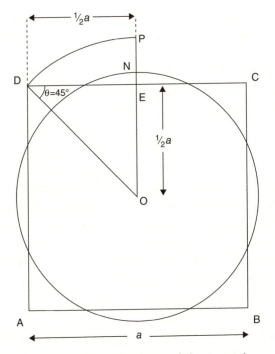

FIGURE 8.6: Turning a square into a circle (or vice versa)

Therefore $OP = \sqrt{2}\,OE$, and so

$$r = OE + \frac{1}{3}(\sqrt{2}\,OE - OE).$$

Since $OE = \frac{1}{2}a$,

$$r = \frac{a}{2} + \frac{\frac{1}{2}a(\sqrt{2}-1)}{3} = \frac{a}{6}(2 + \sqrt{2}).$$

The area of the square is a^2 and that of the circle is πr^2, so

$$a^2 = \frac{\pi}{36}[a(2 + \sqrt{2})]^2,$$

which implies a value of π of 3.088.[10]

All three *Sulbasutras* give the following directions for converting a circle into a square: Divide the diameter into 15 parts and take 13 of these parts as the side of the square. If d is the diameter of the circle and a the side of the required square, then $a = (13/15)d$, which implies a value of π of 3.004.

Irrational Square Roots: An Approximation Procedure

A remarkable achievement of Vedic mathematics is the discovery of a procedure for evaluating square roots to a high degree of approximation. The problem may have originally arisen from an attempt to construct a square altar twice the area of a given square altar. Takao Hayashi has pointed out in a personal communication that the approximation of $\sqrt{2}$ could also be used for constructing a right-angled triangle and a square.

The problem, which the reader may wish to try, is one of constructing a square twice the area of a given square, A, of side 1 unit. It is clear that for the larger square, C, to have twice the area of square A, its side should be $\sqrt{2}$ units. Also, we are given a third square, B, of side 1, which needs to be dissected and reassembled so that by joining cut-up sections of square B to square A, it is possible to make up a square close to the size of square C. Figure 8.7 shows diagrammatically what needs to be done.

The procedure given in the three *Sulbasutras* discussed earlier may be restated as "Increase the measure by its third and this third by its own fourth less the thirty-fourth part of that fourth. This is the value with a special quantity in excess."[11] If we take 1 unit as the dimension of the side of a square, this formula gives the approximate length of the square's diagonal as

$$\sqrt{2} = 1 + \frac{1}{3} + \frac{1}{3 \times 4} - \frac{1}{3 \times 4 \times 34} = 1.4142156....$$

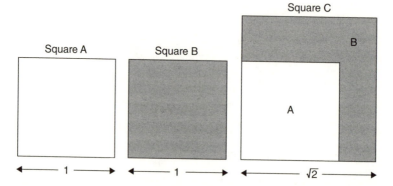

FIGURE 8.7: Doubling a square: Square B is cut into strips and added to Square A to give Square C

The true value is 1.414 213.... A commentator on the *Sulbasutras*, Rama, who lived in the middle of the fifteenth century AD, gave an improved approximation by adding two further terms to the equation:

$$-\frac{1}{3\times 4\times 34\times 33}+\frac{1}{3\times 4\times 34\times 34},$$

which gives a value correct to seven decimal places.

The *Sulbasutras* contain no clue as to how this remarkable approximation was arrived at. Many explanations have been proposed. A plausible one, put forward by Datta (1932a), is as follows. Consider two squares, ABCD and PQRS, each of unit side (see figure 8.8). PQRS is divided into three equal rectangular strips, of which the first two are marked 1 and 2. The third strip is subdivided into three squares, of which the first is marked 3. The remaining two squares are each divided into four equal strips marked 4 to 11. These eleven areas are added to the square ABCD as shown in figure 8.8 to obtain a large square less a small square at the corner F. The side of the augmented square (AEFG) is

$$1+\frac{1}{3}+\frac{1}{3\times 4}.$$

The area of the shaded square is $[1/(3\times 4)]^2$, so that the area of the augmented square AEFG is greater than the sum of the areas of the original squares, ABCD and PQRS, by $[1/(3\times 4)]^2$.

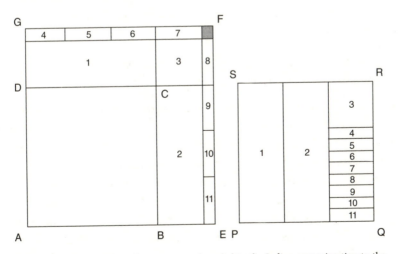

FIGURE 8.8: How doubling the square may have led to the Indian approximation to the square root of 2 (After Datta 1932)

To make the area of the square AEFG approximately equal to the sum of the areas of the original squares ABCD and PQRS, imagine cutting off two very narrow strips, of width x, from the square AEFG, one from the left side and one from the bottom. Then

$$2x\left(1 + \frac{1}{3} + \frac{1}{3 \times 4}\right) - x^2 = \left(\frac{1}{3 \times 4}\right)^{-2}.$$

Simplifying the above expression and ignoring x^2, an insignificantly small quantity, gives

$$x \approx \frac{1}{3 \times 4 \times 34}.$$

The diagonal of each of the original squares is $\sqrt{2}$, which can be approximated by the side of the new square as just calculated:

$$\sqrt{2} \approx 1 + \frac{1}{3} + \frac{1}{3 \times 4} - \frac{1}{3 \times 4 \times 34}.$$

What is particularly appealing about this line of reasoning is that there is other evidence from *Sulbasutra* geometry of the use of this "concrete" mode of argument, which was described as the "out-in" principle in our earlier discussion of Chinese geometry. This mode of demonstration requires neither a well-developed symbolic algebra nor a Greek-style procedure of deductive inference. In this instance, the Indian and Chinese geometric approaches exhibit similarities that may have antecedents in the "geometrical algebraic" approach first found in Mesopotamian mathematics.[12]

The Harappa-Vedic Nexus: Restoring Historical Continuity to Early Indian Mathematics

There is a danger that the magico-religious beliefs surrounding the Vedic rituals may be overemphasized when considering the origins of Indian mathematics. We have already mentioned the role played by the *Agni-cayana* ceremony in generating geometrical concepts and techniques found in the *Sulbasutras*. The rituals associated with the construction of fire altars may be looked at from two standpoints. The first is that the beliefs connect the shapes of altars with specific desires to be fulfilled by their use in the sacrifices. The second is simply technological: how exactly does one construct the altars with specific shapes and sizes, and by using a specific number of bricks, or how does one vary their shapes without affecting their size or area?

It may be argued that the geometry of the *Sulbasutras* has little to do with the first standpoint. Thus, for example, whether a falcon-shaped altar ensures the transport of one's soul to heaven or the annihilation of one's enemies has little relevance to the problem of constructing it to conform to a certain size and shape. As a matter of fact, these construction problems would remain even if the purpose of the construction was to erect an ornamental structure in a garden. In other words, the geometry developed in the *Sulbasutras* was basically needed to solve technological problems involved in construction. It is this geometry, placed in a social context, that should be of primary interest to the historian of mathematics.

Once the *Sulbasutras* are seen primarily as manuals for technicians, the question then arises as to where and when the practical knowledge relating to brick technology was acquired. References to bricks are conspicuous by their absence from the most sacred and earliest of Vedic literature, the *Rig-veda Samhita*. When they do make an appearance in a recension (*Tattiriya Samhita*) of a later Veda, the *Yajur-veda Samhita*, bricks are viewed as marvelous and mysterious entities. In *Tattiriya Samhita*, there is a reference to bricks as "milk cows" (a ready source of income). In *Yajur-veda Samhita*, there are exhortations that "tiles or potsherds" from the ruined, probably Harappan, cities should be gathered for ritual purposes. It is, therefore, likely that the priests were acquainted with the fired bricks from the same sites and would in course of time invest them with magico-religious properties. In one of the last critical revisions to the *Yajur-veda* appears the *Satapatha Brahmana*, in which a discussion of conducting the *Agnicayana* is accompanied by a short discourse on the construction of brick altars of various shapes and sizes. While the discussion lacks the sophistication of the *Sulbasutras*, it is clear that knowledge of brick technology, probably from the Harappan culture, had percolated into the Vedic rituals. Staal's (1978) conclusion that "it is certainly reasonable to suppose that knowledge of the techniques for firing bricks was preserved among the inhabitants of the subcontinent even after the Harappa civilization had disappeared" is particularly apposite. So the geometry embodied in the *Sulbasutras* should be viewed as the outcome of a long and sophisticated tradition of brick technology inherited from the Harappan civilization. If this presumption is correct, the first and earliest of the discontinuities in the chronology of Indian mathematics has been filled with the assistance of bricks.

Early Indian Numerals and Their Development

Three early types of Indian numerals are shown in table 8.2 in chronological order of appearance. The Kharosthi-type numerals, derived from
the Aramaic script, are found in inscriptions dating to a period from the
fourth century BC to the second century AD. Special symbols were used
to show both 10 and 20. Numbers up to 100 were then built up additively;
for larger numbers the multiplication principle came into operation, with
special symbols for higher powers of 10. Following from their West Asian
origins, the Kharosthi numerals were written from right to left. The most
complete example of this type of numerals is the Saka numerals from
around the first century BC. The Brahmi-type numerals were more highly
developed. There were separate symbols for the digits 1, 4 to 9, and the
number 10 and its higher powers. There were also symbols for multiples
of 10 up to 90, and for multiples of 100 up to 900. The number 486, for
example, would be written by using the symbols for 400, 80, and 6. It is
possible that our symbols "2" and "3" are cursive versions of the Brahmi
numerals (i.e., from ⤳ and ⤳ may have evolved 2 and 3).

The earliest trace of Brahmi-type numerals is from the third century
BC, on the Asoka pillars scattered around India, though more detailed
pieces of evidence are found elsewhere later. At the top of Nana Ghat near
Poona in central India is a cave that must once have been a resting place for
travelers; inscribed on the cave walls are numerals representing the signs
for 10 and 7, which date back to 150 BC. Another version of the Brahmi
numerals (shown in table 8.2) is found at Nasik, near present-day Bombay (now Mumbai), from around 100 BC. Both versions resemble each

TABLE 8.2: THREE TYPES OF INDIAN NUMERALS, IN CHRONOLOGICAL
ORDER

	1	*2*	*3*	*4*	*5*	*6*	*7*	*8*	*9*	*10*
Kharosthi	I	II	III	X	IX	IIX	IIIX	XX		ʔ
Brahmi	—	=	≡	⅄	ɾ	ϛ	ʔ	Ϟ	Ϸ	ɑ
Gwalior	٦	ς	३	४	५	८	౨	९	९	९०

Note: The Kharosthi numeral for 9 is not known for certain

other, and it was thought until recently that from them evolved first the Bakhshali number system (c. AD 400–1200) and then the Gwalior system (c. AD 850), which is recognizably close to our present-day number system.[13] In both the Bakhshali and Gwalior number systems, ten symbols were used to represent 1 to 9 and zero. With them it became possible to express any number, no matter how large, by a decimal place-value system.

The earliest appearance of the symbol that we associate with zero in India in a decimal place-value system is in an inscription from Gwalior dated "Samvat 933" (AD 876), where the numbers 50 and 270 are given as 𑀞𑁦 and २७० respectively. Note the close similarity with our notation for 270. For earlier evidence, we have to turn to Southeast Asia when it was under the cultural influence of India. There, three inscriptions have been found bearing dates in the Saka era, which began in AD 78. A Malay inscription at Palembang in Sumatra from AD 684 shows 60 and 606 Saka as 𑀛𑁦 and 𑀛𑁦𑀛 respectively, a Khmer inscription at Sambor in Cambodia from AD 683 gives 605 as 𑀧•𑀜, and an inscription at Ponagar, Champa (now southern Vietnam), from AD 813 represents 735 as ७=𑀜. If, however, the original version of the Bakhshali Manuscript dates from the third century AD, it would be the earliest evidence of a well-established number system with a place-value scale and zero that is also recognizably an ancestor of our present-day number system. In the Bakhshali Manuscript are found the following numbers:

330: ३३०,   846,720: ३ ४ ३ ७ ३ ०,   947: ९ ४ ७.

What we have here is a fully developed decimal place-value system incorporating zero.

The Emergence of the Place-Value Principle

Fascination with numbers has been an abiding characteristic of Indian civilization. Not only large numbers but very small ones as well. Operations with zero attracted the interest of both Bhaskaracharya (b. 1114) and Srinivas Ramanujan (1887–1920). In an elementary class that Ramanujan attended, the teacher was explaining the concept of division (or "sharing") through examples: between three children, each child would get one banana. Similarly, the share would be one banana if four bananas were shared among four children, five bananas among five children, and so on. And when the teacher generalized this idea of sharing x bananas among x boys, Ramanujan asked whether, if x equaled zero, each child would then get one banana! There is no record of the teacher's reply. Ramanujan explained later to his school friends that zero divided by zero could be anything, since the zero of the denominator may be any number of times the zero of the numerator.

Two important features of early numeration may have been of significance in the subsequent development of Indian numerals. Ever since the Harappan period the number 10 may have formed the basis of numeration; there is no evidence of the use of any other base in the whole of Sanskrit literature. Long lists of number-names for powers of 10 are found in various early sources. For example, one of the four major Vedas, the *Yajur-veda*, gives special names for powers of ten from one or 10^0 (*eka*) to one trillion or 10^{12} (*parardha*). In the *Ramayana*, one of the most popular texts of Hinduism and roughly contemporaneous with the later Vedas, it is reported that Ravana, the chief villain of the piece, commanded an army whose total equaled $10^{12} + 10^5 + 36(10^4)$. Facing them was the rival army of Rama, the hero of the epic, which had $10^{10} + 10^{14} + 10^{20} + 10^{24} + 10^{30} + 10^{34} + 10^{40} + 10^{44} + 10^{52} + 10^{57} + 10^{62} + 5$ men! Even though these numbers are fantastic, the very existence of names for powers of ten up to 62 indicates that the Vedic Indians were quite at home with very large numbers. This is to be compared with the ancient Greeks, who had no words for numbers above the myriad (10^4).

And these were by no means the largest numbers ever conceived in ancient India. The Jains, who came after the Vedic Indians, were particularly fascinated by even larger numbers, which were intimately tied up with their philosophy of time and space. This fascination with large numbers is also found in Buddhist literature. In the life of the Buddha, as reported in *Lalita-vistara*, the young Buddha, as part of a competion to win the hand of the princess Gopa, recites a table that includes names for powers of 10 going up to the fiftieth power.[14] (We shall look at the Jaina contribution in detail in a later section.) For units of measuring time, the Jains suggested the following relationships:

1 *purvis* = 756×10^{11} days;

1 *shirsa prahelika* = $(8,400,000)^{28}$ *purvis*.

The last number contains 194 digits!

The early use of such large numbers eventually led to the adoption of a series of names for successive powers of 10. The importance of these number-names in the evolution of the decimal place-value notation cannot be exaggerated. The word-numeral system, later replaced by an alphabetic notation, was the logical outcome of proceeding by multiples of 10. Thus 60,799 is *sasti* (sixty) *sahasra* (thousand) *sapta* (seven) *sata* (hundred) *navati* (nine ten times) *nava* (nine). Such a system presupposes a scientifically based vocabulary of number-names in which the principles of addition, subtraction, and multiplication are used. It requires:

1. The naming of the first nine digits (*eka, dvi, tri, catur, pancha, sat, sapta, asta, nava*)

2. A second group of nine numbers obtained by multiplying each of the first nine digits by ten (*dasa, vimsati, trimsat, catvarimsat, panchasat, sasti, saptati, asiti, navati*)

3. A group of numbers that are increasing integral powers of 10, starting with 10^2 (*sata, sahasra, ayuta, niyuta, prayuta, arbuda, nyarbuda, samudra, madhya, anta, parardha* . . .).

In forming the words of the second and third groups of numbers, the multiplicative principle applies, as in the example quoted: 60,000 is *sastisahasra*. The additive principle is employed when the numbers from the first and second group are used, for example, 27 is *sapta-vimsati*. The subtractive principle may apply occasionally and in a limited way; for example, *ekanna-catvarimsat* indicates $40 - 1 = 39$, where *ekanna* means "one less."

To understand why word-numerals persisted in India, even after the Indian numerals became widespread, it is necessary to recognize the importance of the oral mode of preserving and disseminating knowledge. An important characteristic of written texts in India from time immemorial was the sutra style of writing, which presented information in a cryptic form, leaving out details and rationale to be filled in by teachers and commentators. In short pithy sentences, often expressed in verses, the sutras enabled the reader to memorize the content easily.

As a replacement for the older word-numeral system that consisted of merely names of numbers, a new system (a concrete number system) was devised to help versification and memory. In this system, known as *bhuta-samkhya*, numbers were indicated by well-known objects or ideas. Thus, zero was *shunya* (void) or *ambara akasa* (heavenly space or sky or ether) or other empty things, one was *candra* (moon) or *bhumi* (earth) or other single things, two was *netra* (eyes) or *paksa* (wings of a bird) or other pairs, three was *kala* (time: past, present, and future) or *loka* (heaven, earth, and hell) or other trios, and so on. With multiple words available for each number, the choice of a particular word for a number would be dictated by literary considerations. This form of notation continued for many years in both secular and religious writings because it was aesthetically pleasing and offered an easier way of remembering numbers and rules.

There were two major problems with the *bhuta-samkhya* system. First, there was an "exclusionist" element, in that to decode the words for their numerical values required considerable familiarity with the philosophical and religious texts from which the correspondences were established in the first place. Second, at times the same word stood for two or more different numbers, since some writers had their own preferences when it came to choosing words to correspond to numbers as, for example, when *paksa* was used for 2 as well as 15 and *dik* for 8, 10, and 4.

There are traces of this system of numeration in the *Yavanajataka* (AD 269) of Sphujidhvaja, although the first clearest and detailed evidence of it is found in the works of the astronomer Varahamihira (d. AD 587). Thus, except for the actual symbols themselves, the present-day number system with distinct numerals for the numbers from zero to 9, the place-value principle, and the use of the zero within the decimal base is essentially what we see in this early number system.[15] In a sense, what is used as a symbol for a number, whether it be a letter, a word, or a specially invented squiggle, is of little importance. Indeed, an unduly close association—or even identity—between a number and the symbol used to represent it may even be counterproductive, preventing the strength of the place-value principle from being fully exploited in elementary operations.

A third system of numerical notation originated with Aryabhata (b. 476 AD). In his *Aryabhatiya,* he introduced an alphabetical scheme for representing numerals, based on distinguishing between classified (*varga*) and unclassified (*avarga*) consonants and vowels. The *vargas* fall into five phonetic groups: *ka-varga* (guttural), *ca-varga* (palatal), *ta-varga* (lingual), *ta-varga* (dental), and *pa-varga* (labial). Each group has five letters associated to it, and represented numbers from 1 to 25. There were seven *avargas* consisting of semivowels and sibilants representing numerical values 30, 40, 50, . . . , 190. An eighth *avarga* was used to extend the number to the next place value. The ten vowels denoted successive integral powers of 10 from 100 onward.

This form of representation, closer to the system that preceded *bhuta-samkhya,* has the advantage of brevity and clarity but the disadvantage of having limited potential for formations of words that are pronounceable and meaningful, both necessary requirements for easy memorization. For example, in the Aryabhatan system, the representation of the number of revolutions of the moon in a *yuga* (calculated as 57,753,336 days) is the unpronounceable and meaningless word *cayagiyinusuchlr*!

From a refinement of Aryabhata's alphabet-numeral system of notation emerged the *katapayadi* system, which the legendary founder of the Kerala school of astronomy, Varurici, was believed to have popularized around the fourth century AD. In this system, every number in the decimal place-value system can be represented by words, each letter of the word representing a digit. A vowel not preceded by a consonant stands for zero, but vowels following consonants have no special value. In the case of conjunct consonants (a combination of two or more consonants), only the last consonant has a numerical value. Number-words are read from right to left so that the letter denoting the "units" is given first, and so on.

This was a system devised to help memorization, since memorable words can be made up using different chronograms. For example, if such a system is applied to English, the letters *b, c, d, f, g, h, j, k, l, m* would represent the numbers zero to 9. So would *n, p, q, r, s, t, v, w, x, y*. The last letter, *z*, denotes zero. The vowels, *a, e, i, o, u* are helpful in forming meaningful words but have no numerical values associated with them. Thus, the sentence "I love Madras" represents the numbers 86 and 9,234. To take another example from Kunjunni Raja (1963, p. 123), the number 1,729,133 could be represented by *balakalatram saukhyam* (i.e., the [company] of a young woman is sheer happiness) or *lingavyadhir asahyah* (i.e., the demise of sexual virility is unbearable).

The close relationship between literacy and numeracy, implied by such varied systems of numerical notation, may have its roots in the way that Sanskrit developed in its formative period after its separation from other languages of the Indo-European family. A long tradition of oral communication of knowledge was a characteristic of that period and left a singular mark on the nature and transmission of knowledge, whether religious or scientific, in Indian culture. After many years, as Sanskrit became a written language, three kinds of scientific Sanskrit developed with varying degrees of artificiality: grammatical, logical, and mathematical Sanskrit.

Mathematical Sanskrit remained the least artificial of the three, with the greatest artificiality found in the development of grammatical Sanskrit by Panini and Patanjali, followed five hundred years later by the logical

Sanskrit of *Nyaya*, which culminated a thousand years later in *Navya-Nyaya*. This has important implications for a comparative study of the historical development of Indian and Western mathematics, according to Staal (1995). First, the chronological order of the development of artificial scientific languages in the West was a reversal of the Indian experience. In the West, logic followed mathematics, and linguistics was a late developer. In India, mathematical Sanskrit never quite became an artificial language, although it employed abbreviations and artificial notations outside Sanskrit as shorthand for practical procedures. And logical Sanskrit never became, like its Western counterpart, an important adjunct to "mathematical philosophy."

The Enormity of Zero

The word "zero" comes from the Arabic *al-sifr*.[16] *Sifr* in turn is a transliteration of the Sanskrit word *shunya*, meaning void or empty, which later became the term for zero. Introduced into Europe during the Italian Renaissance in the twelfth century by Leonardo Fibonacci (and by Nemorarius, a less well-known mathematician) as *cifra*, the word emerged in English as "cipher." In French it became *chiffre*, and in German *ziffer*, both of which mean zero.

The ancient Egyptians never used a zero symbol in writing their numerals. Instead they had a stand-alone zero to represent a benchmark value or magnitude. A bookkeeper's record from the Thirteenth dynasty (about 1700 BC) shows a monthly balance sheet for items received and disbursed by the royal court during its travels. On subtracting total disbursements from total income, a zero remainder was left in several columns. This zero remainder was represented by the hieroglyph *nfr*, which also means beautiful or complete in ancient Egyptian. The same *nfr* symbol also labeled a zero reference point for a system of integers used on construction guidelines at Egyptian tombs and pyramids. These massive stone structures required deep foundations and careful leveling of the courses of stone. A vertical number-line labeled the horizontal leveling lines that guided construction at different levels. One of these horizontal lines, often at pavement level, was used as a reference and was labeled *nfr* or zero. Horizontal leveling lines were spaced 1 cubit apart. Those above the zero level were labeled as 1 cubit above *nfr*, 2 cubits above *nfr*, and so on. Those below the zero level were labeled 1 cubit below *nfr*, 2 cubits below, and so forth. Here zero was used as a reference for directed or signed numbers.

It is quite extraordinary that the Mesopotamian culture, more or less contemporaneous to the Egyptian culture, developed a full positional-value number system on base 60 and did not use zero as a number. A symbol for zero as a placeholder appeared late in the Mesopotamian culture. The early Greeks, who were the intellectual inheritors of Egyptian mathematics and science, emphasized geometry to the exclusion of everything else. They did not seem interested in perfecting their number notation system. They simply had no use for zero. In any case, they were not greatly interested in arithmetic, claiming that arithmetic should only be taught in democracies, for it "dealt with relations of equality." On the other hand, geometry was the natural study for oligarchies, for "it demonstrated the proportions within inequality."[17]

In India, zero as a concept probably predated zero as a number by hundreds of years. The Sanskrit word for zero, *shunya*, meant "void" or "empty." The word is probably derived from *shuna*, which is the past participle of *svi*, "to grow." In one of the early Vedas, *Rig-veda*, there is another meaning: the sense of "lack" or "deficiency." It is possible that the two different words were fused to give *shunya* a single sense of "absence" or "emptiness" with the potential for growth. Hence, its derivative, *Shunyata*, described the Buddhist doctrine of "Emptiness," being the spiritual practice of emptying the mind of all impressions. This was a course of action prescribed in a wide range of creative endeavors. For example, the practice of *Shunyata* is recommended in writing poetry, composing a piece of music, producing a painting, or in any activity that comes out of the mind of the artist. An architect was advised in the traditional manuals of architecture (the *Silpas*) that designing a building involved the organization of empty space, for "it is not the walls that make a building but the empty spaces created by the walls." The whole process of creation is vividly described in the following verse from a Tantric Buddhist text:

First the realization of the void [*shunya*],
Second the seed in which all is concentrated,
Third the physical manifestation,
Fourth one should implant the syllable.

The mathematical correspondence was soon established. "Just as emptiness of space is a necessary condition for the appearance of any object, the number zero being no number at all is the condition for the existence of all numbers."

A discussion of the mathematics of the *shunya* involves three related issues: (1) the concept of the *shunya* within a place-value system, (2) the symbols used for *shunya*, and (3) mathematical operations with the *shunya*. Materials from appropriate early texts are used as illustrations below.

It was soon recognized that the *shunya* denoted notational place (placeholder) as well as the "void," or absence of numerical value, in a particular notational place. Consequently all numerical quantities, however great, could be represented with just ten symbols. A twelfth-century text (*Manasollasa*) states:

> Basically, there are only nine digits, starting from "one" and going to "nine." By adding the zeros these are raised successively to tens, hundreds, and beyond.

And in a commentary on Patanjali's *Yogasutra* there appears in the fifth century the following analogy:[18]

> Just as the same sign is called a hundred in the "hundreds" place, ten in the "tens" place, and one in the "units" place, so is one and the same woman referred to (differently) as mother, daughter, or sister.

One of the earliest mentions of a symbol for zero occurs in the *Chandahsutra* of Pingala (fl. third century BC), which discusses a method for calculating the number of arrangements of long and short syllables in a meter containing a certain number of syllables (i.e., the number of combinations of two items from a total of *n* items, repetitions being allowed). The symbol for *shunya* began as a dot (*bindu*), found in inscriptions in India, Cambodia, and Sumatra around the seventh and eighth centuries, and then became a circle (*chidra* or *randhra*, meaning a hole). The association between the concept of zero and its symbol was already well established by the early centuries of the Christian era, as the following quotation shows:

> The stars shone forth, like zero dots [*shunya-bindu*] scattered in the sky as if on a blue rug, [such that] the Creator reckoned the total with a bit of the moon for chalk. (Vasavadatta, c. AD 400)

Sanskrit texts on mathematics/astronomy from the time of Brahmagupta usually contain a section called *shunya-ganita* or computations involving zero. While the discussion in the arithmetical texts (*patiganita*) is limited only to addition, subtraction, and multiplication with zero, the treatment in algebra texts (*bijaganita*) covers such questions as the effect of zero on the positive and negative signs, division with zero, and more particularly the relation between zero and infinity (*ananta*).

Take, as an example, Brahmagupta's seventh-century text *Brahma Sphuta Siddhanta*. In it he treats the zero as a separate entity from the positive (*dhana*) and negative (*rina*) quantities, implying that *shunya* is neither positive nor negative but denotes the boundary between the two kinds, being the sum of two equal but opposite quantities. He states that a number, whether positive or negative, remains unchanged when zero is added to or subtracted from it. In multiplication with zero, the product is zero. A zero divided by zero or by some number becomes zero. Likewise the square and square root of zero is zero. But when a number is divided by zero, the answer is an undefined quantity, "that which has that zero as the denominator."[19] In the twelfth century, Bhaskaracharya stated that if you were to divide by zero you would get a number that was "as infinite as the god Vishnu"!

The Spread of Numeracy in India: A Historical Perspective

A search for the social origins of numeracy must consider the everyday practices and institutions that make the numerals and operations with them familiar to the ordinary person. The structure of Indian mathematics education for all may have been set by a Jaina text, called *Sthananga Sutra*, dating back to about 300 BC. In that, the first two topics out of ten, *parikarma* (number representation and the four fundamental operations of arithmetic) and *vyavahara* (arithmetic problems, including the "rule of three"), came to be referred to as *patiganita* (etymology: "calculation on tablet") and were meant to be studied by all. The other eight topics were plane geometry calculations as carried out with a rope (*rajju*), mensuration of plane figures and solids (*rasi*), advanced treatment of fractions (*kalasavarna*), study of that which is unknown or algebra (*yavat-tavat*), problems involving squares and square roots (*varga*), problems involving cubes and cube roots (*ghana*), problems involving higher powers and higher roots (*varga-varga*), and permutations and combinations (*vikalpa*).

Although being taught at home was the usual practice for the higher-caste males and for all females, all other castes attended schools. There are early British descriptions of indigenous village schools where emphasis on numeracy was an important part of the school curriculum. A report, submitted in 1838 by William Adam, of such schools in certain districts of Bengal and Bihar (Dharampal 1983) is quite illuminating. The period a student spent in an elementary school was divided into four stages. The first stage, when the child first entered school, seldom exceeded ten days. During that time the young child was taught "to form letters of the alphabet on the ground with a small stick or slip of bamboo," or on a sand board, a board on which sand was sprinkled as a writing surface. The second stage, lasting from two and a half to four years, involved pupils being taught to read from and write on palm leaves. During the same period, the pupil was expected to memorize "the Cowrie Table, the Numeration Table as far as 100, the *Katha* Table and the *Ser* Table," the latter two being tables of weights and measures. To help them with this enormous task, different systems of word-numerals were taught. The third stage, lasting from two to three years, was spent on improving their literary skills practiced on plantain leaf, as well as completing the basic course on *patiganita*. In the fourth and final stage, lasting up to two years, pupils were expected to read religious and other texts, both at school and at home, undergo training in commercial and agricultural accounts, and compose letters and petitions. A few would continue their education in institutions or within the household, where Sanskrit was the language of instruction and the teachers and students were predominantly Brahmins.

Apart from numeracy skills, *patiganita* consisted of all the mathematics needed for daily living. The *vyavaharaganita* included problems involving calculation of volumes of grains and heaps, estimating amounts in piles of bricks and timber, construction of roads and building, calculation of the time of the day, interest and capital calculations, barter and exchange, and recreational problems. In modern terminology, this was practical mathematics, which included commercial mathematics. The authors who wrote texts on *patiganita*, such as the unknown author of Bakhshali Manuscript, or Mahavira (fl. 850 AD), or Sridhara (fl. AD 800) began with a review of arithmetic operations, though the extent and detail to which this was done varied with different texts; the earlier the text, the more detailed the treatment.

The level of numeracy in traditional Indian society was high, partly because of the manner in which numeracy was acquired and passed on and partly because of the lack of any institutional, religious, or philosophical inhibitions to the acquisition and practice of numeracy. Yet the absence of a commercial revolution in India meant that the social milieu that nurtured interest in matters scientific in Europe was missing. In particular, no artificial language evolved, and while notations were fun and intellectually distracting, they did little to advance science, which ultimately stagnated.

And practical mathematics, the handmaiden of numeracy, continued to remain at the same level for about a thousand years, eventually to be submerged by the rise of Western mathematics. Even the remnants of indigenous numeracy that exist in subterranean occupations, such as astrology and traditional architecture, may soon become a historical memory.

Jaina Mathematics

The rise of Buddhism and Jainism around the middle of the first millennium BC was in part a reaction to some of the excesses of Vedic religious and social practices. The resulting decline in offerings of Vedic sacrifices, which had played such a central role in Hindu ritual, meant that occasions for constructing altars requiring practical skills and geometric knowledge became few and far between. There was also a gradual change in the perception of the role of mathematics: from fulfilling the needs of sacrificial ritual, it became an abstract discipline to be cultivated for its own sake. The Jaina contribution to this change should be recognized. Unfortunately, sources of information on Jaina mathematics are scarce, though there are enough to show how original the work was.

A number of Jaina texts of mathematical importance have yet to be studied, and what we know of them is based almost entirely on later commentaries. Of particular relevance is the old canonical literature: *Surya Prajnapti, Jambu Dvipa Prajnapti, Sthananga Sutra, Uttaradhyayana Sutra, Bhagavati Sutra,* and *Anuyoga Dvara Sutra.* The first two works are from the third or fourth century BC, and the others are from at least two centuries later. As mentioned in the previous section, the *Sthananga Sutra* gives a list of mathematical topics that were studied at the time. Expressed in their modern equivalents, they were the theory of numbers, arithmetical operations, geometry, operations with fractions, simple equations, cubic equations, biquadratic (quartic) equations, and permutations and combinations. This classification by the Jains was adopted by later mathematicians.

Given the paucity of existing evidence and the little scrutiny it has received, our survey of Jaina mathematics must be rather piecemeal. We shall examine four main areas in which the Jaina contribution was distinctive.[20]

Theory of Numbers

Like the Vedic mathematicians, the Jains had an interest in the enumeration of very large numbers, which was intimately tied up with their philosophy of time and space. The Jaina cosmology involved two suns, two moons, and two sets of stars. One of four suggested subjects of investigation was *ganita-anuyoga* (inquiry into calculation). It was as a part of this inquiry that the Jains developed their interest in the concepts of infinity (very large numbers) and the infinitesimal (very small numbers).

We mentioned earlier that they devised a measure of time, called a *shirsa prahelika*, that equaled $756 \times 10^{11} \times (8,400,000)^{28}$ days! Other examples of the Jaina fascination with very large numbers are these two definitions: a *rajju* is the distance traveled by a god in six months if he covers a hundred thousand *yojana* (approximately a million kilometers) in each blink of his eye; a *palya* is the time it will take to empty a cubic vessel of side one *yojana* filled with the wool of newborn lambs if one strand is removed every century.

The contemplation of such large numbers led the Jains to an early concept of infinity, which, if not mathematically precise, was by no means simpleminded. All numbers were classified into three groups—enumerable, innumerable, and infinite—each of which was in turn subdivided into three orders:

1. *Enumerable*: lowest, intermediate, and highest

2. *Innumerable*: nearly innumerable, truly innumerable, and innumerably innumerable

3. *Infinite*: nearly infinite, truly infinite, and infinitely infinite

The first group, the enumerable numbers, consisted of all the numbers from 2 (1 was ignored) to the highest. An idea of the "highest" number is given by the following extract from the *Anuyoga Dvara Sutra*, from around the beginning of the Christian era:

Consider a trough whose diameter is that of the earth (100,000 *yojana*) and whose circumference is 316,227 *yojana*. Fill it up with white mustard seeds counting one after another. Similarly fill up with mustard seeds other troughs of the sizes of the various lands and seas. Still the highest enumerable number has not been attained. [In this case 10 *yojana* is about 10 kilometers.]

But if and when this highest number, call it N, is attained, infinity may be reached via the following sequence of operations:

$$N + 1, N + 2, \ldots, \qquad (N + 1)^2 - 1,$$

$$(N + 1)^2, (N + 2)^2, ..., \ (N + 1)^4 - 1,$$
$$(N + 1)^4, (N + 2)^4, ..., \ (N + 1)^8 - 1,$$

and so on. Five different kinds of infinity are recognized: infinite in one direction, infinite in two directions, infinite in area, infinite everywhere, and infinite perpetually. This was quite a revolutionary idea in more than one way:

1. The Jains were the first to discard the idea that all infinities were the same or equal, an idea still generally accepted in Europe until the work of Georg Cantor in the late nineteenth century.

2. The highest enumerable number (i.e., N) of the Jains has some resonance with another concept developed by Cantor, aleph-null (the cardinal number of the infinite set of integers $1, 2, \ldots, N$), also called the first transfinite number. It was Cantor who defined the concept of a sequence of transfinite numbers and devised an arithmetic of such numbers.[21]

3. In the Jaina work on the theory of sets (not discussed here, though documentation is given in the reference list), two basic types of transfinite number (i.e., the cardinal numbers of infinite sets) are distinguished. On both physical and ontological grounds, a distinction is made between *asamkhyata* and *ananta*, between rigidly bounded and loosely bounded infinities. With this distinction, the way was open for the Jains to develop a detailed classification of transfinite numbers and mathematical operations for handling transfinite numbers of different kinds. However, unsurprisingly, they did not do so given their limited technical and symbolic compass. (For further details see Jain [1973, 1982] and N. Singh [1987].)

Indices and Logarithms

Without a convenient notation for indices, the laws of indices cannot be formulated precisely. But there are some indications that the Jains were aware of the existence of these laws and made use of related concepts.

The *Anuyoga Dvara Sutra* lists sequences of successive squares or square roots of numbers. Expressed in modern notation as operations performed on a certain number a, these sequences may be represented as

$$(a)^2, (a^2)^2, [(a^2)^2]^2, \ldots$$
$$\sqrt{a}, \sqrt{\sqrt{a}}, \sqrt{\sqrt{\sqrt{a}}}, \ldots$$

In the same *Sutra*, we come across the following statement on operations with power series or sequences: "The first square root multiplied by the second square root [is] the cube of the second square root; the second square root multiplied by the third square root [is] the cube of the third square root." Expressed in terms of a, this says that

$$a^{1/2} \times a^{1/4} = (a^{1/4})^3, \text{ and } a^{1/4} \times a^{1/8} = (a^{1/8})^3.$$

As a further illustration, the total population of the world is given as "a number obtained by multiplying the sixth square by the fifth square, or a number that can be divided by 2 ninety-six times." This gives a figure of $2^{64} \times 2^{32} = 2^{96}$, which in decimal form is a number of 29 digits!

Does this statement indicate that the laws of indices,

$$a^m \times a^n = a^{m+n}, \text{ and } (a^m)^n = a^{mn},$$

were familiar to the Jains? From the period around the eighth century AD, some interesting evidence in the *Dhavala* commentary by Virasenacharya suggests that the Jains may have developed the idea of logarithms to base 2, 3, and 4 without using them for any computational purposes. The terms *ardhacheda, trikacheda,* and *caturthacheda* of a quantity may be defined as the number of times the quantity can be divided by 2, 3, and 4, respectively, without a remainder. For example, since $32 = 2^5$, the *ardhacheda* of 32 is 5. Or, in the language of modern mathematics, the *ardhacheda* of x is $\log_2 x$, the *trikacheda* of x is $\log_3 x$, and so on.[22]

Permutations and Combinations

A permutation is a particular way of ordering some or all of a given number of items. Therefore, the number of permutations that can be formed from a group of unlike items is given by the number of ways of arranging them. As an example, take the letters a, b, and c, and find the number of permutations of two letters at a time. Six arrangements are possible: *ab, ac, ba, ca, be, cb.*

Instead of listing all possible arrangements, we can work out the number of permutations by arguing as follows: the first letter in an arrangement can be any of three, while the second must be either of the other two letters. Consequently, the number of permutations for two of a group of three letters is $3 \times 2 = 6$. The shorthand way of expressing this result is $_3P_2 = 6$.

A *combination* is the number of selections of r different items from n distinguishable items when order of selection is ignored (unlike a *permutation*, where order is taken into account). Therefore, the number of combinations that can be formed from a group of unlike items is given by the number of ways of selecting them. To take the same illustration as above, the number of combinations of two letters at a time from *a*, *b*, and *c* is three: *ab*, *ac*, *bc*. Again, instead of listing all possible combinations, we can work out how many there are as follows: in each combination the first letter can be any of the three, the second letter has two possibilities, and the third letter has just one possibility, so that there are 6 possibilities in total. But if you are not concerned about order of appearance of the two letters (i.e., although *ab* and *ba* are two different permutations, they amount to the same combination), you must divide the total possibilities (6) by 2 to get 3 as the number of combinations. A shorthand way of expressing this result is $_3C_2 = 3$.

Permutations and combinations were favorite topics of study among the Jains. Statements of results, presumably arrived at by methods like the one just discussed, appear quite early in the Jaina literature. The *Bhagavati Sutra* (c. 300 BC) sets forth simple problems such as finding the number of combinations that can be obtained from a given number of fundamental philosophical categories taken one at a time, two at a time, and three or more at a time. Others include calculation of the groups that can be formed out of the five senses, and selections that can be made from a given number of men, women, and eunuchs. The *Bhagavati Sutra* gives the corresponding values correctly for selections of up to three at a time. Expressed in modern mathematical notation, the results are

$$_nC_1 = n, \qquad _nC_2 = \frac{n(n-1)}{1 \times 2}, \qquad _nC_3 = \frac{n(n-1)(n-2)}{1 \times 2 \times 3}.$$

$$_nP_1 = n, \qquad _nP_2 = n(n-1), \qquad _nP_3 = n(n-1)(n-2).$$

Values are given for $n = 2, 3, 4$, and there is then the following observation: "In this way, 5, 6, 7, . . . , 10, etc., or an enumerable, unenumerable, or infinite number of things may be specified. Taking one at a time, two at a time, . . . , ten at a time, as the number of combinations are formed, they must all be worked out." Apart from the generalizations implied, the application of the principle to different kinds of infinities or different dimensions is noteworthy.

Even before the advent of Jainism there was some interest in the notion of permutations and combinations. Sushruta's great work on medicine, mentioned at the beginning of this chapter, contained the statement that sixty-three combinations may be made out of six different tastes (*rasa*)—bitter, sour, salty, astringent, sweet, hot—by taking the *rasa* one at a time, two at a time, three at a time, and so on. This solution of 63 can easily be checked as follows: $_6C_1 + {}_6C_2 + {}_6C_3 + {}_6C_4 + {}_6C_5 + {}_6C_6 = 6 + 15 + 20 + 15 + 6 + 1 = 63$.

Another interesting example from the Vedic period relates to the number of ways of combining different meters (*chandas*) in a poetic composition. In a book titled *Chandahsutra* (Rule of Metrics) from the second century BC, Pingala considered a method of calculating the number of combinations of short (*laghu*) and long (*guru*) sounds (or syllable patterns) in a given poetical composition. During this period, the music of sound variations (*varnasangita*) was based mainly on these two sounds. Pingala considered a three-syllabic meter, for which the following different combinations of the sounds of *guru* and *laghu* could result: three *guru* sounds will occur once, two *guru* and one *laghu* three times, one *guru* and two *laghu* also three times, and three *laghu* sounds once. The rule given in the original sutra is cryptic to the point of incomprehensibility. We have to be dependent on the commentaries. In modern terms, the rule may be expressed thus:

If we represent *guru* by a and *laghu* by b, then the different combinations may be represented by the coefficients of the binomial expansion:

$$(a + b)^3 = a^3 + 3a^2b + 3ab^2 + b^3.$$

For a four-syllabic meter, different combinations of the two sounds can be found by the same representation:

$$(a + b)^4 = a^4 + 4a^3b + 6a^2b^2 + 4ab^3 + b^4.$$

This technique of finding the number of variations of sounds was useful as a means of testing the quality of different meters, and after Pingala it was commonly used for this purpose.

Around the end of the tenth century AD, Halayudha produced a commentary on Pingala's *Chandahsutra* in which he introduced a pictorial representation of different combinations of sounds, enabling them to be read off directly. Figure 8.9 shows Halayudha's *meruprastara* (or pyramidal arrangement) for a binomial expansion of $(a + b)^n$, where $n = 0, 1, 2, 3, 4$. We came across the same triangular array of numbers, Pascal's triangle, in the previous chapter on Chinese mathematics. However, there

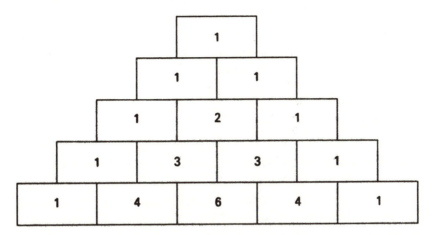

FIGURE 8.9: Halayudha's *meruprastara* (Pascal's triangle)

is no evidence that this triangle was used for any other purpose, such as numerical solutions to higher-order equations, as it was in China. Indeed, there is no evidence that the device was ever incorporated into Indian mathematics.

Sequences and Progressions

Jaina interest in sequences and progressions developed out of the Jains' philosophical theory of cosmological structures. Schematic representations of the cosmos constructed according to this theory contained innumerable concentric rings of alternate continents and oceans, the diameter of each ring being twice that of the previous one, so that if the smallest ring had a diameter of 1 unit, the next largest would have a diameter of 2 units, the next 2^2 units, and so on to the nth ring of diameter 2^{n-1} units.

Arithmetic progressions were given the most detailed treatment. Separate formulas were worked out for finding the first term a, the common difference d, the number of elements n in the series, and the sum S of the terms. This was well explored in a Jaina text titled *Trilokaprajnapti* of Yativrsabha (500 AD). One of its problems is to find the sum of a complicated series consisting of forty-nine terms made up of seven groups, each group itself forming a separate arithmetical progression, and the terms of each group forming another. We shall not attempt a solution here; for details see Bag (1979).

It was the elaborate treatment of mathematical series by Mahavira (c. AD 850) that paved the way for some notable work by medieval mathematicians in this area. We shall examine these developments briefly in the next chapter.

Geometry

The term *rajju* was used in two different senses by the Jaina theorists. In cosmology it was a frequently occurring measure of length, approximately 3.4×10^{21} kilometers according to the Digambara[23] school. But in a more general sense it was the term the Jains used for geometry or mensuration, in which they followed closely the Vedic *Sulbasutras*. Their notable contribution was with measurements of the circle. In Jaina cosmography the earth is a large circular island called the Jambu Island, with a diameter of 100,000 *yojana*. While there are a number of estimates of the circumference of this island, including the rather crude 300,000 *yojana*, an interesting estimate mentioned in both the *Anuyoga Dvara Sutra* and the *Triloko Sara*, from around the beginning the first millennium AD, is 316,227 *yojanna*, 3 *krosa*, 128 *danda*, and 13½ *angula*, where 1 *yojanna* is about 10 kilometers, 4 *krosa* = I *yojanna*, 2,000 *danda* = 1 *krosa*, and 96 *angula* (literally a finger's breadth) = 1 *danda*. This result is consistent with taking the circumference to be given by $\sqrt{10}\,d$, where $d = 100,000$ *yojanna*. The choice of the square root of 10 for the number we call π was quite convenient, since in Jaina cosmography islands and oceans always had diameters measured in powers of 10.

Mathematics on the Eve of the Classical Period

It was in the field of astronomy during the early centuries of the first millennium that India began to make its mark. While written records of such activity have not survived the ravages of time, there is enough to get a flavor of what was happening then. The fundamental objective of mathematical astronomy (as exemplified by the *siddhantas*) was to help to locate the position of the luminaries in the sky as seen from a particular place and at a particular time for the purposes of constructing accurate calendars or to make astrological predictions or to determine geographical directions.

It was soon recognized that this could be achieved by determining the mean position of the celestial body in question, correcting this position for its orbital anomalies, and then using the true position to predict the occurrence of sunrise, new and full moons, conjunctions, eclipses, and so on. A typical *siddhanta* contained not only an explanation of the methods involved but also a discussion of the technical instruments available then for measuring time and angles. In terms of both intellectual and technological transfer across cultures, the *siddhanta* became an important tool for the advancement of mathematical astronomy in India. By the early centuries of the first millennium AD, a synthesis was emerging between indigenous traditions of astronomical and calendrical concepts and computations and the Hellenistic contributions in the form of plane trigonometry of chords and geocentric models involving spherical bodies with planetary eccentrics and epicycles. The period of intellectual exchange seems to have come to an end before the emergence of Ptolemaic astronomy in the third century AD, after which Indian astronomy developed in near isolation for the next few centuries. The first of the major fully preserved astronomical texts that integrated both mathematical methods with astronomical explanations as well as Hellenistic components with indigenous elements is found in the well-known work *Aryabhatiya* of Aryabhata (b. AD 476). A discussion of this work and its author will be found in the next chapter.

However, for an exemplar of early Indian mathematics, we need to look elsewhere. In 1881, near a village called Bakhshali near the northwest border of India, a farmer digging in a ruined stone enclosure came across a manuscript written in an old form of Sanskrit, using Sarada characters, on seventy leaves of birch bark. The find was described as being as fragile as "dry tinder," with a substantial part mutilated beyond repair. What remained was put in order and parts of it translated into English by Rudolph Hoernle; it now resides in the Bodleian Library at Oxford.

G. R. Kaye produced the first translation of this manuscript, published in 1933 together with a commentary by him. Unfortunately, it is partial, incomplete, and fragmentary. There are serious errors in both his translation and his interpretation—errors that have passed into histories of mathematics that cite his work. There has been much controversy over the manuscript's age, and here Kaye's pronouncement has been particularly unfortunate. On the basis of rather dubious literary evidence, Kaye argued that the Bakhshali Manuscript belonged to the twelfth century AD. The general consensus supports Hoernle's assessment that the manuscript is a later copy of a document probably composed sometime in the early centuries of the Christian era. Hoernle's dating is based on a careful consideration of a number of aspects, including the mathematical content, the units of money given in some examples, the use of the symbol + for the negative sign, and the lack of reference to certain topics (especially the solution of indeterminate equations) that appeared in works known to have been written later. Note, however, that Hayashi (1995) claims that the original is probably from the seventh century AD, although the manuscript itself is a later copy made between the eighth and the twelfth centuries. Hayashi has given the first *complete* translation of the manuscript. The section on the Bakhshali Manuscript in this chapter owes a considerable debt to his translation and interpretation.

If we accept Hoernle's dating, the manuscript may therefore be the next substantial piece of evidence, after Jaina mathematics, to bridge the long gap between the *Sulbasutras* of the Vedic period and the mathematics of the Classical period, which began around AD 500. It is also the earliest evidence we have of Indian mathematics free from any religious or metaphysical associations. Indeed, there is some resemblance between the manuscript and the Chinese *Jiu Zhang Suan Shu* from a few centuries earlier, which we examined in chapter 6, both in the topics discussed and in the style of presentation of results. It should, however, be added that the premier Chinese text is far more wide-ranging and "advanced" than the Bakhshali work.

The Bakhshali Manuscript is a handbook of rules and illustrative examples together with their solutions. It is devoted mainly to arithmetic and algebra, with just a few problems on geometry and mensuration. Only parts of it have been restored, so we cannot be certain about the balance between different topics. The arithmetic examples cover fractions, square roots, profit and loss, interest, and the "rule of three," while the algebraic problems deal with simple and simultaneous equations, quadratic equations, and arithmetic and geometric progressions. There is no clue as to who was the author of the work.

The subject matter is arranged in groups of sutras and presented as follows. In a typical case, a rule is stated and then a relevant example is given, first in verse and then in notational form. The solution follows in prose, and finally we have the demonstration or "proof."[24] This method of presentation is quite unusual in Indian mathematics. The few texts arranged in this way are invariably commentaries on earlier works. Since, in terms of its content, the Bakhshali Manuscript is but a prelude to more substantial work in the Classical period, we confine our discussion to a few novel features in this text. We begin with an examination of the system of notation used, as it is a recognizable precursor of later systems.

Notations and Operations

The system of notation used in the Bakhshali Manuscript bears some resemblance to those used by mathematicians such as Aryabhata I (c. AD 476–550), Brahmagupta (b. 598), and even Bhaskaracharya (b. 1114). But there is one important difference. In the Bakhshali text we find that the sign for a negative quantity looks exactly like the present "plus" symbol used to denote addition or a positive quantity. This sign was placed after the number it qualifies. For example, in the manuscript,

$$\begin{array}{|cc|} \hline 15 & 8+ \\ 4 & 3 \\ \hline \end{array}$$

means 15/4 − 8/3. Later, the + sign was replaced by a dot over the number to which it referred. Incidentally, this is one of the clues telling us that the Bakhshali Manuscript must have originated before the twelfth century. Another interesting aspect of the notation shown in the example above is the representation of fractions. It is similar to the present-day representation in that the denominator is placed below the numerator, but the line between the two numbers is missing.

This and other aspects of the notation and operations will be brought out if we take an example from the twenty-fifth sutra. There the following representation appears:

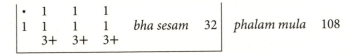

Here the black dot is used very much in the same way as we use the letter *x* to denote the unknown quantity whose value we are seeking. A fraction is denoted by placing one number under another, without a line between them. A compound fraction is shown by placing three numbers under one another; thus the second column of the representation above denotes 1 minus 1/3, or 2/3. (Without the + sign, it would denote 1 plus 1/3.) Multiplication is usually indicated by placing the numbers side by side. Thus the representation above means $(2/3) \times (2/3) \times (2/3)$, or $8/27$. *Bha* is an abbreviation of *bhaga*, meaning "part," and indicates that the number preceding it is to be treated as a denominator; *bha* is thus the symbol for division. The representation above therefore means

$$x = \left[\left(\frac{2}{3}\right)^3\right]^{-1} \times 32 = \left(\frac{27}{8}\right) \times 32 = 108,$$

or, in words, "the remainder (*sesam*) is divided by $1(1 - \frac{1}{3})(1 - \frac{1}{3})(1 - \frac{1}{3})$. The result is 108."

In the Bakhshali Manuscript the dot is also used to represent zero. The use of the same symbol to represent both an unknown quantity and a numeral is interesting. At the time the dot indicated an empty place, as its Sanskrit name shows: *shunya* means "empty," or "void." It is this dual meaning that gives us a clue to the age of the text.

On only two occasions the symbol for addition, which is the abbreviation *yu* (for *yuta*), is used. On almost all occasions, the two numbers to be added are put side by side. The whole operation is enclosed between lines, and the result is set down on the right of *pha*. Thus $3 + 6 = 9$ is represented as

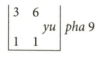

The Rule of Three (Trairasika)

One of the problems from the *Jiu Zhang* (example 6.3, discussed in chapter 6) has a solution that clearly shows a knowledge of the "rule of three." An early statement of this rule is in Aryabhata's *Aryabhatiya*:

> In the rule of three [*trairasika*], multiply the *phala-rasi* [fruit] by the *iccha-rasi* [desire or requisition] and divide by the *pramana* [measure or argument]. The required result *iccha-phala* [or fruit corresponding to desire or requisition] will be thus obtained.

Symbolically, this rule can be expressed thus: given that f, i, p, and m are *phala-rasi*, *iccha-rasi*, *pramana*, and *iccha-phala* respectively, then $m = fi/p$.

This is the first time in Indian mathematics that the technical names for the "rule of three" and for the four numerical quantities involved are given. However, the succinct manner in which the rule is given would indicate that it was already well known and that Aryabhata was merely restating it as a prelude to its use in astronomical computations. The antecedents of this rule have been traced back about a thousand years to a verse in the *Vedanga Jyotisa* and are discussed by Sarma (2002).

Discussion of this rule becomes a standard feature of all texts after Aryabhata. Though normally employed in solving commercial problems, the rule played a more important role in other areas of Indian mathematics and astronomy. In arithmetic, as we will see below, it was used as a means of verification in solving other problems. More importantly, it was employed in astronomical computations, for example, in the computation of the mean position of a planet from the number of its revolutions in a *kalpa* of 4,320,000,000 years. Many of the problems of spherical trigonometry were solved by applying the rule to similar right-angled triangles such as those found in the *aksajaksetra* (figure produced by the latitude).[25] Also, the rule forms the basis for computing trigonometric ratios. Bhaskara I's commentary on Aryabhata's work contains a detailed discussion of the rule and points to how it can be extended to encompass rules of five, seven, and so on. He also introduces the question of the logical sequence in which the three numerical quantities should be set down and the order in which the multiplication (f times i) and division by p should be carried out. Later, Brahmagupta's formulation of the rule became a model for subsequent writers bringing out more explicitly the fact that the three quantities should be set down in such a way that the first and last be of like denomination and the middle one of a different denomination. This is reiterated by Sridhara (c. 800), Mahavira (c. 850), and Aryabhata II (c. 950) without adding much to the principles underlying the rule. However, Bhaskara II (b. 1114) in his *Lilavati* states the important point that nearly the entire arithmetic is based on the "rule of three" and that most of the topics dealt with in *ganita* are but variations of this "rule of three":

> Just as the universe is pervaded by Hari with His manifestations, even so all that has been taught [in arithmetic] is pervaded by the "rule of three" with its variations. (S. R. Sarma 2002, p. 133)

Further, Nilakantha declares in his commentary on the *Aryabhatiya* that the entire mathematical astronomy (*graha-ganita*) is pervaded by two fundamental laws: by the law of relation between the base, perpendicular, and hypotenuse in a right-angled triangle—which goes today under the name of Pythagoras theorem—and by the rule of three.

However, depending on the date of composition of the original Bakh-shali Manuscript, the first recorded application of this rule may very well have been here. The problem to which the rule is applied is of a type famil-iar to schoolchildren today. For example, If 8 oranges cost 92 pence, what will 14 oranges cost? The solution is $(92 \times 14)/8 = £1.61$.

The method suggested in the Bakhshali Manuscript, which is also found in later works, may be stated in the following terms: If p oranges (argu-ment, or *pramana*) yield f pence (fruit, or *phala*), what will i oranges (req-uisition, or *iccha*) yield? It is suggested that the three quantities be set down as follows:

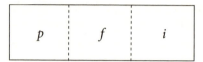

where p and i are of the same denomination and f is of a different denomi-nation. For the required result the middle quantity is to be multiplied by the last quantity and divided by the first, to give the result as fi/p. The fol-lowing example from the Bakhshali Manuscript (example 2 from sutra 53) illustrates the "rule of three," although the solution in the manuscript does not follow the method used here. For the Bakhshali solution, the reader is invited to refer to Hayashi (1995, pp. 385–86).

EXAMPLE 8.1 Two page boys are attendants of a king. For their services one gets 13/6 *dinaras* a day and the other 3/2. The first owes the second 10 dinaras. Calculate and tell me when they have equal amounts.

Suggested Solution

Take the denominators 6 and 2, together with the number 10 that the first has to give. The lowest common multiple of 2, 6, and 10 is 30, so 30 is the *iccha* (requisition). Now apply the rule of three (see table 8.3). Note that if the first page boy gives the second 10 *dinaras*, both will be left with 55 *dinaras*.

TABLE 8.3: INFORMATION FOR EXAMPLE 8.1

	p (day)	f (dinaras)	i (day)	Required result (fi/p)
First page boy	1	13/6	30	65
Second page boy	1	3/2	30	45

Consider a more apposite illustration from sutra 27 of the "rule-and-example" type of exposition, beginning with the statement of the rule, followed by the problem expressed verbally and then in a numerical form, and ending with a verification using the "rule of three." This example is based on a minor modification of what is contained in Hayashi (1995).

EXAMPLE 8.2

Rule

Multiply [the weights of] the gold pieces by [their respective] impurities. Divide their sum by [the weights of] the gold pieces added together. The [result] gives the loss [of gold] per unit [weight of the alloy].

Example

[Four] gold pieces, the quantities of which are one, two, three, and four suvarnas [respectively] are debased by one-half, one-third, one-fourth, and one-fifth of a masa [per suvarna in that order]. [They are melted] and formed into a single alloy. What is the impurity [of that alloy]? [Note: 1 suvarna = 16 masas]

The Problem in Numbers

Weight in *suvarnas*	1	2	3	4
Impurity in *masas*	1/2	1/3	1/4	1/5

Computation

Multiply [the weight of each] gold piece by its own impurity. And then set up the result as given below [see table 8.4]. Add the fractions

TABLE 8.4: CALCULATIONS FOR EXAMPLE 8.2

Type	Pramana (p)	Phala-rasi (f)	Iccha-rasi (i)	Iccha-phala (m) $m = fi/p$
1	10/1	163/60	1/1	163/600 *masas*
2	10/1	163/60	2/1	163/300 *masas*
3	10/1	163/60	3/1	163/200 *masas*
4	10/1	163/60	4/1	163/150 *masas*
Total				1630/600 *masas*

Continued . . .

Continued...

together by reducing each to the same denominator. Result 163/60. Divide this sum by the weights of the gold pieces added together (Result: 163/600). This is the loss [of gold] per suvarna.

The calculation is therefore verified by applying the "rule of three" in obtaining the combined impurity of all four pieces, which amounts to $1,630/600 = 163/60$ masas. In the terms of modern notation: If w_i = weight of the ith gold piece in suvarnas and c_i is the impurity (or the nongold component) of the ith gold piece in masas for $i = 1, 2, \ldots, n$, then the combined impurity of all n gold pieces when melted together is

$$\frac{w_1 c_1 + w_2 c_2 + \ldots + w_n c_n}{w_1 + w_2 + \ldots + w_n}.$$

Extracting Square Roots

The Bakhshali Manuscript extended the work on square roots in the *Sulba-sutras*, which we discussed earlier in this chapter, to give a more accurate formula for finding an approximate value of the square root of a nonsquare number. The relevant sutra (no. 18) may be expressed more comprehensibly in the following terms:

In the case of a nonsquare number, subtract the nearest square number; divide the remainder by twice the nearest square; half the square of this is divided by the sum of the approximate root and the fraction. This is subtracted, and will give the corrected root.

In symbolic form, this rule is:

$$\sqrt{A} = \sqrt{a^2 + r} \approx a + \frac{r}{2a} - \frac{(r/2a)^2}{2(a + r/2a)},$$

where a^2 is the perfect square nearest to A and $r = A - a^2$. For example:

$$\sqrt{41} \approx 6 + \frac{5}{12} - \frac{(5/12)^2}{2(6 + 5/12)} \approx 6.4031 \text{ (to four decimal places)}.$$

(The reader may wish to try using this rule to evaluate $\sqrt{3}$ and $\sqrt{5}$.) The formula was applied in the manuscript to calculate the approximate square root of 481 as 424,642/19,362, which is correct to four decimal places. The formula may be compared with the approximation procedure for finding the square root of a nonsquare integer usually attributed to the Hellenistic mathematician Heron (c. AD 200), although we came across a close antecedent of the procedure in Babylonian mathematics in chapter 4. Heron's formula for finding the square root of A is

$$\sqrt{A} \approx \frac{1}{2}\left(a^* + \frac{A}{a^*}\right),$$

where a^* is a first approximation to the square root of A, and can be a non-integer (this is not possible with the Bakhshali method).[26]

Indeterminate Equations: Their First Appearance

The following is one of a number of similar problems found in the manuscript:

EXAMPLE 8.3 Three persons possess 7 asavs, 9 hayas, and 10 camels, respectively [asavs and hayas are two breeds of horses]. Each gives two animals, one to each of the others. They are then equally well off. Find the price of each kind of animal and the total value of the livestock possessed by each person.

Solution (in Symbolic Terms)

Let x_1, x_2, and x_3 be the prices of an *asav*, a *haya*, and a camel respectively. Then, from the information given in the question,

$$5x_1 + x_2 + x_3 = x_1 + 7x_2 + x_3 = x_1 + x_2 + 8x_3 = k,$$

or

$$4x_1 = 6x_2 = 7x_3 = k,$$

and we seek values of x_1, x_2, x_3, and k that are positive integers.

To get integer solutions, we take k to be any multiple of the lowest common multiple of 4, 6, and 7. In the Bakhshali Manuscript k is taken as $4 \times 6 \times 7 = 168$. Then the price of an *asav* is 42, the price of a *haya* is 28, and the price of a camel is 24. The total value of livestock in the possession of each person is 262.

From these humble beginnings, over the next thousand years there was to be a systematic development of indeterminate analysis, which will be examined in the next chapter.

An Unusual Series

Among the arithmetic series found in the Bakhshali Manuscript, there are some unusual ones. The next problem, reconstructed from a mutilated birch-bark strip, is Hayashi's translation and interpretation of example 7 in sutra N6 (Hayashi 1995, pp. 286 and 395).

EXAMPLE 8.4　O wise man! A certain king gave five horsemen a gift of fifty-seven [monetary unit missing]. Each person in order, I tell [you], obtained twice the amount of his predecessor and one more. What then was obtained by the first person and what by each of the others?

Solution (in Symbolic Terms)

The problem may be expressed thus. Let x_i be the share of the ith horseman. Then

$$x_{i+1} = 2x_i + 1 \, (i = 0, 1, 2, 3, 4), \text{ and } \sum_{i=0}^{4} x_i = 57.$$

The solution is lost in the original manuscript, except for the following tabulation of the numerical data (see table 8.5).

TABLE 8.5: DATA FOR EXAMPLE 8.4

0	1	3	7	15	57
1	1	1	1	1	1
1	2	4	8	16	
1	1	1	1	1	

Expressed in modern notation, the cells of the table are the coefficients of the following equations:

$$x_1 = x_0 + 0, x_2 = 2x_0 + 1, x_3 = 4x_0 + 3, x_4 = 8x_0 + 7, x_5 = 16x_0 + 15.$$

Continued . . .

Continued . . .

The solution is obtained by applying the algorithm given in sutra N6:

$$x_0 = \frac{57 - (1 + 3 + 7 + 15)}{1 + 2 + 4 + 8 + 16} = \frac{31}{31} = 1.$$

Hence

$$x_1 = 1, x_2 = 3, x_3 = 7, x_4 = 15, x_5 = 31$$

are the shares received by the five horsemen.

The state of Indian mathematics at the middle of the first millennium AD, as represented by the Bakhshali Manuscript, may be summarized as follows.

1. Whereas the mathematics of the Vedic age and of the Jains were in part inspired by religion, the mathematics of this period became more practical and secular, being applied to everyday problems. Examples of profit and loss, computation of the average impurities of gold, of wages, or of gifts to be paid to subordinates, and of speeds and distances to be covered, form the subject matter of the Bakhshali Manuscript.

2. Whereas the writers of the Sulbasutras had already devised rules to find approximate values of $\sqrt{2}$, these rules were now more elaborate and were being used to obtain the square root of any number to a greater degree of accuracy.

3. There is some evidence that work on series begun during the Jaina period was continued.

4. This period marks the beginning of the great interest in indeterminate analysis. Such an interest did not arise solely from the demands made by astronomical calculations. Other problems, some of them of a recreational nature, were also included. The examples discussed in the Bakhshali Manuscript and the solutions offered are not difficult, but they mark the beginning of a study that was to reach an advanced level during the so-called Classical period of Indian mathematics.

5. There is evidence of a well-developed place-value number system that included zero (represented by a dot). The ease with which the

system is used in the manuscript suggests that the system predates the document by a few hundred years.

6. In contrast to the vast majority of Indian mathematical works composed before and after this manuscript, the method of exposition follows a systematic order, as illustrated by example 8.2 above: (1) statement of the rule (sutra), (2) example(s) to apply the rule, (3) a solution using the rule, and (4) verification of the correctness of the solution. Most of the other sources of Indian mathematics, until the emergence of Kerala mathematics in the late fourteenth century, contain concise statements of the rules, usually without any attempt at deriving or demonstrating them. These were left to subsequent commentators or teachers to explain.

Notes

1. The beginnings of European awareness of ancient Indian mathematics may be traced to the English translation by Colebrook (1817) of three notable classics of Indian mathematics, namely the *Brahma Sphuta Siddhanta* of Brahmagupta (AD 628) and the *Lilavati* and *Bijaganita* of Bhaskara II (1150). One response was to believe that the mathematics contained in these texts could be traced to the Greeks, who were perceived as the originators of Western mathematics. Cantor (1880–1908), the author of an influential four-volume treatise on the history of mathematics, maintained at a number of places that the Indians had learned algebra through traces of algebra within Greek geometry. And similarly, Brahmagupta's solution to quadratic equations had Greek origins. In response to Cantor, Hankel (1874, p. 204) argued:

> [H]umanist education [has] deeply inculcated prejudice that all higher intellectual culture in the Orient, in particular all science, is risen from the Greek soil and that the only mentally truly productive people have been Greek. This makes it difficult to turn around the direction of interest for one instant.

Heeffer (2007) argues that soon a split arose between the "believers," such as Rodet (1879) and Hankel (1874), in an independent development of Indian mathematics and the "nonbelievers," such as Cantor (1905) and Kaye (1915), who argued that Indian mathematics was derived from the Greeks. The latter group traced the development of Indian indeterminate algebra back to Diophantus, whose algebra in turn was believed to have originated with Pythagoras. For a discussion of the alleged Indian debt to Greek mathematics, see Heeffer (2007).

2. This is not to deny that a case can be made that Hipparchus of Nicaea's (fl. 150 BC) work on chords may have influenced early Indian trigonometry. For details, see Toomer (1973) and Duke (2005).

3. In this chapter and the next, a number of titles of Indian texts are given without translation. This is sometimes because words in the title have a number of possible interpretations, and sometimes because a literal translation would not be particularly meaningful.

4. In recent years, the term "Vedic mathematics" has been used to describe a set of computational algorithms based originally on a book by that name, first published in 1965 and authored by Krishna Tirthaji. The computation rules (given in the form of modern Sanskrit sutras) were claimed to have been "rediscovered" from a *parisista* (appendix) of the *Atharavaveda*. Extensive search for this *parisista* has proved to be fruitless. Hence, in this edition, for the sake of historical accuracy and consistency, it was decided to omit this topic, which nevertheless contains fascinating computational tools. For further elaboration of the methods, see Nelson et al. (1993) and Krishna Tirthaji (1965).

5. The dating of early Indian texts is highly uncertain. The dates given here are rough and conservative estimates of when the first versions of the texts were recorded. It is very likely that, before they were written, an earlier oral tradition kept the contents alive. Copying old texts was a common pursuit of the Indian scholar and student, sanctioned by religion and custom. It is therefore important not to depend on the dates of copies of mathematical texts in assessing the true age of a particular method or technique.

6. This is a conservative estimate. There are varying astrochronological estimates that suggest dates around 4000 BC, 3300BC, and 3000 BC, any of which are possible dates for the composition of Vedic texts or, more plausibly, the origin of the ideas that were later incorporated into them. For further information on these interpretations, see Kak (2005).

7. For further details, see Kuppanna Sastry (1985).

8. These statements are not literal translations of the original texts. For the sake of clarity, modifications have been made. Thus, for example, a more literal translation of the first statement reads: "The rope [equal to] the diagonal of a [square] quadrilateral makes twice the area. It is the 'two-maker' [*dvi-karani* or 'double'] of the square."

9. The *Mahavedi* provides the prototype for a smaller sacrificial altar to the chief of the gods, Indira, with proportions identical to those of the Great Altar but having only one-third of the area. A scaling factor of $1/\sqrt{3}$ is applied to achieve the desired result. It would seem that the *Sulbasutras* exhibit considerable skills in arithmetical manipulations, including those of fractions and their combinations.

10. The rule, stated in the *Sulbasutras*, was as follows:

> Wishing to make a [square] quadrilateral a circle: Bring [a cord] from the center to the corner [of the square]. [Then] stretching [it] toward the side, draw a circle with [radius equal to the half-side] plus a third of the excess [of half the diagonal

over the half-side]. This is definitely the [radius of the] circle. As much as is added [to the edges of the circle] is taken out [of the corners of the square]. (Quoted in Plofker 2009, p. 23)

It would seem that the authors of the *Sulbasutras* were primarily "interested in practical results and show no direct concern with proof procedures as such at all" (Lloyd 1990, p. 104).

11. In the *Katyayana Sulbasutra*, there is a clear recognition that the procedure produces only an approximate result, implied in the term "having a difference" (from the exact value).

12. In chapter 4, on Mesopotamian mathematics, we came across an approximation method for evaluating $\sqrt{2}$. The result, which in sexagesimal notation is 1;24,51,10, is more or less the same as the *Sulbasutra* value, and this prompts the question of whether the *Sulbasutra* procedure was in some sense derived from the Mesopotamian. For speculations regarding the Mesopotamian connections, see Datta (1932a, pp. 192–94) and Neugebauer (1962, p. 34).

13. This would of course depend on the dating of the Bakhshali Manuscript.

14. See Plofker (2009, p. 57).

15. However, the order of the digits is the opposite of what had have today, beginning with the least significant (unit) and moving on to digits representing increasing powers of 10. Thus "earth-eye-sky-time" would be read as 3,021 in our full-fledged decimal place-value system.

16. Or more precisely, "zero" is also known in Arabic by the term *daira saghira* (small circle). It is this small circle that appears in early Latin manuscripts (S. R. Sarma 2009, p. 215).

17. See Han (2002, p. 106) for the source of the quotation.

18. As early as the first century AD, the Buddhist philosopher Vasumitra was using a similar analogy when he compared the varieties of realities to the merchants' counters: the same single clay counter can represent units, hundred, thousands, and so on, depending on its position.

19. In his *Brahma Sphuta Siddhanta*, Brahmagupta gave rules for zero and negative numbers in terms of "fortunes," which represent positive numbers, and "debts," indicating negative numbers. For example, a debt subtracted from zero is a fortune, and a fortune subtracted from zero is a debt. Also the product or quotient of a debt and a fortune

is a debt, while the product or quotient of two fortunes is a fortune. In his attempt to extend these statements to include division by zero, Brahmagupta stated wrongly that zero divided by zero is zero. In terms of modern notation, he also suggested that any number n divided by zero is $n/0$, which is saying very little. However, Brahmagupta is given credit for being one of the earliest mathematicians to tackle this fascinating subject of "calculations with zero" (*shunya ganita*).

20. An interesting area for historical investigation is the connection between the mathematics contained in the Buddhist and Jaina texts and the development of logic. A belief in a truth-functional two-valued logic was denied by both the Buddhists and the Jains. Instead, as expressed by the philosopher Nagarjuna (c. third century AD), the Buddhists had a logic of four alternatives: "Everything is such (X), not such (not X), both such and not such (X and not X), neither such nor not such (neither X nor not X)." It is interesting that Nagarjuna's main contribution to Buddhist philosophy is in the development of the concept of *shunyata*, discussed earlier in this chapter. For further elaboration, see Matilal (1985).

21. It is not practicable to examine this fascinating area of mathematics here. Simple introductions to the concept of transfinite numbers and operations with such numbers are given by Stewart (1981, pp. 127–43) and by Sondheim and Rogerson (1981, pp. 148–59).

22. A cautionary note should be sounded here. The use of the word "logarithm" in this context does not imply that it ever became a computational tool in Indian mathematics.

23. The Digambara are one of the two main sects of Jain monks whose members shun all property and wear no clothes. In their active practice of nonviolence, they use a peacock-feather duster to clear their path of insects to avoid trampling them.

24. It is interesting that the prose explanations do not always reflect the solution procedures offered, possibly suggesting that some of the verses stating the problem may have been recorded earlier than the explanations.

25. I am grateful to Takao Hayashi for providing a precise definition of the term *aksajaksetra*.

26. It is worthy of note that the Babylonian procedure, discussed in chapter 4, permits successive approximation (also not possible with the Bakhshali method). It is easily seen that the Bakhshali formula and Heron's formula produce identical results for the square root of 3, for $a = 1$ and $a^* = 1.5$. In Heron's example in his book, *Metrica 1.8*, $A = 720$, $a^* = 27$, and $A/a^* = 720/27 = 26\frac{2}{3}$ produces a square root approximation for 720 as $(27 + 26\frac{2}{3})/2 = 26 + \frac{1}{2} + \frac{1}{3}$, which is correct to three decimal places.

Chapter Nine

Indian Mathematics: The Classical Period and After

From the previous chapter it is clear that our evidence of mathematical activities after the Vedic period, as represented by Jaina canonical literature and the Bakhshali Manuscript, is imperfect and incomplete. Our knowledge of the development of mathematics and astronomy between the *Sulbasutras* and the period of Aryabhata I (c. AD 500) is therefore fairly sketchy. Yet this hiatus in our knowledge is particularly puzzling given the wealth of evidence we have for the same period in other fields, notably in medicine and chemistry, and in philosophy, where outstanding work was produced by the Nyaya and Mimamsa schools.

Various explanations have been offered for this apparent discontinuity. The virtual disappearance of Vedic sacrifices removed, as it were, the raison d'être for continued interest in geometry. The sheer size of the Indian subcontinent would have restricted communication between different parts of the country, with an adverse effect on the transmission of mathematical ideas, which were widely scattered and normally restricted to certain families. If a particular generation showed little interest or aptitude, the family's mathematical knowledge might be lost forever. Mathematical ideas were transmitted orally in a verse form that could easily be memorized. This sutra form was specially suited for this purpose, but to the uninitiated it required elaborate explanations—without commentaries, the sutras often made little sense. This form of transmitting knowledge had the result of confining mathematical pursuits to a tiny elite. This elitism, born of the caste system, is probably one of the reasons why Indian mathematics floundered for a few centuries after its impressive beginnings. (For other disciplines such as medicine and chemistry, knowledge was concentrated in schools and did not suffer in this way.)

The revival, which came in the middle of the first millennium AD, also established channels of communication both within India, where mathematical work was concentrated in three centers of learning (Kusum Pura and Ujjain in the north and Mysore in the south) and within other cultures, first Persia and later the Islamic world and China. The scene was set for the transmission of Indian mathematical ideas to the West and the incorporation of important Babylonian and Hellenistic ideas, mainly from Alexandria, into Indian astronomy.

The few centuries preceding and following the beginning of the Christian era saw the emergence of a class of astronomical texts called the *Siddhantas*.[1] Contained in them were important changes in astronomical methods and practices. The traditional system, based on tracking the movement of the planets in relation to twenty-seven (or twenty-eight) stars chosen as reference points, was dispensed with and replaced by the twelve signs of the zodiac. More sophisticated mathematical methods were used to determine the periods of planetary revolutions. The mean longitudes were calculated from the number of days that had elapsed since the beginning of the present Kaliyuga era (Friday, February 18, 3102 BC, on the Gregorian calendar). Different measures of the duration of the day and year were correctly determined. Planetary positions were computed using Ptolemaic epicycles and deferents; eclipses were calculated and the results corrected for parallax (the apparent displacement of celestial objects resulting from the changing location of the observer as the earth moves in its orbit). These computations required a wide range of mathematical techniques, including certain innovative methods of plane and spherical trigonometry and applications of indeterminate equations. Some of these methods will be examined in this chapter.

Major Indian Mathematician-Astronomers

Aryabhata I (b. AD 476)

In his best-known work, *Aryabhatiya*, Aryabhata I states that he composed it during the 3,600th year of the Kaliyuga, when he was twenty-three years old.[2] According to the *Suryasiddhanta*, an astronomical treatise that is the basis of all Hindu and Buddhist calendars, Kali Yuga began at midnight on January 23, 3102 BC, according to our present calendar. So he must have been born in AD 476 and completed his work in 499 (i.e., $3601 - 3102 = 499$). There are a number of conjectures about his birthplace, ranging from the south (Kerala, Tamil Nadu, Andhra Pradesh) to the northeast (Bihar, Bengal). But all we know is that he wrote his great work in Kusum Pura, near modern Patna in Bihar. It was then the imperial capital of the Gupta empire, and had been an important center of learning since the Jaina period; the great Jaina metaphysician-scientist Umasvati (c. AD 200) recorded that a famous school of mathematics and astronomy had stood there before his time.

The *Aryabhatiya* is short and concise, and is essentially a systematization of results known earlier and probably contained in the older *Siddhantas*. The mathematical section consists of just thirty-three verses. It is, however, of particular importance not only because of the picture it gives of the state of mathematical knowledge of the period but also for the impetus it gave to the future study of the subject. It contains details of an alphabet-numeral system of notation (discussed in chapter 8), rules for arithmetical operations, and methods of solving simple and quadratic equations and indeterminate equations of the first degree. The book pays some attention to trigonometry and introduces the sine and versine (i.e., $1 - \text{cosine}$) functions—a notable innovation on earlier work both in and outside India. The Indian contribution to trigonometry will be discussed later in this chapter.

Aryabhata hit upon 3.1416 as a close approximation to the ratio of the circumference of a circle to its diameter, a fact mentioned earlier in our discussion of the history of π in chapter 7. He also gave correct general rules for computing the sum of natural numbers, and of their squares and cubes. There was, on the evidence of a later commentator, a nonextant work under the title *Arya Siddhanta*, a more detailed examination of astronomy (and possibly trigonometry), but there are serious doubts about its authorship.

Aryabhata's place as the premier and pioneering mathematician of India will have to be reassessed in the light of recent discoveries about the scope and quality of Jaina mathematics. As the mathematical activities of the post-Vedic and pre-Classical period become better known, Aryabhata will come to be seen mainly as an astronomer who had a great influence on those who came after him. It is therefore appropriate that India's first artificial satellite, designed and built in India, and launched in the former USSR on April 19, 1975, was named *Aryabhata*.

Two astronomers who followed Aryabhata extended his work. As an astronomer, Varahamihira (fl. AD 500) is remembered for his construction of trigonometric tables in his famous astronomical treatise *Pancasiddhantika* and for a revised version of the Indian calendar, which he corrected for the amount of precession that had accumulated since the preparation of *Suryasiddhanta* (one of the widely known earlier *Siddhantas*). Mathematically, his work is interesting for its detailed exposition of trigonometry. It gives a number of relations between three functions, *jya* (Indian sine, or half chord), *kojya* (Indian cosine, or cosine-chord of an arc), and *utkramajya* (Indian versine), which we shall examine in the section on trigonometry later in this chapter. Also, extending the work of Aryabhata I, he gave values of different *jyas* in a quadrant drawn at a fixed interval (i.e., an Indian sine table).

Bhaskara I was one of the most competent exponents of Aryabhata's astronomy. His three major works consist of two treatises on astronomy and a commentary on the *Aryabhatiya* (629). His work reveals not only the depth and breadth of his mathematical understanding but also that he was part of a well-established tradition of investigation into the subject and its foundations. He may be seen as the first illustrious commentator of the Aryabhatan school, who would influence not only the mathematicians who came after him during the Classical period but also the Kerala mathematicians who lived more than seven hundred years later. His notable contributions to mathematics include his solution of indeterminate equations of the first degree and a remarkably accurate approximation formula for calculating the Indian sine of an acute angle without the use of a table, of which the former was to have a significant influence on a later mathematician-astronomer, Brahmagupta.[3]

Brahmagupta (b. AD 598)

After Varahamihira, the best-known mathematician-astronomer of the Ujjain school is Brahmagupta. He wrote two works, the first and more important, *Brahma Sphuta-siddhanta* (Corrected Siddhanta of Brahman), when he was thirty. The book is a comprehensive astronomy text, several chapters of which deal with mathematics. Brahmagupta called the twelfth chapter *Ganita* (Arithmetical Calculation), although it includes a discussion of mathematical series and a few geometric topics (including the well-known "Brahmagupta theorem" discussed in a later section of this chapter). The eighteenth chapter, *Kuttaka* (literally Pulverizer, but broadly translated as Algebra), contains solutions of indeterminate equations of the first and second degree.[4] Scattered through his second book, an astronomical treatise titled *Khanda Khadyaka*, are further developments in trigonometry, including a method of obtaining the sines of intermediate angles from a given table of sines. The method employed is equivalent to the Newton-Stirling interpolation formula up to second-order differences. We shall return to this later.

Brahmagupta holds a special place in the history of mathematics. As we shall see in chapter 11, it was partly through a translation of his *Brahma Sphuta-siddhanta* that the Islamic world, and then the West, became aware of Indian astronomy and mathematics. This was to have momentous consequences for the development of the two subjects.

Sridhara (fl. AD 800)

There remains some controversy over Sridhara's time and place of birth; some scholars suggest he came from Bengal, others from southern India. What is definitely known is that he wrote the partly extant *Patiganita* and the highly influential *Trisatika*. The latter proved to be one of the most popular textbooks on arithmetic before the *Lilavati* of Bhaskaracharya over three centuries later. In it he deals with elementary operations, including extracting square and cube roots, and fractions. Eight rules are given for operations involving zero (but not division). His methods of summation of different arithmetic and geometric series were to become standard references in later works.

Mahavira (fl. AD 850)

Mahavira was the best-known Indian mathematician of the ninth century. A Jain by religion, he was familiar with Jaina mathematics, which he included and refined in his book *Ganita-sara-sangraha*. It is possible that Mahavira knew the works of Aryabhata and Brahmagupta. Unlike his predecessors, Mahavira was not an astronomer—his work was confined to mathematics. He was a member of the mathematical school at Mysore in southern India.

In *Ganita-sara-sangraha* he gives a lucid classification of arithmetical operations and a number of examples to illustrate the rules. His contributions include:

1. A detailed examination of operations with fractions, with some ingenious methods for decomposing integers and fractions into unit fractions (a subject of practical utility for the ancient Egyptians, as we saw in chapter 3)

2. A statement of general rules of operations with zero and positive and negative quantities

3. An extension and systematization of the Jaina work on permutations and combinations, for both of which he provides the well-known

general formulas illustrated with examples involving combinations of flavors, of different precious stones making up a necklace, and of different flowers contained in a garland[5]

4. Solutions of different types of quadratic equations, as well as an extension of his predecessors' work on indeterminate equations

5. Geometric work on right-angled triangles whose sides are rational and, something unusual in Indian mathematics, attempts (albeit unsuccessful) to derive formulas for the area and perimeter of an ellipse

The book was widely used in southern India and translated into Telegu, a regional language, during the eleventh century.

Mahavira's contribution may be looked at in two ways. *Ganita-sara-sangraha* may be seen as the culmination of Jaina work on mathematics (indeed it is the only substantial treatise on Jaina mathematics that we have). Alternatively, Mahavira can be seen as summarizing and extending the mathematical content of the works of his predecessors such as Aryabhata, Bhaskara I, and Brahmagupta. He was very conscious of the debt he owed those who came before him. In the introductory chapter of his book, he wrote:

> With the help of the accomplished holy sages, who are worthy to be worshipped by the lords of the world. . . . I glean from the great ocean of the knowledge of numbers a little of its essence, in the manner in which gems are [picked] from the sea, gold from the stony rock, and the pearl from the oyster shell; and I give out according to the power of my intelligence, the *Sara Sangraha*, a small work on arithmetic, which is [however] not small in importance.

Later than Mahavira was the astronomer Aryabhata II, who lived around the middle of the tenth century. In his major astronomical treatise of eighteen chapters, *Maha-siddhanta*, there is a clear treatment of *kuttaka*, which had by then come to mean the solution of indeterminate equations. Following them were the mathematician-astronomers Sripati (fl. AD 1050), the author of *Ganita-tilaka* (Forehead Mark of Calculation), and Jayadeva, who lived around the middle of the eleventh century. More popular than any of them was Bhaskara II.

Bhaskara II (b. AD 1114)

Bhaskara II or Bhaskaracharya (Bhaskara the Teacher), as he is still popularly known in India, lived in the Sahyadri region in Maharashtra and came from a family of court scholars. Little is known about him, except that his grandson helped to set up a school for the study of his writings. His fame rests on three works: *Lilavati*, *Bijaganita*, and *Siddhanta-siromani*. The last, a highly influential astronomical work, was written in 1150 when he was thirty-six years old.

Lilavati, which is based on the works of Brahmagupta, Sridhara, and Aryabhata II, shows a profound understanding of arithmetic. Bhaskaracharya's work on fundamental operations, his rules of three, five, seven, nine, and eleven, his work on permutations and combinations, and his rules of operations with zero[6] together speak of a maturity, a culmination, of five hundred years of mathematical progress.

In 1587, on the instructions of the Mughal emperor Akbar, the court scholar Fyzi translated *Lilavati* into Persian. Fyzi tells a charming story of the book's origin. Lilavati was the name of Bhaskaracharya's daughter. From casting her horoscope, he discovered that the auspicious time for her wedding would be a particular hour on a certain day. He placed a cup with a small hole at the bottom in a vessel filled with water, arranged so that the cup would sink at the beginning of the propitious hour. When everything was ready and the cup was placed in the vessel, Lilavati suddenly out of curiosity bent over the vessel, and a pearl from her dress fell into the cup and blocked the hole in it. The lucky hour passed without the cup sinking. Bhaskaracharya believed that the way to console his dejected daughter, who now would never get married, was to write her a manual of mathematics!

Bhaskaracharya's *Bijaganita* contains problems on determining unknown quantities, evaluating surds (i.e., square roots that cannot be reduced to whole numbers) and solving simple and quadratic equations, and some general rules that went beyond Sridhara in dealing with the solution of indeterminate equations of the second degree and even equations of the third and fourth degree. Bhaskaracharya's "cyclic" method for solving indeterminate equations of the form $ax^2 + bx + c = y$ was rediscovered in the West by William Brouncker in 1657.

In *Siddhanta-siromani*, Bhaskaracharya demonstrates his knowledge of trigonometry, including the sine table and relationships between different trigonometric functions. Certain preliminary concepts of the infinitesimal calculus and analysis can be traced in his work, concepts that would be taken up in the Kerala school of mathematicians in their work on infinite series some two hundred years later.

He won such a great reputation that his manuscripts were still being copied and commented upon as late as the beginning of the nineteenth century. A medieval temple inscription refers to him in the following terms (and here reappears the imagery of the peacock that provides the inspiration for the title of the present book):

> Triumphant is the illustrious Bhaskaracharya whose feats are revered by the wise and the learned. A poet endowed with fame and religious merit, he is like the crest on a peacock.

It was generally believed until recently that mathematical developments in India came to a virtual halt after Bhaskaracharya. This opinion has had to be revised in the light of recent research on what one could describe as medieval Indian mathematics. A number of the manuscripts of this period are yet to be published, or even subjected to critical scrutiny. But we are able to identify the following notable contributors to mathematics during the medieval period.

Narayana Pandita (fl. AD 1350)

Narayana lived during the reign of Firoz Shah (1355–1388) and composed *Ganita-kaumudi* (Moonlight of Computation), a treatise on arithmetic, and *Bijaganita-avatamsa* (Garland of Algebra), a work on algebra, both of which were heavily influenced by Bhaskaracharya. The topics contained in Narayana's books include laws of signs, mathematical operations with zero, approximation methods for finding the square root of a nonsquare number, detailed investigation of permutations and combinations ("net of numbers"),[7] and a diagrammatic method of representing different mathematical series, to be discussed in the last section of this chapter. His work is also notable for its treatment of magic squares.[8]

Madhava of Sangamagramma (c. AD 1340–1425)

Madhava was probably the greatest of the Indian medieval astronomer-mathematicians, but he has come to the fore only in recent years as a result of growing knowledge of Kerala mathematics. It was Madhava who "took the decisive step onwards from the finite procedures of ancient mathematics to treat their limit-passage to infinity, which is the kernel of modern classical analysis" (Rajagopal and Rangachari 1978, p. 101).

Sangamagramma was a village with a temple dedicated to a deity of the same name and situated near Cochin in Kerala. This place-name is often given when referring to Madhava so as to distinguish him from others such as the astrologer Vidya Madhava. Later astronomers called him Golavid (or Master of Spherics). Of his works that have survived, all are astronomical treatises; for his mathematical contributions we rely on reports by his contemporaries and successors. These contributions, which include infinite-series expansions of circular and trigonometric functions and finite-series approximations, are discussed in chapter 10.

Nilakantha Somayaji (AD 1445–1545)

Nilakantha, who was a student of the eminent astronomer Paramesvara of Vatasreni, lived to the ripe old age of one hundred. He came from a Nambuthri Brahmin family in South Malabar, Kerala. He was a versatile scholar, but, like Madhava, all his surviving works are on astronomy. The mathematical sections of his *Tantra Samgraha* elaborate and extend the contributions that are attributed to Madhava.

The Kerala school of mathematics and astronomy continued for another two centuries, producing detailed commentaries on the works of classical mathematicians such as Aryabhata and Bhaskaracharya as well as continuing the work on trigonometry and infinite series begun by Madhava. In a notable work in Malayalam (the regional language of Kerala) titled *Yuktibhasa*, Jyesthadeva (fl. 1550) provides a detailed summary of the mathematical contributions made by the Kerala school. It is unusual in Indian mathematics since it contains derivations of most of the theorems and formulas stated in the text. Finally, it is worth mentioning a highly influential figure outside the Kerala tradition. Ganesa (b. 1507) was the author of the popular work *Tithi-cintamani* (Thought-Jewel of Lunar Days), a detailed commentary on the *Lilavati* of Bhaskaracharya. He came from a family in Nandigram in Gujerat, who had over several generations earned the reputation of being noted astrologer/astronomers.

In the rest of this chapter we survey the major contributions of the Indian mathematicians of the Classical (from Aryabhata I to Bhaskaracharya) and medieval (from after Bhaskaracharya until about 1600) periods. Our approach will be a thematic one, with the exception of the contribution of the Kerala school, which will be examined in the next chapter. It may help whenever necessary to refer both to the map in figure 8.1 and to the Indian chronology in table 8.1.

Indian Algebra

It was briefly indicated in the previous chapter that the *Sulbasutras* and the later Bakhshali Manuscript contain some early algebra, including the solution of linear, simultaneous, and even indeterminate equations. But it is only from the time of Aryabhata I (fifth to sixth centuries) that algebra grew into a distinct branch of mathematics. Different names were used for this area of mathematics. Brahmagupta (sixth century) has a separate chapter in his book, *Brahma Sphuta-siddhanta*, called *Kuttaka*. The topic was of such importance that he placed it at the beginning of the chapter even before the discussion of basic topics such as the six arithmetic operations involving negative numbers, zero, irrational numbers, and unknown numbers. It was Sridhara (fl. AD 800) who used the term *bijaganita* (computation with seeds) for algebra for the first time. According to Bhaskara II, Sridhara was the author of at least one book on algebra, which is no longer extant. However, his book on arithmetic, referred to earlier, which he called *Patiganita*, brought to the fore a distinction drawn between two major fields in Indian mathematics: *patiganita* and *bijaganita*. Mahavira (fl. AD 850) introduced the term *kuttukara* to describe the procedure for the solution of equations of the first and second degree whether they were determinate or indeterminate. It would therefore seem that terms to describe different branches of mathematics varied over time.

A significant feature of early Indian algebra that distinguishes it from other mathematical traditions was the use of symbols, such as a dot (in the Bakhshali Manuscript) or the letters of the alphabet, to denote unknown quantities. In fact it is this very feature of algebra that one immediately associates with the subject today. The Indians were probably the first to make systematic use of this method of representing unknown quantities. A general term for any unknown was *yavat tavat*, which was shortened to the algebraic symbol *ya*. When Brahmagupta uses the word *avyakta*, he simply means "invisible" or "unknown." However, when he prescribes rules for equations in several unknown numbers, he uses the word *varna*, meaning "color," for indicating them. Thus, "Having subtracted the colors other than the first color [from the opposite side, that side] is divided by the [coefficient of] the first [color]. The result is the value of the first [color]" (*Brahma Sphuta-siddanta*, 18.51).

Simple operations were also indicated by abbreviations or symbols. We saw in chapter 8 that addition was represented a few times in the Bakhshali Manuscript by placing *yu* (which stood for *yuta*, "added" or "increased") between the terms to be added, and subtraction by placing the sign + after the term to be subtracted. Multiplication was indicated by placing *gu* (for *gunita* or multiplied) after the second term (i.e., 3 4 *gu* = 3 × 4), and division by putting *bha* (for *bhaga*) after the two terms (i.e., 3 4 *bha* = 3 ÷ 4) or between the two terms (i.e., 3 *bha* 4 = 3 ÷ 4). A square root was indicated by *mu* (for *mula*) after the term. There were several other similar abbreviations for other operations.

In Prthudakaswami's (fl. AD 864) commentary on Brahmagupta's *Brahma Sphuta-siddhanta* appears the following representation:

yava   0   ya 10 ru 8

yava   1   ya 0 ru 1

Here *ya* is an abbreviation for *yavat tavat* (the unknown quantity, or *x*) and *yava* is an abbreviation for *yavat avad varga* (the square of the unknown quantity, or x^2); *ru* stands for *rupa* (the constant term). In other words, this is what we would now write as $10x + 8 = x^2 + 1$.

Solutions of Determinate Equations

A geometric solution to a linear equation in one unknown (an equation like $3x + 8 = 23$) may be discerned in Baudhayana's *Sulbasutra*, while an algebraic solution appears for the first time in the Bakhshali Manuscript. The method used was an inversion method, whereby one works backward from a given piece of information—an approach particularly favored by Islamic mathematicians five hundred years later, which may have reached them from India. (An illustration of the procedure is given later in this section.)

Quadratic equations make their first appearance in the *Sulbasutras* in the forms $ax^2 = c$ and $ax^2 + bx = c$. No solution is given. For an equation of the form $ax^2 + bx - c = 0$, the Bakhshali Manuscript offers the following solution (in modern notation):

$$x = \frac{\sqrt{b^2 - 4ac} - b}{2a}. \tag{9.1}$$

The first explicit statement of a general rule appears in a work by Sridhara, which is unfortunately lost, though the rule is preserved in quotations by Bhaskaracharya and others. It is:

Multiply both sides [of the equation] by a known quantity equal to four times the coefficient of the square of the unknown; add to both sides a known quantity equal to the square of the coefficient of the unknown; then [extract] the square root.

This solution, obtained by transforming the left-hand side of the quadratic equation

$$ax^2 + bx = c$$

by multiplying both sides by $4a$, adding b^2, and finally taking the square root, is a variant of equation (9.1). There is no evidence that Sridhara used both signs of the radical, but Mahavira was certainly aware of both possibilities. He gave the solution (in modern notation) as

$$x = \frac{-b/a \pm \sqrt{(b/a - 4c/b)\, b/a}}{2}. \tag{9.2}$$

In the works of Mahavira, Bhaskaracharya, and others are found a number of fascinating problems, clearly devised to stimulate the interest of the reader. Let us consider a few of these, about linear and quadratic equations, beginning with one attributed to Aryabhata by his commentator Bhaskara I, whose solution used the method of "algebraic inversion."[9]

EXAMPLE 9.1 O maiden with beaming eyes, tell me, since you understand the method of inversion, what number multiplied by 3, then increased by three-quarters of the product, then divided by 7, then diminished by one-third of the result, then multiplied by itself, then diminished by 52, whose square root is then extracted before 8 is added and then divided by 10, gives the final result of 2?

Solution

The solution offered is elegant and simple. We start with the answer, 2, and work backward. When the problem says divide by 10, we multiply by that number; when told to add 8, we subtract 8; when told to extract the square root, we take the square; and so on. It is precisely the replacement of the original operation by the inverse that gives the method its name of "inversion."

Therefore the original number is obtained thus:

Continued . . .

Continued . . .

$$[(2)(10) - 8]^2 + 52 = 196;$$

$$\sqrt{196} = 14;$$

$$\frac{(14)(3/2)(7)(4/7)}{3} = 28.$$

EXAMPLE 9.2 Out of a certain number of Sarasa birds, one-fourth the number are moving about among the lotus plants; one-ninth together with one-fourth as well as 7 times the square root of the total number of birds are found on a hill nearby; 56 birds remain on the Vakula trees. What is the total number of birds? (From Mahivira's Ganita-sara-sangraha)

Solution

In modern notation the solution is simple. If x is the total number of birds, this gives the equation

$$x = \frac{x}{4} + \frac{x}{9} + \frac{x}{4} + 7\sqrt{x} + 56 \text{ birds,}$$

which solved for x gives 576 birds.

EXAMPLE 9.3 From a swarm of bees, a number equal to the square root of half the total number of bees flew out to the lotus flowers. Soon after, 8/9 of the total swarm went to the same place. A male bee enticed by the fragrance of the lotus flew into it. But when it was inside the night fell, the lotus closed, and the bee was caught inside. To its buzz, its consort responded anxiously from outside. O my beloved! How many bees are there in the swarm? (From Bhaskaracharya's Lilavati)

Solution

Bhaskaracharya's approach is equivalent to solving the following equation:

$$\sqrt{0.5x} + \left(\frac{8}{9}\right)x + 2 = x,$$

Continued . . .

Continued . . .

where x is the total number of bees in the swarm. It is to be assumed from the question that the male bee and his consort were late arrivals from the same swarm—hence the 2 in the above equation. Only $x = 72$ bees is given as an admissible solution. However, in the following problem from *Bijaganita*, Bhaskaracharya admits that more than one solution is valid.

EXAMPLE 9.4 Inside a forest, a number of apes equal to the square of one-eighth of the total apes in a pack are playing noisy games. The remaining 12 apes, who are of a more serious disposition, are on a nearby hill and irritated by the shrieks coming from the forest. What is the total number of apes in the pack?

Solution

The solutions $x = 16$ and $x = 48$ are equally admissible, according to Bhaskaracharya.

In the case of a number of problems, the recreational and poetic elements were dominant. Consider the following example from the *Ganita-sara-sangraha* of Mahavira.

EXAMPLE 9.5 One night in spring, a certain young lady was lovingly happy with her husband on the floor of a big mansion, white like the moon, situated in a pleasure garden full of trees heavy with flowers and fruits. The whole place was resonant with the sweet sounds of parrots, cuckoos, and bees intoxicated with the honey from the flowers in the garden. In the course of a "love quarrel" between the couple, the lady's necklace came undone and the pearls got scattered all around. One-third of the pearls reached the maidservant who was sitting nearby; one-sixth fell on the mattress; one-half of what remained (and one-half of what remained thereafter and again one-half of what remained thereafter and so on, counting six times in all) were scattered everywhere. On the broken necklace, there remained 1,161 pearls. Oh my love, tell me quickly the total number of pearls on the necklace.

Continued . . .

Continued...

The solution is tedious (but not difficult today) and takes us into the realms of fantasy. We will not attempt it here. With the answer as 148,608 pearls, this is truly a fantasy necklace. Such problems were not to be taken too seriously. They reflect a fascination with large numbers referred to previously.

EXAMPLE 9.6 Three merchants find a purse lying on a road. One of them says, "If I keep this purse, I shall be twice as rich as both of you together." "Give me the purse and I will be thrice as rich," says the second, while the third exclaims, "I shall be much better off than either of you if I keep the purse. I shall become five times as rich!" How much money is there in the purse? How much money has each merchant? (From Mahavira's Ganita-sara-sangraha)

Solution

The solution starts by setting up the following relationships:

$$m + x = 2(y + z),$$
$$m + y = 3(x + z),$$
$$m + z = 5(x + y),$$

where m is the amount of money in the purse and x, y, and z are the amounts of money in the possession of the three merchants.

The final solution is given in the form of ratios since there is no unique solution set $m{:}x{:}y{:}z = 15{:}1{:}3{:}5$.

Interest in such indeterminate problems, with no unique solution, has been a characteristic of Indian mathematics ever since the Vedic period.[10]

Indeterminate Equations

It is in Aryabhata I's work, *Aryabhatiya*, that we come across the first unequivocal discussion of the subject of indeterminate analysis. It arose, just as it did in China, in the field of astronomy, where there is a need to determine the orbits of planets. The problem that Aryabhata addresses may be expressed in modern terms as follows:

Find an integer (N) which when divided by another integer (a) leaves the remainder (r_1) and when divided by another integer (b) leaves the remainder (r_2).

And also the general problem:

Find an integer (N^*) which being divided severally by the given numbers $a_1, a_2, \ldots, a_n$ leaves remainders $r_1, r_2, \ldots, r_n$ respectively.

Symbolically, the two problems may be expressed thus:

$$N = ax + r_1 = by + r_2; \tag{9.3a}$$

$$N^* = a_1 x_1 + r_1 = a_2 x_2 + r_2 = \ldots = a_n x_n + r_n. \tag{9.3b}$$

If c denotes the difference between r_1 and r_2 in equation (9.3a), then (9.3a) may be rewritten as

$$ax \pm c = by. \tag{9.3c}$$

It is suggested that c always be kept positive by appropriately labeling r_1 and r_2 such that $r_1 > r_2$.

The following solution is offered in verses 32 and 33. There is some controversy as to how they are to be interpreted.[11]

Divide the greater remainder by the divisor of the smaller remainder. The mutual division [of the previous divisor] by the remainder [is made continuously. The last remainder], having a "clever" [quantity] for multiplier, is added to the difference of the [initial] remainders [and divided by the last divisor]. (Verse 32)

The one above is multiplied by the one below, and increased by the last. When [the result of this procedure] is divided by the divisor of the smaller remainder, the remainder, having the divisor of the greater remainder for multiplier, and increased by the greater remainder, is the [quantity that has such] remainders for the two divisors. (Verse 33)

In Bhaskara I's commentary on the *Aryabhatiya*, the following numerical example is given to illustrate the method:

A quantity divided by 12 leaves a remainder of 5. Furthermore, if such a quantity divided by 31 leaves a remainder of 7, what should one such quantity be?

Bhaskara I's explanation of Aryabhata's solution procedure is clear and concise when translated into modern notation.[12] In Bhaskara I's example, integers N, x, and y are sought such that $N = 12y + 5 = 31x + 7$. The "mutual division" is understood to mean continued divisions recasting the original equation between two unknown quantities with smaller and smaller coefficients until it is reduced or "pulverized" into a form that can be solved by inspection.[13] Thus

$$y = \frac{31x + 2}{12} = 2x + w,$$

$$x = \frac{12w - 2}{7} = 1w + v,$$

$$w = \frac{7v + 2}{5} = 1v + u,$$

$$v = \frac{5u - 2}{2}.$$

At this point, a "clever" integer solution is found by inspection showing that $u = 2$ and $v = 4$. By working our way to the top through a chain of substitutions, we find that the minimum solution set is ($x = 10$, $y = 36$, and $N = 317$).

This method of solution came to be known as *kuttaka*. The word is derived from *kutt*, meaning to "crush," "grind," or "pulverize," and describes a successive process of breaking something down into smaller and smaller pieces, in this case making the values of the coefficients a and b in equation (9.3a) smaller and smaller. All the great mathematicians of the Classical period dealt with the *kuttaka*, and it is one of the very few topics in Indian mathematics to be made the subject of a special monograph, titled *Kutta-kara Siromani*, written by a commentator on Aryabhata I named Devaraja.

However, the climax of Indian work in this area is the solution of indeterminate equations of the second degree. Brahmagupta considered the following two equations, the second of which is a special case of the first:

$$ax^2 \pm c = y^2, \tag{9.4a}$$

$$ax^2 + 1 = y^2, \tag{9.4b}$$

where a and c are known as the multiplier and augment, and x and y as the smaller and larger roots. Equation (9.4b) is a form of Pell's equation, wrongly named by Euler after the English mathematician John Pell

(1610–85). Brahmagupta was probably the first mathematician to give solutions to both equations (9.4a) and (9.4b) in rational integers. His approach is ingenious and general; here, though, we give a simple example to illustrate his approach. The algebraic intricacies are gone into by Bag (1979, pp. 216–17).

EXAMPLE 9.7 Solve the equation $8x^2 + 1 = y^2$.

Solution

Brahmagupta's method may be expressed in the following way.[14]

From inspection, it is obvious that the smallest integral solution (root) for x is 1, and the y that corresponds to this solution is 3, the minimum solution (root) for y.

Now arrange this information as follows:

SMALLER ROOT	LARGER ROOT	AUGMENT
1	3	1
1	3	1

Multiply crosswise as indicated by arrows and add the products. Thus

$3 + 3 = 6 = x$, so $y = 17$.

Arrange the old and new sets of values of x and y together with the augment the following way:

SMALLER ROOT	LARGER ROOT	AUGMENT
1	3	1
6	17	1

Multiply crosswise the first two columns and add to obtain $x = 35$. The corresponding value for y is found by substituting into the original equation, which gives $y = 99$.

Proceeding along these lines, we can construct the following sequence of diagrams to obtain larger and larger solution sets:

Continued . . .

Continued...

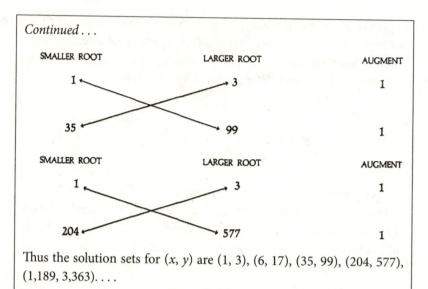

SMALLER ROOT	LARGER ROOT	AUGMENT
1	3	1
35	99	1

SMALLER ROOT	LARGER ROOT	AUGMENT
1	3	1
204	577	1

Thus the solution sets for (x, y) are $(1, 3)$, $(6, 17)$, $(35, 99)$, $(204, 577)$, $(1,189, 3,363)$....

The last column of the diagrams in example 9.7 comes into play when, in the process of calculation, we no longer obtain perfect squares, as in the next example.

EXAMPLE 9.8 Solve the equation $11x^2 + 1 = y^2$.

Solution

Follow the same procedure as before. Take the smaller root $x = 1$.

The left-hand side of the equation is not a perfect square. If, however, the augment is -2 rather than 1, the left-hand side becomes 9, a perfect square.

The diagrammatic representation is then as follows:

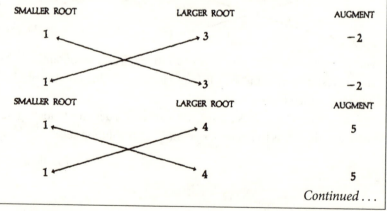

SMALLER ROOT	LARGER ROOT	AUGMENT
1	3	-2
1	3	-2

SMALLER ROOT	LARGER ROOT	AUGMENT
1	4	5
1	4	5

Continued...

Continued . . .

Take the first diagram, multiply crosswise, and add to get $x^* = 6$. Find the product of the smaller roots (i.e., 1×1), multiply the product by 11 (1×11), and add to it the product of the larger root (3×3) to give

$$11 + 9 = 20 = y^*.$$

But $x^* = 6$ and $y^* = 20$ satisfy the equation

$$11x^2 + 4 = y^2,$$

which has an augment of 4 and is not the same as the equation we started with. Now, the product of the assumed augment is $-2 \times -2 = 4$. Dividing 4 by 4 gives an augment of 1. Thus to obtain the values of x and y that correspond to the original augment 1, divide x^* and y^* by 2 to give one solution set for (x, y) as $(3, 10)$.

If we begin with the smaller root $x = 1$ and the larger root $y = 4$, the augment is 5. Following exactly the same procedure as before, and operating with the second diagram, the resulting solution set for (x, y) is $(8/5, 27/5)$.

To generate another solution set, proceed as before but use the following diagram:

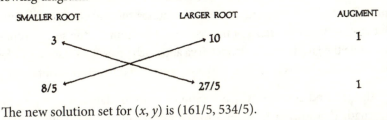

The new solution set for (x, y) is $(161/5, 534/5)$.

It is worth noting that this method was first used by Brahmagupta as early as the seventh century AD, though it is usually attributed to Euler, who named it *theorem elegantissimum*. The sheer ingenuity and versatility of the approach is also highlighted by the fact that it was not until 1767 that Lagrange gave a complete solution to Pell's equation, using continued fractions.

Jayadeva (c. 1000) was one of the first to point out that, while Brahmagupta's approach would easily produce an infinite number of solutions with an augment of ± 1, ± 2, or ± 4, with all other augments a trial-and-error process was necessary. In Udayadivakara's eleventh-century commentary on Bhaskara I's *Laghu Bhaskariya*, titled *Sundari*, Jayadeva's twenty verses are quoted, which constitute a general method for solving indeterminate equations of the kind just discussed. The method was refined by Bhaskara-charya about a hundred years later. The Jayadeva-Bhaskaracharya method was known as the *chakravala*, or "cyclic," method because the same set of operations is repeated over and over again. It bears a close resemblance to the so-called "inverse cyclic method" based on continued-fraction expansions that attracted the attention of European mathematicians of the caliber of Pierre de Fermat (1601–1665), Leonhard Euler (1707–1783), Joseph Lagrange (1736–1813), and Évariste Galois (1811–1832). The method must be regarded as a purely Indian creation, for there is no record of it at all in Chinese mathematics. For details of the Indian cyclic method, see Bag (1979, pp. 217–24) and Selenius (1975).

There is a problem of considerable historical interest for which Bhaskara II offers the first complete solution. The problem is to solve

$$61x^2 + 1 = y^2$$

for minimum x and y.

He gives the solution $x = 226{,}153{,}980$ and $y = 1{,}766{,}319{,}049$. It was precisely this problem that Fermat set as a challenge to his friend Frénicle de Bessy in 1657. We do not know whether Frénicle de Bessy took up the challenge; the problem was finally solved by Lagrange about a hundred years later. A comparison between Lagrange's and Bhaskaracharya's methods is quite illuminating. Lagrange's method requires the calculation of twenty-one successive convergents of the continued fraction for the square root of 61, while the Jayadeva-Bhaskaracharya approach gives the solution in a few easy steps.[15] Selenius's (1975, p. 180) assessment of the method is interesting:

> The method represents a best approximation algorithm of minimal length that, owing to several minimization properties, with minimal effort and avoiding large numbers always automatically produces the [best] solutions to the equation. . . . The *chakravala* method . . . anticipated the European methods by more than a thousand years. But no European performances in the whole field of algebra at a time much later than Bhaskara's, nay nearly up to our times, equaled the marvelous complexity and ingenuity of *chakravala*.

Indian Trigonometry

The origins of trigonometry are obscure. There are certain problems in the Ahmes Papyrus (c. 1650 BC) relating to measuring the steepness of the face of a pyramid by the ratio of the "run" to the "rise" (the horizontal departure of the oblique face from the vertical per unit height). As we mentioned in chapter 3, this ratio (known as the *seked* of the pyramid) would be considered today as equivalent to the cotangent of the angle made by the face of the pyramid and its base. This angle was kept constant at around 52° in the Great Pyramid at Gizeh and many other Egyptian pyramids. There is also the conjecture, discussed in chapter 4, that a column of numbers contained in a Babylonian cuneiform tablet, Plimpton 322, is a table of secants, but this must be considered a far-fetched idea, especially since more plausible explanations exist. Also, there is no evidence that the Babylonians of that period were familiar with the concept of an angle. However, we cannot be so dismissive about the possibility that the Babylonians of the "New" period may have constructed a form of prototrigonometry for astronomical purposes. The Babylonian astronomers of the first millennium BC were known to have accumulated a large number of observations that survived to provide the Greeks and then the Alexandrians with an impetus for early work on trigonometry.

The beginnings of a systematic study of the relationships between the angles (or arcs) of a circle and the lengths of chords subtending them are usually attributed to the Alexandrian Hipparchus (c. 150 BC), who was also credited with a twelve-part treatise dealing with the construction of a table of chords of arcs of a circle. Ptolemy (c. AD 100) constructed a table of his own that gave the lengths of the chords of all central angles of a given circle in half-degree intervals from $\frac{1}{2}^{\circ}$ to 180°. The radius of a circle was divided into 60 equal parts, and the chord lengths were then expressed sexagesimally in terms of one of these parts as a unit. Ptolemy's table has entries like crd 36° = 0;37,4,55, which means that the length of the chord of a central angle of 36° (see figure 9.1) is equal to 37 small parts of the radius (37/60) plus $4/60^2$ of one of these small parts and $55/60^3$ more of one of the small parts. The division of a circumference into 360° goes back to the period of the Mesopotamians. Again, it was the sexagesimal system that led Ptolemy to subdivide the diameter of his trigonometric circle into 120 parts, each of these in turn being split into 60 minutes and each minute into 60 seconds.

From the earliest surviving works of Indian astronomy, especially Aryabhata's *Aryabhatiya* (c. AD 500) and Varahamihira's *Pancasiddhantika* (c. AD 550), it is clear that the astronomers were interested in finding answers to the same questions that engaged the earlier civilizations: determining solar, lunar, and planetary positions, predicting eclipses, and so on. From a combination of indigenous and Babylonian procedures, they were

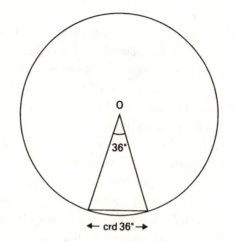

FIGURE 9.1: The chord of an angle

able to derive planetary positions to intervals of up to a billion years since the epoch. The calculations that led to such numbers may have spurred Indian interest in the mathematics of indeterminate equations discussed earlier. And, in a short time, Indian astronomy was able to cut itself off from the coattails of its predecessors and achieve a degree of autonomy, no doubt helped by its singular method of recording past knowledge. Unlike the Greek texts, for example, Indian texts contained in most cases merely prescriptions for calculations given in cryptic verses to aid memorization. And since verses are often open to different interpretations, controversies were bound to arise.

The Sources of Indian Trigonometry

While the work of the Alexandrians Hipparchus (c. 150 BC), Menelaus (c. AD 100), and Ptolemy (c. AD 150) in astronomy laid the foundations of trigonometry, further progress was piecemeal and spasmodic. From about the time of Aryabhata I (c. AD 500), the character of the subject changed, and it began to resemble its modern form. Subsequently it was transmitted to the Arabs, who introduced further refinements. From the Islamic world, the knowledge spread to Europe, where a detailed account of existing trigonometric knowledge first appeared under the title *De triangulis omni modis*, written in 1464 by Regiomontanus.

In early Indian mathematics, trigonometry formed an integral part of astronomy. References to trigonometric concepts are found in the *Surya-siddhanta* (c. AD 400), Varahamihira's *Pancasiddhantika* (c. AD 500), and

Brahmagupta's *Brahma Sphuta-siddhanta* (AD 628). A detailed and system-atic study of the subject was made by Vatesvara (b. AD 880) in the *Vatesvara Siddhanta* and then by Bhaskaracharya in his *Siddhanta Siromani*. He felt that the title *acharya* (i.e., master or teacher) in astronomy could be given only to those who possessed sufficient knowledge of trigonometry. Infinite expansions of trigonometric functions, building on Bhaskaracharya's work, are found in the work of Madhava and Nilakantha, discussed chapter 10.

The Development of Trigonometric Functions

On account of their shapes, the arc of a circle (e.g., the arc ACB in figure 9.2) was known as the "bow" (*capa*) and its full chord (e.g., the line seg-ment AMB in figure 9.2) as the "bow string" (*samastajya*). In their study of trigonometric functions, Indian mathematicians more often used the half chord (e.g., the segment AM or MB). The half chord was known as *ardhajya* or *jyardha*, later abbreviated to *jya* to become the Indian sine.[16] Three functions were developed, whose modern equivalents are defined here with reference to figure 9.2:

$jya\ \alpha$ = AM = $r \sin \alpha$,

$kojya\ \alpha$ = OM = $r \cos \alpha$,

$utkramajya\ \alpha$ = MC = OC − OM = $r - r \cos \alpha = r(1 - \cos \alpha)$
$\qquad\qquad$ = r versin α.

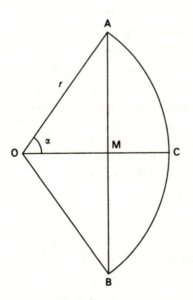

FIGURE 9.2: The Indian sine

To calculate *jya α* ($r \sin α$) for angles $α \leq 90°$, Varahamihira in his *Panca-siddhantika* suggested the following formulas:

$$jya\ 30° = \tfrac{1}{2}r, \quad jya\ 60° = \tfrac{1}{2}\sqrt{3}r, \quad jya\ 90° = r.$$

With the help of these formulas he calculated the values of $r \sin α$ ranging $3° 45'$ in twenty-four multiples to 90°.

Bhaskara I (c. 600), in his *Maha Bhaskariya*, gave the following approximate formula for calculating the Indian sine of an acute angle without the use of a table:

$$r \sin α \approx \frac{4r(180 - α)α}{[40,500 - α(180 - α)]},$$

which is equivalent to

$$\sin β \approx \frac{16β(π - β)}{5π^2 - 4β(π - β)},$$

where $β$ radians correspond to $α$ degrees. If now $β = π/3$ and then $π/7$,

$$\sin(π/3) \approx 0.8643 \ldots \text{ and } \sin(π/7) \approx 0.4314 \ldots,$$

which are both correct to the second decimal place.[17]

A similar degree of accuracy is achieved when the values of $\sin π$, $\sin(π/2)$, and $\sin(π/4)$ are obtained from the above approximation formula. Bhaskara I ascribed this formula to Aryabhata I. It occurs in Brahmagupta's *Brahma Sphuta-siddhanta* and in several later works.

Some other trigonometric relations found in the astronomical texts of the Classical period are shown below, together with the names of the author in whose work they first appear. (For ease of expression we take $r = 1$, so that the Indian sine [usually denoted as Sine with capital S] becomes equal to the modern sine.)

$\sin(n + 1)α - \sin nα =$
$\sin nα - \sin(n - 1)α - (1/225)\sin nα$ Aryabhata I

$\cos α = \sin(\tfrac{1}{2}π - α)$ Varahamihira

$\sin^2 α + \cos^2 α = 1$ Varahamihira

$\sin^2 α = \tfrac{1}{4}(\sin^2 2α + \text{versin}^2 2α) = \tfrac{1}{2}(1 - \cos 2α)$ Varahamihira

$1 - \sin^2 α = \cos^2 α = \sin^2(\tfrac{1}{2}π - α)$ Brahmagupta

$$\sin^2(\tfrac{1}{4}\pi + \tfrac{1}{2}\alpha) = \sqrt{\tfrac{1}{2}(1 \pm \sin \alpha)} \qquad\qquad \text{Aryabhata II}$$

$$\sin(\alpha \pm \beta) = \sin \alpha \cos \beta \pm \cos \alpha \sin \beta \qquad\qquad \text{Bhaskara II}$$

In 1658, the astronomer Kamalakara in his *Siddhanta Tattvaviveka* gave the formulas for Sin 2α, Cos 2α, Sin 3α, Cos 3α, and increasing multiples generated by specifying $n\alpha = \alpha + \alpha + \alpha + \dots$ and then applying the sum formula repeatedly. Thus, for example,

$$\text{Sin } 3\alpha = \text{Sin } \alpha\left[3 - \frac{(\text{Sin } \alpha)^2}{(\text{Sin } 30^\circ)^2}\right],$$

which is equivalent to $\sin 3\alpha = 3 \sin \alpha - 4 \sin^3 \alpha$.

The above expression is important since it could be used to get an estimate of the seed value for sin 1° from sin 3°, which provides the starting point in the construction of sine tables. The method was first applied by the Islamic astronomer al-Kashi to obtain a highly accurate value of sin 1° using an iterative procedure to solve cubic equations. Al-Kashi's work will be discussed in chapter 11. Kamalakara approached the problem differently. He used a triple angle approximation formula, particularly suitable for small angles, to obtain more accurate seed values.[18]

The Construction of Sine Tables

Various relationships between the sine of an arc and its integral and fractional multiples were used to construct sine tables for different arcs lying between 0° and 90°. These tables were used for astronomical calculations, for example, to compute exact locations of planets. The general formula by the name of Aryabhata I above was used to compute tables of half chords in a quadrant divided into twenty-four equal parts, so that the smallest arc is 3° 45′ (or 225′). It is worth noting that this formula is too crude to be used in the construction of the twenty-four sines as given in table 9.1. However, if we use Aryabhata's formula as interpreted in Nilakantha's *Aryabhatiyiabhasya*, we get the values of the twenty-four Sines shown.[19] Since the Indian sines are not the ratios of the corresponding half chords and the radius, but represent the half chords themselves, their values obviously depend on the length of the radius chosen.

Many Indian Sine tables used $r = 3{,}438$. This follows from the fact that if the circumference is measured in minutes (60 minutes = 1 degree), then the total circumference is $360 \times 60 = 21{,}600$ minutes, and the corresponding radius is $3{,}437.7467\dots$, or approximately $3{,}438$ minutes. This value has

TABLE 9.1 INDIAN SINES: VALUES GIVEN BY VARAHAMIHIRA AND
ARYABHATA I, AND THEIR MODERN EQUIVALENTS

Angle, θ	Varahamihira		Aryabhata		Modern value of sin θ
	r sin θ (r = 120')	Computed sin θ	r sin θ (r = 3438')	Computed sin θ	
3°45′	7′51″	0.06542	225′	0.06545	0.06540
7°30′	15′40″	0.13056	449′	0.13060	0.13053
11°15′	23′25″	0.19514	671′	0.19517	0.19509
15°	31′ 4″	0.25889	890′	0.25962	0.25882
18°45′	38′34″	0.32139	1105′	0.32141	0.32143
22°30′	45′56″	0.38278	1315′	0.38249	0.38268
26°15′	53′ 5″	0.44236	1520′	0.44212	0.44229
30°	60′	0.50000	1719′	0.50000	0.50000
33°45′	66′40″	0.55556	1910′	0.55556	0.55556
37°30′	73′ 3″	0.60875	2093′	0.60878	0.60876
41°15′	79′ 7″	0.65931	2267′	0.65910	0.65935
45°	84′51″	0.70708	2431′	0.70710	0.70711
48°45′	90′13″	0.75181	2585′	0.75189	0.75184
52°30′	95′13″	0.79347	2728′	0.79348	0.79335
56°15′	99′46″	0.83139	2859′	0.83159	0.83147
60°	103′56″	0.86611	2978′	0.86620	0.86602
63°45′	107′38″	0.89694	3084′	0.89703	0.89687
67°30′	110′53″	0.92402	3177′	0.92408	0.92388
71°15′	113′38″	0.94694	3256′	0.94706	0.94693
75°	115′56″	0.96611	3321′	0.96597	0.96593
78°45′	117′43″	0.98097	3372′	0.98080	0.98079
82°30′	119′	0.99167	3409′	0.99156	0.99144
86°15′	119′45″	0.99792	3431′	0.99796	0.99786
90°	120′	1.00000	3438′	1.00000	1.00000

Adapted from table 3.4 in Bose et al. (1971, p. 200)

a great advantage in that the Sines of small arcs are almost equal to the arcs
themselves. For example, in Aryabhata's Sine table, which has a step size of
3°45′, the Sine of the smallest value (3°45′) is equal to 225′. Whether this
was a deliberate choice of the largest convenient arc at which Sin $x = x$, or
because it would result in a natural division of the right angle into twenty-
four (21,600/90) segments, is a moot point. If it was the latter, an indebted-
ness to Hipparchus (second century BC) becomes a possibility.[20] The values
for the radius adopted by Varahamihira, Aryabhata I, and Brahmagupta

were 120′, 3,438′, 3,270′, and 150′. 120′ was used by Varahamihira, 3,438′ by Aryabhata, 3,270′ and 150′ by Brahmagupta. 900′ is the value of a unit arc when a quadrant is divided into six equal parts. The multiplicity of the values taken for the radius has some interesting historical implications. For example, Varahamihira's odd choice of $r = 120$ and the resulting Sine table has been linked with Ptolemy's table of chords. This has led to the suggestion that Varahamihira's values were taken from Ptolemy. It could equally be argued that Ptolemy's *Almagest* and Varahamihira's *Pancasiddhantika* were both dependent on an earlier source. The jury is still out on whether this earlier source was of Greek or Indian origin.

The first known variant of a sine table is Ptolemy's. He gave a table of chords within a circle of radius 60 units and expressed it in sexagesimal units. The arc ranges from $\frac{1}{2}°$ to 180° at half-degree intervals. With the help of this table, the corresponding length of the chord can be calculated when the length of the arc is known, and vice versa. There is some controversy as to the source of the Indian sine table. What seems likely is that both Ptolemy and the Indian astronomers were indebted to an earlier source, possibly Hipparchus.[21]

However, a uniquely Indian approach to constructing a sine table soon appeared. Beginning from the assumption that the first entry in the table is sin 225′ = 225, various procedures were tried for the successive computation, one at a time, of the remaining twenty-three sine values. An early such attempt is found in verses 11 and 12 of Aryabhata I's *Aryabhatiya*.[22] The values of the twenty-four Sines given in the *Pancasiddhantika* and *Aryabhatiya* are given together with equivalent modern values in table 9.1. The accuracy of Aryabhata's Sines is quite impressive.

While these methods were ingenious, a sine table that contained entries only for every 225 minutes would seem to be rather limited. However, in the twelfth century, the sine of 18° entered Indian mathematics in an appendix titled *Jyotpatti* to Bhaskara II's astronomical treatise *Siddhanta Siromani*. In terms of modern notation, *Jyotpatti* gives the formula for the Indian sine (r sin):

$$r \sin 18° = \frac{\sqrt{5r^2 - r}}{4},$$

which is equivalent to

$$\sin 18° = \frac{\sqrt{5} - 1}{4} \text{ and } \sin 36° = \frac{\sqrt{5 - \sqrt{5}}}{8}.$$

Two more building blocks complete Bhaskara II's attempt to build a fuller sine table. The first is to obtain an accurate approximation to Sin 1°, which is relatively simple since, as we saw earlier, sin (225') ≈ 225 and, even more accurately, sin (60' = 1°) ≈ 60. The second is to find a method of calculating the sines of sums and differences of angles. The rule is given in *Jyotpatti* as follows:[23]

> The Sines of the two given arcs are crossly multiplied by [their] Cosines and [the products are] divided by the radius. Their [i.e., the quotients obtained] sum is the Sine of the sum of the arcs; their difference is the Sine of the differences of the arc.

The above may be expressed in modern notation as

$$\text{Sin}(\alpha \pm \beta) = \frac{\text{Sin } \alpha \text{ Cos } \beta}{r} \pm \frac{\text{Cos } \alpha \text{ Sin } \beta}{r}.$$

The equivalent of this formula for the modern sine was given earlier. The analogous formula for the cosine of sums and differences [cos (α ± β) = cos α cos β ± sin α sin β] is rarely shown in Indian mathematical texts since, as stated in a seventeenth-century commentary on *Jyotpatti*, once Sin (α ± β) is known, Cos (α ± β) can be more easily calculated using the Pythagorean rule. A variety of derivations of the sine and cosine addition rules based on both geometrical and other methods are to be found in Indian mathematics. One of the more interesting derivations involves the use of indeterminate analysis, a favorite subject in Indian mathematics.[24]

In the year 665, when Brahmagupta was sixty-seven years old, he wrote an astronomical treatise titled *Khanda Khadyaka*. In its ninth chapter he shows how to interpolate the sines of intermediate angles from a sine table. Brahmagupta's rule may be stated as follows:

> Multiply half the difference of the *gata khanda* [tabular differences passed over] and *bhogya khanda* [the difference to be passed over] by the residual arc [*h* in minutes] and divide by 900. The result is added to and subtracted from half the sum of *gata khanda* and *bhogya khanda* according to whether this half sum is less than or greater than the tabular difference to be crossed. The result obtained is the true functional difference to be crossed. [Brahmagupta takes the radius as 150' and the interval to be 900'.]

We can use Brahmagupta's interpolation formula to find the sine of 67°:

the interval is $h = 15°$ or $900'$,

the residual angle is $\Delta\theta = (67 - 60) = 7°$ or $420'$,

the relevant tabulated values are $D_p = 24, D_{p+1} = 15$,

where D_p and D_{p+1} are the corresponding functional differences, *gata khanda* and *bhogya khanda* respectively, and are taken from table 9.2.

The value given by the interpolation formula is then

$$\frac{\Delta\theta}{h}\left(\frac{D_{p+1} + D_p}{2} + \frac{\Delta\theta}{h} \times \frac{D_{p+1} - D_p}{2}\right).$$

Applying the Brahmagupta interpolation formula gives

$$\frac{7}{15}\left(\frac{15 + 24}{2} + \frac{7}{15} \times \frac{15 - 24}{2}\right) = 8.12.$$

Hence

$$jya\ 67° = 130 + 8.12 = 138.12,$$

TABLE 9.2: CALCULATION OF SINES USING BRAHMAGUPTA'S INTERPOLATION FORMULA

Angle (degrees)	Indian sine (jya)	First difference, D_p	Second difference, $D_{p+1} - D_p$
0	0		
		39	
15	39		−3
		36	
30	75		−5
		31	
45	106		−7
		24	
60	130		−9
		15	
75	145		−10
		5	
90	150		

Adapted from Bag (1979, p. 257)

which is close to 150 sin 67° = 138.08, the modern value, and good enough for most astronomical purposes. Given that Brahmagupta began with a sine table with just six entries, the accuracy that he got for any given arc is very impressive. The reader is invited to estimate 150 sin 78° using the Brahmagupta interpolation formula and to check how close the approximation is to the modern value of 146.72.

The rule is equivalent to the Newton-Stirling interpolation formula to second-order differences, expressed as

$$f(a + xh) = f(a) + \frac{x[\Delta f(a) + \Delta f(a - h)]}{2} + \frac{x^2 \Delta^2 f(a - h)}{2!},$$

where Δ is the first-order forward-difference operator [i.e., $\Delta f(a) = f(a + h) - f(a)$, or the column D_p values given in table 9.2], Δ^2 is the second-order difference operator (i.e., the column of $D_{p+1} - D_p$ values in table 9.2), and $x = \Delta\theta/h$. This Brahmagupta scheme for approximating sines could very well be the earliest use of finite difference interpolation.

Two centuries after Brahmagupta, the astronomer Govindasvamin (fl. 800–850), an early Kerala commentator on the works of Aryabhata and Bhaskara I, produced a rule for second-order interpolation to compute intermediate functional values. This proved to be a particular case (up to second order) of the general Newton-Gauss interpolation formula:

$$f(a + xh) = f(a) + x\Delta f(a) + \tfrac{1}{2}x(x - 1)[\Delta f(a) - \Delta f(a - h)].$$

Vatesvara was a notable but neglected figure who showed great understanding of trigonometric concepts and computations. Born in 880, he composed his *Siddhanta* in 904. In verses 2–51 of this text he gives us a list of the values of ninety-six Indian sines and versed sines at intervals of 56.25 minutes.[25] This list is interposed with verses indicating the relationships between sines, cosines, and versed sines in various quadrants; several methods for computing desired sines from given arc and tabular values; and different methods of first- and second-order interpolation and inverse interpolation procedures for finding desired arcs from given sine and tabular values.[26]

The culmination of the Indian effort in the construction of trigonometric tables is found in the *Golasara* of Nilakantha Somayaji, a small text on spherical astronomy consisting of fifty-six verses. The author points out that he was computing sines and cosines because they were required for a discussion of the motion of planets in their respective orbits on the stellar sphere. The text begins by providing a geometrical method of computing successively the values of sines of half angles starting from $30° = 1,800'$ along the lines indicated in the *Aryabhatiya*. The method is then repeatedly applied to find the sine of $1,800/2''$ for $n = 1, 2, 3, 4....$ Next, taking the value of sine of $1,800/2^m$ for some chosen m as the first sine value, a sine table of length $l = 3 \times 2^m$ could be constructed.

Sine tables of lengths 3, 6, 12, 24, 48, 96, 192, 384,... are easily computed using this *Golasara* algorithm. It is quite interesting in this context to note that Nilakantha has referred to the last and the first sine differences by the terms *antya* and *adi khanda* without mentioning that the last sine is the twenty-fourth. So it may be inferred that Nilakantha's rule for the determination of the sines successively gives a general method for constructing sine and cosine tables. Cosine tables may also be constructed similarly.[27]

Other work on computing sine and cosine functions, mainly in Kerala, produced expressions that are similar to the modern Taylor approximations to second order, and predate Taylor by more than three hundred years. We will examine in the next chapter this work of the Kerala mathematicians from about the fourteenth to the seventeenth centuries.

Other Notable Contributions

It is clearly impossible, given the scope of this book, to examine the whole range of subjects covered by Indian mathematicians over a period of two thousand five hundred years. In this section we consider three areas where contributions were notable or unusual, though they may have had only a limited impact outside India. We begin with medieval approaches to mathematical series, and then discuss briefly some special topics in Indian geometry. We conclude by assessing the preliminary notions of the infinitesimal calculus to be found in the work of Bhaskaracharya.

Geometric Representation of Arithmetic Series

Interest in number sequences that follow particular laws has been shown by several mathematical cultures, beginning—as we saw in chapter 3—with the Egyptians. In India, it was not until Mahavira's time (c. AD 850) that a systematic examination of the properties of different series was first attempted. We concentrate here on one memorable aspect of Indian work in this area: the study of the properties of different arithmetic series through diagrams (or *sredhiksetras*), which aroused interest well into the fifteenth century.

In Nilakantha's commentary on the *Aryabhatiya*, mentioned earlier, an arithmetic series is represented by piling rectangular strips, of unit width and of lengths equal to the number of units in each term of the series, on top of each other, with the shortest strip at the top and the longest at the bottom (see figure 9.3a).[28] If two of these *sredhiksetras* are joined together, one inverted to fit with the other, as shown in figure 9.3a, the resulting figure will be a rectangle with height n units (the number of terms, or rectangular strips) and length $a + f$, where a and f are the first and last terms of the series, respectively. Since the area of this rectangle is $n(a + f)$ square units, the area of one of the *sredhiksetras* will be the sum of the series, $\frac{1}{2}n(a + f)$.

Nilakantha proceeds to demonstrate the relationship

$$\Sigma \frac{n(n + 1)}{2} = \frac{n(n + 1)(n + 2)}{6}$$

by taking six *sredhiksetras*, representing the sum of n natural numbers, and combining pairs of them to form rectangular strips in the shape of figure 9.3b having adjacent sides of n, $n + 1$, and $n + 2$ units and thickness 1 unit. If one of these strips is placed flat on the ground, as in figure 9.3c, and the other two are held vertically touching the edge of the first, so that the sides of the section along the top of the strip are $n + 1$ and $n + 2$, then the three strips define a rectangular block of sides n, $n + 1$, and $n + 2$. The inside of the block is filled with the set of rectangles formed by joining three pairs each of the *sredhiksetras* representing the sums of $n - 1$, $n - 2$, $n - 3$, ..., 2, 1 natural numbers. A solid cuboid measuring n by $n + 1$ by $n + 2$ is the final result. The volume of this cuboid is equal to $n(n + 1)(n + 2)$. This may be equated with the "contents" of the cuboid, known to be $6\Sigma\frac{1}{2}n(n + 1)$, so that each of the *sredhiksetras* occupies one-sixth of this:

$$6\Sigma\tfrac{1}{2}n(n + 1) = n(n + 1)(n + 2),$$

so

$$\Sigma \frac{n(n + 1)}{2} = \frac{n(n + 1)(n + 2)}{6}.$$

A similar geometrical representation, which has the great advantage of being immediately convincing, is found in the demonstration of a number of other results for mathematical series. In *Kriyakramakari* there is the intriguing statement that a demonstration similar to the one above is

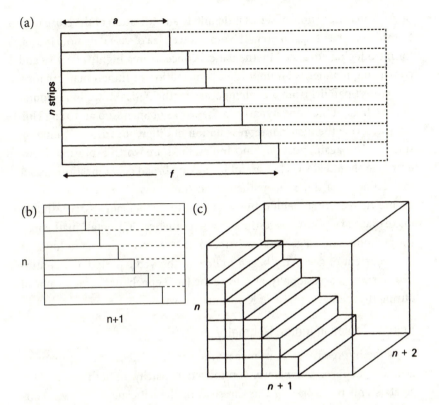

FIGURE 9.3: Piling rectangular strips

possible for arithmetic series that lead beyond the three-dimensional cube. It would indeed be interesting to see how such a demonstration would have proceeded to show that

$$\sum \frac{n(n+1)(n+2)(n+3)}{4!} = \frac{n(n+1)(n+2)(n+3)(n+4)}{5!}.$$

The lack of explanation by the authors of *Kriyakramakari* may have been because of the difficulty on the part of the reader to conceive of spaces having more than three dimensions. After all, at the beginning of the book it is stated that the book has been composed for the benefit of the less intelligent! The formulas of the sum of squares and sum of cubes of natural numbers are also treated diagrammatically by Nilakantha and Sankara Variyar.

The *sredhiksetra* method of representing mathematical series is an interesting feature of Indian mathematics and probably a legacy from the Vedic constructions. The terminology like *citi* (pile) or *ghana* (solid content) used by Aryabhata for sums of series is definitely an indication of the close relationship of series to geometry. The innovative idea of visual demonstration for introducing advanced mathematical concepts in a highly effective and convincing manner is an important contribution of Indian mathematics. The closest parallel from a contemporary mathematical tradition is found in the China of the Song dynasty, in the works of Shen Kuo and Yang Hui. They describe the pictorial representation of different series as "piling up stacks." Here again, possible Sino-Indian links are worth exploring. However, one should always be careful in making broad generalizations about the character of different mathematical traditions. The subtle nature of geometric reasoning behind this approach to mathematical series should make one wary of any suggestions of a hypothetical "Oriental" mathematics, predominantly algebraic in character. It is just possible that in India an undercurrent of geometry began to flow in the Vedic period and continued, surfacing occasionally, as during the Jaina epoch, but came to a head during the medieval phase of Kerala mathematics.[29]

Special Topics in Indian Geometry

After the impressive start in mensuration in the *Sulbasutras*, subsequent geometrical developments were on the whole patchy. In both Jaina mathematics and the works of the Classical mathematicians there was considerable emphasis on simple rules of mensuration but little sign of the sophistication found in Chinese geometry (see chapters 6 and 7). But in one area of geometry, the Indian contribution was notable: the study of the properties of a cyclic quadrilateral (i.e., a quadrilateral inscribed in a circle). In chapter 12 of the *Brahma Sphuta-siddhanta*, Brahmagupta gives the following two results:

1. The area of a cyclic quadrilateral is given by the product of half the sums of the opposite sides, or by the square root of the product of four sets of half the sum of the sides (respectively) diminished by the sides.

2. The sums of the products of the sides about the diagonal should be divided by each other and multiplied by the sum of the opposite sides. The square roots of the quotients give the diagonals of a cyclical quadrilateral. (Verse 28)[30]

In modern notation, and with reference to figure 9.4, these rules may be expressed as follows. Let *a*, *b*, *c*, and *d* be the sides of a cyclic quadrilateral

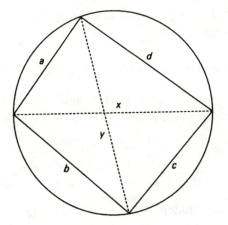

FIGURE 9.4: A cyclic quadrilateral

of area A; let $s = \frac{1}{2}(a + b + c + d)$ be the semiperimeter, and x and y the diagonals. Then

$$A = \sqrt{(s - a)(s - b)(s - c)(s - d)},$$

$$x = \sqrt{\frac{(ab + cd)(ac + bd)}{ad + bc}}, \qquad y = \sqrt{\frac{(ad + bc)(ac + bd)}{ab + cd}}.$$

The first statement of the expressions for the diagonals in Western mathematics is found in 1619 in the work of Willebrord Snell, some thousand years later.

The derivations of these results are first referred to in a tenth-century commentary on Brahmagupta's work but find their full expression in the sixteenth-century Kerala text *Yuktibhasa*. This contains a detailed discussion of the properties of a cyclic quadrilateral and how it is used to arrive at various trigonometric results. It makes use of Ptolemy's theorem, which states that the product xy of the diagonals of a cyclic quadrilateral is equal to the sum of the products of the two pairs of opposite sides, $ac + bd$. Bag (1979) and Sarasvati (1979) give details of the proofs.

Notable extensions in this area are contained in Narayana Pandita's *Ganita Kaumadi* in the fourteenth century and Paramesvara's *Lilavati Bhasya*, a detailed fifteenth-century commentary on Bhaskaracharya's *Lilavati*. The cyclic quadrilateral was an important device used by the Kerala school for deriving a number of important trigonometric results, including

$$\sin^2 A - \sin^2 B = \sin(A + B) \times \sin(A - B)$$

and

$$\sin A \times \sin B = \sin^2 \tfrac{1}{2}(A + B) - \sin^2 \tfrac{1}{2}(A - B).$$

A new rule is found in the work of Paramesvara for obtaining the radius r of the circle in which a cyclic quadrilateral of sides a, b, c, and d is inscribed:

$$r = \sqrt{\frac{(ab + cd)(ac + cd)(ad + bc)}{(a + b + c - d)(b + c + d - a)(c + d + a - b)(d + a + b - c)}}.$$

A detailed demonstration of this result is found in a later commentary on Bhaskaracharya's *Lilavati*, from the Kerala school, titled *Kriyakramakari* (Gupta 1977). This result makes its first appearance in European mathematics in 1782 in the work of l'Huilier.

The Beginnings of the Calculus

One of the most important problems of ancient astronomy was the accurate prediction of eclipses. In India, as in many other countries, the occasion of an eclipse had great religious significance, and rites and sacrifices were performed. It was a matter of considerable prestige for an astronomer to demonstrate his skills dramatically by predicting precisely when the eclipse would occur.

In order to find the precise time at which a lunar eclipse occurs, it is necessary first to determine the true instantaneous motion of the moon at a particular point in time. The concept of instantaneous motion, known as *tatkalika-gati* in Indian astronomy, is found in the works of Aryabhata I and Brahmagupta. They calculated this quantity from the formula (in modern notation)

$$u' - u = v' - v \pm e(\sin w' - \sin w), \tag{9.5}$$

where u, v, and w denote the moon's true longitude, mean longitude, and mean anomaly at a particular time; u', v', and w' are these same quantities after a specific interval of time; and e is the eccentricity, or sine of the greatest equation of the orbit. The use of sine tables and interpolation formulas would then yield values of the sines of angles over very short intervals.

Manjula (c. AD 930) was the first Indian astronomer to recognize that equation (9.5) could also be expressed as

$$u' - u = v' - v \pm e(w' - w)\cos w, \qquad (9.6)$$

since $(\sin w' - \sin w) = (w' - w)\cos w$.

In modern notation, we would write equation (9.6) as

$$\delta u = \delta w \pm e \cos w \, \delta u.$$

Bhaskaracharya extended this result to obtain the differential of sin w as

$$d(\sin w) = \cos w \, dw.$$

Bhaskaracharya proceeded to use this equation to work out the position angle of the ecliptic, the other quantity required for predicting the time of an eclipse.

This result in itself was a notable technical achievement in the astronomy of the period, but it may well be much more than this. It may seem a far-fetched claim, on this evidence alone, that Bhaskaracharya was one of the first mathematicians to conceive of the differential calculus, but there is further evidence to be found in his *Siddhanta Siromani*:[31]

1. In computing the instantaneous motion of a planet, the time interval between successive positions of the planet was no greater than a *truti*, or 1/33,750 of a second, and his measure of velocity was expressed in this "infinitesimal" unit of time.

2. Bhaskaracharya was aware that when a variable attains the maximum value, its differential vanishes.

3. He also showed that when a planet is either at its farthest from the earth or at its closest, the equation of the center[32] vanishes. He therefore concluded that for some intermediate position the differential of the equation of the center is equal to zero.

In the third observation above there are traces of the "mean value theorem," which today is usually derived from Rolle's theorem (1691).

Later mathematicians, particularly the Kerala school, continued the work of Bhaskaracharya. Nilakantha (1443–1543) derived an expression for the differential of an inverse sine function, and Acyuta Pisarati (c. 1550–1621) gave the rule for finding the differential of the ratio of two cosine functions. As we will see in the next chapter, a number of other ideas of the Kerala school—notably those relating to approximating very small arc segments to their sines and making adjustments to infinite series—anticipate seventeenth-century European work on infinitesimal calculus. The idea of using the integral calculus to find the value of π, and the areas of curved surfaces and the volumes enclosed by them, is implicit in the method of exhaustion that we examined in earlier chapters. Such ideas and their development are also found in the works of Bhaskaracharya, Narayana Pandita (c. 1350), and Jyesthadeva (c. 1550), but there are few novel features in the Indian treatment of the subject.

Outside Europe, there has been only one other country apart from India in which some form of calculus developed. In Japan, during the seventeenth century, Seki Kowa (or Seki Takakazu) developed a form of calculus called *yenri* that was primarily used in circle measurements (Mikami 1913; Smith and Mikami 1914). In India itself, where the concept of differentiation was understood from the time of Manjula, differential calculus was applied only to astronomy and certain problems in mensuration and did not spread across the broad spectrum of mathematics. This spread has been an important factor in promoting the phenomenal development of modern mathematics during the last few hundred years. The crucial concept of the "limit" of a function or a sum is essentially a modern idea, not to be found in Indian or any premodern mathematics. But its absence should not make one ignore the advances made during the Classical and medieval phases of Indian mathematics.

Indian Mathematics from Persian and Arabic Sources

The arrival of the Islamic scientist al-Biruni in India in AD 1018 as a prisoner of the invading army of Muhammad Ghaznavi was an important landmark in the scientific contact between the two cultures. During his enforced stay in India, which he profitably occupied by learning the language and culture of the country, al-Biruni translated from Arabic into Sanskrit Euclid's *Elements*, Ptolemy's *Almagest*, and his own work on the construction of the astrolabe. None of these Sanskrit translations are extant today. However, al-Biruni's example was followed by Sultan Firus Tughlaq, who occupied the throne of Delhi between 1351 and 1388. He ordered the astronomical text *Brhatsamhita* of Varahamihira to be translated into Persian, while his court astronomer, Mahendra Suri, wrote *Yantraraja*, which introduced astronomical principles and practices from central and western Asia into India. Unfortunately, the ideas proposed in the book of welding together principles from the *Siddhantas* and the Persian-Arabic system did not get a favorable reception. However, ideas from the West continued to

flow into India, including those of Nasir al-Din al-Tusi (1201–1274). His
Kitab Zij-Ilkani, based on the observations made at the Marghah Observatory, became the model for future *zij* (astronomical tables) in the Islamic
world. Two of his texts, one each in Arabic and Persian, became compulsory reading for students of astronomy in Indian *madrassahs* (mosque
schools) and appear to have inspired the astronomer Muhammad Jaunpuri
to question the validity of the Ptolemaic planetary model.

Geometry has always been an integral part of astronomy. However, it has
developed on different lines in different cultural areas. In India, as we saw
in chapter 8, it originated to serve the needs of rituals and cosmographic
speculations. In central and western Asia, under Islamic rule, Euclidean
geometry, which developed as a deductive science, became dominant. As
a result, a thorough knowledge of Euclidean geometry was required before
students of mathematics and astronomy, taught in Arabic-Persian language
schools in India, took up the study of Ptolemy's *Almagest,* or Archimedes'
On the Sphere and Cylinder, or Appollonius's *Conics.* The most widely used
Arabic translation of Euclid's *Elements* was Nasir al-Din al-Tusi's *Tahrir
Uqlidis,* which in turn went into a number of Persian translations. In 1732,
seven centuries after al-Biruni's attempt to introduce Euclid to India, Jag-
annatha Samrata translated the Persian text into a Sanskrit version of the
Elements titled *Rekhaganita.* Only one hundred years later, five chapters
of Hutton's *Euclidean Geometry* were rendered from English into Sanskrit
by Yogadyana Misra in Calcutta. By then, English-language education had
become the norm in a number of schools in major cities of India.

The major difference between the two streams of mathematical activity, namely those working within the Sanskritic tradition and others within
the Arabic-Persian tradition, is well brought out in the research preoccupations of the two groups. The Kerala mathematicians, with their work
on infinite series, were inspired by Aryabhata and his school. The Indian
mathematicians working within the Greek-Arabic-Persian tradition, whose
interests were primarily in Greek geometry and Ptolemaic astronomy, offered another model. An interesting illustration of the difference between
the preoccupations of the two traditions is shown by the work of Ghulam-
Hussain Jaunpuri (b. 1790), one of the notable Indian mathematicians from
the Arabic-Persian tradition. In *Jame-i-Bahadur Khani,* composed in 1833,
Juanpuri tackles the problem of trisecting an angle, which had engaged
mathematicians from the Greek-Arabic-European tradition over a long
period of time, including notable names such as Archimedes, al-Biruni,

Thabit ibn Qurra, and François Viete. Various approaches were tried out, including those involving conic sections, transcendental curves, circles, and *neusis* (insertion of a static line). Jaunpuri used the last method to achieve a construction that is sound and practicable. The details of his method are found in Rizvi (1983) and are of little importance within the context of this discussion. What would have been very unlikely is that such a problem would have engaged the interest of an Indian mathematician from the Sanskrit tradition.

The two parallel traditions met in a few cases involving astronomy in the courts of the Tughluq and Mughal emperors at Delhi and later in the court of Jai Singh in Jaipur, but hardly ever on matters relating to pure mathematics. The Islamic astronomical table (*zij*) and astronomical instruments (particularly the astrolabe, known in Sanskrit as the *yantraraja* or "king of instruments") had a significant impact on mathematical astronomy in India. The practical advantage of referring to a table rather than calculating planetary positions using the rules from a *siddhanta* became immediately obvious. Indian astronomers soon took to these tables, and apart from their usefulness for working out planetary positions, they became indispensable aids for synchronizing time units in calendars.[33] There were translations of texts such as *Lilavati* and *Bijaganita* into Persian under the patronage of Mughal emperors Akbar and Shah Jahan and dictionaries compiled to help this process. However, such cross-cultural exchanges were exceptional: a missed opportunity that has had considerable repercussions for the development of Indian mathematics. But that is another story.

Notes

1. This sentence, without further comment here, would be oversimplifying the development of Indian astronomy from the post-Vedic times onward. The emergence of the early Classical Indian astronomy may be discerned from later summaries contained in the *Pancasiddhantika* (the Five *Siddhantas*) composed in the fifth century AD by Varahamihira. The five named are *Paulisa-siddhanta* (the Text of Paulisa), *Romaka-siddhanta* (the Text of Romans), *Vasistha-siddhanta* (the Text of the Sage Vasistha), *Surya-siddhanta* (the Text of the Sun), and *Paitamaha-siddhanta* (the Text of Pitamaha, or the deity Brahman). These texts were composed at varying times during the first half of the second millennium AD. On the basis of similarities detected between the *Pancasiddhantika* and Hellenistic astronomical texts, it has been conjectured that a transmission occurred from the Hellenistic world to India during the period. In particular, similarities between Indian sines and the Greek chords have led some to hypothesize

on the Greek origins of the Indian sine. Irrespective of whether such a transmission took place, it should be borne in mind that the Indian astronomers were the first to replace the chord geometry of triangles inscribed in a semicircle with the geometry of sines of right-angle triangles in the quadrants of a circle. The subject will be taken up in a later section of this chapter.

2. It may be argued that the word "states" in this context is somewhat imprecise since it cannot help in fixing the precise year of the composition of *Aryabhatiya*. The relevant verse gives the year of the birth of Aryabhata as AD 476: "When sixty times sixty years and three quarter *yugas* (of the current *yuga*) had elapsed, twenty-three years had then passed since my birth" (Verse 3.10, *Aryabhatiya* 499, translated by Shukla and Sarma 1976, p. 95).

3. A recent translation and discussion of Bhaskara I's commentary on the mathematical chapter of the *Aryabhatiya* will be found in the two volumes by Keller (2006).

4. A useful summary of the topics contained in chapter 18 is found in Plofker (2007, pp. 428–34).

5. The following problem in chapter 6 of Mahavira's book is quite deep and throws an interesting light on the social context of his time.

> Five men are enamored by a courtesan, of whom only three she finds attractive. However, to each one of them separately, she says: "You are my only beloved." How many of her statements are true?

The solution offered by Mahavira: Multiply the total number of men (5) by the number of those found attractive plus one (3 + 1). Diminish this product by twice the number found attractive (2 × 3) and you will get the number of false statements. The square of the number of men (5^2) minus the number of false statements gives the number of true statements. Or

Number of false statements = $(5 \times 4) - (2 \times 3) = 14$;

Number of true statements = $25 - 14 = 11$.

It is important to recognize here that the rule concerns the truth values of a set of n explicit statements and $(n^2 - n)$ implicit statements. For further details, see Plofker (2007, pp. 446–47).

6. Operations with zero (*sunya-ganita*) had been a characteristic of Indian mathematics texts from the time of Brahmagupta. While the discussion in the arithmetical texts (*patiganita*) was limited only to addition, subtraction, and multiplication with zero, the treatment in algebra texts (*bijaganita*) covered such questions as the effect of zero on the positive and negative signs, division with zero, and more particularly the relation between zero and infinity (*ananta*). In the *Lilavati* of Bhaskaracharya, the eight

operations involving zero—addition, subtraction, multiplication, and division with zero as well as the square, square root, cube, and cube root of zero—are listed. A number divided by zero is given, like Brahmagupta, as "zero-divided" or "that which has zero as the denominator." For further details, see Joseph (2002a).

7. At the beginning of the thirteenth chapter of *Ganita-kaumudi*, Narayana writes: "I will briefly describe the net of numbers which causes enjoyment for mathematicians, in which those who are jealous, depraved and poor mathematicians fall down. . . . It is applied to dance and music, metrics, medicine, garland-making, and mathematics as well as architecture. Knowledge of these [subjects] is [indeed acquired] by means of numbers" (Plofker 2007, 499).

8. For further details on Narayana's work on magic squares, see P. Singh (1982, 1986).

9. Note that example 9.1 is taken from Bhaskaracharya's *Lilavati*.

10. For example, in the *Baudhyana Sulbasutra* (c. 800 BC), there appears the problem of designing a *Garuda Chayana* altar (an altar in the shape of an eagle with outstretched wings). The altar should have five layers of bricks, with each layer containing 200 bricks of four different sizes covering an area of $7\frac{1}{2}$ square *purushas*. In terms of modern notation, if x, y, z, w represent the numbers of the four sizes of bricks in any layer, and the bricks of each of those sizes have the area of $1/m, 1/n, 1/p, 1/q$ respectively, then what is required is the solution of the indeterminate equations

$$x + y + z + w = 200;$$

$$x/m + y/n + z/p + w/q = 7\tfrac{1}{2}.$$

The solution set that Baudhayana accepts is ($x = 24, y = 120, z = 36, w = 20; m = 16, n = 25, p = 36, q = 100$).

11. The interpretations include those of Rodet (1879), Kaye (1908), Heath (1910), Majumdar (1911–12), Sengupta (1927), Ganguli (1929), Clark (1930), and Datta (1932a). The translations of Rodet and Kaye are now accepted as faulty. But the damage persisted with the adoption of Kaye's interpretation by certain Western and Indian historians of mathematics, notably Heath and Majumdar. Sengupta's interpretation is based on Brahmagupta and Clark's on Paramesvara. Datta's and Ganguli's refer to Bhaskara I, whose authority is now acknowledged to be the more plausible one.

12. A translation of the full verbal explanation of the procedure is given in Keller (2006, vol. 1, p. 131). The explanation of this procedure in modern notation that follows is based on the exposition by Plofker (2007, pp. 416–17).

13. To illustrate the process of mutual division: Taking $y = (31x + 2)/12$, we divide to get $2x + [(7/12)x + 1/6]$. Setting $w = (7/12)x + 1/6$, which reduces to $7x = 12w - 2$, we continue with the process of mutual division. Bhaskara I, while commenting on this solution procedure, called it *kuttaka*.

14. Brahmagupta uses the principle of composition (*samasabhavana*) to arrive at a solution. In algebraic symbols, assume that for conveniently chosen values of c_1 and c_2, (a_1, b_1) and (a_2, b_2) is a set of solutions of $Nx^2 + c_1 = y^2$ and $Nx^2 + c_2 = y^2$ respectively. Then $x = a_1 b_2 \pm a_2 b_1$ and $y = b_1 b_2 \pm N a_1 a_2$ will satisfy the equation $Nx^2 + c_1 c_2 = y^2$.

15. Bill Farebrother points out in a personal communication that using a Pascal program, Bhaskaracharya's result for $d = 61$ was obtained from evaluating the square root of 974 quadratic expressions. A contemporary adaptation of Lagrange's method required the evaluation of as many as 226,153,980 square roots before alighting upon the correct result!

16. Van Brummelen (2009), following Datta and Singh (1962), conjectures that the replacement of the chord function by the sine in even the most ancient of extant Indian texts started as a "time-saving" device when "some early Indian astronomer (having to) repeatedly double arcs and (halve) the resulting chords" realized that he could save time by tabulating the half chords (or *ardhajya*).

17. One radian is the angle subtended at the center of the circle by an arc whose length is equal to the radius of the circle. Since an arc of $2\pi r$ (i.e., the circumference) subtends an angle of 360° at the center of the circle, it would follow that $360° = 2\pi$ radians or 1 radian $= 360/2\pi \approx 57$ degrees. Bhaskara I merely stated this rule without providing any explanation. However, a number of explanations have been proposed, including those based on the assumption that Bhaskara expressed the sine function as a ratio of two quadratics and then proceeded to solve for the coefficients by substituting known sine values. For further details on the various explanations, see Gupta (1967, 1986). Note that by substituting $\pi = 180°$ into the approximate formula expressed in terms of radians, the formula for degrees can be derived.

18. For a more detailed account of Kamalakara's work on multiple angle formulas, see Gupta (1974a).

19. For further details, see Hayashi (1997).

20. For further details, see Van Brummelen (2009).

21. For further details, see Duke (2005).

22. The relevant verses are:

A quadrant of the circumference of a circle is divided and from the [right] triangles and quadrilaterals as many r sines (*jya-ardhas*) of equal arcs as desired are found for any given half diameter.

The r sine of the first arc, divided by itself and lessened by the quotient, gives the second r sine difference. That first r sine diminished by all the quotients obtained by dividing each of the preceding r sines by the first r sine gives the remaining r sine differences.

Expressed in modern notation, the $(n + 1)$th r sine-difference (d_{n+1}) is given by

$$d_{n+1} = r \sin \theta - \left[\frac{r \sin \theta}{r \sin \theta} + \frac{r \sin 2\theta}{r \sin \theta} + \frac{r \sin 3\theta}{r \sin \theta} + \dots + \frac{r \sin (n\theta)}{r \sin \theta} \right],$$

or

$$r \sin \times [(n + 1)\theta] - r \sin (n\theta) = r \sin \theta - \left[\frac{r \sin \theta + r \sin 2\theta + r \sin 3\theta + \dots + r \sin (n\theta)}{r \sin \theta} \right].$$

With these formulas it is possible to generate the r sine-differences successively from the preceding ones and also to find the r sines, which then become the entries in table 9.1. For further details, see Mallayya and Joseph (2009b).

23. The source of this quotation is Van Brummelen (2009) from Gupta (1974b, p. 165).

24. For further details, see Gupta (1974a).

25. It is interesting in this context that the radius used by Vatesvara in computing his sines was the more accurate 3,437'44" rather than Aryabatha's 3,438'. Govindasvamin's radius was even more accurate at 3437'44"19''' while Madhava, the founder of the Kerala school of mathematics and astronomy, whose links with Govindasvamin have been established, must have used 3437'44"48''' for his sine computation. Since the radius of a circle can be estimated from 360/2π, it would follow that the relative accuracy of π in decimal places (dp) of the four mentioned are: Aryabhathan (3 dp), Vatesvara (4 dp), Govindasvamin (4 dp), and Madhava (6 dp).

26. A detailed discussion of Vatesvara's *Siddhanta* is found in Mallayya (2008).

27. For further details, see Mallayya (2004) and Mallayya and Joseph (2009).

28. A similar geometrical treatment is also found in Sankara Variyar and Narayana's *Kriyakramakari*, which is a commentary on Bhaskaracharya's *Lilavati*.

29. Further details of the *sredhiksetra* geometry are given by Sarasvati (1963, 1979) and Mallayya (2002).

30. The flavor of the original text is captured by the following more literal translation of verse 28, which reads: "One should multiply the sum of the products of the arms adjacent to the diagonals, after it has been mutually divided on either side, by the products of the arms and the counterarms. For an unequal [cyclical quadrilateral] the two square roots are the two diagonals."

31. It may be argued quite legitimately that the core of the method of calculus is not its ability to deal with sines and cosines but its universal power to approach all sorts of different functions. It is stretching a point to suggest that the application of the calculus

technique to a small subset of functions (such as sines and cosines) does qualify as calculus as we understand it today.

32. This is a measure of how far a planet is from the position it is predicted to be in by assuming it to move uniformly. The predicted and actual positions differ because planetary orbits are elliptical, whereas uniform motion implies a circular orbit.

33. Since the Indian calendar is a lunisolar calendar, there is need for true lunar months to be synchronized with true solar years, and this is most efficiently achieved with the help of tables. One of the best-known examples of such a text is *Tithi-cintamani* by Ganesa Daivajna (1525). For further details see Ikeyama and Plofker (2001).

Chapter Ten

A Passage to Infinity: The Kerala Episode[*]

Along the southwest coast near the tip of the Indian peninsula lies a strip of land known as Kerala. It has figured prominently in history, not only as a stopover for travelers and explorers such as ibn Battuta (b. 1304) and Vasco da Gama (b. 1460) arriving from across the Arabian Sea, but as a center of maritime trade, with its variety of spices greatly in demand even as early as the time of the Mesopotamians. While most of India was in political upheaval during the first part of the second millennium AD, Kerala was a place of relative tranquillity, sheltered by the high mountains of the Western Ghats to the east and the Arabian Sea to the west. In recent years, Kerala has played a central role in the reconstruction of medieval Indian mathematics.

The Actors

Two powerful tools contributed to the creation of modern mathematics in the seventeenth century: the discovery of the general algorithms of calculus, and the development and application of infinite-series techniques. When introduced to calculus, one is often told that the names normally associated with the development of the subject are Newton and Leibniz. The other, less well-known stream, the discovery and applications of infinite series, is often downplayed despite its importance in the development of modern mathematics. Historically, the two streams tended to reinforce each other in their simultaneous development by each extending the range of application of the other.

---

[*]While the third edition was being prepared for the press, a book on this subject came out that contains a detailed examination of this remarkable episode in the history of mathematics. Readers may wish to consult Joseph (2009).

It is generally assumed that modern calculus developed from ideas and techniques inspired by ancient Greek mathematics culminating with the "method of exhaustion" deployed by Archimedes (287–212 BC).[1] After a period of more than eighteen centuries these techniques were rediscovered in Europe with the translation of the works of Archimedes in the sixteenth century, and then developed further by a chain of European mathematicians including Roberval (1602–1675), Cavalieri (1598–1647), and Fermat (1601–1665), culminating in the consolidation of calculus by Leibniz (1646–1716) and Newton (1643–1727). This version of history takes for granted that no significant developments took place between the time of Archimedes and the seventeenth century that could have had a bearing on the all-European chain of transmission.

However, it is now generally recognized that the origin of the analysis and derivations of certain infinite series, notably those relating to the arctangent, sine, and cosine, are not to be found in Europe but in an area in South India that now falls within the state of Kerala. From a region of about a thousand square kilometers north of Cochin, during the period between the fourteenth and sixteenth centuries, there emerged discoveries in infinite series that predate similar work of James Gregory, Newton, and Leibniz by at least two hundred years.

There are a number of questions worth asking about the activities of this group of mathematician-astronomers (referred to hereafter as the Kerala school[2]), apart from those relating to the mathematical content of their work. The questions include specific ones relating to the social and historical landscape in which the Kerala school developed as well as to the motivation underlying their work. Figure 10.1 provides a useful point of reference for the notable members of the Kerala school.

There are six texts that constitute the main evidence of the work of the Kerala school. They are *Aryabhatiyabhasya* (A Commentary on *Aryabhatiya*) and *Tantrasamgraha* (A Digest of Scientific Knowledge) of Nilakantha (1443–1544); *Yuktibhasa* (An Exposition of the Rationale) of Jyesthadeva (fl. 1500–1610); *Kriyakramakari* (Operational Techniques) of Sankara Variyar (c. 1500–1560) and Narayana (c. 1500–1575); *Karanapaddhati* (A Manual of Performances in the Right Sequence) of Putumana Somayaji (fl. 1660–1740); and *Sadratnamala* (A Garland of Bright Gems) of Sankara Varman (1800–1838).

An important feature of these texts is their claim to have derived their principal ideas from Madhava of Sangamagrama (c. 1340–1425), who was

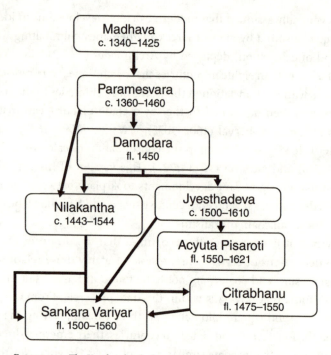

FIGURE 10.1: The Kerala school of mathematics and astronomy

mentioned earlier. We have little information of his family background, except that he belonged to a subcaste of Brahmins known as the Empranatiri, who were not originally Nambutiri Brahmins (the highest-ranking Brahmins in Kerala) but had over the years attempted to enter this group. His only surviving works are in astronomy. We know, from the reports of those who came after him, of Madhava's contribution to the development of Kerala mathematics. He was frequently referred to as Golavid or "One Who Knows the Sphere." His fame rests on his discovery of the infinite series for circular and trigonometric functions, notably the Gregory series for arctangent, the Leibniz series for π, and the Newton power series for sine and cosine. There are also some remarkable approximations attributed to him based mainly on incorporating "correction" terms for these slowly converging series.

Madhava's distinguished student was Paramesvara, whom we came across in the previous chapter. Born in 1360 into a Nambutiri Brahmin family of Vedic scholars in the village of Alattur, he was reputed to have learned his mathematics and astronomy from Madhava. Alattur had become a famous center of learning and scholarship some centuries earlier during the period of Jain and Buddhist dominance. Paramesvara wrote a number of commentaries including ones on Aryabhata's *Aryabhatiya*, on Bhaskara I's *Mahabhaskariya* and *Laghubhaskariya*, and on Bhaskara II's *Lilavati*.

The foundation laid by Paramesvara heralded the emergence of the major figure of Nilakantha Somayaji, born in 1443 into a Nambutiri Brahmin family of *somatiris* or *somayajis* (those who performed the *soma* sacrifice) in Trikkantiyur. He stayed and studied in the house of Damodara, the son of Paramesvara, where he was probably taught by Paramesvara. His major works include a commentary on three chapters of *Aryabhatiya*, and the seminal text *Tantrasamgraha*.

The latter consists of eight chapters, containing 432 verses, dealing with various topics connected with astronomical calculations, including the setting up of a sundial, calculations of the meridian, the method(s) of determining the latitude,[3] and the prediction of eclipses.

In the *Tantrasamgraha*, Nilakantha carried out a major revision of the Aryabhatan model for the interior planets, Mercury and Venus, arriving at a more accurate specification of the equation of the center[4] for these planets than any other that existed in Islamic or European astronomy before Kepler (born about 130 years after Nilakantha). In *Aryabatiyabhasya*, Nilakantha developed a computational scheme for planetary motion more efficient than that of Tycho Brahe in that it correctly takes account of the equation of center and latitudinal motion of the interior planets. This computational scheme implied a heliocentric model of planetary motion in which the five planets (Mercury, Venus, Mars, Jupiter, and Saturn) move in eccentric orbits around the mean sun which, in turn, goes round the earth. This model is similar to the one suggested by Brahe when he revised Copernicus's heliocentric model. It is significant that all astronomers of the Kerala school who followed Nilakantha accepted his planetary model.[5]

The other works that he wrote late in life were either commentaries on his earlier texts, such as those on his *Chandrachayaganita* and *Siddhantadarpana* (a short work in thirty-two verses dealing with certain important astronomical constants and the theory of epicycles); or works such as *Golasara*, mentioned earlier, a book in three chapters on spherical astronomy; or *Sundararaja Prasnottara*, a work no longer extant but mentioned elsewhere as giving answers to questions raised by a Tamil astronomer called Sundararaja. The last work is important because it provides rare evidence of the spread of the influence of Kerala mathematics and astronomy to other areas of South India.

One of Nilakantha's students was Citrabhanu (fl. 1475–1550). He was a Nambutiri and came from the village of Sivapuram (situated in present-day Trissur). His work, *Karanamrta*, contained four chapters dealing with advanced astronomical calculations. He was also the author of *Ekavimsati Prasnottara* (Twenty-one Questions and Answers), in which he offered solutions for each of a set of twenty-one pairs of simultaneous equations in two unknowns. The twenty-one pairs arose from taking, at a time, any two of the following seven quantities (*a* to *g*) given on the right side of the following equations:

$$x + y = a; \; bx - y = b; \; xy = c; \; x^2 + y^2 = d; \; x^2 - y^2 = e; \; x^3 + y^3 = f; \; x^3 - y^3 = g.$$

The solutions to fifteen of the twenty-one pairs ($7C_2$) are fairly straightforward, while the remaining six are not.[6]

A student of Citrabhanu, Narayana (c. 1500–1575) completed one of the major texts of the Kerala school, *Kriyakramakari*. A commentary on Bhaskara II's *Lilavati*, it was begun by Sankara Variyar (c. 1500–1560), a student of both Nilakantha and Citrabhanu. The Variyars were a group of non-Brahmin temple officials who assisted the Brahmin priests in their religious rituals. A number of them were skilled in astrology, and many were fluent in Sanskrit. The text *Kriyakramakari* is important in the history of Kerala mathematics and astronomy for its detailed discussion of the works of earlier writers, some of which are not extant, and for providing rationale and proof for a number of earlier results.[7]

Another student of Nilakantha was a Nambutiri from the Alattur village (the birthplace of Paramesvara), Jyesthadeva (fl. 1500–1610). He was the author of the seminal text of the Kerala school, *Yuktibhasa*. There are at least three versions of this text, of which the Malayalam version became well known throughout Kerala.[8] Based on Nilakantha's *Tantrasamgraha*, it is unique in Indian mathematical literature for giving detailed rationale, proofs, or derivations of many theorems and formulas in use among the astronomer-mathematicians of that time.

A student of Jyesthadeva came from the Pisarati community. They were not Brahmins but performed traditional functions as cleaners and suppliers of flowers and plants for the temple. They were also employed by some Nambutiri families to give instructions to family members on the calculation of the astrological calendar (*panchanga*) and on time reckoning.

Acyuta Pisarati (c. 1550–1621) was a versatile scholar who made a mark not only in astronomy but also in literature and medicine. His major contribution is found in his work *Sphuta-nirnaya,* where he introduced for the first time in Indian astronomy a correction called "reduction to the ecliptic," around the same time as Tycho Brahe did in Western astronomy.

In Charles Whish's 1832 paper appears the passage: "The author of the *Karanapaddhati* whose grandson is now alive in his seventieth year was Putumana Somayaji, a Nambutiri Brahmana of Trisivapur [Trissur] in Malabar." An influential work in the dissemination of Kerala mathematics and astronomy not only in Kerala but also in the neighboring areas of present-day Tamilnadu and Andhra Pradesh, *Karanapaddhati* was written in 1732, almost two hundred years after Jysthadeva's *Yuktibhasa.*

After Acyuta Pisarati, little in the way of original work was done, although the tradition of providing corrections and contributing to the preparation of astronomical ephemerides for the daily needs of faithful observers and practitioners continued for a long time. About one hundred years after *Karanapaddhati* came the last of the known texts of the Kerala school, *Sadratamala.* The author of this book, Sankara Varman, belonged to a minor royal family and was a contemporary of Charles Whish. *Sadratnamala,* written in 1823, contains many of the results of the Kerala school, but given without the rationales or derivations found in the earlier texts. Whish met him and described him as "a very intelligent man and acute mathematician." He died six years after Whish's article on Kerala mathematics and astronomy appeared in 1832.

The authors mentioned above form part of a tradition of continuing scholarship in Kerala over a period of four hundred years, from the birth of Madhava in 1340 to the probable death of Putumana Somayaji in 1740. The current level of knowledge of source materials means that it is difficult to assign many of the developments to any particular person. The results should be seen as produced by members of a school spread over several generations.

The Social Background

To understand the context in which mathematics developed in Kerala, there is a need to take a careful look at the social landscape of medieval Kerala and, in particular, the structure of medieval Kerala society, the pivotal role of the Kerala temple, and the mode by which scientific knowledge was acquired and disseminated. Each of these topics could well provide

sufficient subject matter for a whole chapter. Instead, we will confine ourselves to making a few observations.

It is clear, from the discussion so far, that the members of the Kerala
school were mostly Nambutiri Brahmins. Within a mainly two-tier caste
system in Kerala, consisting of Brahmins and Nairs, two institutions operated to strengthen and sustain the economic and social dominance of the
Nambutiris to a degree not known elsewhere in India: a system of feudalism (the *janmi* system of landholding) headed by the Nambutiris, and their
control of vast areas of arable land owned by temples.

There were also certain social factors that strengthened the Nambutiri dominance over the Kerala society of that time. The Nairs practiced
the *marumakkattayam* (matrilineal) system of descent outside the formal
institution of marriage. Sexual alliances between Nair women and Nambutiri men were permitted, indeed sometimes encouraged, with children
of such unions remaining the sole responsibility of their mother's family.
At the same time, the Nambutiris operated a system of patrilineal descent
(*makkatayam*), with a form of primogeniture that allowed only the eldest
son to inherit property and to marry Nambutiri women. The eldest son
was also required by custom to provide for the material needs of his sib-
lings, consisting of younger brothers and unmarried sisters (of whom there
were a large number, given the way that the system operated).

It is known that the pursuit of activities such as studying mathematics
and astronomy did not traditionally confer high status, which was reserved
for those who carried out ceremonial and ritualistic duties. The most notable member of the Kerala school after Madhava, Nilakantha, belonged
to the highest rank among the Nambutiris. His high social status arose
from the fact that he was a *somayaji*, one of the select subcastes among the
Nambutiris who carried out the *soma* sacrifices. In the traditional *soma*
sacrifice, the preparation and consumption of the hallucinatory juice of the
soma plant played a central role. Yet Nilakantha single-mindedly pursued
his interests in mathematics and astronomy. There were other members of
the Kerala school who were not even Brahmins. There was, for instance,
Sankara Variyar: the name Variyar indicates that he belonged to the *Am-
bilavasis*, a caste of temple servants, as does the name of Acyuta Pisarati.
This would suggest that the Kerala school were a mixed group, probably
brought together by their interest in mathematics and astronomy, undertaking pursuits that did not have great social status—a group that cut
across caste lines to an extent, and which probably had strong contacts

with the temple personnel.[9] The temple fulfilled an important purpose as an institution for acquiring and disseminating scientific knowledge. It was an influential organization since it combined religious power with secular power, its Nambutiri members being in many cases powerful landlords in their own right. The temple served as a medium through which the Nambutiris asserted their power and kept other groups in check.

Another aspect of the social background, for present speculation and future research, is whether a number of the Nambutiri members of the Kerala school were younger sons. If that was so, we would have a group of Nambutiris freed of all economic and family responsibilities, a truly leisured class with religious duties confined to a few, and not very demanding, rituals. Some in such a situation whiled away their time writing erotic poetry, and many were engaged in other less demanding pursuits. But a few pursued their interest in astronomy and mathematics consistently over a period of several centuries, sustained by the institution of the *guru-sisya* (teacher-disciple) relationship, which was characteristic of the educational system then.

While this explanation for the emergence and continued existence of the Kerala school might appear attractive, particularly to those who are disposed to seeking major explanations of any Indian phenomena in the caste system, it seems somewhat unconvincing. First, it does not account for the presence and the role of non-Brahmins in the Kerala school. Second, this explanation ignores the symbiotic nature of the relationship between the traditional *jyotisa* (astronomer/astrologer), who often came from the lowly Kaniyan caste, and the Nambutiris. Third, the *granthaveri* (or village records) of Kerala of this period contain ample evidence of the metrical precision of a number of artisans and craftsmen (such as the carpenter, the trader, the builder, and the architect). These records show among these artisans some awareness of the developments taking place in astronomy and mathematics during that period. The *granthaveri* and temple records remain a good but relatively untapped source of information about the "calculating people" of the period.

Further study of the social context of Kerala mathematics may yield an unexpected bonus. There is a deeply entrenched notion, as mentioned in relation to Egyptian and Mesopotamian mathematics, that all non-European mathematics is utilitarian. A number of scholars have fallen into the same trap. Their search in astronomy, navigation, and other practical pursuits for the motivation behind Kerala mathematics can offer only a partial answer. One should, of course, never ignore the practical motivation. After all, many of the members of the Kerala school were both mathematicians and astronomers. The texts of that period cover both subjects. However, a lot of the works on infinite series do not have any direct applications to astronomy. So what led them on in their pursuit of knowledge? I have a vision of a group of pure mathematicians in Kerala between the fourteenth and sixteenth centuries (like Ramanujan, Hardy, and Littlewood at the University of Cambridge early in the twentieth century) indulging in their passion and probably proud of the fact that the mathematics that they did was of no use to anyone! Some members of the Kerala school must have taken delight in long and tedious calculations, such as the one reportedly undertaken by Madhava in calculating the sine tables to twelve decimal places. Such fascination with numbers and delight in calculation has been a characteristic of Indian mathematics over the ages, as we saw in chapter 8.

The Motivation and Method

An important "mathematical" motivation for the Kerala school may be traced to a verse in *Aryabhatiya* that explains how, for a given diameter, the circumference of a circle is calculated:

> Add 4 to 100, multiply by 8, and add 62,000. The result is approximately the circumference of a circle whose diameter is 20,000. (Verse 10)

Some historians of mathematics have argued, partly on the basis of such quotations, that the Indians were not aware of the fact that the circumference of a circle (and therefore π) could never be exactly determined. The confusion may have risen because of the mistranslation of the word *asanna* as "approximate" or "rough value," as in the quotation above. The word is subtler than that. What it conveys is the notion of "unattainability," that is, something that cannot be reached.

This is illustrated by a passage from Nilakantha's commentary on *Aryabhatiya*:

> Why is only the approximate value (of circumference) given here? Let me explain. [It is approximate] because the real value cannot be obtained. If the diameter can be measured without a remainder, the circumference measured by the same unit [of measurement] will leave a remainder. Similarly, the unit that measures the circumference without a remainder will leave a remainder when used for measuring the diameter. Hence, the two measured by the same unit will never be without a

remainder. Though we try very hard we can reduce the remainder to a small quantity but never achieve the state of "remainderlessness." This is the problem.

What the passage shows is that Nilakantha and others understood the "irrational" nature of the ratio we now represent as π. So the question arose as to what could be done as a result. The following passage from Sankara Variyar and Narayana's *Kriyakramakari* contains a strategy:

> Thus even by computing the results progressively, it is impossible theoretically to come to a final value. So, one has to stop computation at that stage of accuracy that one wants and take the final result arrived at by ignoring the previous results [obtained along the way].

In applying the infinite-series approach to estimate the circumference, the Kerala mathematicians came across a serious difficulty: the special case of the Madhava-Gregory series (discussed in a later section) converges very slowly. The problem was tackled by Kerala mathematicians in two directions:

1. Obtain rational approximations by applying corrections to partial sums of the series.

2. Obtain more rapidly converging series by transforming the original series.

An unusual aspect of the Kerala approach to the derivation of a number of infinite series is their use of the method of direct rectification. The method of direct rectification of an arc of a circle involves summation of very small arc segments and reducing the resulting sum to an integral. This is an interesting geometric technique different from the "method of exhaustion" used in Islamic and European mathematics. In the Kerala case, you are subdividing an arc into *unequal* parts, whereas in the Islamic and European case there is a subdivision of the arc into *equal* parts. The different technique used in Kerala does not indicate that the method of exhaustion was unknown to the Indians. Indeed, it is likely that Aryabhata preferred the octagon method rather than the hexagon method used by Greek and Islamic mathematicians to compute his accurate estimate of the circumference of the circle.[10] The method of exhaustion was probably avoided because it involved working out the square roots of numbers at each stage of the calculation, a tedious and time-consuming task.

Astronomy provided an important motive for the study of infinite-series expansions of π and rational approximations for different trigonometric functions. For astronomical work, it was necessary to have both an accurate value for π and highly detailed trigonometric tables.[11] In this area Kerala mathematicians made the following discoveries:

1. The power series for the inverse tangent, usually attributed to Gregory

2. The power series for π, usually attributed to Leibniz, and a number of rational approximations to π

3. The power series for sine and cosine, usually attributed to Newton, and approximations for sine and cosine functions (to the second order of small quantities), usually attributed to Taylor; this work was extended to a third-order series approximation of the sine function, usually attributed to Gregory

Apart from the work on infinite series, there were extensions of earlier work, notably of Brahmagupta and Bhaskara II, already discussed in Chapter 9.

The Madhava-Gregory Series for the Inverse Tangent

The power series for $\tan^{-1} x$ is

$$\tan^{-1} x = x = \frac{x^3}{3} + \frac{x}{5} - \dots \text{ for } x \leq 1 \tag{10.1}$$

and is generally known as the Gregory series for the inverse tangent after the Scottish mathematician James Gregory, who derived it in 1667. Madhava is credited with the following rule found in various texts, including the *Yuktibhasa* and the *Kriyakramakari*. The first of these two sources gives the rule as follows.

The first term is the product of the given Sine and radius of the desired arc divided by the Cosine of the arc. The succeeding terms are obtained by a process of iteration when the first term is repeatedly multiplied by the square of the Sine and divided by the square of the Cosine. All the terms are then divided by the odd numbers 1, 3, 5, ... The arc is obtained by adding and subtracting [respectively] the terms of odd rank and those of even rank. It is laid down that the [Sine of the arc] or that of its complement whichever is smaller should be taken here [as the given Sine]. Otherwise, the terms obtained by this above iteration will not tend to the vanishing magnitude.

The use of capital letters in Sine and Cosine in this extract indicates that we are dealing with the Indian sine and cosine, where $\text{Sin } \theta = r \sin \theta$ and $\text{Cos } \theta = r \cos \theta$, r being the radius. The condition given at the end of this rule may be interpreted as ensuring that $r \sin \theta$ is less than $r \cos \theta$, or that $\tan \theta$ (i.e., x in equation 10.1) should be less than 1 to ensure absolute convergence of the series. Thus Madhava's rule given above may be written as

$$r\theta = \frac{r(r \sin \theta)}{1(r \cos \theta)} - \frac{r(r \sin \theta)^3}{3(r \cos \theta)^3} + \frac{r(r \sin \theta)^5}{5(r \cos \theta)^5} - \cdots$$

or

$$\theta = \tan \theta - \frac{\tan^3 \theta}{3} + \frac{\tan^5 \theta}{5} - \cdots \qquad (10.2)$$

This is equivalent to the Gregory series (10.1) for the inverse tangent.

When $x = 1$ in equation (10.1) or $\theta = 45° = \pi/4$ in equation (10.2), the Madhava-Gregory series reduces to the Liebniz series:

$$\frac{\pi}{4} = 1 - \frac{1}{3} + \frac{1}{5} - \frac{1}{7} + \cdots \qquad (10.3a)$$

Or, expressed in terms of the circumference (C) and diameter (d) of a circle, we have the usual form found in Kerala mathematics:

$$C = 4d - \frac{4d}{3} + \frac{4d}{5} - \frac{4d}{7} + \cdots \qquad (10.3b)$$

Steps in the derivation of this series are found in Mallayya and Joseph (2009b), Plofker (2009), Roy (1990), Sarasvati (1979), Srinivasiengar (1967), and other publications, based on the original explanations in the *Yuktibhasa* and *Kriyakramakari*. The method used corresponds to what is known today as the method of expansion and term-by-term integraion.[12]

It was soon realized that the infinite-series expansion for the circumference (C) given in equation (10.3b) was not particularly helpful in obtaining accurate estimates of the circumference for a given diameter (i.e., for estimating π), because of the slowness of the convergence of the series. To stop the computation of the C at any desired stage, as the quotation given earlier from *Kriyakramakari* recommends, the infinite series has to be truncated, and such truncation produces some error in the estimate of C. The objective was to minimize this truncation error (or to compensate for the loss of terms because of truncation). This gave impetus to developments in two directions: (1) obtaining rational approximations by applying corrections to partial (or truncated) sums of the series; and (2) obtaining more rapidly converging series by transforming the original series. There was considerable work in both directions, and the details are discussed in both the *Yuktibhasa* and the *Kriyakramakari*.[13] What the work exhibits is a measure of understanding of the concept of convergence, of the notion of rapidity of convergence, and an awareness that convergence can be speeded up by transformations.

As an illustration of the remarkable efficiency of some of the corrections introduced, consider the following examples. In the *Kriyakramakari*, the discussion starts with a quotation from Bhaskara's *Lilavati*.

> When a diameter is multiplied by 3,927 and divided by 1,250, [this is] a very accurate circumference. Or when [a diameter] is multiplied by 22 and divided by 7, [the result is] crude and for practical use.

It is then pointed out that the diameter and the circumference given here can be obtained by dividing *Aryabhatiya*'s original estimate of circumference and diameter given in verse 11, and quoted earlier, by 16. However, a more accurate estimate of the circumference is given as 355 for a diameter of 113, which corresponds to π correct to six decimal places.

Consider another example from the *Yuktibhasa*. What is required is to evaluate the circumference of a circle with a diameter of 10^{11}. Without the correction and using infinite series (10.3b) above with the number of terms on the right-hand side as nineteen, the circumference is about 3.194×10^{11}. However, incorporating one of the corrections gives the circumference as $3.1415926529 \times 10^{11}$, which is correct to nine places.[14] And the interest in increasing the accuracy of the estimate seems to have continued for a long time, so that as late as the nineteenth century Sankara Varman, the author of *Sadratnamala*, estimated the circumference of a circle corresponding to a diameter measure of 1 *parardha* (10^{17}) as 314,159,265,358,979,324—correct to seventeen places.

A second approach to achieving a greater degree of accuracy is by transforming the original series to a more rapidly converging series. For $x = 1/\sqrt{3}$ or $\theta = 30°$, that is, for an arc that is 1/12th of the circumference, we get a transformation attributed to Madhava and found in both the *Yuktibhasa* and the *Kriyakramakari*:

$$C = \sqrt{12}\,d\left[1 - \frac{1}{3 \times 3} + \frac{1}{3^2 \times 5} - \frac{1}{3^3 \times 7} + \ldots\right].$$

It is likely that Madhava used one of these approximation formulas when he estimated correctly to eleven decimal places the circumference of a circle of diameter 9×10^{11}. Madhava's calculated value of the circumference is 2,827,433,388,233 units (implying a value for π of 3.14159265359), as reported in the *Kriyakramakari* (Sarma 1972, p. 26).

A translation of Madhava's verse statement of the circumference gives a flavor of how numbers were recorded in verse. In the notational system known as *bhuta samkhya*, discussed in chapter 8, certain objects were traditionally used to represent numerals, either singly or in pairs, reading from right to left. Thus the circumference 2,827,433,388,233 was recorded as

Gods (33), eyes (2), elephants (8), serpents (8), fires (3), three (3), qualities (3), Vedas (4), *naksatras* (27), elephants (8), and arms (2)—the wise say that this is the measure of the circumference when the diameter of a circle is nine *nikharva* [10^{11}].

These approximations (for obtaining accurate estimates of the circumference for a given diameter) are not to be found in any other mathematical literature until much later.[15] They are unique to Kerala.

The Madhava-Newton Power Series for the Sine and Cosine

In a commentary on Nilakantha's *Tantrasamgraha* by an unknown student of Jyesthadeva, the author of *Yuktibhasa*, are found the following descriptions of the power series for sine and versine without any derivations.

(A) The arc is repeatedly multiplied by the square of itself and divided (in order) by the square of each and every even number increased by itself and multiplied by the square of the radius. The arc and the terms obtained from these repeated operations are to be placed one beneath the other in order, and the last term subtracted from the one above, the remainder from the term then next above, and so on, to yield the [*bhuja*] *jya* [or Indian Sine] of the arc.

(B) The radius is repeatedly multiplied by the square of the arc and divided [in order] by the square of each and every even number diminished by itself and multiplied by the square of the radius, with the first term involving only 2. The resulting terms are placed one beneath the other in order, and the last term subtracted from the one above, the remainder from the term next above and so on, to yield *utkamajya* or *sara* [Indian versine] of the arc.

Expressed symbolically, where r is the radius and a the length of the given arc, the first three are

$$\frac{a \times a^2}{1!(2^2 + 2)r^2} = \frac{a^3}{3!r^2},$$

$$\frac{a^3 \times a^2}{3!(4^2 + 4)r^4} = \frac{a^5}{5!r^4},$$

$$\frac{a^5 \times a^2}{5!(6^2 + 6)r^6} = \frac{a7}{7!r^6}.$$

Hence

$$\text{Indian Sine} = r \sin \theta = a - \frac{a^3}{3!r^2} + \frac{a^5}{5!r^4} - \frac{a^7}{7!r^6} + \dots$$

Substituting $a/r = \theta$ gives

$$\sin \theta = \theta - \frac{\theta^3}{3!} + \frac{\theta^5}{5!} - \frac{\theta^7}{7!} + \dots \tag{10.4}$$

Using the above notation and denoting Indian versine (*sara*) by ($r - r \cos \theta$), the first three even numbers given in (B) can be written as

$$\text{Indian versine}(r - r \cos \theta) = \frac{a^2}{2!r} - \frac{a^4}{4!r^3} + \frac{a^6}{6!r^6}.$$

Substituting $a/r = \theta$ and simplifying gives

$$\cos \theta = 1 - \frac{\theta^2}{2!} + \frac{\theta^4}{4!} - \frac{\theta^6}{6!} + \dots \tag{10.5}$$

The series given in (10.4) and (10.5) are usually named after Newton. They make their first appearance in European mathematics in a letter from Newton to Oldenburg in 1676 and are then elaborated on a firmer

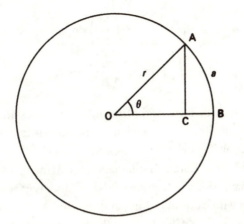

FIGURE 10.2: The Indian sine and cosine

algebraic basis by De Moivre (1708–1738) and Euler (1748). They should be more appropriately named after Madhava, to whom the series are usually attributed by the later members of the Kerala school.

The rules given for generating the two power series given in (10.4) and (10.5) may be explained in terms of figure 10.2 as

Indian sine $= r \sin \theta = $ AC;

Indian cosine $= r \cos \theta = $ OC.

Madhava's approach is to express $r \sin \theta$ and $r \cos \theta$ in terms of r and a. (We shall not attempt to follow the long and complex steps in the derivation of the sine and cosine series in the *Yuktibhasa*. Details of the original explanation involving the use of the method of "expansion and term-by-term integration" are given by Rajagopal and Venkataraman [1949], Sarasvati [1963], and K. V. Sarma [2008]).[16]

These power series were probably used to construct accurate sine and cosine tables for astronomical calculations. In a table of values of half-sine chords reportedly calculated by Madhava for twenty-four arcs drawn at equal intervals for a quadrant of a given circle, the values are correct in almost all cases to the eighth or ninth decimal place. Such an accuracy was not achieved in Europe for another two hundred years.[17]

Approximations for sine and cosine functions to the second power of small quantities are also attributed to Madhava. In modern notation, Madhava's results can be written as

$$\sin(x + h) \approx \sin x + \left(\frac{h}{r}\right) \cos x - \left(\frac{h^2}{2r^2}\right) \sin x,$$

$$\cos(x + h) \approx \cos x - \left(\frac{h}{r}\right) \sin x - \left(\frac{h^2}{2r^2}\right) \cos x,$$

where h is the small quantity and r the radius. These results are but special cases of one of the familiar expansions in mathematics, the Taylor series, named after Brook Taylor (1685–1731).

In our brief look at Kerala mathematics, the name that keeps recurring is that of Madhava of Sangamagrama. His brilliance is generously acknowledged by those who came after him, and the effects of his teaching on the works of Paramesvara, Nilakantha, Jyesthadeva, and others are there to see. It would be quite in keeping with Indian tradition if, in holding him in such awe, his successors were to have credited him with more than his share of discoveries. Of his teachers we know nothing. Madhava's outstanding contributions, in the area of infinite-series expansions of circular and trigonometric functions and finite series approximations to them, predate European work on the subject by two hundred to three hundred years.

We may consider Madhava to be the founder of mathematical analysis. Some of his discoveries in this field show him to have possessed extraordinary intuition, making him almost the equal of a more recent intuitive genius, Ramanujan (1887–1920), who spent his childhood and youth at Kumbakonam, not very far from Madhava's birthplace. Ramanujan also showed a considerable intuitive grasp of infinite-series expansions, particularly of trigonometric and circular functions, as his *Notebooks* (1985) now testify. Do we see in these notes the vestiges of a "hidden" indigenous mathematics not submerged by the influx of modern mathematics from the West ?

There is an interesting Chinese connection that merits further investigation. In the middle of the eighteenth century Ming Antu wrote a book on geometry that contained power-series expansions for trigonometric functions and π. The derivations of the Gregory and Newton formulas contained in his book bear an uncanny resemblance to the work of the Kerala school. The inference is usually that the Chinese were introduced to these results by the Jesuit missionary Pierre Jartoux at the beginning of the eighteenth century, and that the proofs were subsequently arrived at independently. However, the Kerala-China link should also be examined. We mentioned in the historical introduction to chapter 6 that the early

Ming period saw considerable maritime contact between China and parts of Asia and Africa. Kerala has a long history of trade contacts with China. One piece of tangible evidence of technology having been transferred from China to Kerala is the Chinese fishing net, which is still in use today. Would it be far-fetched to suggest that the contacts between the two areas also had a less tangible dimension?[18]

Transmission of Kerala Mathematics

Establishing Transmissions: A Digression

The question to be addressed here, in the case of cross-cultural transmission of mathematical ideas, is how do we establish that an item of knowledge was actually transmitted into, say, Europe and was not an independent discovery within Europe even though it was known earlier in another culture?[19]

Translations provide direct, indubitable evidence of transmissions. So do written acknowledgments of debts owed to particular mathematical traditions. The transmission of Indian or Greek mathematics and astronomy via Islamic scholars to Europe has been established by such *direct* evidence. We know that Indian astronomy was transmitted westward to Baghdad, by a translation into Arabic of the *Siddhantas* around 760, and into Spain, by a translation into Latin of the same work in 1126. This transmission was not just westward, for there is documentary evidence of Indian mathematics and astronomy being imported to China, Thailand, Indonesia, and other Southeast Asian regions from the seventh century onward.

Table 10.1 provides a list of topics on which there has been similar work in Indian, Islamic, and European mathematics. In some of these cases, there is documentary evidence to establish transmission. However, it is often the case that *direct* evidence based on translations is unavailable. Then, we have to turn to *circumstantial* evidence to support transmission claims. Notably, in the exact sciences, the identification of methodological, algorithmic, and epistemological similarities—especially where these similarities may involve duplications in the recipient culture *of incorrect results or of approximations* from the original source—would support the case for transmission. Correct results could have been independently discovered. But incorrect results and approximations that match are less likely to have been independently discovered. Furthermore, showing the existence of an

Table 10.1: Common Mathematical Pursuits in India and Elsewhere

Topic	First appears in	Repeated or used by
Method of calculating values of the chord	Ptolemy AD 150	Islamic mathematicians
Method of construction of 24 Sine values	Aryabhata AD 499	Regiomontanus 1533
Radian measure	Aryabhata AD 499 (Hipparchus 150 BC?)	European mathematians late 19th century AD
Square and cube root extraction	Aryabhata AD 499 (China 1st century BC)	Islamic mathematics 10th century
Methods in algebra including summation of finite series	Aryabhata AD 499 (Archimedes and Diophantus)	Italian mathematics: Bombelli 16th century
Number system with 0 as number and as positional symbol	Indians c. 6th century AD	Al Khwarizimi 9th century
Generalized arithmetic of fractions	Bhasakara I AD 600	Islamic mathematics 9th century
Prime factorizing method involving differences of squares	Narayana AD 1356	Fermat 17th century
Arithmetic involving zero	Brahmagupta 7th century AD	Arab mathematics 9th century, European mathematics 13th century
Elements of the calculus including derivatives of sine and cosine	Bhaskara II 12th century AD	Renaissance mathematics 17th–18th centuries
Continued fractions formulas	Bhaskara II 12th century AD	John Wallis 1655
Pells equation: $61x^2 + 1 = y^2$	Brahmagupta 7th century AD, Bhaskara II 12th century AD, $x = 226{,}153{,}980$, $y = 1{,}766{,}319{,}049$	Fermat challenge problem 17th century, solved by Euler 18th century

continued

Table 10.1: Continued

Topic	First appears in	Repeated or used by
A proof of the "Pythagorean" theorem	Bhaskara II 12th century AD	John Wallis 1655
Infinite series for π	Madhava 15th century AD	Leibniz, James Gregory, 17th century
Infinite series for sine and cosine functions	Madhava 15th century AD	Newton 17th century

accessible corridor of communication between the two cultures involved, as well as an appropriate chronology of the transmission process, could strengthen the case. However, using a legal analogy, such an approach cannot establish a case for transmission beyond reasonable doubt; it can only establish a case for transmission as better than its no-transmission competitor on the balance of probabilities.[20]

Using such approaches, historians have argued for transmissions of ideas from Europe to cultures outside. It was by thus inferring methodological similarities that Van der Waerden (1983) claimed that Aryabhata's trigonometry was borrowed from the Greeks. Earlier in 1976, combining this approach with his much-criticized "hypothesis of common origin," he made claims that Bhaskara II's work on Diophantine equations could be traced to an unknown Greek manuscript that was available to Bhaskara and his students. Van der Waerden concluded his study of the Greek origins of the works of Aryabhata and Bhaskara by stating that "in the history of science independent inventions are exceptions: the general rule is dependence." Neugebauer (1962) used "priority, accessible communication routes and methodological similarities" to establish his conjecture about the Greek origins of the astronomy contained in the Indian Siddhantas. What we see from these examples is that a case for claiming the transmission of knowledge from Europe to places outside does not necessarily rest on direct documentary evidence. In certain circumstances, priority, communication routes, and similarities appear to establish transmission from West to East as more plausible, on the balance of probabilities, than independent discovery in the East. However, when it comes to East-to-West transmissions, there seems to be a complete change of orientation. The criterion for establishing transmission is no longer the comparative notion of "balance of probabilities" but the absolute notion of "beyond all reasonable doubt." This double standard makes it possible to sustain a case for Eurocentric histories against their dialogical competitors, even in those situations where an across-the-board application of the principle of the balance of probabilities would make a stronger case for East-to-West transmission.

So how can our conjecture of transmission of Kerala mathematics possibly be established? The tradition in Renaissance Europe was that mathematicians did not always reveal their sources or give credit to the original source of their ideas. However, the activities of the monk Marin Mersenne from the early 1620s to 1648 suggest some attempt at gathering scientific information from the Orient. Mersenne corresponded with the leading Renaissance mathematicians such as Descartes, Pascal, Fermat, and Roberval.[21] Though a minim monk, Mersenne had had a Jesuit education and maintained ties with the Collegio Romano. Mersenne's correspondence reveals that he was aware of the importance of Goa and Cochin (in a letter from the astronomer Ismael Boulliaud to Mersenne in Rome[22]). He also wrote of the knowledge of Brahmins and "Indicos"[23] and took an active interest in the work of orientalists such as Erpen. Regarding Erpen, he mentions his collection of manuscripts in Arabic, Syriac, Persian, and Indian languages.[24]

It is our conjecture that between 1560 and 1650 knowledge of Indian mathematical, astronomical, and calendrical techniques accumulated in Rome and diffused to neighboring Italian universities like Padua and Pisa, and to wider regions through Cavalieri and Galileo, and through visitors to Padua like James Gregory. Mersenne may have also had access to knowledge from India acquired by the Jesuits in Rome and, via his well-known correspondence, helped to diffuse this knowledge throughout Europe. Certainly the way James Gregory acquired his geometry after his four-year sojourn in Padua, where Galileo taught, supports this possibility.

All this is *circumstantial*. To make the case stronger for the transmission of Kerala mathematics to Europe, we require documentary evidence to show that the Jesuits acquired and comprehended mathematics of Kerala and disseminated this information among those in Europe who had the necessary background to assimilate this information. Failing the availability of such *direct* evidence, the *indirect* route for establishing transmission needs to be strongly delineated. In addition to the Neugebauer's criteria of priority, communication routes, and methodological similarities, we propose to test the hypothesis of transmission on the grounds of *motiva-tion* and *evidence of transmission activity by Jesuit missionaries*. In the next few sections all these aspects will be discussed.

The Case for Transmission: Applying the Neugebauer Criteria

The priority of the Kerala work over that of Europe is now beyond doubt. Madhava (1340–1425) is credited with the original ideas in Kerala mathematics. These ideas led to derivation of the infinite series for π, which we illustrated earlier, and to infinite series for a range of trigonometric functions. These developments, therefore, precede the late-seventeenth-century work of Gregory, Newton, and Leibniz by at least 250 years.

A corridor of communication between the South of India and the Arabian Gulf (via the port of Basra) had been in existence for centuries. The arrival of the Portuguese Vasco da Gama to the Malabar coast in 1499 heralded a direct route between Kerala and Europe via Lisbon. Thus, after 1499, despite its geographical location, which prevented easy communication with the rest of India, Kerala was linked with the rest of the world and, in particular, directly to Europe.

While the two aspects of priority and communication routes are readily established, the existence of methodological similarities requires further examination. It is beyond the scope of this book to do so; the interested reader may refer to the literature listed in endnote 19 in the present chapter. However, consider the intriguing similarity in a key result in the *Yuktibhasa* and one adopted by Fermat, Pascal, and Wallis among others in European mathematics.[25] In the *Yuktibhasa*, the following result, expressed in modern notation, is proved:

$$\lim_{n \to \infty} \frac{1}{n^{k+1}} \sum_{i=1}^{n} i^k = \frac{1}{k+1}, \quad k = 1, 2, 3, \dots.$$

The same result was adopted in Europe in the seventeenth century to evaluate the area under the parabola $y = x^k$ or, equivalently, calculate $\int x^k \, dx$. At this point it should be pointed out that Wallis used reasoning somewhat similar to that given in the *Yuktibhasa*.[26] Methodological similarities between the mathematics of the Aryabhata school, upon which Kerala mathematics is based, and the works of the renaissance European mathematicians are not infrequent. Apart from the similarities between their infinitesimal methods, both traditions shared a common interest in quadratures and rectifications for different reasons.[27]

The Case for Transmission: Applying the Legal Standard of Motivation and Opportunity

The primary *motivation* for Europeans to import knowledge from India was navigation: the need for more accuracy in computation, a better calendar, and more advanced astronomy.[28] By the middle of the sixteenth century there was an error in the calculations that formed the basis of the existing Julian calendar. The true solar year was around 11.25 minutes shorter than the assumed 365.25 days, thus causing a cumulative error that was offsetting the date of Easter appreciably. For example, the vernal equinox was scheduled by the calendar to take place on March 21, but it actually took place on March 11—thus, without correction, Easter would eventually take place in summer rather than in spring. An awareness of this inadequacy in relation to other calendars may be inferred from the calculation involving the *tithi*[29] measure in Viete's critique of the Gregorian calendar reform (Bein 2007). In astronomy, the remarkable similarities between the planetary model devised by the Kerala mathematician Nilakantha and the later one by Tycho Brahe, and the adoption for a time by Kepler of the tenth-century Indian lunar model of the astronomer Munjala, are worthy of note.[30]

The arrival of Francis Xavier in Goa in 1540 heralded a continuous presence of the Jesuits in the Malabar till 1670. The early Jesuits were interested in learning the vernacular languages to further their work of religious conversion, but the later Jesuits, who arrived after 1578, were of a different mold. The famous Matteo Ricci was in the first batch of Jesuits, trained in the new mathematics curriculum introduced in the Collegio Romano by Clavius. Ricci was an accomplished mathematician. He also studied cosmography and nautical science in Lisbon prior to his arrival in India in 1578. Ricci's arrival in Goa was significant in respect to Jesuit acquisition of local knowledge. His specialist knowledge of mathematics, cosmography, astronomy, and navigation made him an eminent candidate for "discovering the knowledge of the colonies and he had specific instructions to investigate the sciences of India" (Bernard 1973, p. 38).

Subsequently several other scientist Jesuits trained by Clavius or Grienberger (Clavius's successor as mathematics professor at the Collegio Romano) were sent to India. Most notable of these, in terms of their scientific activity in India, were Johann Schreck and Antonio Rubino. The former had studied with the French mathematician Viete, well known for his work in algebra and geometry. At some point in their stay in India these Jesuits went to the Malabar region, including the city of Cochin, the epicenter of developments in Kerala mathematics. The Jesuits were keen to acquire local knowledge, especially relating to the calendar and navigation. They had, as mentioned earlier, an interest in the calendar that stemmed from the Church's desire to reform the erroneous dating of Easter and other festivals. Clavius was a member of the commission that ultimately reformed the Gregorian calendar in 1582. Also, as discussed earlier, improving navigational skills became a matter of vital importance, as shown by the large prizes offered by various governments in Europe.[31] All this has led to the conjecture that these Jesuits took part in an interchange of scientific ideas between Europe on the one hand and India and China on the other.[32]

With regard to the earlier Jesuits on the Malabar Coast, we observe that several references in the historical works of Wicki indicate that they were interested in the arithmetic, astronomy, and timekeeping of the region. Indeed, it appears that the Jesuits tried to augment their knowledge of indigenous sciences by including subjects such as *jyotisa* (astronomy/astrology) in the curriculum of the Jesuit colleges on the Malabar Coast. They were also active in the transmission of local knowledge back to Europe. Evidence of this knowledge acquisition is contained in the manuscript collections Goa 38, 46, and 58 to be found in the Jesuit historical library in Rome (ARSI). The last collection mentioned contains the work of Father Diogo Gonsalves on the judicial system, the sciences, and the mechanical arts of the Malabar region. This work of knowledge acquisition started from the very outset of the Jesuit presence in Kerala.[33] The translation of the local sciences into European languages prior to transmission to Europe was epitomized by Garcia da Orta's popular *Colloquios dos simples e drogas he cousas mediçinas da India*, published in Goa in 1563. There may have been other publications of this type that remain inaccessible, possibly because of linguistic and nationalistic reasons.

If the earlier and later Jesuits were involved in learning the local sciences, then—given the academic credentials of the Jesuits such as Ricci, Schreck, and Rubino of the middle period—it is a plausible conjecture that this work continued and with greater intensity. There is fragmentary documentary evidence that this did happen. It is known that Ricci made inquiries about the Indian calendar—in a letter to Maffei he states that he requires the assistance of an "intelligent Brahmin or an honest Moor" to help him understand the local ways of recording and measuring time.[34] Then there is de Menses, who, writing from Kollam in 1580, reports that, on the basis of local knowledge, he has detected inaccuracies in European maps.[35] There were other later Jesuits who reported on scientific findings on such diverse things as calendrical sciences and inaccuracies in the astronomical tables. Antonio Rubino wrote in 1610 about inaccuracies in European mathematical tables for determining time.[36] Then there is the letter from Schreck, in 1618, on astronomical observations intended for the benefit of Kepler—the latter had requested the eminent Jesuit mathematician Paul Guldin to help him to acquire these observations from India to support his theories.[37] Although this does not establish the fact that these Jesuits obtained manuscripts containing Kerala mathematics, it does establish that their scientific investigations about local astronomy and calendrical sciences could have led them to an awareness of this knowledge. There are some reports that the Brahmins were secretive and unwilling to share their knowledge. However, this was not an experience shared by many others. For example in the mid–seventeenth century Fr. Diogo Gonsalves, who learned the local language Malayalam well, was able to write a book about the administration of justice, sciences, and mechanical arts of the Malabar. This book is to be found in the manuscript Goa 58 collection in the Jesuit historical library (ARSI), Rome. Also there is a report of a Brahmin who spent eight years translating Sanskrit works for Fr. Frois during the same time.[38]

The information gathering and transmission activities of the Jesuit missionaries are thus not in doubt. In addition, after the 1580 annexation of Portugal by Spain and subsequent loss of funding from Lisbon, the rationale for transmission acquired another dimension, that of profit. Whatever the nature of the profit, intellectual or material, the motivation may have been sufficient for the learned Jesuits to have acquired (or at the very least read and understood) the relevant manuscripts containing Kerala mathematics.

A Conjecture on the Mode of Acquisition of Manuscripts by the Jesuits

The question arises as to how the Jesuits might have obtained key manuscripts of Indian astronomy such as the *Tantrasangraha* and the *Yuktibhasa*.[39] It would require the Jesuits being in close contact with scholars who had access to such manuscripts. We know that at least one scholarly Brahmin was working for the Jesuits.[40] In addition, as we shall now show, the Jesuits were in communication with the members of the Court of Cochin, whose scholarship and authority may have enabled them to help the Jesuits to acquire Kerala mathematics.

The rulers of Cochin came from a scholarly family who were reputed to be knowledgeable about the mathematical and astronomical works of medieval Kerala, at least to the extent that they were aware of the methods for astrological prediction and of the manuscripts that contained these methods. They were in possession of a large number of manuscripts in mathematics and astronomy and were known for a tradition of helping other scholars. "Thus, the royal family could itself have been a possible source of knowledge for the Jesuits. Indeed, the Jesuits working on the Malabar Coast had close relations with the royal court of Cochin. Furthermore, around 1670, they were granted special privileges by Raja Rama Varma who, despite his misgivings about the evangelical work of the Jesuits, permitted members of his household to be converted to Christianity. The close relationship between the Raja of Cochin and the foreigners from Portugal was cemented by Rama Varma's appointment of a Portuguese as his tax collector" (The Aryabhata Group 2002, p. 47). Given this close relationship with the kings of Cochin, the Jesuits' desire to acquire local knowledge, and the royal family's contiguity to the works on Indian astronomy, it is quite possible that the Jesuits may have gained access to key manuscripts of the Kerala school via the royal household.

In conclusion, it is worth noting that we have focused so far on documentary evidence of direct transmissions of Kerala ideas to Europe and pointed to certain conjectures based on circumstantial evidence. It should be emphasized that a painstaking trawl of the mass of manuscripts and other materials mentioned earlier in this chapter and discussed in Almeida and Joseph (2009) has yielded *no direct evidence of the conjectured transmission.* Therefore, on the basis of the evidence in documents studied so far, we should conclude that the European Renaissance developments of prototypical calculus were independent of the developments in that subject in Kerala some centuries earlier. Baldini (2009), who has provided some of the more cogent arguments against transmission, concludes:

[U]nless new evidence is found and some basically new circumstance is established, the only possible deduction seems to be that not only no information exists on a Jesuit mathematician having managed to study some advanced Indian text (not to say to transmit it, or its content, to Europe), but no serious clue appears of a scientific interchange not purely superficial and more than occasional. (p. 288)

The debate is by no means over. As pointed out by Bala (2009), the transmission of the discoveries of Kerala mathematics could have been as "know-how" and computation techniques through the channel of craftsmen and technicians. This may explain the absence of direct documentary evidence in Jesuit communications. Even if there exists documentary evidence on the use of approximate series derived from the discoveries of the Kerala school in sixteenth- and seventeenth-century European manuals on navigation, map making, and calendar construction, it would hardly have been directly communicated to European mathematicians. After all, craftsmen oriented to practical rather than theoretical concerns would have been unlikely to write to leading mathematical figures or to be taken seriously if they did so. This raises the question as to whether it was possible to transmit the knowledge of infinite series through computations and calculations contained in navigation charts and similar aids, and if so what would be the precise nature of the calculation with series that could be transmitted. Only a closer look at ship records and other practical manuals would help to finally resolve the validity of Bala's hypothesis.[41]

Notes

1. See Aaboe and Berggren (1996, pp. 295–316) for further details. The paper also argues that Archimedes *may* have used infinitesimals to produce his results.

2. In this book, the term "school" is used in two different ways. In this instance, the "Kerala school," as shown in figure 10.1, describes the institution of *guru-parampara* or a sequence of direct transmissions from teachers to students living in the same locality, even if this happens over a number of generations. In the other sense, the "Aryabhata school" is a term used to describe those who were influenced by Aryabhata even if the influence lasted over a thousand years.

3. The traditional Indian method for determining latitude (discussed in the *Laghu Bhaskariya* of the seventh century AD) involved measuring solar altitude at noon (α), which is in turn a function of the latitude (λ) and the solar declination (i.e., the angular distance of the sun north or south of the equator) denoted by δ. At any given place, α varies with δ at different days of the year. The relationship between the three variables, according to the *Laghu Bhaskariya*, can be expressed as

$$\sin \delta = \sin \lambda \sin \alpha.$$

Thus to determine the latitude correctly, an accurate calendar giving the number of days that elapsed since the last equinox becomes an absolute requirement. The traditional

Indian day-count system (*ahargana*) achieved this purpose well, in contrast to the Julian calendar.

4. The *equation of the center* is equal to the difference between the actual angular position in the elliptical orbit and the position the orbiting body would have if its angular motion was uniform. It arises from the ellipticity of the orbit, and is zero at the pericenter (i.e., the point on the orbit nearest to the center) and the apocenter (i.e., the point on the orbit that is farthest from the center). The difference is at its greatest approximating midway between these points.

5. For further details, see Ramasubramanian et al. (1994).

6. For further details, see Hayashi and Kusuba (1998).

7. While other commentaries (such as those of Bhaskara I and Nilakantha on *Aryabhatiya*) contain some form of rationale and verification of the results, the proofs offered in *Yuktibhasa* and *Kriyakramakari* use complicated geometrical constructions to be found in no other mathematical traditions.

8. In Kerala at the time of the *Yuktibhasa*, "advanced" mathematics were used mainly in two areas, astronomy and building science (known as *vastu vidya* or *thachu sastra*). A contemporary of Jyesthadeva, Thirumangalath Neelakantan, had written a very popular text on traditional building science, *Manushyalaya Candrika*.

9. It would seem that there existed some caste flexibility that allowed the Brahmins and the non-Brahmins (i.e., *Ambalavasis* such as Variyars and Pisaratis) to share Sanskritic knowledge and collaborate on more or less equal terms.

10. The "octagon method" appears as one of the earliest methods of estimating the area of a circle. It is found in the Ahmes Papyrus of ancient Egypt and discussed in chapter 3. The "hexagon method" is of Greek origin and is usually associated with the "method of exhaustion." This method is discussed in chapter 7.

11. It should be noted that the main purpose of developing the mathematics of the sine and cosine power series was to provide a sounder basis for mathematical astronomy. To construct a sine table accurate even to seven decimal places, which is what Madhava achieved, a better value for π than that implied by the traditional Indian radius of 3,438′ was necessary. Madhava seems to have realized this. And it was his search for a more accurate estimate of the circumference of a circle in terms of its diameter that probably led him to the infinite series corresponding to that measure.

12. Chapter 6 of the *Yuktibhasa* discusses another procedure for estimating the circumference of a circle by approximating it to regular polygons. The method involves approximating the circle to regular polygons having an increasing number of sides by

an iterative process of finding at each stage the length of the sides of regular polygons of $2n$ sides from a regular regular polygon of n sides. Thus we could derive the length of the side of a regular octagon from a square, the length of the side of a sixteen-sided regular polygon from a regular octagon, of a thirty-two-sided polygon from a sixteen-sided polygon, and so on, such that when n becomes very large, the resulting polygon will approximate a circle. Known in the *Yuktibhasa* as the "square-square root method," it makes use of the relation between the sides of a right-angled triangle. For a useful discussion of this method and its derivation, as stated in the *Yuktibhasa*, see Rajasekhar (2009, pp. 113–36) and Sarma (2008, pp. 46–49, 180–83).

13. The following corrections (in ascending order of accuracy) are suggested for incorporation as the last term in equation (10.3b):

(1) $F_1(n) = \dfrac{1}{4n}$,

(2) $F_2(n) = \dfrac{n}{4n^2 + 1}$,

(3) $F_3(n) = \dfrac{n^2 + 1}{4n^3 + 5}$,

where n is the number of terms on the right-hand side of (10.3b).

   Historians have argued that continued factor approximations to the errors produced by successive partial sums lie behind Madhava's procedure. If that is so, Madhava was not only a skilled geometer but also showed remarkable numerical intuition reminiscent of Ramanujan five hundred years later, who had a lot in common with him. For a discussion of the possible derivations of the various correction terms, see Hayashi et al. (1990), Gupta (1992), and Sarma (2008, vol. 1, pp. 72–82).

14. One can only assume that of the three corrections given in note 15, the one used in this instance, is

$$F_2(n) = \frac{n}{4n^2 + 1}.$$

15. Sankara states that Madhava's value for the ratio of circumference to diameter is more accurate than the traditional value of 355/113 given in *Lilavati*.

16. For some of the more recent work outside India, see the relevant articles in M. Anderson et al. (2004).

17. Note that the trigonometric table relating to sines was constructed by a contemporary of Madhava, Ulugh Beg (1393/4–1449) of Samarkand, who achieved a similar level of accuracy to eight decimal places.

18. The role of the Jesuit conduit in India and China during the period of the Ming dynasty has become an interesting area of study in recent years. In this book, we identify two specific cases: a conjecture regarding possible transmission of the Chinese solution

to the mathematical problem of equal temperament in music through the agency of the Jesuits, discussed briefly in chapter 7; and, in the section that follows, a more detailed examination of possible transmission of mathematical and astronomical ideas from Kerala to Europe again through the Jesuits.

19. This section is based on the findings of a research project funded by the Arts and Humanities Research Board, UK, undertaken by Dennis Almeida and the author. For further details, see the Aryabhata Group (2002), Almeida et al. (2001), Almeida and Joseph (2004, 2007, 2009), and Bala (2009). For a short period, C. K. Raju was a member of the Aryabhata Group, which had been modeled along the lines of Nicolas Bourbaki, a collective nom de plume of mainly French mathematicians who aimed to publish an ambitious *Éléments de mathématiques*, a text in many volumes in which the fundamental structures of modern mathematics were to be treated in a rigorous fashion. The Aryabhata Group was dissolved in 2000, before the author began his collaboration with Dennis Almeida, the origins of which may be traced back to a discussion between us in July 1997 in South Africa of a Jesuit role in the spread of Kerala mathematics to Europe. Raju (2007) has recently brought out a book that contains his own interpretation of the "transmission of the calculus from India to Europe in the 16th century." It also makes claims regarding the nonexistence of a historical Euclid and his *Elements* before the tenth century AD and expresses similar skepticism about Archimedes and his works. In writing this section of the book, the author owes a considerable debt to the Aryabhata Group, whose work (2002) is extensively referred to.

20. In all common-law systems, the standard of proof in civil cases is on a "balance of probabilities," while in criminal cases, where the prosecution bears the burden of proof, the standard is proof "beyond reasonable doubt."

21. These letters have been published in eighteen volumes (Mersenne 1945–).

22. See Mersenne (1945–), vol. 13, p. 267.

23. See Mersenne (1945–), vol. 13, pp. 518–21.

24. See Mersenne (1945–), vol. 2, pp. 103–15. It is interesting in this context that Mersenne's classical work on musical theory explored the relationship between combinatorics and music theory, a topic covered by Pingala (third century AD) in *Chandasutra*, as stated in chapter 8. Further, as discussed in chapter 7, about thirty years before Mersenne set down the mathematical basis for the concept of equal temperament in music, the same theory was propounded in 1584 by Zhu Zaiyu. Mersenne's work came out sixteen years *after* the first reference to Zhu's work in Europe.

25. The mathematical representation that follows is taken from the paper by the Aryabhata Group (2002, p. 42), who in turn credit it to a paper presented by C. K. Raju in 1999 at the National Seminar on Applied Sciences in Sanskrit Literature, Various Aspects of Utility, Agra, February 20–22.

26. That is, Wallis replaces the term n^2 by $n(n + 1)$, implying by that, as n tends to ∞, $(n + 1)$ can be replaced by n. (Scott 1981, p. 30).

27. *Quadrature* involves the process of determining the area of a plane geometric figure by dividing it into a collection of shapes of known area and then finding the limit (as the divisions become ever finer) of the sum of these areas. A similar process called *rectification* is used in determining the length of a curve. The curve is divided into a sequence of straight-line segments of known length, and the sum of a large number of these segments gives an estimate of the length of the curve.

28. The navigational problems in Europe related to difficulties of obtaining (1) an accurate measurement of latitude (the prerequisites being an accurate calendar and accurate trigonometric values, as explained in note 3 above); and (2) an accurate method of determining longitude (the prerequisite before the invention of the marine chronometer being an accurate measurement of the size of the earth). In the absence of (1) and (2) the problem was one of tracing the path of a ship along a constant course (a rhumb line or loxodrome), given a compass that was often unreliable, a chart that was often incomplete and inaccurate, and the lack of any identifiable star or celestial body (such as the fixed polestar, which would disappear from the horizon once the ship reached the equator). For further explanation of the difficulties faced by European navigators during the fifteenth and sixteenth centuries, see Waters (1958).

29. In traditional Indian timekeeping, a *tithi* is a lunar day (i.e., the time it takes for the longitudinal angle between the moon and the sun to increase by 12°). *Tithis* begin at different times of day and vary in duration from approximately 19 to 26 hours. The *tithi* is a basic measure of the Indian lunisolar calendar (i.e., a calendar whose date shows both the moon phase and the time of the solar year).

30. "The eccentric version of [Manjula's] *Laghumanasam* model has an interesting subsequent history. Essentially the same model, with the small epicycle rotating through the angle 2ψ, was used by Kepler at an intermediate stage of his lunar research, and then abandoned" (Duke 2007, p. 157).

31. These offers by governments included the Spanish prize of 1567, the Dutch prize of 1636, the French prize of 1666, and the English prize of 1711. Portugal had earlier instituted a special post for navigational studies in Lisbon, first occupied by Pedro Nunes as early as 1529. And among the first tasks that the infant Royal Society of London and the Royal Academy of Paris set for themselves was improving the navigational methods of their respective countries.

32. Baldini (1992, p. 70) wrote: "It can be recalled that many of the best Jesuit students of Clavius and Geienberger (beginning with Ricci and continuing with Spinola, Aleni, Rubino, Ursis, Schreck, and Rho) became missionaries in the Oriental Indies. This made them protagonists of an interchange between the European tradition and those

of India and China, particularly in mathematics and astronomy, which was a phenomenon of great historical meaning."

33. Ferroli (1939, p. 402) wrote: "In Portuguese India, hardly seven years after the death of St. Francis Xavier the fathers obtained the translation of a great part of the 18 *Puranas* and sent it to Europe. A Brahmin spent eight years in translating the works of *Veaso* [*Vyasa*].... several Hindu books were got from Brahmin houses, and brought to the Library of the Jesuit college. These translations are now preserved in the Roman Archives of the Society of Jesus" (*Goa*, 46).

34. Letter by Matteo Ricci to Petri Maffei dated December 1, 1581 in Josef Wicki, *Documenta Indica*, vol. 12, pp. 472–77 (p. 474).

35. "I have sent Valignano a description of the whole world by many selected astrologers and pilots, and others in India, which had no errors in the latitudes, for the benefit of the astrologers and pilots that every day come to these lands, because the maps of theirs are all wrong in the indicated latitudes, as I clearly saw" Josef Wicki, *Documenta Indica*, vol. 11, p.185.

36. " . . . comparing the real local times with those inferable from the ephemeridis [tables] of Magini, he [Rubino] found great inaccuracies and, therefore, requested other ephemeridi" Baldini (1992, p. 214).

37. For details, see Iannaccone (1998, p. 58).

38. For details, see Ferroli (1939, vol. 2, p. 402).

39. This section takes as a starting point the discussion of this conjecture in the paper by the Aryabhata Group (2002, pp. 46–47), which should be referred to for further details.

40. See note 32 above.

41. There is the further question as to how necessary it was to obtain increasingly accurate values for sines, cosines, and π from a practical point of view. Opinions are divided as to whether accuracy up to ten or more decimal places was required for navigational purposes. Of course, it is always possible that a "delight in accurate calculation" may have driven the Kerala mathematicians to attempt increasingly accurate approximations.

Chapter Eleven

Prelude to Modern Mathematics:
The Islamic Contribution

Historical Background

The year AD 622 is a momentous one in world history. It was then that
the Prophet Muhammad fled from Mecca and took refuge in Yathrib (now
Medina) about 350 kilometers away. He had incurred the wrath of pilgrims
who had come to worship at a shrine called the K'aba—a shrine then dedicated to many gods. Muhammad's preaching of a monotheistic faith, which
he claimed had been directly revealed to him by the Archangel Gabriel,
had aroused considerable hostility, contributing to his decision to flee his
birthplace. Eight years later he returned at the head of an army, and two
years after that he died. But he had already created a whirlwind that would
eventually lead to the establishment of Islamic rule[1] over areas stretching
from North Africa in the south to the borders of France in the west, right
across Persia and the central Asian plains to the borders of China in the
east, and down to Sindh in northern India. Figure 11.1 shows the extent of
the empire at its height, and the location of places referred to in the course
of this chapter. Much of this vast territory was brought under Islamic rule
in less than a hundred years.

Such a rapid expansion was possible for two reasons. First, there was
something quite irresistible about the passion and egalitarian character of
early Islam that fired the imagination and won the devotion of many who
came across it for the first time. And second, in a number of areas through
which the forces passed, local rulers were so unpopular with their subjects that the conquering armies were welcomed as liberators. With the
physical conquest completed, the government of this vast territory passed
into the hands of *khalifa* (caliphs), who were the Prophet's deputies and
whose role included duties such as leading the army into battle or solving
legal disputes. They belonged to one of two dynasties: the Umayyad and

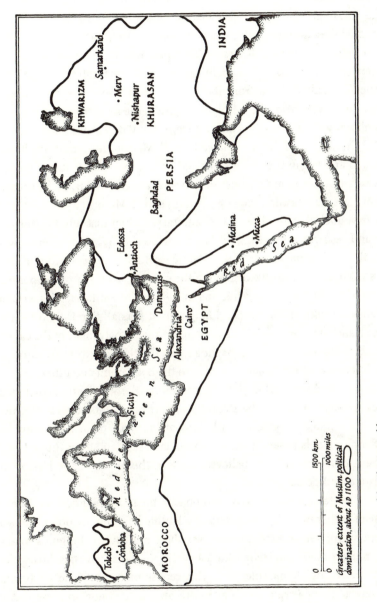

FIGURE 11.1: Map of the Islamic world

the Abbasid. The Umayyads were the early rulers of the Eastern empire, with their capital at Damascus, but in the year 750 they were overthrown and power passed to the Abbasids. This was not the end of the Umayyad dynasty, though. Among the few who escaped was a young man of twenty named Abd al-Rahman, who reached Spain and reestablished Umayyad power there. For the next three centuries Spain was to be the center of Muslim power in the West, with its political and intellectual capital at Cordoba.

The Abbasids differed from the Umayyads in one important respect. While both came from the Arabian Peninsula, the former were more cosmopolitan, welcoming new converts from many different ethnic groups. In 762 the second of the Abbasid caliphs, al-Mansur, moved his capital to Baghdad and began the process of building it into a new center of power. This ambitious program of construction was carried out during the caliphates of Harun al-Rashid (786–809) and his son al-Ma'mun (809–833), and may have included an observatory, a library, and an institute for translation and research named Bait al-Hikma (House of Wisdom), which was to be the intellectual center of the Islamic world for the next two hundred years.[2] Within its walls lived some of the greatest scientists of the period. It housed translators, busy rendering into Arabic scientific classics written in Sanskrit, Pahlavi (the classical language of Persia), Syriac, and Greek. These translations were often carried out under the patronage of the caliph himself, or notable families in Baghdad, and the classic texts were acquired in different ways, by scholars traveling to the Byzantium and other territories or by caliphs as part of peace negotiations. Early collections included Greek manuscripts from the Byzantium. Some accounts suggest that the caliphs were able to obtain translations of Mesopotamian astronomy by the Syriac schools based in Antioch and Damascus and even the remains of the Alexandrian library believed to be in the hands of the Nestorian Christians at Edessa.

It is a mistake, however, to overemphasize the role of Baghdad at the expense of earlier pre-Islamic centers of scientific learning. By doing so, two questions are left unresolved: why was there a greater willingness to accept Indian rather than Hellenistic astronomy and mathematics during the earlier period of Islamic rule? And why were so many of the early scholars in Baghdad from a region such as Khurasan, which is now in present-day Iran and Afghanistan?

The answers to these apparently rhetorical questions are simpler than one might think. The researchers' interest focused at the beginning on

Sasanian astrology, which drew on Hellenistic as well as Sanskrit sources; the Iranian base of the Abbasids and their Iranian astrologers explain the preference for this kind of astrology and its astronomical aspects. Closely related to this interest was the translation of Sanskrit sources as far as they were done in Baghdad. Other translations were made in Sind and maybe Kashmir, which were all parts then of eastern Iran. The answer to the second question is that the Abbasids recruited their power base in eastern Iran and many of their followers came with them to Iraq and settled there. Because of their Iranian military and ideological setup, people from other parts of Iran joined them.

Long before the Islamic conquest, there were scientific and translation centers in Syria as well as in Sasanian Iran. In astronomy, for example, the first translation of Ptolemy's *Almagest* and its important commentary by Theon of Alexandria (c. AD 300) was from Greek into Syriac. Three important components of Persian astronomy were carried into the Islamic period:

1. Syriac astronomy, inspired mainly by Hellenistic influences and notably Ptolemy

2. Pre-Sasanian astrology, dating back to the Babylonian astronomy of the Seleucid period and earlier

3. Indian astronomy transmitted to central Asia, probably during the first and second centuries AD, when such regions as Parthia and Bactria, as well as northwestern India, were part of the great Kushan empire

Against this background was composed the first *zij* of Sasanian Iran. The word *zij*, probably a distortion of the Pahlavi word *zeh* (bowstring), came later to be applied to Islamic works on calendar construction and tables of movement of the sun, moon, and planets, as well as trigonometric and geographical tables.[3]

It is possible to distinguish three main influences that went into the creation of medieval Islamic mathematics. The first was Greek mathematics, and notably the geometrical works of Euclid, Apollonius, and Archimedes, followed chronologically by the work of Diophantus with his solutions of indeterminate equations and concluding with Heron, who wrote on practical mathematics, and Pappus, whose best-known work, *Synagoge*, is a compendium of mathematics of which eight volumes survive. The second

influence was that of the Indians, with their ingenious arithmetical calculations based on a numeral system consisting of nine symbols and a dot for
zero. The Indians also contributed ideas on algebraic notations and solutions, an early trigonometry of sines and cosines that extended the work of
Ptolemy, and finally methods of solid geometry useful in solving problems
in astronomy.[4] The third influence, often ignored in historical discussion,
is the mathematics of practitioners such as surveyors, architects, builders,
merchants, and government officials. Their mathematics often formed a
part of an oral tradition that transcended ethnic and linguistic divisions
and constituted the common heritage of the Islamic world and beyond. By
the very nature of oral transmission, it is difficult to trace the origins and
impact of such traditions on the mathematical culture of the day.

Thus the scientific culture that developed in Baghdad arose from an interaction of these different traditions. By far the greatest contribution of
the Islamic culture, as we shall see, was to continue this creative synthesis.
This the scholars pursued with an openness of mind and a clearer understanding than had been shown by any of the earlier scientific cultures of
the need to balance empiricism and theory in mathematics and other sciences. The process of synthesis was aided by the creative tension between
two main traditions of astronomy and mathematics represented in Baghdad, even from the early years of Islamic rule. One tradition was derived
directly from Indian and Iranian sources and is best exemplified in the
astronomical tables and the algebraic approach to mathematics. One of
the greatest exponents of this tradition, who left an indelible mark on the
subsequent development of Islamic mathematics, was al-Khwarizmi. To
him mathematics had to be useful and help with practical concerns such as
determining inheritances, constructing calendars, or informing religious
observances. The other tradition looked to Hellenistic mathematics, with
its strong emphasis on geometry and deductive methods. A well-known
proponent of this school was Thabit ibn Qurra, who was both an outstanding translator of Greek texts and an original contributor to geometry and
algebra. That the two traditions eventually merged is evident in the work of
later Islamic mathematicians such as Omar Khayyam and al-Kashi.

Our knowledge of medieval Islamic mathematics is mainly from documents recorded with pen and paper in the Arabic language. Chinese prisoners taken at the battle of Atlakh in AD 751 had shown their captors how
to manufacture paper, and this technology had spread rapidly throughout the Islamic world. Extant collections of these documents are found

not only in those countries which formed part of the medieval Islamic world but also in countries of Europe that had at some time exerted colonial domination. The manuscripts contain not only prose compositions but also a large number of tables of numbers, often computed for astronomical purposes and illustrating certain mathematical procedures. Apart from this documentary evidence, artifacts in the form of mathematical or astronomical instruments have increasingly become important sources of evidence of medieval Islamic mathematics.

We shall now proceed to examine the contributions of medieval Islamic mathematicians in three areas in which the results of their creativity and synthesis are most apparent: the introduction and popularization of our present-day numerals, the bringing together of the geometric and algebraic approaches to the solution of equations, and the first systematic treatment of trigonometry. First, though, let us look at the lives and achievements of the mathematicians themselves.

Major Medieval Islamic Mathematicians

Muhammad ibn Musa al-Khwarizmi (c. 780–850)

Abu Jafar Muhammad ibn Musa al-Khwarizmi (to give him his full name, which means Muhammad, the father of Jafar and the son of Musa, from Khwarizm) was born in about 780. The name "al-Khwarizmi" suggests that either he or his family came from Khwarizm, east of the Caspian Sea in present-day Uzbekistan. Little is known of his early life. There is a reference to al-Khwarizmi as "al-Majusi" in a book titled *History of Envoys and Kings* by al-Tabari, an Islamic historian who lived between the ninth and tenth centuries. Now, a Zoroastrian was sometimes referred to as a *magos* (i.e., belonging to the Magi, a Median tribe). From that word is derived our words "magi" and "magician." There is, therefore, the view that al-Khwarizmi may have been of Zoroastrian descent and acquired his early knowledge of Indian mathematics and astronomy from Zoroastrian clergy, some of whom were reputed to be well acquainted with these subjects. However, there is some uncertainty about his origins. For example, there is a report by the historian Tabari that al-Khwarizmi came from a town not far from Baghdad.

In about 820, he was invited by Caliph al-Ma'mun to move to Baghdad, where he was appointed first astronomer and then head of the library at the House of Wisdom. He continued to serve other caliphs, including al-Wathiq during his short rule from 842 to 847. There is a story, told by the historian al-Tabari, that when al-Wathiq lay seriously ill, he asked al-Khwarizmi to cast his horoscope and find out whether he would live. Al-Khwarizmi assured the caliph that he would live another fifty years, but al-Wathiq died within ten days. Whether this story illustrates al-Khwarizmi's highly developed sense of survival or his ineptness as a fortune-teller, it is difficult to say. We know little else of his later life except that he probably died before 850.

However, we do have information on his scientific work. A bibliographer, Ibn al-Nadim, lists four astronomical works: the *Zij al-Sindhind* (an astronomical handbook according to the *Sindhind*), a treatise on the sundial, and two works on the astrolabe. Of these, the first is no longer extant in Arabic but available in Latin translation; the second seems to be extant, as are fragments of a work on the astrolabe. Of his mathematical works, the two most influential were *Hisab al-jabr w'al-muqabala* (Calculation by Restoration and Reduction), and *Algorithmi de numero indorum* (Calculation with Indian Numerals). The original Arabic version of the latter no longer exists, and so we have it only in Latin translation. The first, hereafter known as the *Algebra*, was the starting point for Islamic work in algebra, and indeed gave the subject its name. It is an interesting blend of a variety of mathematical traditions including the Mesopotamian, Indian, and Greek.[5] The second book, which we shall call the *Arithmetic*, served to introduce the decimal positional number system developed in India a few hundred years earlier. It was also the first book on arithmetic to be translated into Latin, and gave currency to the word "algorithm," derived from the name of the author and frequently used today to denote any systematic procedure for calculation.[6]

Al-Khwarizmi also constructed a *zij* (i.e., a set of astronomical tables) that was to remain influential in astronomy for the next five centuries.[7] The antecedents of this *zij* are interesting, for they are indicative of the shadowy path through which Indian mathematics and astronomy entered the Islamic world. The first Arabic translations of Indian astronomical texts were made in the later Umayyad and early Abbasid caliphates in Sind and Kashmir, although little is known about the impact of these translations. A historian, al-Qifti (c. 1270), reported that in the year AH 156 (or AD 773)[8] a man well versed in astronomy, by the name of Kanaka (or possibly Ganaka, meaning an astrologer or a calculator) came to Baghdad as a member of a diplomatic mission from Sind, in northern India. It is possible that he brought with him Indian astronomical texts, including *Surya Siddhanta* and the works of Brahmagupta. Caliph al-Mansur ordered that some of these texts be translated into Arabic and, according to the principles given in them, that a handbook be constructed for use by the court astronomers. The task was delegated to al-Fazari, who produced a text that came to be known by later astronomers as the *Great Sindhind*.[9] The word *sindhind* is derived from the Sanskrit word *siddhanta*, meaning a "doctrine or teaching." It was mainly on the basis of such texts, as well as some other elements from Babylonian and Ptolemaic astronomy, that al-Khwarizmi constructed his *zij*. Unfortunately, the original Arabic text is no longer extant. But a Latin translation, made in 1126 from an edited version produced by Maslama al-Majriti (a Spanish astronomer who lived in Cordoba in about the year 1000), became one of the most influential astronomical texts in medieval Europe and elsewhere.[10]

Finally, there is al-Khwarizmi's geographical work, in particular his contribution to cartography. He was believed to have been a member of a team that was given the following tasks by Caliph al-Ma'mun:[11]

1. To measure the length of one degree of longitude at the latitude of Baghdad (the result obtained was quite accurate, at 91 kilometers in modern measurement)

2. To use astronomical observations to find the latitude and longitude of 1,200 important places on the earth's surface, including cities, lakes, and rivers

3. To collate the personal observations of travelers on the physical features of different areas of the caliphate and traveling times between them.

Al-Khwarizmi incorporated some of these in his book *The Image of the Earth*, which contains substantial parts of Ptolemy's *Geography* with many non-Ptolemaic coordinates and place-names. He corrected Ptolemy's overestimate of the length of the Mediterranean Sea by the simple device of moving the prime meridian by ten degrees, and provided detailed and accurate descriptions of the geography of Asia and Africa.[12]

In other fields, al-Khwarizmi wrote a history of the Islamic caliphates that contained horoscopes of prominent persons. A number of minor works on topics such as the astrolabe, the sundial, and the Jewish calendar showed his versatility.[13] His mathematical texts are still recommended reading in some Islamic countries, not for their mathematical content but for their legal acumen. His book on algebra contains an analysis of property relations, the distribution of inheritance according to Islamic law, and rules for drawing up wills. He was but the first of a succession of remarkable scientists who contributed to some of the most significant scientific discoveries of all time. These included Thabit ibn Qurra (836–901), al-Razi (c. 865–901), Ibn al-Haytham (c. 965–1039), al-Biruni (973–1051), Ibn Sina (980–1037), Umar al-Khayyami or Omar Khayyam (c. 1048–1126), al-Tusi (1201–1274), and al-Kashi (d. 1429), some of whom are better known in the West by their Latin names. Let us look very briefly at one of these who was not primarily a mathematician but whose contributions to mathematics will be noted in later sections.

Ibn al-Haytham, known in Latin as Alhazen, wrote more than two hundred books on mathematics, physics, astronomy, and medicine, and commentaries on Aristotle and Galen. But his major work was in optics. It included an early account of refraction, the mathematics of finding the focal point of a concave mirror, and a refutation of the theory put forward by both Euclid and Ptolemy that human vision works by the eye sending out rays to the object observed. Ibn al-Haytham's work had a significant impact on Roger Bacon and Johannes Kepler.

Thabit ibn Qurra (c. 836–901)

Hasan Thabit ibn Qurra Marwan al-Harrani was born in Harran in al-Jazira (northern Mesopotamia), probably in 836, and died in 901. Little is known of his early life except that when he reached adulthood he became a money changer. Thabit belonged to a religious sect that was believed to be descended from Babylonian star worshippers and that produced eminent scholars in both astronomy and mathematics. Members of the sect called themselves Sabeans (after a Chaldean sect that was later designated as "the People of the Book") to avoid being persecuted as polytheists. Either his unorthodox religious beliefs or a quarrel with his community led him to leave Harran and head for Baghdad. He had been befriended by a member of a wealthy and influential Baghdad family, the Banu Musa (Sons of Musa), who invited him to come to Baghdad, where he joined a circle of scholars and translators. Thabit's command of languages, namely Arabic, Greek, and Syriac, soon established him as one of the foremost translators in Baghdad. His translations (as well as corrections of translations by

Hunayun ibn Ishaq) included Greek mathematical texts such as Euclid's *Elements*, several works by Archimedes, parts of Apollonius's *Conics*, and Ptolemy's *Almagest*. They were in turn rendered into Latin by Gherardo of Cremona and Adelard of Bath in the twelfth century, in which form they were to have a momentous impact on medieval Europe.

Thabit's passion for translation led him to set up a school for translators in Baghdad, the members of which included the son of his distinguished collaborator, Hunayun ibn Ishaq, and some of the greatest translators of the day. Two of his grandsons who were physicians followed the family tradition and became mathematicians in their own right. Ibrahim ibn Sinan ibn Thabit ibn Qurra's commentary extending Archimedes' work on the quadrature[14] of the parabola has been described as one of the most innovative approaches known before the emergence of the novel technique of integral calculus. Whether it was the range of translations that he undertook or the unusual breadth of mind possessed by the scholars of the period, Thabit himself became highly competent in a number of subjects that included—apart from mathematics—medicine, astronomy, philosophy, theology, and meteorology. However, it is for his work in mathematics that he is best remembered.

His notable contributions in mathematics included the rule for discovering pairs of "amicable numbers" (discussed in a later section); a "dissection" proof of the Pythagorean theorem, strangely reminiscent (no "connection" intended!) of the Chinese approach, which we discussed earlier; work on spherical trigonometry; an attempt to prove Euclid's parallel postulate;[15] and his work on conic sections and mensuration of parabolas and paraboloids, which some look upon as providing the essential link between Archimedes and later European mathematicians such as Cavalieri, Kepler, and Wallis. As a geometer he had few equals in the Islamic world, and it is clear from the above list that his algebraic strength was also considerable. Clearly, we shall not be able to explore all the contributions of this remarkable polymath, but we shall consider briefly some of his geometric work and his rule for generating amicable numbers.

Omar Khayyam (c. 1040–1123)

The *Ruba'iyat of Omar Khayyam*, a number of quatrains (verses of four lines) freely translated into English by Edward Fitzgerald in the middle of the nineteenth century, is one of the best-known and most translated books in world literature. But what is not widely known outside the Islamic world is that the poet was also a distinguished mathematician, astronomer, and philosopher.

Abul-Fath Umar ibn Ibrahim al-Khayyami was born in about 1040 at Nishapur in Khurasan, now part of Iran and Afghanistan. This region had already produced two distinguished figures—Firdausi (c. 940–1020), a poet, and ibn Sina. It is quite possible that these two men in different ways had a considerable influence on the young Omar. The name al-Khayyam would indicate that either Omar or his family were tent makers. Little else is known about his childhood or his youth.

In 1070, he wrote his great work on algebra. In it he classified equations according to their degree and gave rules for solving quadratic equations, which are very similar to the ones given by his predecessors. The true importance of his algebra lies in his geometric theory of cubic equations. (We shall be looking at Omar's solution of cubic equations in a later section.) He also wrote on the triangular array of binomial coefficients known as Pascal's triangle.

In 1074, Omar was appointed by Sultan Malik Shah as one of the eight learned men involved with the task of revising astronomical tables and reforming the calendar. They produced a new calendar, according to which eight out of every thirty-three years were made into leap years. This treatment produces a more accurate measure of a solar year than does our Gregorian calendar year.[16]

Three years later, Omar wrote *Sharh ma ashkala min musadarat kitab Uqlidis* (Explanations of the Difficulties in the Postulates of Euclid). An important section of this book is concerned with Euclid's famous parallel postulate, which had also attracted the interest of Thabit ibn Qurra. Al-Haytham too had attempted a demonstration of the postulate; Omar's attempt was also unsuccesful, although it marked a distinct advance on his predecessors.[17]

Omar Khayyam died in Nishapur in 1123. Unlike the image of him we may get from the *Rubai'yat*, of a hedonist who lived only for the present, he was a scholar, a poet, a Sufi, and a gnostic. He was that rare combination—an outstanding poet and a mathematician.

Jamshid al-Kashi (b. unknown–1429)

Ghiyath al-Din Jamshid al-Kashi was born at Kashan, a town in Iran not far from Isfahan, in the latter half of the fourteenth century. A short period of turmoil had followed the Mongol invasion under Hulegu Khan, grandson of Genghis Khan. The caliphate at Baghdad was destroyed in 1258, and a century later a new empire was created under Timur (also known as Tamburlaine or Timur the Lame). But mathematical activities continued throughout this period, probably because of the patronage offered by Hulegu Khan to astronomers of the caliber of al-Tusi, and by Ulugh Beg, grandson of Timur, to a later group that included al-Kashi.[18]

Little is known of al-Kashi's life until 1406, when he began a series of observations of lunar eclipses from his birthplace, Kashan. At Samarkand, Ulugh Beg had established an observatory and a madrassa (a school of advanced study in science or theology), and it is probable that al-Kashi was invited to join a group of scientists there. We know that in 1414 he revised a set of astronomical tables produced by al-Tusi and dedicated it to Ulugh Beg, who was a knowledgeable astronomer himself. Samarkand under the rule of Ulugh Beg had become an intellectual center where, as al-Kashi observed in a letter to his father, "the learned are gathered together, and teachers who hold classes in all the sciences are at hand, and the students are all at work on the art of mathematics."

Al-Kashi's strength lay in prodigious calculations. His approximation for π, correct to sixteen decimal places, was obtained by circumscribing a circle by a polygon having 3×2^{28} (805,306,368) sides. His best-known work, *Miftah al-hisab* (The Key of Arithmetic), completed in 1427, provides us with a compendium of the best of Islamic arithmetic and algebra. Its contents include the first systematic exposition of decimal fractions; a method of extracting the nth root of a number, similar to the so-called Horner's method and thus probably derived from the Chinese; and the solution of a cubic equation to obtain a value for the sine of one degree. On al-Kashi's death in 1429, Ulugh Beg praised the mathematician's achievements, in the preface to his own *zij*, calling him "the admirable mullah known among the famous of the world, who had mastered and completed the science of the ancients, and who could solve the most difficult problems."

Medieval Islam's Role in the Rise and Spread of Indian Numerals

In chapter 8 we saw how the Indian numerals evolved and spread to Southeast Asia; we now take up the story of their spread westward. The Islamic world contained the leading actors in this drama. The first evidence of the westward migration of Indian numerals is found in the following (rather aggrieved) passage from the fragments of a book in Syriac. It was written in 662 by a Nestorian bishop, Severus Sebokht, who came from Keneshra in the upper reaches of the Euphrates. He had written previously on both geography and astronomy, and hurt by the arrogance of some Greek (or Byzantine) scholars who looked down on his people, he wrote:

> I will omit all discussion of the science of the Indians, a people not the same as the Syrians; of their subtle discoveries in astronomy, discoveries that are more ingenious than those of the Greeks and the Babylonians; and of their valuable methods of calculation which surpass description. I wish only to say that this computation is done by means of nine signs. If those who believe, because they speak Greek, that they have arrived at the limits of science, [would read the earlier texts], they would perhaps be convinced, even if a little late in the day, that there are others also who know something of value.

This supports the view that, even before the beginning of Islamic rule, knowledge of Indian numerals had spread westward, probably as a result of widespread interest in Indian astronomy. Christian sects, particularly the Nestorian and Syrian Orthodox denominations, needed to calculate an accurate date for Easter, and various astronomical texts were examined with this problem in mind. (It was a problem that continued to occupy mathematicians, including Gauss, down to the nineteenth century.) There is also the possibility, given the thriving commercial relations between Alexandria and India, that the Indian numeral system had reached the shores of Egypt as early as the fifth century AD. It would have been regarded as a useful commercial device rather than a system that might become more widely used or accepted; it would not have been adopted for scientific and astronomical calculations by Alexandrian scientists, who used the Mesopotamian sexagesimal system.

After the Islamic conquest, Indian numerals probably arrived at Baghdad in 773 with the diplomatic mission from Sind to the court of al-Mansur. Around 825, al-Khwarizmi wrote his famous *Book of Addition and Subtraction according to the Indian Calculation*, the first text to deal with the new numerals. As mentioned earlier, although the original Arabic text is now lost, it gave rise to a whole genre of works in Arabic, Latin, and Greek. Europe came to know it only through several partial Latin translations undertaken five hundred years later by John of Seville and Robert of

Chester. The text contained a detailed exposition of both the representation of numbers and operations using Indian numerals. Al-Khwarizmi was at pains to point out the usefulness of a place-value system incorporating zero, particularly for writing large numbers.

The earliest extant texts to examine arithmetical operations with Indian numerals are Abul Hassan al-Uqlidisi's *Kitab al-fusul fil-hisab al-Hindi* (The Book of Chapters on Indian Arithmetic, 952) and Kushyar ibn Labban's *Usul hisab al-Hind* (Principles of Indian Reckoning, c. 1010). Operations are given in both the Indian decimal place-value system and the Babylonian sexagesimal system. Al-Uqlidisi's book is particularly notable for the first use of decimal fractions in computing with the new numerals; Ibn Labban introduces "Indian" reckoning as part of a general discussion of Indian methods used in astronomical calculations. Ibn Labban's work became an influential arithmetic textbook in the Islamic world. Some aspects of both works will be highlighted in the next section.

Other references to Indian numerals are found in works by later writers. The opinions of the tenth-century polymath Abu Rayhan al-Biruni are particularly valuable since he lived in India and knew Sanskrit. Two of his books, *Risalah* (Book of Numbers) and *Rasum al-Hind* (Indian Arithmetic) contain an assessment of Indian numeration as well as some corrections to earlier works on the subject.

A minor incident from the autobiography of Ibn Sina (c. 980–1037) shows us how the use of the Indian numerals was spreading. When he was about ten years old a group of missionaries belonging to a small Islamic sect came to Bukhara from Egypt, and it was from these people that Ibn Sina learned "Indian arithmetic." There is also a story of the young Ibn Sina being taught "Indian calculation" and algebra by a vegetable vendor. What these stories illustrate is that by the beginning of the eleventh century Indian numeration was being used from the borders of central Asia to the southern reaches of the Islamic empire in North Africa and Egypt—and not just by scholars.

In the transmission of Indian numerals to Europe, as with almost all knowledge obtained from the Islamic world, Spain and (to a lesser extent) Sicily played the role of intermediaries, being the two areas in Europe that had been under Islamic rule for many years. (This was one of the important aspects emphasized in our examination of the spread of mathematical knowledge in chapter 1.) So it is not surprising to find that the oldest record of Indian numeration in Europe, dating from the year 976, is found in a monastery in northern Spain. This manuscript, known as the *Codex Vigilanus*, is now kept in a museum in Madrid. The relevant passage reads:

> So with computing symbols. We must realize that the Indians had the most penetrating intellect, and other nations were way behind them in the art of computing, in geometry, and in other free [? probably meaning natural] sciences. And this is evident from the nine symbols with which they represented every rank of number at every level.

There follows a set of symbols, now known as the West Arabic or *Ghubar* (Gobar) numerals, from which our present numerals derive. The shapes of these numerals are shown in figure 11.2, which outlines the evolution of our numerals from some of the earlier forms. The Indian numerals from which the two main forms of Arabic numerals (East and West) were derived quite likely resembled those found in the Gwalior inscription of 876, which we discussed in chapter 8. The western version of the Arabic numerals that stemmed from Indian figures were called *Ghubar* numerals—presumably because, as the word *Ghubar* suggests, these symbols were written on a sand board containing dust, a practice that was popular in India. The *Ghubar* numerals were widely used in the western part of the Islamic empire, including Spain and Sicily; indeed, they are still found in parts of North Africa. The eastern Arabic numerals may have come to the Islamic world by a more indirect route that included Persia. In the early years the differences between the two types of Arabic numerals were slight, but they grew with the passage of time. A striking feature of the evolution of Indian numerals from the Gwalior script to the present form, as shown in figure 11.2, is how little they have changed on passing through one culture after another. In several instances what changed was the orientation of a symbol, not its form.

The oldest date to appear in the new numerals in Europe is on a Sicilian coin from the reign of the Norman king Roger II. On it the year is expressed as AH 533 (AD 1138). The use of the Muslim date is not surprising, since Roger II encouraged the pursuit of Islamic learning in his kingdom.

We now come to a landmark in the spread of Indian numerals: the appearance of one of the most influential mathematical texts in medieval Europe, the *Liber Abaci* (Book of the Abacus) by Fibonacci (1170–1250). The young Fibonacci grew up in North Africa, where his father was in charge of a customshouse. He was first introduced to Indian numerals by his Islamic teachers there. As a young man, he traveled extensively around

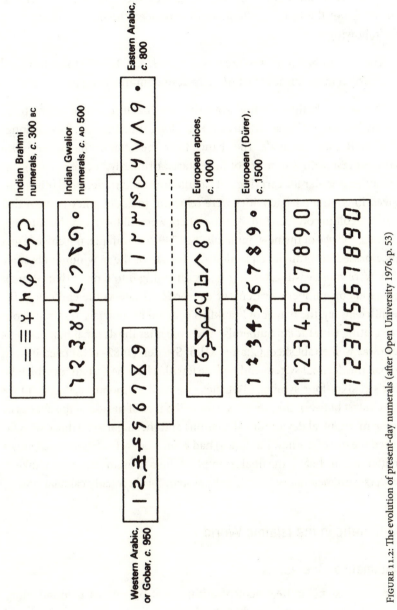

FIGURE 11.2: The evolution of present-day numerals (after Open University 1976, p. 53)

the Mediterranean, visiting Egypt, Syria, Greece, Sicily, and southern France, observing the various computational systems used by merchants, particularly in the Islamic world. He quickly recognized the enormous advantage of the Indian system, and introduced the new numerals with the following words:

> The nine Indian numerals are 9, 8, 7, 6, 5, 4, 3, 2, 1. With these nine and 0, which in Arabic is called *sifr*, any desired number can be written.

It was mainly through this work that the Indian numerals came to be widely known in Christian Europe. For a long time they were used alongside the Roman numerals. The change from the latter to the former was a slow process with a number of false starts, primarily because the abacus remained popular for carrying out calculations, and traders and others engaged in commercial activities were reluctant to adopt a new system that was difficult to comprehend. At times there were diktats from above to discourage the use of the new numerals. In 1299, for example, the city of Florence passed an ordinance prohibiting the use of the new numerals since they were more easily altered (e.g., by changing 0 to 6 or 9) than Roman numerals or numbers written out in words. As late as the end of the fifteenth century, the mayor of Frankfurt ordered his officials to refrain from calculating with Indian numerals. And even after the decimal numeral system was well established, Charles XII of Sweden (1682–1718) tried in vain to ban the decimal system and replace it with a base 64 system for which he devised sixty-four symbols! But these were all temporary setbacks. Once the contest between the "abacists" (those in favor of the use of the abacus or some mechanical device for calculation) and the "algorists" (those who favored the use of the new numerals) had been won by the latter, it was only a matter of time before the final triumph of the new numerals, with bankers, traders, and merchants adopting the system for their daily calculations.

Arithmetic in the Islamic World

Arithmetical Operations

The first systematic treatment of arithmetical operations is found in al-Khwarizmi's *Arithmetic*, in which he discusses the place-value system and rules for performing the four arithmetical operations. In later works, notably those of Ibn Labban and al-Uqlidisi referred to in the previous section, there are computational schemes for carrying out operations, such as long multiplication, which were reproduced in medieval Latin texts. Al-Kashi's *Key of Arithmetic* presents a comprehensive treatment of arithmetical methods, including operations with decimal fractions (to be discussed later).

We have not the space here to examine in any detail how the mathematicians of the Islamic world carried out the basic operations; in any case, their pratices were not very different from ours. However, we shall look at a method they popularized for long multiplication known as the "sieve" or "lattice" method, which is of historical significance. Ibn Labban discusses the method, probably of Indian origin, in his *Principles of Indian Reckoning*. Traces of the method are found later in the "grating" methods explained in the *Treviso Arithmetic* of 1478, in the mechanical device known as "Napier's rods" or "Napier's bones" after its inventor, John Napier (1550–1617) of Scotland, and the nineteenth-century Japanese Sokuchi method.[19] Even today the "lattice" provides a useful diversion in learning long multiplication.

Figure 11.3 shows how the lattice method is used to multiply 1,958 by 546. The numbers to be multiplied are entered as shown. The number of paths from (a) to (g) is 7, the sum of the digits in the two numbers to be multiplied. The product to be entered in each cell is obtained by multiplying the numbers of the row and column in which it lies and arranging the

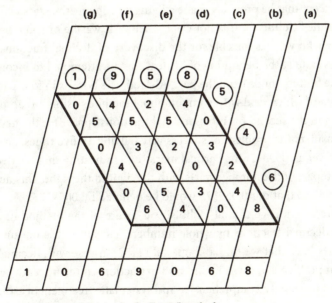

FIGURE 11.3: Multiplication by the "lattice" method

result with the "units" digit below and the "tens" digit above the cell's diagonal. For example, in the top right-hand cell of the square is entered 40, the product of 8 and 5, arranged so that 4 is above the diagonal and 0 below it. After all the cells have been filled, additions are made along the diagonal paths from (a) to (g), starting at (a) and continuing to (g), carrying over if necessary, and the results written at the bottom. For example, summing the numbers in path (c) gives $0 + 3 + 0 + 3 + 4 = 10$, so 0 is written at the bottom and 1 carried over to (d). The answer appears at the bottom of figure 11.3 as 1,069,068.

The reader is invited to multiply 1,990 by 365 using this method.

Decimal Fractions

In previous chapters we have seen the versatility of the Babylonian sexagesimal system for representing fractions, and the Chinese facility in manipulating fractions using their rod numerals. However, neither the Mesopotamians nor the Chinese had a device or symbol for separating integers from fractions. The credit for such a symbol must go to our Islamic precursors.

Decimal fractions make their first appearance in Islamic mathematics in *The Book of Chapters on Indian Arithmetic*, written in Damascus in the year 952 or 953 by Abul Hassan al-Uqlidisi. We know little about the author, except that he probably made a living by copying the works of Euclid (hence the "Uqlidisi"). It is not clear whether it was he or some previous scholar who was responsible for the discovery of decimal fractions, since he does state at the beginning of his text that he attempted to incorporate in it the best methods from the past. Al-Uqlidisi's book is in four parts, the first two of which deal with computational methods using Indian numerals. It is in the second part that decimal fractions appear for the first time. He considers the problem of successively halving 19 five times, and gives the answer as 0̇59375, the vertical mark on 0 indicating that the decimal-fraction part of the number starts with the digit to the right. This notation is general, so that our 0.059375 would be denoted by 0̇059375.

There is evidence that al-Uqlidisi was aware of the method of multiplying decimal fractions by whole numbers. However, it is not until two centuries later that we find al-Samawal (1172) at ease with decimal fractions in problems of division and root extraction. In the fifteenth century, al-Kashi provided a comprehensive and systematic treatment of arithmetic operations with fractions. Devices for separating integers from fractions now included not only the vertical mark placed on the last digit of the integer part but also the use of different colors, or a numerical superscript giving the number of decimal places specified: thus $36\overset{2}{}$ would indicate the decimal fraction 0.36. Al-Kashi's multiplication procedure is identical to the one we use today.

During the fifteenth century this method of representing decimal fractions came to be known outside the Islamic world as the Turkish method, after a Turkish colleague of al-Kashi known as Ali Qushji, who provided an explanation. Knowledge of the method then spread to Vienna, where in 1562 it appeared in a collection of Byzantine problems. It is quite likely that the Dutch mathematician Simon Stevin (1548–1620), who is often credited with the first systematic exposition of decimal fractions, may have learned of the so-called Turkish method from this Byzantine text or a similar source. In place of the short vertical line over the last digit of the integer part of the number, which was the original notation of al-Uqlidisi, Stevin used a cipher, so that the number 6.8145 would be represented as $6\overset{0}{}8145$. To his contemporary John Napier we owe the present convention of using a decimal point to separate integer and fractional parts.

Another contribution of al-Uqlidisi's was to adapt the Indian sand board techniques of computation to methods suitable for pen and paper. It was a common practice to perform arithmetical calculations by writing number symbols in sand or dust, rubbing out intermediate steps as one proceeded. Al-Uqlidisi had strong reservations about this procedure, not for any shortcomings in the method itself but because dust-board calculations were carried out by street astrologers and other "good-for-nothings" to earn their livelihood! He suggested the use of pen and paper so as not to be associated with such company. Indeed, the "pen and paper methods" for carrying out multiplication and division that found their way into medieval Latin works owe more to al-Uqlidisi than to al-Khwarizmi.

Mathematics in the Service of Islamic Law: Problems of Inheritance

The second half of al-Khwarizmi's *Algebra* contains a series of problems about the Islamic law of inheritance. These laws are fairly straightforward in that when a woman dies her husband receives one-quarter of her estate, and the rest is divided among the children such that a son receives twice as much as a daughter. However, if a legacy is left to a stranger, the division gets more complicated. The law on legacies states that a stranger cannot receive more than one-third of the estate without the permission of the natural heirs. If some of the natural heirs endorse such a legacy but others do not, those who do must between them pay pro rata, out of their own shares, the amount by which the stranger's legacy exceeds one-third of the estate. In any case, the legacy to the stranger has to be paid before the rest is shared out among the natural heirs. Clearly, it is possible to construct problems of varying degrees of complexity that illustrate different aspects of the law. Here are two simple examples considered by al-Khwarizmi:

EXAMPLE 11.1 A woman dies leaving a husband, a son, and three daughters. She leaves a bequest consisting of $1/8 + 1/7$ of her estate to a stranger. Calculate the shares of her estate that go to each of her beneficiaries.

Solution

The stranger receives $1/8 + 1/7 = 15/56$ of the estate, leaving $41/56$ to be shared out among the family. The husband receives one-quarter of what remains, that is, $1/4$ of $41/56 = 41/224$. The son and the three daughters receive their shares in the ratio 2:1:1:1; that is, the son's share is two-fifths of the estate after the stranger and the husband have been given their bequests. So, if the estate is divided into $5 \times 224 = 1{,}120$ equal parts, the shares received by each beneficiary will be

Stranger: $15/56$ of 1,120 or 300 parts

Husband: $41/224$ of 1,120 or 205 parts

Son: $2/5$ of $(1{,}120 - 505)$ or 246 parts

Each daughter: $1/5$ of $(1{,}120 - 505)$ or 123 parts

EXAMPLE 11.2 A man dies, leaving his mother, his wife, and two brothers and two sisters by the same father and mother as himself. He (also) bequeaths to a stranger one-ninth of his estate. Calculate the shares going to each of the beneficiaries.

The original text reads as follows:

Calculation

You work out their shares by taking them out of forty-eight parts. You know that if you take one-ninth from the capital, eight-ninths

Continued . . .

Continued . . .

will remain. Add now to the eight-ninths one-eighth of the same and to the forty-eight also one-eighth of them, namely, six, in order to complete your capital. This gives fifty-four. The person to whom one-ninth is bequeathed receives six out of this, being one-ninth of the whole capital. The remaining forty-eight will be distributed among the heirs, proportionally to their legal share.

The reader may wish to work this out in terms of modern mathematics.

It is important, however, not to exaggerate the role of mathematics in the service of Islam. Undoubtedly, religious requirements such as implementing the law of inheritance, determining the direction of Mecca for daily prayer, or identifying the beginning and end of the period of fasting (Ramadan) gave a special impetus to the development of certain areas of mathematics. But there is a limit to the usefulness of mathematics for such purposes, and to the stimulus that such activities provided for the further development of mathematics.

The Theory of Numbers

In the theory of numbers, as in other fields, the mathematicians of the Islamic world managed to produce a creative synthesis of the ideas they obtained from different traditions—notably India and the Hellenistic world. This is best exemplified in the work of Ibn Sina (or Avicenna, as he came to be known in Europe). Although he is widely known for his *Canon of Medicine*, a standard text used for centuries in medieval Europe, his mathematical work is little appreciated outside the Islamic world. His *Kitab al-Shifa* (Book of Healing) contains sections on arithmetic. They begin with a discussion, based on Greek and Indian sources, of different types of number (e.g., odd, even, deficient, perfect, and abundant numbers[20]) and an explanation of different arithmetical operations, including the rule for "casting out nines."[21] He then proceeds to state the two rules of summation. The first is for summing a square array of odd numbers:

If successive odd numbers are placed in a square table, the sum of the numbers lying on the diagonal will be equal to the cube of the side; the sum of the numbers filling the square will be the fourth power of the side. (Al-Daffa and Stroyls 1984, pp. 77–78)

9	7	5	3	1
19	17	15	13	11
29	27	25	23	21
39	37	35	33	31
49	47	45	43	41

FIGURE 11.4: A square array of odd numbers

Ibn Sina illustrates this rule by the square shown in figure 11.4. The diagonals of the square add up to

$$9 + 17 + 25 + 33 + 41 = 125 = 1 + 13 + 25 + 37 + 49,$$

which is equal to the cube of the "side," 53. The total sum of the numbers of the square is $625 = 5^4$, the fourth power of the "side." There is the clear implication here that Ibn Sina knew that the sum of successive odd numbers starting with 1 is equal to the square of the number of odd numbers being added. For example, $1 + 3 + 5 + 7 + 9 + 11 = 36$, which is the square of 6, the number of odd numbers added.

The second rule is for summing a triangular array of odd numbers:

> If successive odd numbers are placed in a triangle, the sum of the numbers taken from one row equals the cube of the [row] number. (Al-Daffa and Stroyls 1984, p. 78)

Such a triangular array of odd numbers from 1 to 30 is shown in figure 11.5. It is easily seen that the sum of the numbers in, say, the third row is 27, the cube of the row number.

Figurate numbers were studied by several Islamic mathematicians, particularly those who favored a geometric rather than an algorithmic approach to numbers; al-Khwarizmi represented the latter, Thabit ibn Qurra the former approach. Thabit was one of the first Islamic mathematicians to recognize the value of a geometric interpretation of an algebraic problem. However, his most notable contribution to the theory of numbers

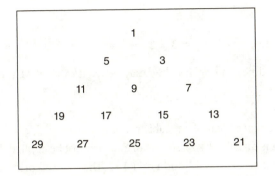

FIGURE 11.5: A triangular array of odd numbers

apparently had no geometric motivation—this was his derivation of a formula for generating pairs of amicable numbers. This appears in a text on amicable numbers written after his translation into Arabic of Nicomachus's *Introduction to Arithmetic.*

After explaining that the Pythagoreans and other ancient Greek mathematicians had made use of two main kinds of numbers in illustrating their philosophy, namely perfect numbers and amicable numbers, Thabit goes on to define these and other "imperfect" (i.e., abundant and deficient) numbers. He then points out that neither Euclid nor Nicomachus paid any attention to the "theory" of amicable numbers. The theory that he suggests may be expressed in the following modern form.

A pair of natural numbers, M and N, is defined as amicable if each is equal to the sum of the proper divisors (i.e., all the divisors of a number, including 1, but not itself) of the other. The smallest pair of amicable numbers is 220 and 284. The proper divisors of 220 are 1, 2, 4, 5, 10, 11, 20, 22, 44, 55, 110, the sum of which is 284. Similarly, the proper divisors of 284 are 1, 2, 4, 71, 142, the sum of which is 220.

Thabit then proceeds to provide a verbal rule that can be translated into the following formula for deriving pairs of amicable numbers. Let $p, q,$ and r be distinct prime numbers given by

$$p = 3 \times 2^{n-1} - 1, \quad q = 3 \times 2^n - 1, \quad r = 9 \times 2^{2n-1} - 1,$$

where the integer n is greater than 1. Then M and N will be a pair of amicable numbers, given by

$$M = 2^n pq, \quad N = 2^n r.$$

For $n = 2$,

$$p = 3 \times 2 - 1 = 5, \quad q = 3 \times 2^2 - 1 = 11, \quad r = 9 \times 2^3 - 1 = 71.$$

As p, q, and r are all prime, they may be used to yield M and N as

$$M = 2^2 \times 5 \times 11 = 220, \quad N = 2^2 \times 71 = 284,$$

which is the smallest pair of amicable numbers. For $n = 3$, $q = 287$, which is not prime (it is divisible by 7), so the formula cannot be applied. For $n = 4$, it is found that $p = 23$, $q = 47$, and r $= 1,151$. These are all primes, and so the next pair of amicable numbers generated from Thabit's formula is

$$M = 2^4 \times 23 \times 47 = 17,296, \quad N = 2^4 \times 1,151 = 18,416.$$

Thabit obtained only the first pair of amicable numbers from his rule. It was a later Islamic mathematician, Ibn al-Banna (1256–1321), who found the above pair corresponding to $n = 4$. Some six hundred years after Thabit, Fermat rediscovered this rule and the pair for $n = 4$, and shortly after this Descartes too rediscovered the rule and set $n = 7$ to yield 9,363,584 and 9,437,056. It turns out that Thabit's rule generates pairs of amicable numbers for $n = 2, 4$, and 7, but for no other value of n below 20,000. Fortunately there are other lines of attack. Euler found more than sixty pairs, using methods he developed himself—methods that still form the basis for present-day search techniques. Over a thousand pairs are now known. Curiously, the second-smallest pair, 1,184 and 1,210, was overlooked by all the famous "amicable number chasers"—it was discovered in 1866 by an Italian schoolboy!

To complete the story it is worthwhile mentioning that Abu Mansur al-Baghdadi (d. 1027), in his book the *Book of Completion on Reckoning*, discusses a notion related to amicable numbers, the property of "balanced numbers." Take the example given by al-Baghdadi:

> We wish to know two numbers such that [the sum of] the aliquot[22] parts of each gives 57. So we subtract 1 from it, and there remains 56. We divide this into two prime parts, [say] 3 and 53, and we multiply the one by the other; the result is 159 and summing its aliquot parts gives 57. Next, we again divide 56 into two other prime parts, say 13 and 43, and we multiply the one by the other: [the result is 559]. Then the sum of the aliquot parts of the result is 57. (Quoted from Berggren 2007, p. 563)

It is interesting to note that apart from the pair of balanced numbers 159 and 559, given above, there is a third number, 703 (19 × 37).

Extraction of Roots

The Arabic word for "root," *jidhir*, was the term introduced by al-Khwarizmi to denote the unknown in an equation. With the help of the terms *mal* (second power, literally "wealth") and *kab* (cube), he was able to describe equations of first or second degree and exponents of degree higher than third. More precisely, al-Khwarizmi described only solutions to equations of first and second degree, but referred to higher exponents and higher reducible equation. Thus, for example:

$$mal\ mal = x^4, \quad mal\ kab = x^5, \quad kab\ kab = x^6, \text{and so on.}$$

There are close parallels between the way the Islamic mathematicians extracted square and cube roots (by seeking numerical solutions to quadratic and cubic equations) and the methods used in other mathematical traditions. Both the Mesopotamian (c. 1800 BC) numerical method of extracting square roots, which we examined in chapter 4, and the closely related method found in the *Sulbasutras* (c. 500 BC), discussed in chapter 8, probably came to be known to the Islamic mathematicians through work of the Alexandrian mathematician Heron (c. first century AD). We also saw, in chapters 6 and 7, that the Chinese had developed quite sophisticated procedures for obtaining approximate solutions of $x^n = A$, for any integral of n and to any degree of accuracy, based on variants of Horner's method using the binomial coefficients of Pascal's triangle. Later Islamic mathematicians worked with methods very similar to those used by the Chinese to extract roots of the second and higher orders.[23]

Algebra in the Islamic World

Perhaps Islamic mathematics should be best remembered for its synthesis of geometry (mainly from Euclid) and algebra from the Eastern world, beginning with al-Khwarizmi's geometric solution of quadratic equations and culminating in Omar Khayyam's geometric solution of cubic equations.

The word *al-jabr* appears frequently in mathematical texts that followed al-Khwarizmi's influential *Hisab al-jabr w'al-muqabala* (Compendium on Calculation by Completion and Reduction), written in the first half of the ninth century.[24] There were two meanings associated with *al-jabr*. The more common was "restoration," as applied to the operation of adding equal terms to both sides of an equation so as to remove negative quantities, or to "restore" a quantity that is subtracted from one side by adding it to the other. Thus an operation on the equation $2x + 5 = 8 - 3x$ that led to $5x + 5 = 8$ would be an illustration of *al-jabr*. There was also another, less common meaning: multiplying both sides of an equation by a certain number to eliminate fractions. Thus if both sides of the equation $(9/4)x + 1/8 = 3 + (5/8)x$ were multiplied by 8 to give the new equation $18x + 1 = 24 + 15x$, this too would be an instance of *al-jabr*. The common meaning of *al-muqabala* is the "reduction" of positive quantities in an equation by subtracting equal quantities from both sides. So for the two equations above, applying *al-muqabala* would give

$$5x + 5 = 8,$$
$$5x + 5 - 5 = 8 - 5,$$
$$5x = 3,$$

and

$$18x + 1 = 24 + 15x,$$
$$18x - 15x + 1 - 1 = 24 - 1 + 15x - 15x,$$
$$3x = 23.$$

The words *al-jabr* and *al-muqabala*, linked by *wa*, meaning "and," came to be used for any algebraic operation and eventually for the subject itself. Since the algebra of the time was almost wholly confined to the solution of equations, the phrase meant exactly that.

A Geometric Approach to the Solution of Equations

Al-Khwarizmi distinguished six different types of equation by using certain word conventions mentioned earlier. The unknown quantity (which we now denote by x) was referred to as "root" or "thing," and the constant was known as "number." So the six different types of equation were

1. roots equal squares: $bx = ax^2$

2. roots equal numbers: $bx = c$

3. squares equal numbers: $ax^2 = c$

4. squares and roots equal numbers: $ax^2 + bx = c$

5. roots and numbers equal squares: $bx + c = ax^2$

6. squares and numbers equal roots: $ax^2 + c = bx,$

where a, b, and c are positive integers. Al-Khwarizmi provided rules for solving these equations and in a number of cases the geometric rationale for these solutions. Let us take one example of a type 4 equation (squares [*mal*] and roots equal numbers) that is interesting historically since it recurs in later texts.

EXAMPLE 11.3 One square [mal] and 10 roots of the same equals 39 direhems. [Or, in modern notation, solve $x^2 + 10x = 39$.]

Suggested Solution

AL-KHWARIZMI'S EXPLANATION	EXPLANATION IN MODERN NOTATION
1. You halve the "number" of roots: result 5.	$x^2 + 10x = 39$.
2. This you multiply by itself: result 25.	
3. Add this to 39: result 64.	$(x + 5)^2 = 39 + 25 = 64$.
4. Take the square root of this: result 8.	$x + 5 = 8$.
5. Subtract from 8 the result given in step 1: result 3.	$x = 3$.

This is the root of the square you sought [the square itself is 9].

(The negative root, $x = -13$, is ignored.)

Variants of this rule are found in both Mesopotamian and Indian mathematics, and there is every likelihood that this algorithm came from either or both of these sources.

The real novelty of the Islamic approach is contained in the following statement of al-Khwarizmi's. After giving numerical solutions for all six types of equation, he goes on:

We have said enough, so far as numbers are concerned, about the six types of equation. Now it is necessary to demonstrate geometrically the truth of the same problems which we have explained in numbers.

We can illustrate al-Khwarizmi's geometric approach by returning to the above exercise. In figure 11.6, ABCD is a square of side *x*. AD and AB

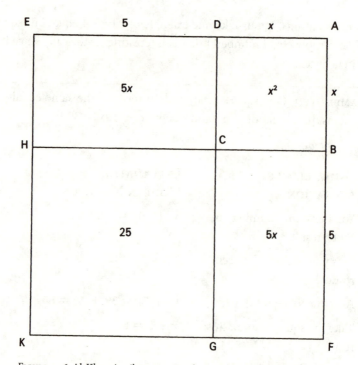

FIGURE 11.6: Al-Khwarizmi's geometric solution of a quadratic equation

are extended to E and F such that DE = BF = 5. The square AFKE is completed, and DC is extended to G and BC to H. From the diagram it is clear that the area of AFKE is equal to

$$x^2 + 10x + 25, \quad \text{or} \quad (x + 5)^2.$$

Adding 25 to both sides of the equation $x^2 + 10x = 39$ gives

$$x^2 + 10x + 25 = 39 + 25 = 64,$$

from which one of the sides of AFKE, say EK, is found to be $x + 5 = 8$, and so EH = $x = 3$.

Al-Khwarizmi's geometric demonstration for each type of equation is based on a specific example. Thabit ibn Qurra presented the first general demonstration, using two of Euclid's theorems, in a short work titled *Problems of Algebra through Geometrical Demonstrations*, of which only a single manuscript has survived. We can illustrate this demonstration for the type 4 equation discussed above—squares and roots equal numbers, or $x^2 + bx = c$.[25]

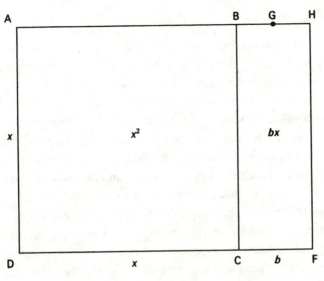

FIGURE 11.7: Thabit's general solution of quadratic equations

In figure 11.7,

ABCD + BHFC = AHFD = $x^2 + bx$.

The application of a result from Euclid's *Elements* (Book II, proposition 5) on the equivalence of areas gives

AHFD + BG^2 = AG^2

if G is the midpoint of BH. Now,

AHFD = $x^2 + bx = c$,   BG$^2 = (\frac{1}{2}b)^2$, AG$^2 = c + (\frac{1}{2}b)^2$,

and therefore

$$x^2 + bx + (\tfrac{1}{2}b)^2 = c + (\tfrac{1}{2}b)^2. \tag{11.1}$$

Applying another of Euclid's propositions (*Elements*, Book II, proposition 4) to the left-hand side of equation (11.1) gives

$$x^2 + bx + (\tfrac{1}{2}b)^2 = (x + \tfrac{1}{2}b)^2. \tag{11.2}$$

Substituting (11.2) in (11.1) gives

$$(x + \tfrac{1}{2}b)^2 = c + (\tfrac{1}{2}b)^2.$$

Therefore

$$x = \sqrt{c + (\tfrac{1}{2}b)^2} - \tfrac{1}{2}b.$$

The validity of this result can be checked by applying it to example 11.2 to solve for x. Thabit also provided a similar geometric proof for type 6 equations, squares and numbers equal to roots, or $x^2 + c = bx$.

To sum up, the Islamic mathematicians' work on solving quadratic equations is yet another illustration of their ability to bring together two strands of mathematical thinking—the geometric approach that had been carefully cultivated by the Greeks, and the algebraic/algorithmic methods that had been used to such effect by the Mesopotamians, Indians, and Chinese. The Islamic mathematicians went far beyond the ingenuity and calculating skills of the Mesopotamians, of which we have seen ample evidence in chapter 4. By devising an efficient system of classifying equations, they, starting with al-Khwarizmi, reduced all equations to six main types. For each type they offered solutions as well as a geometric rationale, thereby laying the foundation of modern algebra. Thus were the *ahl al-jabr* (the "algebra people"), in Thabit's words, and the "geometry people" brought together.

Changing Algebra: Al-Khwarizmi to Abu Kamil to al-Karaji

Around AD 830, Muhammad ibn Musa al-Khwarizmi composed the earliest-known Islamic treatment of algebra, beginning a preoccupation that continued for several centuries. As discussed earlier, his treatise, *Hisab al-jabr wa'l-muqabala*, began with a discussion of the algebra of first- and second-degree equations, proceeding to practical applications of this algebra to questions of mensuration and legacies. Different influences and sources went into the making of this treatise. The use of geometry to justify algebraic manipulation has been traced to Euclid's *Elements*, although Chinese and Indian mathematics, discussed in earlier chapters, contain similar strands. His interest in equations of a second degree and the rhetorical style of his presentation show vestiges of both Mesopotamian and Indian influences.

A generation later, Abu Kamil ibn Aslam (c. 850–930) wrote his own text, *Kitab fi al-jabr wa'l-muqabala*, based on al-Khwarizmi's work. In it, Abu Kamil not only quotes directly from al-Khwarizmi but also incorporates almost half of al-Khwarizmi's forty examples into his work, with numerical changes. However, the mathematical environment in which Abu Kamil's thought developed had undergone significant changes, notably in that the Greek influence had become more manifest. Compare, for example, Abu Kamil's proof of the solution of the equation $x^2 + 10x = 39$ with the proof of al-Khwarizmi.[26] What Abu Kamil did was to use the under-pinnings of Greek mathematics without destroying the concrete base of al-Khwarizmi's algebra to create an algebra based on practical realities. This approach remained unchanged in other commentaries on al-Khwarizmi's work until the middle of the tenth century.

The next change may have been a result of the gradual impact of the ninth-century Arabic translation of Diophantus's *Arithmetica* by Qusta b. Luqa, which introduced the Diophantine approach to the solutions of determinate and indeterminate equations. It is important, in this context, to distinguish between the approach to and the actual solution of indeterminate equations, since it is likely that the Islamic mathematicians had come across Indian and possibly Chinese work on the subject.

One mathematician whose work reflected this change in approach was Abu Bakr al-Karaji (al-Karkhi) (c. 1000). In *al-Fakhri fi'l-jabr wa'l-muqabala* (Glorious Work on Algebra[27]), al-Karaji solved indeterminate equations of degrees two and three in up to three unknowns by methods clearly influenced by the *Arithmetica* while retaining the geometrical approach of al-Khwarizmi. At the same time, he rejected aspects of both approaches, in particular al-Khwarizmi's complete reliance on geometry for proof and Diophantus's syncopated notation.

Omar Khayyam's Geometric Solution of Cubic Equations

Omar Khayyam's work may be seen as the culmination of the geometric approach to the solution of equations, in particular general cubic equations. It is almost certain that he was unaware of the arithmetic solutions we have found in Chinese mathematics. He bemoaned the fact that he could not find an "algebraic" solution for cubics, as had been done for quadratics, and hoped that his successors would be able to do so. Instead, he explored the possibility of using geometric methods, in particular whether parts of intersecting conics[28] could be used to solve cubic equations. Traces of such an approach are found in the works of earlier writers such as Menaechmus (c. 350 BC) and Archimedes (287–212 BC), and Omar's near-contemporary Ibn al-Haytham (c. 965–1039). They had observed, for quantities a, b, c, and d, that if

$$\frac{b}{c} = \frac{c}{d} = \frac{d}{a},$$

then

$$\left(\frac{b}{c}\right)^2 = \left(\frac{c}{d}\right)\left(\frac{d}{a}\right) = \frac{c}{a},$$

or

$$c^3 = b^2 a.$$

Now, if $b = 1$, the cube root of a can be evaluated as long as c and d exist such that

$$c^2 = d \quad \text{and} \quad d^2 = ac. \tag{11.3}$$

Omar Khayyam's great contribution was to discover the geometrical argument embedded in this algebra. For the Greek mathematicians who were the inspiration behind Omar's work, the equations leading to 11.3 can hardly be described as "algebra." For them, the equations were the solutions of certain ratio problems, such as finding two mean proportionals between two given quantities, that could be found by dealing with conic sections.[29]

If we think of c and d as variables and of a as a constant, then equations (11.3) are the equations of two parabolas with perpendicular axes and the same vertex. This is illustrated in figure 11.8. The two parabolas, whose construction is explained in example 11.4 later in this chapter, have the same vertex B, with axes AB ($= a$) and CB ($= b = 1$), and they intersect at E. In the rectangle BDEF, BF = DE = c, and BD = FE = d. Since AB is a line segment and the point E lies on the parabola with vertex B and axis AB, the rectangle BDEF has the property that

$$(\text{FE})^2 = \text{AB} \times \text{BF}, \quad \text{or} \quad d^2 = ac. \tag{11.4}$$

Similarly, for the other parabola with vertex B and axis BC,

$$(\text{BF})^2 = \text{CB} \times \text{BD}, \quad \text{or} \quad c^2 = bd = d. \tag{11.5}$$

From (11.4) and (11.5) we get

$$c^2 = a. \tag{11.6}$$

Therefore, DE = BF is a root of equation (11.6).

Applying similar reasoning, Omar extended his method to solve any third-degree equation for positive roots. He discussed nineteen types of cubic equations (expressed with only positive coefficients). Five of these could easily be reduced to quadratic equations. Each of the remaining fourteen he solved by means of conic sections. It is possible to classify these

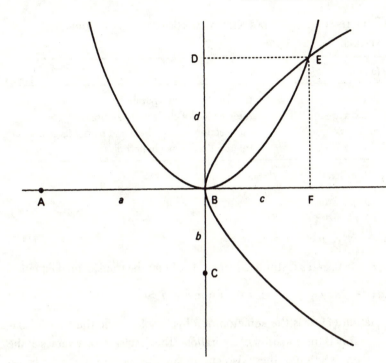

FIGURE 11.8: Omar Khayyam's solution of cubic equations

fourteen cubics, using modern notation, into four main types as in table 11.1. To illustrate this reduction, take a cubic of the form

$$z^3 + pz^2 + qz + r = 0,$$ (11.7)

where the coefficients p, q, and r can be positive, negative, or zero. Setting $z = x - p/3$ in equation (11.7) gives an equation of the form

$$x^3 + gx + h = 0,$$

where the coefficients g and h are again positive, negative, or zero. It can be shown that

$$g = q - \frac{1}{3}p^2 \quad \text{and} \quad h = \frac{2}{27}p^3 - \frac{1}{3}pq + r.$$

There is evidence of Omar Khayyam's facility with other cubics, especially those that can be transformed to one of the standard types given in table 11.1. Take for example the cubic

TABLE 11.1: SOME OF OMAR KHAYYAM'S SOLUTIONS OF CUBIC
EQUATIONS

Type (a > 0, c > 0)	Method
1. $x^3 + c$	Intersection of two parabolas
2. $x^3 + ax = c$	Intersection of circle and parabola
3. $x^3 \pm c = ax$	Intersection of hyperbola and parabola
4. $x^3 = ax + c$	Intersection of two hyperbolas

Note: In each case one positive root was found.

$$x^3 + ax^2 + b^2x + c = 0 \tag{11.8}$$

for $a, b, c > 0$, and substitute $2dy$ for x^2 to obtain the transformed equation

$$2dxy + 2ady + b^2x + c = 0, \quad \text{for } a, b, c, d > 0. \tag{11.9}$$

Equation (11.9) is the equation of a hyperbola, while the transformation $x^2 = 2dy$ is the equation of a parabola. The abscissas, or x values at the point of intersection of these two curves, are the roots of the cubic equation (11.8).

Without the language of modern mathematics, in particular its symbolic notation, Omar's task of exposition was much more difficult. In his approach, a, b, c, and x are line segments, and the problem is:

Given a, b, and c, to construct a line segment x such that equation (11.8) holds.

Omar begins by declaring that a line segment cannot be constructed by using only a straightedge and compass; at some point in the construction conic sections must be introduced. His knowledge of conic sections was derived mainly from the *Conics*, the work of the Hellenistic mathematician Apollonius of Perga (c. 200 BC). This is one of the more difficult works of Alexandrian geometry. It is a measure of the level of sophistication of Islamic mathematics in the tenth century that the *Conics* together with Archimedes' *On the Sphere and Cylinder* constituted the two pillars of Islamic geometry. Omar's solutions for each type of cubic listed in table 11.1 are too long and involved for us to discuss here. His own book, *Al-jabr w'al-muqabala* (*Algebra*, translated into English and edited by Kasir, 1931), gives an indication of the breadth of coverage of his approach to different cubics.

Omar's achievement is typical of Islamic mathematics in its application (in a systematic fashion) of geometry to algebra. While Omar made no addition to the theory of conics, he did apply the principle of intersecting conic sections to solving algebraic problems. In doing so he not only exhibited his mastery of conic sections but also showed that he was aware of the applications of what was then a highly abstruse area of geometry.

Also important was Omar's systematic classification of cubic equations, with his demonstration of a geometric solution for each type. Despite the constraints imposed by the character of the mathematical language of the time (he used either geometric magnitudes or numbers capable of geometric interpretation), the clarity of Omar's presentation is striking, and the cases he cannot demonstrate are relatively few. He was aware that sometimes there was more than one positive solution, sometimes none at all (for nonintersecting conic sections). His neglect of negative or imaginary roots is perfectly understandable given the mathematical climate of the time. But, with hindsight, this does not imply that the methods he used were not adequate for the purpose of extracting negative roots. In fact, referring to table 11.1, it can be shown, first, that the absolute value of the negative roots of type 1 is identical to the positive root of type 2, and vice versa; and second, that the absolute values of the negative roots of type 4 correspond to the positive roots of type 3, and vice versa. Indeed, Omar's geometrical methods were the only ones available until the algebraic methods were developed by the Italian algebraists, notably Girolamo Cardano and Niccolò Tartaglia.

What Omar Khayyam and those who came after him failed to do was find an algebraic solution of cubic equations. In an earlier chapter we examined the numerical procedures the Chinese used to solve equations of any degree, but the Chinese showed little interest in the algebra of these solutions. Sharaf al-Din al-Tusi's work (Rashed 1986) contains an interesting application of what we would describe today as the determination of maxima and minima to the solutions to cubic equations. However, it should be noted that Sharaf al-Din al-Tusi found the positive solutions to his equations by a numerical procedure similar to the Chinese procedure discussed in chapter 7. For an equation $x^3 + c = ax^2$ expressed in the form

$$x^2(a - x) = c, \tag{11.10}$$

al-Tusi notes that whether it has a positive solution depends on whether the expression on the left-hand side reaches c. He is aware that, for any value of x lying between 0 and a,

$$x^2(a - x) \leq \left(\frac{2a}{3}\right)^2\left(\frac{a}{3}\right).$$

We can easily see with the help of elementary calculus that a (relative) maximum occurs for x in equation (11.10) when $x_0 = 2a/3$. But there is no indication as to how al-Tusi found this value. He proceeds to make the following inferences:

If $4a^3/27 < c$, no positive solution exists.

If $4a^3/27 = c$, only one positive solution ($x = 2a/3$) exists.

If $4a^3/27 > c$, two positive solutions (x_1 and x_2) exist,

where $0 < x_1 < 2a/3$ and $2a/3 < x_2 < a$. However, there is no evidence that al-Tusi actually found the two positive solutions (x_1 and x_2).

Islamic Algebra and Its Influence on Europe

Tracing the lines of Islamic influence on European mathematics is a difficult task at the best of times, although there are one or two lines of which we are reasonably certain. Al-Khwarizmi's *Algebra* is generally recognized, through its Latin translations, as having been highly influential in the development of European algebra. Abu Kamil (c. AD 900), popularly known as the "Egyptian Calculator," wrote a commentary on al-Khwarizmi's work in which he systematically treated the fundamental rules of algebraic operations and solution of equations, including nonlinear simultaneous equations. This work influenced Fibonacci, whose impact on medieval European mathematics cannot be overstated. There was a third Islamic mathematician and scientist whose geometric theory of the solution of equations, particularly as applied to problems in optics, had a direct impact on Europe—Ibn al-Haytham. It is one of the ironies of history that the works of Thabit and Omar, two of the greatest Islamic mathematicians who turned to geometry for rigorous derivations of results, were less well known than those just mentioned.

It is interesting that, in the westward movement of algebra, it was mainly the twelfth-century Latin translations of al-Khwarizmi's text, by Robert of Chester and Gerard of Cremona, that shaped the mathematical environment of medieval Europe. The same applied in the case of Fibonacci, who came across al-Khwarizmi's work either from reading one or both translations or from learning algebra during his youth in North Africa or during his travels as an adult in Egypt, Syria, Greece, and Sicily. The fact remains that al-Khwarizmi's ideas on the theory of quadratic equations figured prominently in the fifteenth and final chapter of Fibonacci's influential book, *Liber Abbaci* (1202).

Not until the 1850s did Omar's work begin to be mentioned in the standard Western histories of mathematics, when Woepcke's translation of his *Algebra* appeared, though Kasir (1931, pp. 6–7) produces evidence of European interest in Omar's work from over a hundred years before. And there is no evidence that Thabit ibn Qurra may have directly influenced the development of mathematics in Europe, yet there are pieces of circumstantial evidence that are rare but quite suggestive. The texts translated by Gherardo of Cremona (c. 1175), through which more Islamic science entered Europe than in any other way, may have indirectly drawn on Thabit's works. Furthermore, we may question the originality of John Wallis's *Treatise on Angular Sections* (1685), in which there are echoes of the works of Ptolemy (c. AD 100) and of Thabit's generalization of the Pythagorean theorem. (Thabit's work in this area will be discussed in a later section.) Given the steady stream of Arabic manuscripts that flowed into Europe in the seventeenth century and the appointment of professors of Arabic in some of the major universities of Europe who were conversant with mathematical and astronomical works of the Islamic world, it is important to keep an open mind on possible transmissions, about which more may emerge from detailed research on how the Islamic world influenced medieval Europe.[30] It was not all that long ago that the Islamic scholars were dismissed as mere custodians, or at best pale imitators, of Greek science and philosophy.

Geometry in the Islamic World

Throughout the Islamic world are to be found buildings decorated with intricate geometric designs. These are a common feature of Islamic art, which has an ornamental tradition since the Islamic religion has generally discouraged the portrayal of living things. Such superb craftsmanship in various media, including wood and tile, would have required considerable geometric skills in construction. Islamic mathematicians were particularly interested in geometrical constructions, both Euclidean using straightedge and compass as well as more advanced constructions using conic sections. Around the year 960, Abu al-Wafa al-Buzjani (940–997) is believed to have written a book (now lost), titled *Knowledge of Geometry*

Necessary for the Craftsman, that provided a number of constructions, many of which could be achieved with just a straightedge and compass. These included construction of polygons, inscribing (or circumscribing) circles in (or around) various polygons, dividing the surface of a sphere into given shapes, decomposing a square into a given number of squares or constructing a square equal to a given number of squares, and so on. Consider one example of such a construction.

EXAMPLE 11.4 To inscribe an equilateral triangle in a square so that its angles touch its sides.

Method

We make a square ABCD [see figure 11.9], and we extend AB to E to make BE equal to AB. And on AE we construct the semicircle ECA. Then with A as center and distance BA [as radius] we mark G [on the semicircle ECA] and with E as center and distance EG we mark F [on line AB]. And we make CH equal to AF. Then we draw DH, DF, and FH. Then triangle DHF is equilateral, and it was constructed in the square ABCD.

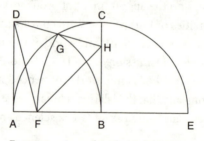

FIGURE 11.9: An equilateral triangle in a square

[A proof is provided to show that the triangle DHF is equilateral with its angles touching the sides of the square.][31]

However, it was in the construction of conic sections that the Islamic mathematicians made their distinctive contribution.

In *Rasm al qutu as salasa* (On Drawing the Three Conic Sections) by Ibrahim ibn Sinan (d. 946), a grandson of Thabit ibn Qurra, there are detailed instructions on how to construct a parabola and an ellipse, and three different ways of constructing a hyperbola. Ibn Sinan begins with a cautionary note. "When we found that it was difficult to draw these three sections with a compass or other instruments, we tried hard to draw numerous points to which [a] man can add as many as he wants and such that these points will be on one of the three sections. Everything [that was] determined that way proved how these sections, along with others, are generated from the circle" (Berggren 2007, p. 565).

To illustrate, let us consider Ibn Sinan's constructions of a parabola and a hyperbola. (The reader may like to try drawing these figures with just a ruler and compass to appreciate the sheer ingenuity and geometric "sense" displayed here.)

EXAMPLE 11.5 To construct a parabola.

Method

Draw a line AB [see figure 11.10], and construct a perpendicular CE cutting AB at D. On the line segment DB, mark a number of points G,

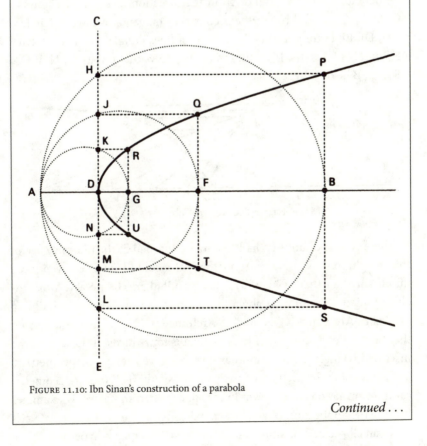

FIGURE 11.10: Ibn Sinan's construction of a parabola

Continued . . .

Continued . . .

F, ... Next, construct circles with diameters AB, AF, AG, . . . , which intersect CE at H and L, J and M, K and N, . . . , respectively. Through H and L, draw lines parallel to AB, and through B draw a line parallel to CE. Let these lines through H, L, and B meet at P and S. Similar lines drawn through J, M, and F, and through K, N, and G, intersect at points Q and T, and R and U, respectively. Ibn Sinan provides a proof that all such points of intersection lie on a parabola. The parabola has turning point at D, axis AB, and parameter AD.

EXAMPLE 11.6 To construct a hyperbola.

Method

AB is a line segment [see figure 11.11] that is also the diameter of a semicircle, center O. Extend AB in the direction of B. Choose points C, D, E, . . . , and through them construct tangents of the semicircle: CH, DI, EJ. From points H, I, J construct lines parallel to AB such that HR = HC, IQ = ID, JP = JE. Ibn Sinan shows that points M, N, P, Q, R, . . . , B lie on a hyperbola.

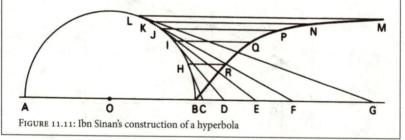

FIGURE 11.11: Ibn Sinan's construction of a hyperbola

There is little doubt that had Ibn Sinan not died at the untimely age of thirty-eight, his contribution to mathematics could have been even more important than that of his illustrious grandfather. His notable work included quadrature of a parabola using a method of integration building on that of Archimedes.[32] Following Archimedes, he gave an elegant proof that the area of a segment of the parabola is four-thirds of the area of the inscribed triangle. And in another text, he solved forty-one geometrical problems using the method of analysis and synthesis.[33] Indeed, geometrical transformations of different kinds appear often in his work. Examples include the application of an orthogonal compression to transform a circle to an ellipse and of an oblique compression to map a hyperbola into a second hyperbola. In another work, he uses a transformation that maps figures keeping invariant the ratio between their areas.[34]

It is difficult to establish the origins of the mathematician, mentioned earlier, considered by many to have been al-Khwarizmi's natural successor. The difficulty arises from a characteristic of Arabic writing: sometimes letters are distinguished, not by their different shapes, but by the location of a dot near the letter. So whether this mathematician was al-Karkhi (which places his origins in Iraq) or al-Karaji (which places his origins in Iran) depends on whether the dot was placed above the relevant Arabic letter or below it; both versions are recorded. Whatever his origins, we shall refer to him as al-Karaji (i.e., one who was born in the town of Karaj, near Tehran). All we know about him is that he lived in Baghdad around the year 1000, where he dedicated a book on algebra to a vizier Fakhr al-Din. His contributions include his development of the binomial coefficients and the "Pascal triangle," deriving rules of operations with exponents, solving equations of higher degree, and an elaboration of the "Indian calculation."[35] Al-Karaji gives an interesting geometric construction.

EXAMPLE 11.7 Construct a circle whose area is equal to a given fraction $(1/n)$ of the area of a given circle.

Method

[In figure 11.12], AOB is the diameter of the given circle, with center O. Draw TA perpendicular to AB. Extend BA to C such that CA $= (1/n)$AB.

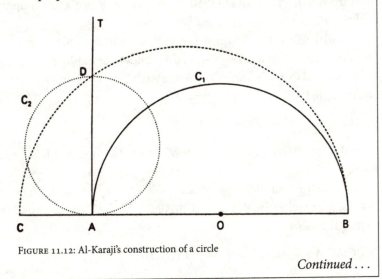

FIGURE 11.12: Al-Karaji's construction of a circle

Continued . . .

Continued...

Construct a circle with CB as diameter. The point at which this circle cuts TA is denoted by D. AD is then the diameter of the required circle, equal to $(1/n)$AB, so the area of this circle, C_2, is $1/n$ of the given circle, C_1.

Proof

The proof follows easily from two propositions of Euclid's on intersecting chords (*Elements*, Book II, propositions 12 and 13), which give

$$CA \times AB = DA^2 = \left(\frac{d}{n}\right)d,$$

where AB $= d$, CA $= (1/n)$AB. Therefore DA $= d/\sqrt{n}$. Now, the circle with diameter AB is C_1 and the circle with diameter AD $= d/n$ is C_2. So the ratio of the areas of C_1 and $C_2 = d^2:d^2/n = 1:1/n$.

Thabit ibn Qurra's Generalization of the Pythagorean Theorem

In a letter to a friend, Thabit expressed disappointment with an existing (so-called Socratic) proof of the Pythagorean theorem because it applied only to isosceles right-angled triangles. He then proceeded to give three results, of which the third is a generalization of the Pythagorean theorem applicable to all triangles, whether right-angled or not. We shall look at the third result, which Sayili (1960, p. 35) has described as an "important contribution" to the history of mathematics.

Consider figure 11.13, which is constructed in the following manner. From the vertex A of a triangle ABC drop lines intersecting the base BC at B′ and C′ and forming angles AB′B and AC′C respectively, each of which equals angle BAC. We wish to show that

$$AB^2 + AC^2 = BC(BB' + CC').$$

Thabit ibn Qurra provides no proof, except to say that it follows from Euclid.

A reconstruction of the proof that makes use of similar triangles is as follows. It can easily be shown that BAC, BAB′, and CAC′ are similar triangles. Therefore

$$BC:AC:AB = AC:CC':AC' = AB:AB':BB'.$$

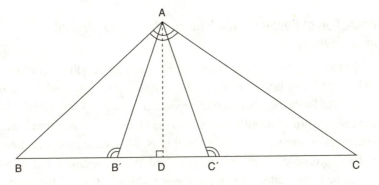

FIGURE 11.13: Thabit's generalized Pythagorean theorem

Then

$$AB/BB' = BC/AB, \text{ so } AB^2 = (BC)(BB'); \qquad (11.11)$$

and

$$AC/CC' = BC/AC, \text{ so } AC^2 = (BC)(CC'). \qquad (11.12)$$

Adding equations (11.11) and (11.12) gives Thabit's generalization of the Pythagorean theorem:

$$AC^2 + AB^2 = BC(BB' + CC').$$

Figure 11.13 shows an obtuse triangle, with the angle at A greater than a right angle. Thabit also considers an acute triangle, for which B′ and C′ lie outside BC, but for which the above proof (with minor modifications) still applies; and he considers the Pythagorean right-angled triangle, for which B′ and C′ coincide at D.

Thabit's work on this theorem was discovered as late as 1953 in the library of the Aya Sofia Museum in Turkey. However, it made its first appearance in European mathematics in 1685, when John Wallis's proof of the theorem was published in his *Treatise on Angular Sections*. It is a reasonable conjecture that Wallis was aware of Thabit's work in this area, since he was sufficiently acquainted with Islamic mathematics to know of al-Tusi's work on the parallel postulate. Indeed, Scriba (1966, p. 67) is of the opinion that the *Treatise on Angular Sections* is based on Thabit ibn Qurra's generalization of the Pythagorean theorem and Ptolemy's work in this area.

Summation of Powers in the Islamic World: The Work of Ibn al-Haytham

Ibn al-Haytham (965–1040), known by his Latin name Alhazen in Europe, was born in Basra but spent most of his adult life in Egypt, where he had been invited by the caliph al-Hakim to work on a project to control the Nile. His most important scientific work was his *Optics*, translated into Latin in the thirteenth century and studied in Europe for several centuries thereafter. It was probably his interest in optics and his close examination of what came to be known after him as "Alhazen's problem"[36] that led him to prove a set of results on the sum of powers of whole numbers, which then became an integral part of his measurement of the volume of a certain kind of paraboloid.[37] Rather than consider how Ibn al-Haytham proceeded to measure the volume of the solid of revolution formed by rotating a parabola around a line at a right angle to its axis, we will examine some of the results that he stated and proved relating to the sums of powers of whole numbers.[38]

> *Result 1.* If one has a sequence of natural numbers, beginning with one, and one takes half the largest and half of one, adds these halves and multiplies this sum by the largest number, one has the sum of all given numbers.

Or in modern notation

$$\sum_{i=1}^{n} i = \left(\frac{1}{2}n + \frac{1}{2}\right)n = \frac{n^2}{2} + \frac{n}{2} = \frac{1}{2}n(n+1).$$

> *Result 2.* One has again the same sequence of numbers. One takes the third part of the largest and third part of one, adds these parts, and multiplies the sum by the largest numbers. Then one adds to the largest numbers the half of one and multiplies this sum by the former product. One has the sum of the squares of the given numbers.

Or in modern notation

$$\sum_{i=1}^{n} i^2 = \left(\frac{1}{3}n + \frac{1}{3}\right)n\left(n + \frac{1}{2}\right) = \frac{n^3}{3} + \frac{n^2}{2} + \frac{n}{6}.$$

> *Result 3.* One is again given the same sequence of numbers. One takes the fourth part of the largest and the fourth part of one, adds these parts,

and multiplies the sum by the largest number. One then adds one to the largest number, multiplies the sum by the largest number, and multiplies this product by the former product. One then has the sum of the cubes of the given numbers.

Or in modern notation

$$\sum_{i=1}^{n} i^3 = \left(\frac{1}{4}n + \frac{1}{4}\right)n(n+1)n = \frac{n^4}{4} + \frac{n^3}{2} + \frac{n^2}{4}.$$

Result 4. One is again given the same sequence of numbers. One takes the fifth part of the largest and the fifth part of one, adds these parts, and multiplies the sum by the largest number; then one adds to the largest number half of one, and multiplies this sum by the former product. Now one adds one to the largest number, multiplies this sum by the largest number, subtracts from the product the third part of one, and multiplies this result with the previous product. One then has the sum of the fourth powers of the given numbers.

Or in modern notation

$$\sum_{i=1}^{n} i^4 = \left(\frac{1}{5}n + \frac{1}{5}\right)n\left(n + \frac{1}{2}\right)\left[(n+1)n - \frac{1}{3}\right] = \frac{n^5}{5} + \frac{n^4}{2} = \frac{n^3}{3} - \frac{n}{30}.$$

Ibn al-Haytham provides proofs for these results, which will not be discussed here. The approach used by Ibn al-Haytham may be extended further for any positive integral k (Katz 1993). Inductively, it can be established from the results above that

$$\sum_{i=1}^{n} i^k = \frac{n^{k+1}}{k+1} + \frac{n^k}{2} + p(n),$$

where $p(n)$ is a polynomial in n of degree $< k$. As Katz (1995, pp. 168–69) points out, this formula for the sum of fourth powers appears in works of other mathematicians in the Islamic world, including Ibn Haydur (d. 1413), Ibn Ghazi (1437–1514), and in *The Key of Arithmetic* of al-Kashi (d. 1429). As discussed in the previous chapter, a variant of this result occurs in Kerala mathematics in the fifteenth century. It was used by European mathematicians, including Fermat, Roberval, and Pascal, in the seventeenth century to evaluate the area under the parabola. This brings us back to the mathematics of Ibn al-Haytham, a true pioneer of early modern mathematics.

Trigonometry in the Islamic World

As with so many other areas of mathematics, the Islamic scholars selected Hellenistic and Indian concepts of trigonometry and combined them into a distinctive discipline that bore little resemblance to its precursors. It then became an essential component of modern mathematics. We shall consider three aspects of Islamic trigonometry:

1. The introduction of six basic trigonometric functions, namely sine and cosine, tangent and cotangent, secant, and cosecant

2. The derivation of the sine rule and establishment of other trigonometric identities

3. The construction of highly detailed trigonometric tables with the aid of various interpolation procedures

Introduction of Trigonometric Functions

Basic to modern trigonometry is the sine function. It was introduced into the Islamic world from India, probably through the famous Indian astronomical text *Suryasiddhanta*. This was one of the texts brought to the court of al-Mansur during the eighth century by a diplomatic mission from Sind. We saw in an earlier chapter that there were two types of trigonometry: one based on the geometry of chords (see figure 9.1) and best exemplified in Ptolemy's *Almagest*, and the other based on the geometry of semichords (see figure 9.2), which was an Indian invention.

In the Indian scheme, the length of the semichord (AM in figure 9.2) that corresponded to the semiangle at the center of the circle (of radius 3,438′, where each minute was a unit of length equal to 1/60 of the length of 1° of arc on the circle) was given at intervals of 3° 45′: effectively, a sine table. The only difference between this table and a modern one is that it gives the Indian sine, or *jya*, of the angle α:

$$jya\ \alpha = r \sin \alpha = 3,438 \sin \alpha.$$

From the tenth century onward, starting with the work of Abu Nasr Mansur (c. 960–1036), Islamic mathematicians brought the sine function closer to its modern form with a few defining it for the first time in terms of a circle of unit radius, although it remained defined for an arc of a circle rather than the angle subtended at the center.

The etymology of the word "sine" is instructive, for it shows what can happen as a result of imperfect linguistic and cultural filtering. The Sanskrit term for sine in an astronomical context was *jya-ardha* (half chord), which was later abbreviated to *jya*. From this came the phonetically derived Arabic word *jiba*, which, following the usual practice of omitting vowels in Semitic languages, was written as *jyb*. Early Latin translators, coming across this word, mistook it for another word, *jaib*, which had among its meanings the opening of a woman's garment at the neck, or bosom; *jaib* was translated as *sinus*, which in Latin had a number of meanings, including a cavity in facial bones (whence sinusitis), bay, bosom, and, indeed, curve. And hence the present word "sine."

There are two other functions that Islamic mathematicians may have derived from the Indians. *Kojya* (i.e., $r \cos \alpha$, or OM in figure 9.2) and *ukramajya* (i.e., r vers $\alpha = r(1 - \cos \alpha)$, or MC in figure 9.2) were trigonometric functions commonly used in Indian astronomy during the period of contact between the Indians and the Islamic world. But the tangent and cotangent functions are of Islamic origin.

During the ninth century the Islamic astronomer Habash al-Hasib examined the length of the shadow of a rod of unit length horizontally mounted on a wall when the sun was at a given angle to the horizontal. It is easily shown (figure 11.14a) that the length s of the shadow on the wall can be calculated as

$$s = \frac{\sin \alpha}{\cos \alpha} = \tan \alpha,$$

where α is the angle of elevation of the sun above the horizon. The length t of the shadow cast by a vertical rod (see figure 11.14b) is

$$t = \frac{\cos \alpha}{\sin \alpha} = \cot \alpha.$$

Al-Hasib also contructed the first table of tangents and cotangents.

The secant and cosecant functions seldom appeared in Islamic trigonometric tables. They were first mentioned without special names by Abu al-Wafa (940–997), who was also one of the first to construct a table of tangents. But such tables were of little practical use until navigational tables were computed in the fifteenth century. The first printed table of secants appeared in European mathematics in the work of Georg Joachim Rhaeticus (1514–1576), who was a younger contemporary of Copernicus

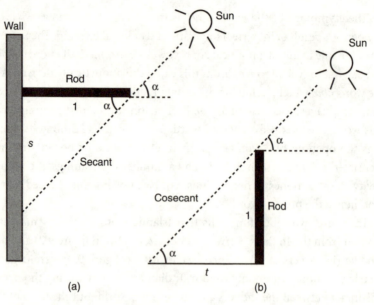

FIGURE 11.14: A "shadow" problem resolved

(1473–1543) and later his disciple. However, where they were referred to in Islamic mathematics, the secant and the cosecant were known respectively as the "hypotenuse of the shadow" (the distance from the top of the horizontal rod to the tip of the shadow in figure 11.14a) and the "hypotenuse of the reversed shadow" (the distance from the top of the vertical rod to the tip of the shadow in figure 11.14b). A long-standing trigonometric tradition based on shadow lengths is found in both Indian and Islamic mathematics.[39]

Derivation of Trigonometric Relationships

Abu al-Wafa's work on trigonometry contains more than a systematic treatment of the six functions. In his *Zij almagesti* he gives a rule for calculating the sine of the sum of two arcs and the sine of their difference when each of them is known:

> Multiply the sine of each of them by the cosine of the other, expressed in sixtieths, and we add the two products if we want the sine of the sum of the two arcs, but take the difference if we want the sine of their difference.

Expressed in modern notation, this rule becomes the familiar

$$\sin(\alpha \pm \beta) = \sin \alpha \cos \beta \pm \cos \alpha \sin\beta.$$

The reference to sine and cosine functions expressed in sixtieths shows that calculations were carried out in sexagesimal fractions. Al-Wafa provides a proof of this result in terms of arcs of a circle of unit radius (Berggren 1986, pp. 136–38).

The sine rule in its modern version is sometimes wrongly attributed to Nasir al-Din al-Tusi, although even the spherical version of the rule was known at least 250 years before him. In spherical trigonometry, the ratio of the sines of any two angles is equal to the ratio of the sines of the great arcs forming the sides opposite the angles. In plane trigonometry the ratio of the sines of any two angles is equal to the ratio of the two opposite sides. This result for a spherical triangle was discovered almost simultaneously by Nasir ibn Iraq and Abu al-Wafa, and a long controversy ensued on the question of priority. Abu al-Wafa's proof is the one that is better known and is contained in his astronomical text, the *Almagest*. Unfortunately, a discussion of the significant contributions of Islamic mathematicians to spherical trigonometry is well beyond the scope of this book.[40]

The sine rule in plane trigonometry may be stated in the following way. Given any triangle ABC (see figure 11.15),

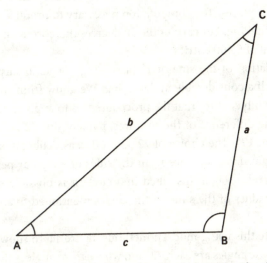

FIGURE 11.15: Al-Tusi's procedure for calculating the sides of a triangle

$$\frac{b}{c} = \frac{r \sin B}{r \sin C},$$

where r is taken to be 60 units. Al-Tusi provides a proof for this rule (Beggren 1986, pp. 138–39) and proceeds to consider how the result could be used to calculate the dimensions of a triangle, given a knowledge of different combinations of angles and sides. For example, knowing the values of one angle (say B) and two sides (b and c), the other angle (C) can be calculated by first using the rule given above and then looking up the angle in a sine table. This would immediately yield the other angle (A), and the sine rule could then be used to obtain the length of side a.

Construction of Trigonometric Tables

The interest of Islamic mathematicians in trigonometry was triggered by their discovery of sine tables in the Indian *Siddhantas*. They soon realized that trigonometric calculations, whether applied to astronomy or to geometry, required detailed and accurate tables, and they proceeded to construct tables that were more accurate than any before. Al-Hasib (c. 850) constructed the first sine and tangent tables at intervals of $1°$, accurate to three sexagesimal (five decimal) places. Subsequent work concentrated on reducing the intervals and increasing the accuracy of these tables. Thus, in the works of the astronomer-king Ulugh Beg in 1440 are tables for the two functions at intervals of 1/60 of a degree correct to five sexagesimal (nine decimal) places. The computation necessary to produce such a table is quite breathtaking. For each of the 90 degrees there would be 60 entries, making a total of 5,400 entries.

The calculation of the sine of $1°$ (assuming, for simplicity, a unit radius) was itself a considerable undertaking. We know from Abu al-Wafa's work in the tenth century that the procedure was to apply the formula for the sine of the difference of the two arcs, namely sin $(72° − 60°)$, which would give sin $12°$. The choice of $72°$ and $60°$ was deliberate and derives from Ptolemy's *Almagest*, since from the sides of a regular pentagon and of an equilateral hexagon inscribed in a circle it is possible to work out the required values of the sines of the angles mentioned to any degree of accuracy.

To illustrate this using modern methods, figure 11.16 shows a triangle ABC whose base angles are each $72°$ and whose third angle is therefore $36°$. It can be shown that[41]

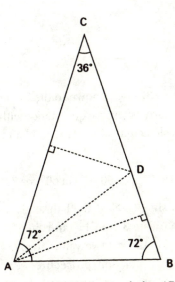

FIGURE 11.16: Equation 11.13 is established by drawing the line AD bisecting angle CAB, then dropping perpendiculars from A and D

$$2 \cos 36° = 2 \cos 72° + 1. \tag{11.13}$$

From the double-angle formula for the cosine (i.e., $\cos 2x = 2\cos^2 x - 1$), equation (11.13) can be expressed as

$$2c = 2(2c^2 - 1) + 1, \quad \text{where } c = \cos 36°.$$

Therefore $4c^2 - 2c - 1 = 0$, the solution of which, for $c > 0$, is

$$c = \frac{2 \pm \sqrt{4 + 16}}{8} = \frac{1 + \sqrt{5}}{4},$$

or $\cos 36° = \frac{1}{4}(1 + \sqrt{5})$.

Now, to calculate $\sin 72°$, since $\sin 72° = \cos 18°$, we set $c = \cos 18$. Then

$$2c^2 - 1 = \frac{1 + \sqrt{5}}{4}$$

and

$$c^2 = \frac{5 + \sqrt{5}}{8},$$

so

$$\cos 18° = \sin 72° = \sqrt{\frac{5 + \sqrt{5}}{8}}.$$

The fact that $\sin 60° = \sqrt{3}/2$ was known to the Indians as early as AD 500 probably explains the early Islamic acquaintance with the result.[42] The difference between the calculated values of $\sin 72°$ and $\sin 60°$ is taken and used with the expression

$$\sin(72° - 60°) = \sin 72° \cos 60° - \sin 60° \cos 72°$$

to obtain the value of $\sin 12°$. Next the half-angle formula[43] is applied to yield, successively, $\sin 6°$, $\sin 3°$, $\sin 1\frac{1}{2}°$ and $\sin \frac{3}{4}°$; and then some linear interpolation is applied to the last two values to give an estimate of $\sin 1°$.

Different types of interpolation procedure were experimented with, particularly when it was recognized that while linear interpolation would work well over small intervals, where the growth was uniform, it was not appropriate for large intervals, or for the upper bounds of a tangent function, where the value of the tangent of an angle approaches infinity as the angle approaches 90° (i.e., the function has a vertical asymptote at 90°). This recognition was already implicit in Indian mathematics before it came to be known in the Islamic world. One of the greatest Islamic astronomers, Ibn Yunus, who lived during the first half of the tenth century, had devised a second-order interpolation procedure and used it for constructing his sine table. New ground was broken by al-Kashi. In his book *Risala al-watar wa'l-jaib* (Treatise on the Chord and Sine) he treats the problem of obtaining an accurate estimate of $\sin 1°$ in a different manner, devising an iterative procedure involving the solution of cubic equations. His method highlights some of the similarities between the methods used for this purpose by the Chinese and the Islamic mathematicians.

The approximation to $\sin 1°$ was based on two pieces of information known to Islamic mathematicians at least three centuries before al-Kashi's time. (For the sake of simplicity we shall use base 10 here, not the sexagesimal base with which al-Kashi worked. Also, we shall use the modern sine function whereas al-Kashi's sine function is 60 times the modern one.) The two relationships are, first, that for any given angle α,

$$\sin 3\alpha = 3 \sin \alpha - 4 \sin^3 \alpha,$$

so that

$$\sin 3° = 3 \sin 1° - 4 \sin^3 1°. \tag{11.14}$$

Second, by the method discussed earlier, we calculate that

$$\sin 3° = 0.052335956. \tag{11.15}$$

Combining equations (11.14) and (11.15), and denoting the unknown value $\sin 1°$ by x, we get the cubic equation

$$x^3 - 0.75x + 0.013083989 = 0.$$

To solve this cubic equation al-Kashi used an iterative method that, expressed in modern terms, proceeds as follows. Given an equation of $y = f(x)$, choose an arbitrary value x_0 as a first approximation to the root. Then, by using the relation $x_n = f(x_{n-1})$ for $n = 1, 2, 3, \ldots$, a sequence of values $x_1, x_2, \ldots$, is obtained that approximates more and more closely to the solution, irrespective of what was chosen as x_0, as long as $\lim x_0$ exists. In numerical analysis this procedure is known as "fixed point" or "direct" iteration.

By this method al-Kashi computed the value of $60 \sin 1°$ correct to nine sexagesimal (sixteen decimal) places—a remarkable exhibition of computational skill, even by today's standards. The reader who wishes to know about al-Kashi's method will find Aaboe (1954) illuminating. Values of sines for $\frac{1}{2}°$, $\frac{1}{4}°$, $\frac{1}{8}°$, and so on were obtained by applying the half-angle formula; other fractions or finer divisions were achieved by applying some appropriate interpolation formula. Variants of al-Kashi's method were used by astronomers, mainly working in Samarkand, including his patron Ulugh Beg, who paid him the tribute quoted earlier. An iterative procedure similar to al-Kashi's was used by the German astronomer Johannes Kepler (1571–1630).

Mathematics from Related Sources

Mathematics from Hebrew Sources

The oldest mathematical text in Hebrew is the *Mishnat ha-Middol*. Of unknown authorship and uncertain date, it gives practical rules for mensuration, including that of the Tabernacle that the Jews constructed in the desert. There is a section on the solution of quadratic equations with the geometric analogues of these equations, which may have been a later inclusion since its composition was strongly influenced by al-Khwarizmi's algebra from a later period. It was only in the twelfth century that significant

Hebrew mathematics emerged, in the persons of Abraham bar Hiyya (c. 1065–1145) and Abraham ibn Ezra (1092–1167).

Abraham bar Hiyya spent most of his life in Barcelona in Christian Spain, although it is likely that he was educated in the Muslim kingdom of Zaragoza. His major text, written in Hebrew and titled *Hibbur ha-Meshiha we ha Tisboret* (The Composition on Geometrical Measures), has an extensive coverage that includes algebra akin to that in al-Khwarizmi's work, mensuration on plane figures and solids including the truncated pyramid, some trigonometry of chords, and a lost fragment of Euclid's work on the "division of figures." The text was translated into Latin as *Liber embadorum* (1145) by Plato of Tivoli and would play a significant role in the spread of mathematical knowledge into Europe.

The contribution of Abraham ibn Ezra is more difficult to assess. Born in Aragon and imbued with Islamic culture and learning, he was a versatile scholar and poet. His mathematical reputation is essentially based on one book, *Sefer ha-Mispar* (The Book of Numbers). Written before 1160, it is one of the first expositions in Europe of the Indo-Arabic numerals and arithmetical operations with them. It was an important text in the diffusion of Indian arithmetic among Hebrew readers. He also wrote another book, *Sefer ha-Ehad* (The Book of the One), that was highly prized by some Jewish scholars of the period for the "mathematical hints" (or numerology) contained in it.

The role of the two Abrahams was to make basic mathematical knowledge available to Jewish communities whose members were either unfamiliar with Arabic sources or did not have direct access to these sources. A scientific language close to biblical Hebrew became the vehicle for the transmission of such knowledge. Translations of scientific and philosophical texts from Arabic into Hebrew gathered momentum. For example, Jacob Anatoli (c. 1194–1256), living in Italy under the patronage of Frederic II of Sicily, translated Ptolemy's *Almagest*, Ibn Rushd's *Compendium on Astronomy*, and Euclid's *Elements* into Hebrew. Moses ben Samuel ben Yehuda of Montpellier (fl. 1240–1283) carried out a number of translations of mathematical texts from Arabic, including works of Euclid, al-Farabi (Commentary on the *Elements*), ibn al-Haytham (Commentary on the *Elements*) and al-Hassar (The Arithmetic). Other major translators of mathematical literature from Arabic into Hebrew included Jacob ben Makhir of Montpellier (c. 1236–1305), Yehuda ben Solmon ha-Kohen (b. 1215), and Qalonymos ben Qalonymos (b. 1287). The activities of these and other translators meant that, by the middle of the thirteenth century, there were Hebrew versions of the major Greek texts of Euclid, Archimedes, and Ptolemy, and of works of Islamic mathematicians, such as Abu Kamil (*Algebra*) and Ibn al-Haytham (*Astronomy*).

An outstanding example of the resulting creativity was Levi ben Gershom (1288–1344), better known in Europe by his Latin name Gersonides. He was one of the most versatile Jewish scholars of the European Middle Ages, and wrote on philosophy, logic, mathematics, and astronomy. His treatise *Ma'asheh Hoshev* (Computer's Manual) contains studies of permutations, combinations, and summations of series as well as introductions to arithmetic and algebra. He also wrote commentaries on the *Elements*, including a treatise on the "parallel postulate," and on Thabit ibn Qurra's *Risala fi Shakl al-qatta* (On the Secant Figure). However, his most lasting contribution is contained in his book on astronomy, *Sefer Tekhunah*, which is preserved in both the Hebrew and Latin versions. A lengthy work, divided into 136 chapters, it is notable for its emphasis on astronomical observations and its critical assessment of Ptolemy and the Islamic astronomer al-Bitruji.[44] Using his own solar and lunar models, he calculated sine tables, spherical astronomical parameters, and solar and lunar mean motions and corrections. His practical bent led him to the invention of a transversal scale that helped to reduce the random errors introduced into observations when reading off minutes from the linear scale of an astrolabe calibrated in degrees. His writings on harmonic numbers (now extant only in Latin translation), combinatorial analysis, and the geometry of Euclid have been studied in depth by Tony Levy (1996) for their Arabic sources, and for their impact on later work. As mentioned in chapter 1, there is the possibility of a transmission of the basic formulas for finding permutations and combinations from Gershom's *Computer's Manual* to a Cardano manuscript, which in turn bore great similarities to the treatment in Mersenne's classical book on music theory, *Harmonie Universelle* (1636).

In this short survey of mathematicians from the Hebrew tradition, we have omitted discussion of one of the most influential figures in Jewish intellectual history, Moses Maimonides (d. 1204). His early education in Andalusian Spain introduced him to the scientific treatises of fellow Andalusians Jabir ibn Aflah and al-Mu'tamir ibn Hud. He wrote in both Arabic and Hebrew, and his notable works in the former language include commentaries on Apollonius's *Conics* and Ibn al-Haytham's *Completion of the Conics* (Langermann 1984).

A study of mathematics from Hebrew sources is an instructive exercise. Like Islamic mathematics, of which it was an integral part at times, it went through different phases: translation, assimilation, synthesis, creation, and transmission. Mathematical ideas and practices contained in Arabic texts were translated into Hebrew and then assimilated by Jewish communities unfamiliar with Islamic sources. From their ranks came notable scholars who synthesized and developed the subject. Their writing became an important channel for the diffusion of ideas and texts before Europe discovered the original Greek texts.

Mathematics of the Maghreb

There is a tendency, in studying Islamic mathematics, to confine one's attention to the activities around the Middle East and Spain, and ignore the work in North and Northwest Africa, a large region referred to as the Maghreb, in which there occurred mathematical activity within the framework of the Islamic civilization. The starting point of this activity was the link between Andalusia in Spain and the Maghreb, where the close political, economic, and cultural ties meant that it was difficult to separate the two regions of the Muslim West.

From the information available, mathematical activity began in Ifriqya (present-day Tunisia) around the end of the eighth century, when there is a record of a scholar known as Yahya al-Kharraz, who wrote a book on metrology. For the next two centuries information about mathematical activity in the region is scarce, although there is some evidence of knowledge moving westward from Baghdad. Abu Sahl al-Qayarawani, who may have spent some time in Baghdad, is the first known mathematician from the Maghreb. He wrote a book titled *Kitab fi al-hisab al-hindi* explaining Indian arithmetic. Around the same period, the ruler Ibrahim II (875–902) established a Bail al-Hikma (House of Wisdom) in Raqqada, modeled on the lines of the one in Baghdad, to promote the study of mathematics, astronomy, and other subjects. This institution survived as a scientific center until the establishment of the Fatimid caliphate in Egypt. The patronage of learning was reestablished during the reign of Fatimid caliph al-Mu'izz, and although we know that the study of mathematics and astronomy flourished during this period, there is little information on the scholars and the type of work that they undertook.

There is considerably more information available on mathematical activity in the Maghreb during the Almohad period (twelfth and thirteenth centuries). Of the three mathematicians who have been identified and studied, Ibn Mun'im's (d. 1228) work is the most innovative. In his only extant work, *Fiqh al-hisab*, dealing with combinatory problems, he states the rule for determining all possible combinations of n colors p times and establishes, inductively, the resulting arithmetic triangle of the relationship:

$$^nC_p = {}^{n-1}C_{p-1} + {}^{n-2}C_{p-2} + \ldots + {}^{p-1}C_{p-1}.$$

He applies similar formulas for permutations with and without repetitions using the Arabic alphabet for illustration purposes. One of the more remarkable aspects of his work is the use of combinatorial reasoning, the earliest attempt in Islamic mathematics to do so.

With the advent of the fourteenth century the quantity of mathematical writing increased substantially, although many of these texts took the form of commentaries and summaries of previous work. However, this century produced a mathematician of exceptional ability: Ibn al-Banna (c. 1256–1321). Born in Marrakesh (Morocco), he became a versatile scholar, having made a study of the Arabic language, the Qur'an, astronomy, mathematics, and medicine. His fame rests mainly on his work in mathematics. His best-known book, *Talkis a'mal al-hisab* (Summary of Arithmetical Operations) is still extant. This was widely known in the Islamic world because of its clarity and conciseness. He also wrote two other texts on calculation that have survived: *al-Qanun al-hisab* (Manual of Mathematics) and *Raf al-hijab* (Lifting of the Veil). Other texts include an introduction to Euclid, a book on algebra, a treatise on geometry, and a popular astronomical almanac. In the *Raf al-Hijab* there are some interesting mathematical ideas and results. It contains one of the earliest expressions of the continued fractions used to compute square roots. In the section on summing series Ibn al-Banna obtains the results for n terms as

$$1^3 + 3^3 + 5^3 + \ldots + (2n - 1)^3 = n^2(2n^2 - 1);$$

$$1^2 + 3^2 + 5^2 + \ldots + (2n - 1)^2 = \frac{2n(2n - 1)(2n + 1)}{6}.$$

However, Rashed (1994) considers Ibn al-Banna's work on binomial coefficients to be his most innovative work; it is even more fundamental than the Pascal triangle results given by al-Karaji and al-Samawal.

The Maghreb tradition carried on for a few centuries, but the quality of the work declined. Mathematicians were reduced to teaching or mundane activities such as advising on the determination of times for prayer or distributing inheritances or using astronomical instruments. Yet sparks of curiosity and inventiveness appeared occasionally. Muhammad ibn Muhammad al-Fullani al-Kishnawi was a Fulani from northern Nigeria. He traveled to Egypt and, in 1732, wrote a manuscript (in Arabic) of procedures for constructing magic squares up to order eleven. As words of encouragement to the reader, he writes: "Do not give up, for that is ignorance and not according to the rules of this art. . . . Like the lover, you cannot hope to achieve success without infinite perseverance." He died in Cairo in 1741.

General methods for constructing magic squares first appeared in the Islamic world, including the Maghreb, during the ninth century. From the thirteenth century, recreational and divinatory applications began to replace mathematical study. However, interest in the mathematics of construction survived, as shown by the work of Muhammad ibn Muhammad al-Fulani al-Kishnawi.

The Islamic Contribution: A Final Assessment

In chapter 1 we contrasted modern Eurocentric attitudes toward the Islamic contribution with the seminal role Islamic innovators played in transmitting mathematics to Western Europe, setting the stage for the development of modern mathematics. It should be clear from the present chapter that the traditional view of the medieval Islamic world as a mere custodian of Greek learning and passive transmitter of knowledge is both a partial view and a distorted one. We have seen how original were the Islamic scholars' contributions to algebra and trigonometry, and how crucial was the role they played in bringing together two different mathematical strands—the algebraic and arithmetic traditions so evident in the mathematical cultures of Mesopotamia, India, and China; and the geometric traditions of Greece and the Hellenistic world. The intertwining of these strands had already begun with later Alexandrian mathematicians such as Heron, Diophantus, and Pappus, who had absorbed some of their mathematics from Mesopotamia and Egypt, but there remained the constraints imposed by the straitjacket of the Greek mathematical tradition.[45] It was left to the Islamic scholars to bring together the best of both traditions. In doing so, they provided us with an efficient system of numeration in which calculations were no longer tied to mechanical devices, an algebra that was practical and rigorous, a geometry that was no longer an intellectual pastime, and a trigonometry freed from its ties to astronomy to eventually be used in fields as diverse as optics and surveying.

The Islamic approach to mathematics was no doubt helped in its early years by the existence of a creative tension between the "algebra people" and the "geometry people," best exemplified by al-Khwarizmi and Thabit ibn Qurra, respectively. Each group remained open to influences from the other group, as shown by al-Khwarizmi's geometric approach to the solution of quadratic equations and Thabit's discovery of a rule for generating amicable numbers. As mathematics developed, work on "pure" geometry, such as attempts to prove or modify Euclid's parallel postulate, continued alongside the development of skillful numerical methods for extracting roots and solving higher-order equations. Indeed, one of the main reasons why modern mathematics moved away so substantially from the spirit and methods of Greek mathematics was the intervention of the Islamic scholars. Perhaps, if the lessons had been absorbed earlier and if the works of the notable figures of Islamic mathematics such as Omar Khayyam and Thabit ibn Qurra had been better known than they were, the period of painful transition and the repetitive nature of some medieval European mathematics could have been avoided altogether.

Both history and religion in Europe conspired to stem the flow of ideas from the Islamic world at a time when Europe was rousing itself from its long slumber and taking its first confident steps into the realm of ideas. Increasingly, Europe was exposed only to the Greek vision as represented by various translations into Latin and other languages from Arabic texts. In a search for their roots, Europeans bypassed their Islamic and non-European heritage and homed in on Greece and Rome. Greece thereby became the fount of their intellectual and cultural heritage, while for their religious roots they looked toward Rome and the Byzantium. Here they eventually rewove a synthetic Christianity from some of the various strands into which doctrinal controversies had split the original Pauline faith, although the result was in some ways far removed from its Eastern and Judaic origins. The history of the last five hundred years has tended to strengthen these ties, partly as a consequence of European dominance and partly under the impetus of "Classical" scholarship, much of which regards Greece as the sole source of knowledge and culture.

In mathematics, the glorification of ancient Greece during the Renaissance led to a concentration on Hellenistic texts. The medieval Islamic world also admired the Hellenistic contribution, particularly in geometry.

One wonders, especially with Thabit ibn Qurra, whether some of the time and effort spent translating Greek works might have been put to better use in developing the translators' own, very promising algebra. Nevertheless, the Islamic world remained open to other influences as well.

There is no denying that the Greek approach to mathematics produced remarkable results, but it also hampered the subsequent development of the subject. The strengths of the Greek approach have been discussed extensively; any standard textbook on the history of mathematics deals with this, so there is little point in going over the positive aspects again. But the limiting effect of the Greek mode of thought is another matter. The Greek preoccupation with geometry until the infiltration of the Mesopotamian and Egyptian influences in the later Hellenistic period was a serious constraint. Great minds such as Pythagoras, Euclid, and Apollonius spent much of their time creating what were essentially abstract, idealized constructs; how they arrived at a conclusion was in some way more important than any practical significance. There were in fact two different geometries coexisting at the time: the "pure" geometry of the Greeks, whose validity was determined wholly by its internal consistency and coherence, and the "applied" geometry of other mathematical traditions, whose validity was judged solely by its ability to describe physical reality. (It is interesting to speculate what a Euclid who had absorbed the arithmetic and algebra of the Mesopotamians and had sympathy with their analytic/algebraic approach to geometry might have created with his particular brand of deductive reasoning.) Apollonius's *Conics* seemed to be a product of a Greek abstract geometry that had reached a level of refinement with no further progress in sight. Only with the emergence of the Islamic mathematics were the works of the period rescued and given a new direction. However, the pioneers of modern mathematics in the post-Renaissance period found themselves compelled to undergo a sometimes painful distancing from the Greek geometric approach their predecessors had too readily espoused, unleavened as it was with the Islamic spirit.

We conclude this chapter by returning to the question of transmissions, which we have touched on in several previous chapters. The impact of the Islamic world on the intellectual life of Europe is better chronicled than most other cross-cultural influences. The spread of Indian numerals, the growth of algebra, the introduction of trigonometry, the dissemination of Greek geometry, and the Islamic extensions to it—these are all well authenticated and recognized in the more recent histories of mathematics.

The possible transmission of certain techniques through third parties for which written records are nonexistent or incomplete is more problematic. We have looked at several examples, including solutions of higher-order equations by the Horner-Ruffini method and Thabit's rule for generating amicable numbers (sometimes credited to Fermat and Euler). As a further instance, it is now known that Nicolaus Copernicus (1473–1543) owed a considerable debt to the Islamic mathematician-astronomers Nasir al-Din al-Tusi (1201–1274) and Mu'ayyad al-Din al-'Urdi (d. 1266), some of whose ideas are incorporated in the Copernican solar system (though it was Copernicus who put the sun at the center, thus reviving an idea that dates back to the Greek Aristarchus of Samos in the third century BC) (Saliba 2007, pp. 193–232). There is clearly a need for further examination of known medieval European sources and for a search for other archival material, especially in Arabic and Ottoman sources.[46]

Apart from transmissions to the West, there are two other links that require further elaboration.

1. There is the whole question of possible transmissions of mathematical ideas between the Islamic world and China. We have remarked in this chapter on how Chinese methods of solving numerical equations of higher order may have influenced Islamic mathematics; at the end of chapter 7, on Chinese mathematics, we briefly considered the likelihood that Islamic trigonometry and arithmetic may have reached China. In looking for channels along which such transmissions could have occurred, we must take into account the political and social climate of the first half of the second millennium AD. There is evidence, from the time of the Song dynasty, of political and cultural contacts between the two societies. The Mongol empire stretched across a good part of central Asia. We have seen how receptive some of the rulers were to scientific ideas from lands they had conquered. Hulegu Khan and Ulugh Beg were not only patrons but practitioners of science. The intriguing question remains: did they fulfill the same role as the caliphs of Baghdad by encouraging contacts between their scientists? Here again, more research needs to be done before we can provide any definite answers.

2. Indian mathematics and astronomy absorbed much from Islamic and Persian sources; astronomy was the main beneficiary. In 1370 Mahendra Suri, the astronomer at the court of Firoz Tughlaq, published his work *Yantraraja*, which introduced Persian and Islamic astronomy

into the Sanskrit *Siddhanta* tradition. This flow of astronomical ideas, as well as instruments, continued into the seventeenth century, providing the basic materials for those training in the Ptolemaic system. However, attempts to synthesize the two systems, which had such promising beginnings with Mahendra Suri, proved unsuccessful. Over a period of time there developed two distinct schools in mathematical astronomy: the old Sanskrit school and the new Islamic school. Occasionally they came together, usually under the patronage of an enlightened ruler such as the Mughal emperor Akbar, or Raja Jai Singh of Rajasthan. The latter left as his monument the large masonry astronomical instruments at Delhi, Jaipur, Ujjain, and Varanasi (Benares).

• • •

This is the end of our story. We have traveled around the world in search of our "hidden" mathematical heritage, and in the rich tapestry of early human experience we have discovered mathematics in bones, strings, and standing stones. No society, however small or remote, has ever lacked the basic curiosity and "number sense" that is part of the global mathematical experience. The need to record information that gave birth to written language also brought a variety of number systems, each with its own strengths and peculiarities. And yet if there is a single universal object, one that transcends linguistic, national, and cultural barriers and is acceptable to all and denied by none, it is our present set of numerals. From its remote beginnings in India, its gradual spread in all directions remains the great romantic episode in the history of mathematics. It is hoped that this episode, together with other non-European mathematical achievements highlighted in this book, will help to extend our horizons and dent the parochialism that lies behind the Eurocentric perception of the development of mathematical knowledge.

Notes

1. For the present edition, I decided to substitute the phrase "medieval Islamic mathematics" or, in short where the context allowed, "Islamic mathematics" for "Arab Mathematics." This allows for the accommodation of other groups such as Iranians, Egyptians, North and West Africans. A further qualification then becomes necessary. Although Islam is a religion, the term can be used to describe a culture tha includes significant communities of other religious groups—notably Christians and Jews. During

the period (hereafter referred to as the medieval period) that we are concerned with in this book (from AD 750 to 1450), these groups played an important part. There is a brief section on the contribution of those mathematicians who wrote in Hebrew. However, as we will find out in this chapter, the majority of the mathematicians discussed were Muslims of different denominations living in societies where Islam was the dominant religion.

2. This view is now hotly contested. An alternative view, espoused by Balty-Guesdon (1992) and Gutas (1998) among others, sees Bait al-Hikma as little more than a library where a few translations were carried out.

3. For a detailed examination of the meaning of the term *zij* from a large number of primary sources in different languages, including Persian, Arabic, and Sanskrit, see Mercier (2004, pp. 451–60).

4. To understand the probable debt to Indian algebra, it is important to remember that in the context of algebraic notation a distinction may be made between two different categories of things to be represented: unknown quantities and types of numbers. In Indian algebra, this distinction is present from early times. For example, colors (black [*calaca*], blue [*nilaca*], yellow [*pitaca*], red [*lohitaca*], . . .) and later letters (*ca, ni, pi, loh,* . . .) were used to represent unknown quantities. At the same time, names for the type or species of numbers such as *rupa, varga,* and *ghana* for absolute numbers, squares, and cubes respectively were also used. This is well summed up in the following passage from Bhaskaracharya's *Bijganita*, translated by Colebrooke (1817):

> When absolute number and colour or letter are multiplied one by the other, the product will be colour or letter. When two or more homogenous quantities [meaning those of the same color] are multiplied together, the product will be the square, cube or other [power] of the quantity. But if unlike quantities be multiplied, the result is their (*bhavita*) 'to be' product or factum. (p. 140; my insertions in square brackets)

In al-Khwarizmi's *Algebra*, a distinction exists between "types of numbers that appear in the calculation" (e.g., treasures [*mal*], roots [*jidr*], . . .) and unknown quantities (e.g., thing [*shay*]) but only for the purpose of expressing arithmetic operations rhetorically. Later Islamic mathematicians, notably al-Karaji and Omar Khayyam, extended this closer to our usage: *shay* and *mal* were identified without x and x^2 respectively. But this is another story.

5. More precisely, these consisted of a large dollop of Indian and unknown quantities of Babylonian and Greek ingredients.

6. Algorithm, algorism, or *augrim* originally referred to systematic calculation with decimal numbers. It was referred to as such in Dr. Johnson's 1755 *Dictionary of the English Language.*

7. More strictly, al-Khwarizmi's *zij* became influential after the thirteenth century, except in al-Andulus, where it gained favor from the eleventh century. Its influence on the later *zijes* compiled in medieval Europe is not insignificant, although not as important as the so-called Toledan and Alphonsine tables. I am grateful to an anonymous reader for drawing my attention to this point.

8. The Muslim era begins in the year that Muhammad fled from Mecca (AD 622, the year of the Hejira). The Muslim year is a lunar year and therefore about eleven days shorter than the Western calendar year, so an AH date ("after Hejira") cannot be converted to an AD date simply by adding 622.

9. For further details, see Pingree (1970).

10. Al-Khwarizmi's *Zij al-Sindhind* remains an important "transmission" document in the history of mathematics. It highlighted the role of pre-Islamic Indian astronomy in the birth of Islamic astronomy. The astronomer Ibn al-Adami described al-Khwarizmi's *Zij* as an abridgement of the one that al-Fazari had prepared for Caliph al-Ma'mun soon after the Sind mission during the second half of the eighth century. The fame of al-Khwarizmi's tables had spread among the astronomers not only in Baghdad but also in central Asia and Andalusia. Al-Biruni wrote three commentaries on this work, in one of which he defended al-Khwarizmi against attacks by al-Ahwazi. And as late as the nineteenth century, al-Khwarizmi's *Zij* was being copied in Egypt. For further details and references, see Brentjes (2007).

11. Here again it is difficult to separate fact from fiction. For more recent details and references, see King (2000).

12. For further details, see Berggren (1986).

13. Al-Khwarizmi's treatise on the Jewish calendar contains rules for calculating the mean longitude of the sun and the moon based on this calendar and for determining on what day of the Muslim week the first day of the new year would fall. It also discusses more fanciful subjects such as the lapse of time between the beginning of the Jewish era (i.e., the creation of Adam) and the beginning of the Seleucid era. For further details, see Kennedy (1964).

14. Quadrature is the process of determining the area of a plane geometric figure by dividing it into a collection of shapes of known area.

15. The status of Euclid's parallel postulate has been a source of great controversy in the history of mathematics. The question first raised by Greek writers was whether Euclid had made a postulate out of what amounted to a theorem. Attempts to prove the postulate continued attracting the attention of Islamic mathematicians of the caliber of Omar Khayyam and Nasir al-Din al-Tusi and those who came after.

16. A comparison between the two calendars indicates that, in the case of Omar's, 96 out of 396 years were made into leap years, while the corresponding figure for the Gregorian calendar is 97 out of 400. It is estimated that the Gregorian calendar had an error of one day in 3,330 years, while Omar's (*Jalali*) calendar had an error of one day in 3,770 years.

17. The Islamic work on the parallel postulate is outside the scope of this book. For further details, see Rosenfeld (1988).

18. Courtly patronage offered to scholars by the Mongols can only be a partial explanation for the continuity of mathematical studies during this period, for after all they ruled only Iran, parts of Iraq, and Anatolia. Other explanations should be sought in the widespread interest in mathematics, particularly since the subject was taught also in madrassas all over the Islamic world.

19. These multiplication methods are discussed by Smith (1923–25).

20. The number 6 is perfect since the sum of its proper divisors is $1 + 2 + 3 = 6$.

The number 8 is deficient since the sum of its proper divisors is $1 + 2 + 4 < 8$.

The number 12 is abundant since the sum of its proper divisors is $1 + 2 + 3 + 4 + 6 > 12$.

21. "Casting out nines" is an originally Indian method of checking addition and multiplication. It uses the well-known property that the sum of the digits of any natural number when divided by 9 produces the same remainder as when the number itself is divided by 9. For example, in checking that the product of 436 and 659 is 287,324,

   1. Add the digits of 436 to get 13, whose digits are then added to get 4.

   2. Add the digits of 659 to get 20, whose digits are then added to get 2.

   3. Add the digits of the product 287,324 to get 26, whose digits add to 8.

So "casting out nines" leaves remainders of 4, 2, and 8 respectively, and since $4 \times 2 = 8$, the multiplication is probably correct.

22. An aliquot is any divisor of a given number other than the number itself. A prime number has only one aliquot part: the number 1. Numbers 1, 2, 3, 4, and 6 are all aliquot parts of 12.

23. For a useful discussion of the historical background to extraction of higher roots in the medieval Islamic culture and the mathematics of this procedure, which results in the generation of the numbers in the Pascal's triangle, see Berggren (1986, pp. 53–63). The procedure is equivalent to the Ruffini-Horner method for extracting a fifth root. Al-Kashi (d. 1429), the Persian mathematician, in his book *The Key of Arithmetic* illustrates this procedure by working out the fifth root of the number 44,240,899,506,197.

24. Oakes and Alkhateeb (2007) have argued that words *al-jabr* and *al-muqabala* were used in everyday Arabic long before being "appropriated" into arithmetic and algebra. What is uniquely algebraic is the phrase *al-jabr wa'l-muqabala*. It began as a shorthand way of saying "by *al-jabr* and/or *al-muqabala*." Different Italian books have *argibra*, *alcibra*, *algebra*, and even *aliabraa argibra*. The English word "algebra" derives from the Italian spelling.

25. For further details of Thabit's approach, see Rashed (1994).

26. A detailed step-by-step comparison is found in Joseph (1994d, 65–67) and will not be discussed here. The difference between Al-Khwarizmi's and Abu Kamil's approaches is that the latter makes *explicit* use of propositions from Euclid's *Elements* in solving the quadratic equation discussed. However, it is worth noting that Abu Kamil gave essentially the same proofs for solving quadratic equations as Thabit ibn Qurra, who lived around the same time.

27. Since the book was dedicated to a vizier by the name of Fakhr al-Din, the title may well reflect that fact.

28. A conic is a curve formed by intersecting a cone with a plane. In 200 BC, the Greek mathematician Apollonius of Perga undertook a systematic study of the genesis of five type of conics: circle, hyperbola, ellipse, parabola, and rectangular hyperbola. For example, the circle and the ellipse arise when the intersection of cone and plane is a closed curve. And the circle is a special case of the ellipse: it arises when the plane is perpendicular to the axis of the cone.

29. I am grateful to Victor Katz for this observation.

30. For some of the most recent evidence on the Islamic influence, see Bala (2007) and Saliba (2007).

31. For a statement of this proof and the original diagram on which figure 11.9 is based, see Berggren (2007, p. 579).

32. In the *Quadrature of the Parabola*, Archimedes proved that the area enclosed by a parabola and a straight line is 4/3 multiplied by the area of a triangle with equal base and height. He expressed the solution to the problem as a geometric series that summed to infinity with the ratio 1/4.

33. The method of analysis and synthesis goes back to the Greeks and is a useful approach to solving problems. Analysis is a method of geometrical demonstration that proceeds from the solution and retraces the path of solution to an original set of givens that is known to be true. The ensuing synthesis moves from the known set of givens toward the solution. Ibn Sinan's work exhibits two main preoccupations. First, to obtain the magnitude of certain parts of geometrical figures, assuming that other parts are

given so that, for example, given certain chords and arcs of a circle, the diameter of the circle can be determined. Second, to investigate the construction of certain geometric figures, of which the most interesting are the so-called "contact" problems, such as, for example, how one constructs a circle that is tangent to given lines and circles and/or passes through certain specified points.

34. For more details of Ibn Sinan's innovative work in geometry, see Rashed (1996).

35. Unfortunately, al-Karaji's original work on binomial coefficients and the Pascal triangle is no longer extant, and so we have to depend on the account of al-Samawal (1125–1174). For a brief discussion of al-Karaji's work on binomial coefficients, see Rashed (1994, pp. 66–67).

36. "Alhazen's problem" involves locating a point or points reflecting on the surface of a concave or convex spherical mirror, given that the two points are related to one another as are the eye and the visible object. In Book 5 of his *Optics*, Ibn al-Haytham lays down the solution for a variety of surfaces—spherical, cylindrical, and conical. He does so based on certain results (lemmas) that he has proved for geometrical constructions.

37. A paraboloid is a type of surface in three dimensions. It can be shaped like an oval cup, in which case it is known as an elliptic paraboloid, or shaped like a saddle, in which case it is a hyperbolic paraboloid. Depending on the way a parabola is rotated, we obtain paraboloids of different shapes.

38. The discussion that follows is based on Berggren (2007, pp. 588–92) and Katz (1998). It should be remembered that al-Karaji gave one of the earliest-known proofs of the formula for the sum of cubes using a method of induction. However, as early as AD 500, the Indian mathematician Aryabhata had stated the result. It was rediscovered by Uthman al-Qabisi in tenth-century Baghdad and again in early fourteenth-century France by Levi ben Gershom. The Indian mathematician/astronomer Nilakantha, whose work we referred to in earlier chapters, gave a visual proof in 1500 that captured the essence of al-Karaji's approach.

39. It is interesting in this context that a contemporary of al-Hasib, known as al-Battani, gave the following rule for finding the elevation of the sun above the horizon (θ) in terms of the length s of the shadow cast by a vertical gnomon of height h:

$$s = \frac{h \sin (90° - \theta)}{\sin \theta}.$$

40. The interested reader may wish to consult the relevant sections of Van Brummelen (2009).

41. The result follows from: Let ABCDE be a regular pentagon of side 1 unit. M, P, and N are points on BE such that AP, CM, and DN are all perpendicular to BE.

(a) Find $\triangle$ABP and $\triangle$CBM.

(b) By considering $\triangle$ ABE, prove that BE = 2 cos 36°.

(c) By considering quadrilateral BCDE, prove that BE = 2 cos 72° + 1.

(d) Therefore BE = 2 cos 36° = 2 cos 72° + 1.

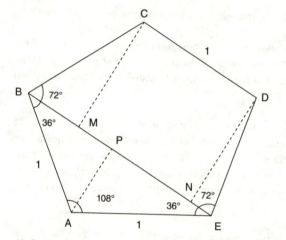

42. It may be argued that the Islamic mathematicians also obtained this value from Ptolemy's work, which contains the value corresponding to the sine of 36° (and therefore the values of 18° and 72°).

43. The half-angle formula follows from the identities $\sin^2 x + \cos^2 x = 1$ and $\cos 2x = \cos^2 x - \sin^2 x$. Thus

$$\sin x = \sqrt{1 - \cos^2 x} \quad \text{and} \quad \sin \frac{1}{2}x = \sqrt{\frac{1 - \cos x}{2}}.$$

44. A well-known translation of Gershom's astronomy is Goldstein (1985).

45. This statement may be seen by some as "incendiary and vague." The "constraints" referred to relate to a view that mathematics is a system of axiomatic/deductive truths inherited from the Greeks and is concerned with a search for infallible eternal truths and modes of establishing them. There is a growing awareness that the formal deductive format found in mathematics texts is a serious obstacle to understanding, leaving many students with no clear idea of what is being talked about. It is now also clearly recognized that the development of mathematical analysis in Europe became possible only when the Greek canon of logical rigor and the Greek mode of establishing mathematical truths was given up during the heyday of the development of "infinitesimal calculus." As a notable historian of calculus writes: "Although the Greek bequest of deductive rigour is the distinguishing feature of modern mathematics, it is arguable that, had all the succeeding generations also refused to use real numbers and limits

until they fully understood them, the calculus might never have been developed and mathematics might now be a dead and forgotten science" (Edwards 1979, p. 79).

46. More importantly, a reevaluation of the roots of the Copernican Revolution is called for. It is generally recognized that the Copernican Revolution transformed the European conception of the universe from an earth-centered vision to one in which the earth became only one planet among others orbiting the sun, and this is often seen as the key event that triggered the birth of modern science. What is often ignored in the Eurocentered history of science is the significant non-European contribution to this revolution. According to Bala (2007), these include Arabic optics and astronomy, Indian mathematics, and Chinese cosmological ideas, acknowledged in the pioneering studies of Needham (1954–) and others who followed him.

References

GENERAL

Aaboe, A., and J. L. Berggren. 1996. Didactical and other remarks on some theorems of Archimedes and infinitesimals. *Centaurus* 38 (4): 295–316.

Alatas, S. H. 1976. *Intellectuals in developing societies*. London: Frank Cass.

Ang Tian Se. 1978. Chinese interest in right-angled triangles. *Historia Mathematica* 5:253–66.

Bala, A. 2006. *The dialogue of civilisations in the birth of modern science*. Basingstoke: Palgrave Macmillan.

Bala, A., and G. G. Joseph. 2007. Indigenous knowledge and western science: The possibility of dialogue. *Race and Class* 49 (1): 39–61.

Bell, E. T. 1940. *Development of mathematics*. New York: McGraw-Hill.

Bernal, J. D. 1969. *Science in history*. London: Penguin.

Bernal, M. 1987. *Black Athena*. London: Free Association Books.

Bonala, R. 1955. *Non-Euclidean geometry*. New York: Dover.

Boyer, C. B. 1944. Fundamental steps in the development of numeration. *Isis* 35:153–68.

———. 1968. *A history of mathematics*. New York: John Wiley. Reprinted by Princeton University Press, Princeton, NJ, 1985.

Brohman, J. 1995a. Universalism, Eurocentrism, and ideological bias in development studies: From modernisation to neoliberalism. *Third World Quarterly* 16 (1): 121–40.

———. 1995b. Economism and critical silences in development studies: A theoretical critique of neoliberalism. *Third World Quarterly* 16 (2): 297–318.

Bukert, W. 1972. *Lore and science in ancient Pythagoreanism*. Trans. E. L. Miner. Cambridge, MA: Harvard University Press.

Bullen, P. S. 2003. *Handbook of means and their inequalities*. 2nd ed. Mathematics and their Applications. New York: Springer.

Burton, D. M. 2005. *The history of mathematics: An introduction*. New York: McGraw-Hill Science.

Cantor, M. 1894. *Vorlesungen über Geschichte der Mathematik*. 4 vols. Leipzig: Teubner.

Cooke, R. 1997. *The history of mathematics: A brief course*. New York: Wiley-Interscience.

Cuomo, S. 2001. *Ancient mathematics*. London and New York: Routledge.

Davidson, B. 1987. The ancient world and Africa. *Race and Class* 29 (2): 1–16.

Denevan, W. M. 1992. *The native population of the Americas in 1492.* 2nd ed. Madison: University of Wisconsin Press.

Dold-Samplonius, Y., et al., eds. 2002. *From China to Paris: 2000 years transmission of mathematical ideas.* Stuttgart: Franz Steiner Verlag.

Edwardes, M. 1971. *East-west passage.* London: Cassell.

Edwards, C. H. 1979. *The historical development of the calculus.* New York: Springer-Verlag.

Eves, H. 1983. *An introduction to the history of mathematics.* 5th ed. Philadelphia: Saunders.

Fahim, H. 1982. *Indigenous anthropology in non-Western countries.* Durham, NC: Carolina Academic Press.

Farebrother, R. 1999. *Fitting linear relationships: A history of the calculus of observations, 1750–1900.* New York: Springer.

Fauvel, J., and J. Gray, eds. 1987. *The history of mathematics: A reader.* London: Macmillan.

Flegg, G. 1983. *Numbers: Their history and meaning.* London: Andre Deutsch.

———. 1989. *Numbers through the ages.* Houndsmills: Macmillan Education.

Fowler, D. H. 1987. *The mathematics of Plato's Academy: A new reconstruction.* Oxford: Oxford University Press.

———. 1999. Inventive interpretations. *Revue d'Histoire des Mathématiques* 5:149–53.

Gillespie, C. C., ed. 1969–. *Dictionary of scientific biography.* 15 vols. New York: Charles Scribner's Sons.

Goodall, H. 2000. *Big bangs: The story of five discoveries that changed musical history.* London: Chatto & Windus.

Gray, J. 1979. *Ideas of space: Education, non-Euclidean, and relativistic.* London: Oxford University Press.

Halsted, G. B. 1912. *On the foundation and technic of arithmetic.* London: The Open Court Company.

Han, B. 2002. *Foucault's historical project: Between the transcendental and the historical.* Trans. E. Pile. Stanford, CA: Stanford University Press.

Harris, M. 1997. *Common threads: Women, mathematics and work.* Oakhill: Trentham Books.

Heath, T. L. 1921. *A history of Greek mathematics.* Oxford: Clarendon Press. Reprinted by Dover, New York, 1981.

Hobson, J. M. 2004. *The Eastern origins of Western civilisation.* Cambridge: Cambridge University Press.

Hodgkin, L. 2005. *A history of mathematics: From Mesopotamia to modernity.* Oxford: Oxford University Press.

Ifrah, G. 1985. *From one to zero: A universal history of numbers.* New York: Viking Penguin.

Joseph, G. G. 1994a. The politics of anti-racist mathematics. *European Education Journal,* July, 67–74.

———. 1997a. Foundations of Eurocentrism in mathematics. In *Ethnomathematics: Challenging Eurocentrism in mathematics*, ed. M. Frankenstein and A. Powell. 61–82. Albany, NY: SUNY Press.

———. 1997b. Mathematics. In *Encyclopedia of the history of science, technology and medicine in non-Western cultures*, ed. H. Selin. Dordrecht: Kluwer Academic Publishers.

———. 2003. What is a square root? In *The changing shape of geometry: Celebrating a century of geometry and geometry teaching*, ed. C. Pritchard, 100–114. Cambridge: Cambridge University Press.

———, ed. 2009. *Kerala mathematics: History and its possible transmission to Europe*. Delhi: B. R. Publishing Corporation.

Kanigel, R. 1991. *The man who knew infinity: A life of the genius Ramanujan*. New York: Charles Scribner's Sons.

Katz, V. 1998. *A history of mathematics: An introduction*. 2nd ed. Reading, MA: Addison Wesley.

———, ed. 2007. *The mathematics of Egypt, Mesopotamia, China, India, and Islam: A sourcebook*, 385–514. Princeton, NJ, and Oxford: Princeton University Press.

Kline, M. 1953. *Mathematics in Western culture*. New York: Oxford University Press. Reprinted by Penguin, London, 1972.

———. 1962. *Mathematics: A cultural approach*. Reading, MA: Addison Wesley.

———. 1972. *Mathematical thought from ancient to modern times*. New York: Oxford University Press.

Lach, D. F. 1965. *Asia in the making of Europe*. 2 vols. Chicago: University of Chicago Press.

Lakatos, I. 1976. *Proofs and refutations*. Cambridge: Cambridge University Press.

Lancy, D. F. 1983. *Cross studies in cognition and mathematics*. New York: Academic Press.

Lloyd, G.E.R. 1990. *Demystifying mentalities*. Cambridge: Cambridge University Press.

Maruzi, A. 2009. Foreword: Seven biases of Eurocentrism; a diagnostic introduction. In *The challenge of Eurocentrism: Global perspectives, policy, and prospects*, ed. R. Kannepalli Kanth, xi–xix. New York: Palgrave Macmillan.

Menninger, K. 1969. *Number words and number symbols: A cultural history of numbers*. Cambridge, MA: MIT Press.

Mersenne, Marin. 1945–. *Correspondance du P. Marin Mersenne*. 18 vols. Paris: Universitaires de France.

Midonick, H. O. 1965. *The treasury of mathematics*. 2 vols. London: Peter Owen. Reprinted by Penguin, London, 1968.

Nelson, D., G. G. Joseph, and J. Williams. 1993. *Multicultural mathematics*. Oxford: Oxford University Press.

Neugebauer, O. 1962. *The exact sciences in antiquity*. New York: Harper & Row. Reprinted by Dover, New York, 1969.

Nunes, T., A. Schliemann, and D. Carraher. 1993. *Street mathematics, school mathematics*. New York: Cambridge University Press.

Palter, R. 1998. Black Athena, Afro-centrist, and the history of science. *History of Science* 31:227–87.

Resnikoff, H. L., and R. O. Wells, Jr. 1984. *Mathematics in civilization*. New York: Dover Publications.

Rosenfeld, B. A. 1988. *A history of non-Euclidean geometry: Evolution of the concept of a geometric space*. New York: Springer-Verlag.

Rouse Ball, W. W. 1908. *A short account of the history of mathematics*. New York: Macmillan. Reprinted by Dover, New York, 1960.

Said, E. 1978. *Orientalism*. New York and London: Vintage.

Sarton, G. 1972. *Introduction to the history of science*. Vol. 1. Baltimore: Williams & Williams Co.

Schmandt-Besserat, D. 1992. *Before writing*. 2 vols. Austin: University of Texas Press.

——. 1999. *The history of counting*. New York: Morrow Junior Books.

Scott, J. 1981. *The mathematical work of John Wallis*. New York: Chelsea.

Seidenberg, A. 1961. The ritual origins of geometry. *Archives for History of Exact Sciences* 1:488–522.

——. 1983. The geometry of the Vedic rituals. In *Agni: The Vedic ritual of the fire altar*, vol. 2, ed. F. Staal, 95–126. Berkeley: Asian Humanities Press.

Singmaster, D. 1994. Recreational mathematics. In *Companion encyclopedia of the history and philosophy of the mathematical science*, ed. I. Grattan-Guinness, vol. 2. London: Routledge.

Smith, D. E. 1923–25. *History of mathematics*. 2 vols. Boston: Ginn & Co. Reprinted by Dover, New York, 1958.

Sondheim, E., and A. Rogerson. 1981. *Numbers and infinity*. Cambridge: Cambridge University Press.

Stewart, I. 1981. *Concepts of modern mathematics*. London: Penguin.

Struik, D. J. 1965. *A concise history of mathematics*. London: Bell.

Suzuki, J. 2001. *A history of mathematics*. Upper Saddle River, NJ: Prentice Hall.

Teresi, D. 2002. *Lost discoveries: The ancient roots of modern science—from the Babylonians to the Maya*. New York: Simon & Schuster.

Toomer, G. J. 1973. The chord table of Hipparchus and the early history of Greek trigonometry. *Centaurus* 18:6–28.

Van Brummelen, G. 2009. *The mathematics of the heaven and the earth: The early history of trigonometry*. Princeton, NJ, and London: Princeton University Press.

Van der Waerden, 1961. *Science awakening*. New York: Oxford University Press.

Van Sertima, I. 1983. *Blacks in science*. New Brunswick, NJ: Transaction Books.

Waters, D. W. 1958. *The art of navigation in Elizabethan and Stuart times*. London: Hollis and Carter.

Wittfogel, K. 1957. *Oriental despotism*. New Haven, CT: Yale University Press.

Mathematics of Egypt

Abdulaziz, A. A. 2008. On the Egyptian method of decomposing $2/n$ into unit fractions. *Historia Mathematica* 35: 1–18.

Bard, K. A. 1994. *From farmers to pharaohs: Mortuary evidence for the rise of complex society in Egypt*. Sheffield: Sheffield Academic Press.

Bernal, M. 1992. Animadversions on the origins of Western science. *Isis* 83:596–607.

———. 1994. Response to Palter. *History of Science* 32:445–64.

Bruckheimer, M., and Y. Salamon. 1977. Some comments on R. J. Gillings' analysis of the 2/*n* table in the Rhind Papyrus. *Historia Mathematica* 4:445–52.

Clagett, M. 1999. *Ancient Egyptian science*. Vol. 3, *Ancient Egyptian mathematics*. Philadelphia: American Philosophical Society.

Fischer-Elfert, H. W. 1986. *Die satirische Streitschrift des Papyrus Anastasi: Übersetzung und Kommentar*. Ägyptologische Abhandlungen 44. Wiesbaden: Otto Harrassowitz.

Friberg, J. 2005. *Unexpected links between Egyptian and Babylonian mathematics*. Singapore: World Scientific.

Gerdes, P. 1985a. Three alternative methods of obtaining the ancient Egyptian formula for the area of a circle. *Historia Mathematica* 12:261–68.

Gillings, R. J. 1962. Problems 1 to 6 of the Rhind mathematical papyrus. *The Mathematics Teacher* 55:61–69.

———. 1964. The volume of a truncated pyramid in ancient Egyptian papyri. *The Mathematics Teacher* 57:552–55.

———. 1972. *Mathematics in the time of the pharaohs*. Cambridge, MA: MIT Press.

Glanville, S.R.K. 1927. The mathematical leather roll in the British Museum. *Journal of Egyptian Archaeology* 13:232–38.

Gunn, B., and T. E. Peet. 1929. Four geometrical problems from the Moscow mathematical papyrus. *Journal of Egyptian Archaeology* 15:167–85.

Herodotus. 1984. *The histories*. London: Penguin.

Imhausen, A. 2002. The algorithmic structure of the Egyptian mathematical problem texts. In *Under one sky: Astronomy and mathematics in the ancient Near East*, ed. J. M. Steele and A. Imhausen, 147–66. Proceedings of the conference held in the British Museum, London, June 25–27, 2001. Alter Orient and Altes Testament 297. Münster: Ugarit Verlag.

———. 2003. Egyptian mathematical texts and their contexts. *Science in Context* 16 (3): 367–89.

———. 2006. UC 32107A verso: A mathematical exercise? In *The UCL Lahun Papyri*, vol. 3, ed. M. Collier and S. Quirke, 288–301. Oxford: Archaeopress.

———. 2007. Egyptian mathematics. In *The mathematics of Egypt, Mesopotamia, China, India, and Islam: A sourcebook*, ed. V. Katz, 7–56. Princeton, NJ, and Oxford: Princeton University Press.

Imhausen, A., and J. Ritter. 2004. Mathematical fragments: UC32114, UC32118, UC32134, UC32159–UC32162. In *The UCL Lahun Papyri: Religious, literary, legal, mathematical and medical*, British Archaeological Reports International Series 1209, ed. M. Collier and S. Quirke, 71–96. Oxford: Archaeopress.

Katz, V. K., and B. Lumpkin. 1995. The relationship of Egyptian to Greek mathematics. *History and Pedagogy of Mathematics Newsletter* 35:10–13.

Kemp, B. 1991. *Ancient Egypt: Anatomy of a civilization*. London: Routledge.

Lichtheim, A. 1975. *Ancient Egyptian literature*, 1:169–84. Berkeley: University of California Press.

Lumpkin, B. 2000. The mathematical legacy of ancient Egypt: A response to Robert Palter. In *Debating black Athena*, ed. M. Bernal. Durham, NC: Duke University Press.

———. 2002. The zero concept in Egypt. In *Proceedings of the International Seminar and Colloquium on 1500 Years of Aryabhateeyam*, 161–68. Kochi: Kerala Sastra Sahitya Parishad.

———. 2004. The mathematical legacy of ancient Egypt—a response to Robert Palter. Manuscript accessed with the permission of the author from www.ethnomath.org/resources/lumpkin1997.pdf.

Oldfather, C. H., trans. 1989. *Diodorus Siculus's Universal History*. Cambridge, MA: Harvard University Press.

Palter, R. 1993. Black Athena, Afro-centrism, and the history of science. *History of Science* 31:227–287

Parker, R. A. 1972. *Demotic mathematical papyri*. Brown Egyptological Studies 7. Providence: Brown University Press.

Peet, T. E. 1923. Arithmetic in the Middle Kingdom. *Journal of Egyptian Archaeology* 9:91–95.

———. 1931. A problem in Egyptian geometry. *Journal of Egyptian Archaeology* 17:100–106.

Ritter, J. 1995. Measure for measure: Mathematics in Egypt and Mesopotamia. In *A history of scientific thought: Elements of a history of science*, ed. Michel Serres, 44–72. Oxford and Cambridge, MA: Blackwell.

———. 2000. Egyptian mathematics. In *Mathematics across cultures: The history of non-Western mathematics*, ed. Helaine Selin, 115–36. Dordrecht/Boston/London: Kluwer.

———. 2002. Closing the eye of Horus. In *Under one sky: Astronomy and mathematics in the ancient Near East*, ed. J. M. Steele and A. Imhausen, 297–323. Proceedings of the conference held in the British Museum, London, June 25–27, 2001. Alter Orient Altes Testament 297. Münster: Ugarit Verlag.

Robins, G., and C. Shute. 1989. *The Rhind mathematical papyrus*. London: British Museum Publications.

Roero, C. S. 1994. Egyptian mathematics. In *Companion encyclopedia of the history and philosophy of the mathematical sciences*, ed. I. Grattan-Guinness, 30–35. London: Routledge.

Scharff, A. 1922. Ein Rechnungsbuch des Kniglichen Hofes aus der 13. Dynastie (Papyrus Boulaq Nr. 18). *Zeitschrift ägyptische Sprache und Alterumskunde* 57:58–59.

MATHEMATICS OF MESOPOTAMIA

Anderson, M., V. J. Katz, and R. J. Wilson. 2004. *Sherlock Holmes in Babylon and other tales of mathematical history*. Washington, DC: Mathematical Association of America.

Bahrani, Z. 1998. Conjuring Mesopotamia: Imaginative geography and a world past. In *Archaeology under fire: Nationalism, politics and heritage in the eastern Mediterranean and Middle East*, 159–74. London: Routledge.

Baqir, T. 1950. An important mathematical text from Tell Harmal. *Sumer* 6:39–54.

———. 1951. Some more mathematical texts from Tell Harmal. *Sumer* 7:28–45.

———. 1952. Foreword. *Sumer* 18:5–14.

Bruins, E. M. 1955. Pythagorean triads in Babylonian mathematics: The errors in Plimpton 322. *Sumer* 11:117–21.

Chace, A. B. 1979. *The Rhind mathematical papyrus.* Reston, VA: National Council of Teachers of Mathematics.

Friberg, J. 1981. Methods and traditions of Babylonian mathematics: Plimpton 322, Pythagorean triples and the Babylonian triangle parameter equations. *Historia Mathematica* 8:277–318.

———. 1993. On the structure of cuneiform metrological table texts from the 1st millennium. In *Die Rolle der Astronomie in den Kulturen Mesopotamiens,* ed. H. D. Galter, 383–405. Grazer Morgen-ländische Studien 3. Graz: R. M. Druck und Verlagsgessellschaft.

———. 1996. Pyramids and cones in ancient mathematical texts: New hints of a common tradition. *Proceedings of the Cultural History of Mathematics* 6:80–95.

———. 1999. A Late Babylonian factorisation algorithm for the computation of reciprocals of manyplace regular sexagesimal numbers. *Baghdader Mitteilungen* 30:139–161.

———. 2000. Mathematics at Ur in the Old Babylonian period. *Revue d'Assyriologie* 94:97–188.

———. 2005. *Unexpected links between Egyptian and Babylonian mathematics.* Singapore: World Scientific.

———. 2007a. *A remarkable collection of Babylonian mathematical texts.* New York: Springer.

———. 2007b. *Amazing traces of a Babylonian origin in Greek mathematics.* Singapore: World Scientific.

Hilprecht, H. V. 1906. *Mathematical, metrological and chronological tablets from the temple library of Nippur.* Philadelphia: University of Pennsylvania Press.

Høyrup, J. 1994. Babylonian mathematics. In *Companion encyclopedia of the history and philosophy of the mathematical sciences,* ed. I. Grattan-Guinness, 21–29. London: Routledge.

———. 2001. The Old Babylonian square texts BM 13901 and YBC 4714: Retranslations and analyses. In *Changing views on ancient Near Eastern mathematics,* ed. J. Høyrup and P. Damerow, 122–218. Berliner Beitrage zum Vorderen Orient 19. Berlin: Dietrich Reimer Verlag.

———. 2002. *Lengths, widths, surfaces: A portrait of Old Babylonian algebra and its kin.* New York: Springer.

Jones, A. 1991. The adaptation of Babylonian methods in Greek numerical astronomy. *Isis* 82:441–53.

———. 1996. Babylonian astronomy and its Greek metamorphoses. In *Traditions, transmission, transformation: Proceedings of two conferences on pre-modern science held at the University of Oklahoma,* ed. F. J. Rageb and S. Rageb, 139–55. Leiden: Brill.

Neugebauer, O. 1935–37. *Mathematische Keilschrift-Texte Quellen und Studien.* 3 vols. Berlin: Springer-Verlag.

Neugebauer, O., and A. Sachs. 1945. *Mathematical cuneiform texts.* New Haven, CT: Yale University Press.

Nissen, H., P. Damerow, and R. Englund. 1933. *Archaic bookkeeping.* Chicago: University of Chicago Press.

Powell, M. A. 1976. The antecedents of Old Babylonian place notation and the early history of Babylonian mathematics. *Historia Mathematica* 3:417–439.

———. 1990. The anecdotes of Old Babylonian place notation and the early history of Babylonian mathematics. *Historia Mathematica* 3:417–39.

Proust, C. 2002. Numération centésimale de position à Mari. *Florilegium Marianum* 6:513–516.

Robson, E. 1999. *Mesopotamian mathematics, 2100–1600 BC: Technical constants in bureaucracy and education.* Oxford: Clarendon Press.

———. 2001a. The Tablet House: A scribal school in Old Babylonian Nippur. *Revue d'Assyriologie* 95:39–57.

———. 2001b. Neither Sherlock Holmes nor Babylon: A reassessment of Plimpton 322. *Historia Mathematica* 26:167–206.

———. 2002. More than metrology: Mathematics education in an Old Babylonian scribal school. In *Under one sky: Astronomy and mathematics in the ancient Near East,* ed. J. M. Steele and A. Imhausen, 325–65. Proceedings of the conference held in the British Museum, London, June 25–27, 2001. Alter Orient Altes Testament 297. Münster: Ugarit Verlag.

———. 2005. Influence, ignorance or indifference? Rethinking the relationship between Babylonian and Greek mathematics. *BSHM Bulletin* 4:1–17.

———. 2007. Mesopotamian mathematics. In *The mathematics of Egypt, Mesopotamia, China, India, and Islam: A sourcebook,* ed. V. Katz, 58–186. Princeton, NJ, and Oxford: Princeton University Press.

———. 2008. *Mathematics in ancient Iraq: A social history.* Princeton, NJ: Princeton University Press.

Rochberg, F. 1993. The cultural locus of astronomy in Late Babylonia. In *Die Rolle der Astronomie in den Kulturen Mesopotamiens,* ed. H. D. Galter, 31–47. Grazer Morgen-ländische Studien 3. Graz: R. M. Druck und Verlagsgessellschaft.

Thureau-Dangin, F. 1938. *Textes mathématiques Babyloniens.* Leiden: Springer-Verlag.

THE MATHEMATICS OF CHINA

Berezkina, E. I. 1957. Matematika v devyati knigakh. *Istoriko-matematicheskie issledovaniay* 10:427–584.

Chemla, K., and S. Guo. 2004. *Les Neuf chapitres: Le classique mathématique de la Chine ancienne et ses commentaries.* Paris: Dunod.

Cho, G. J. 2003. *The discovery of musical equal temperament in China and Europe in the 16th century.* Lewiston, NY: Edwin Mellen Press.

Cullen, C. 1996. *Astronomy and mathematics in ancient China: The Zhou bi suan jing.* Cambridge: Cambridge University Press.

Dauben, J. W. 2007. Chinese mathematics. In *The mathematics of Egypt, Mesopotamia, China, India, and Islam: A sourcebook,* ed. V. Katz, 186–384. Princeton, NJ, and Oxford: Princeton University Press.

———. 2008. *Suan Shu Shu* (A Book on Numbers and Computations), English translation with commentary. *Archives for History of Exact Sciences*, forthcoming.

Dreyer, E. L. 2006. *Zheng He: China and the oceans in the early Ming, 1405–1433*. Library of World Biography Series. London: Longman.

Fukagawa, H., and T. Rothman. 2008. *Sacred mathematics: Japanese temple geometry*. Princeton, NJ: Princeton University Press.

Gillon, B. S. 1977. Introduction, translation and discussion of Chao Chung Ching's "Notes to the Diagrams of Short Legs and Long Legs and of Squares of Circles." *Historia Mathematica* 4:253-93.

Hart, R. 1997. Quantifying ritual: Political cosmology, courtly music, and precision mathematics in seventeenth-century China. Preliminary draft of a paper to be presented at the Disunity of Chinese Science conference, University of Chicago, May 10–12, 2002. http://uts.cc.utexas.edu/~rhart/papers/quantifying.html.

Hidetoshi, F., and T. Rothman. 2008. *Sacred mathematics: Japanese temple geometry*. Princeton, NJ: Princeton University Press.

Hoe, J. 1977. *Les systèmes d'équations-polynômes dans le Siyuan yujian (1930)*. Mémoires de l'Institute des Hautes Etudes Chinoises, vol. 6. Paris: Collège de France, l'Institute des Hautes Etudes Chinoises.

Joseph, G. G. 1994b. Tibetan astronomy and mathematics. In *Companion encyclopedia of the history and philosophy of the mathematical sciences*, ed. I. Grattan-Guinness, vol. 1. London: Routledge.

Keightley, D. N. 1985. *Sources of Shang history: The oracle-bone inscriptions of Bronze Age China*. 2nd ed. Berkeley: University of California Press.

Kim, Y. W. 1994. Korean mathematics. In *Companion encyclopedia of the history and philosophy of the mathematical sciences*, ed. I. Grattan-Guinness, 111–117. London: Routledge.

Lam Lay Yong. 1970. The geometrical basis of the ancient Chinese square root method. *Isis* 61:96–102.

———. 1974. Yang Hui's commentary on the Ying Nu chapter of the *Chiu Chang Suan Shu*. *Historia Mathematica* 1:47–64.

———. 1977. *A critical study of the* Yang Hui Suan Fa. Singapore: Singapore University Press.

———. 1986. The conceptual origins of our numeral system and the symbolic form of algebra. *Archive for History of Exact Sciences* 36:183–99.

———. 1987. Linkages: Exploring the similarities between the Chinese rod numeral system and our numeral system. *Archive for History of Exact Sciences* 37:365–92.

Lam Lay Yong and Ang Tian Se. 1992. *Fleeting steps: Tracing the conception of arithmetic and algebra in ancient China*. Singapore: World Scientific.

Lam Lay Yong and Shen Kangsheng. 1984. Right-angled triangles in ancient China. *Archive for History of Exact Sciences* 30:87–112.

Li Yan and Du Shiran. 1987. *Chinese mathematics: A concise history*. Trans. J. N. Crossley and A. W.-C. Lun. Oxford: Clarendon Press.

Libbrecht, U. 1973. *Chinese mathematics in the thirteenth century*. Cambridge, MA: MIT Press.

Liu Dun. 1994. 400 years of the history of mathematics in China: An introduction to the major historians of mathematics since 1592. *Historia scientiarum* 4:103–111.

Martzloff, J.-C. 1994. Chinese mathematics. In *Companion encyclopedia of the history and philosophy of the mathematical sciences*, ed. I. Grattan-Guinness, 93–103. London: Routledge.

———. 1997a. *A history of Chinese mathematics*. Trans. S. S. Wilson. New York: Springer-Verlag.

———. 1997b. Mathematics in Japan. In *Encyclopedia of the history of science, technology, and medicine in non-Western cultures*, ed. H. Selin, 641–43. Dordrecht: Kluwer Academic Publishers.

Mikami, Y. 1974. *The development of mathematics in China and Japan* (1913). New York: Chelsea.

Murata, T. 1975. Pour une interprétation de la destinée du Wasan—aventure et mésaventure de ces mathématiques. In *Proceedings of the 14th International Congress of the History of Science*, 2:184–99. Tokyo: Science Council of Japan.

———. 1980. Wallis' *Arithmetica Infinitorum* and Takebe's *Tetsujutsu Sankei*—what underlies their similarities and dissimilarities? *Historia Scientiarum* 19:77–100.

———. 1994. Indigenous Japanese mathematics. In *Companion encyclopedia of the history and philosophy of the mathematical sciences*, ed. I. Grattan-Guinness, vol. 1. London: Routledge.

Needham, J. 1954. *Science and civilisation in China*. Vol. 1. Cambridge: Cambridge University Press.

———. 1959. *Science and civilisation in China*, 3:1–168. Cambridge: Cambridge University Press.

Ohashi, Y. 1997. Astronomy in Tibet. In *Encyclopedia of the history of science, technology, and medicine in non-Western cultures*, ed. H. Selin, 136–39. Dordrecht: Kluwer Academic Publishers.

Robinson, K. 1980. *A critical study of Chu Ti-yü's contribution to the theory of equal temperament in Chinese music*. Wiesbaden: Fran Steiner Verlag.

Rogers, D. G. n.d. *Cutting it fine: Euclid's problem on slicing the circle*. Accessed from http://www.ics.mq.edu.au/~gerry/slices.pdf on December 7, 2009.

Sang-Woon Jeon. 1974. *Science and technology in Korea*. Cambridge, MA. MIT Press.

Shen Kangsheng. 1988. Historical development of the Chinese remainder theorem. *Archives for History of Exact Sciences* 38 (4): 285–305.

Shen Kangsheng, J. N. Crossley, and Antony W.-C. Lun. 1999. *The* Nine Chapters on the Mathematical Art: *Companion and commentary*. Oxford and Beijing: Oxford University Press and Science Press.

Shigeru Jochi and B. Rosenfeld. 1997. Seki Kowa. In *Encyclopedia of the history of science, technology, and medicine in non-Western cultures*, ed. H. Selin, 890–92. Dordrecht: Kluwer Academic Publishers.

Sivin, N. 1995. Science and medicine in Chinese history. In *Science in ancient China: Researches and reflections*. Brookfield, VT: Variorum.

Smith, D. E., and Y. Mikami. 1914. *A history of Japanese mathematics*. Chicago: Open Court.

Swetz, F. 2008. *Legacy of the Luo Shu: The 4,000 year search for the meaning of the magic square of order three*. Wellesley, MA: A. K. Peters.

Swetz, F., and T. I. Kao. 1977. *Was Pythagoras Chinese? An examination of right triangle theory in ancient China*. University Park: Pennsylvania State University Press.

Tenzer, M., ed. 2006. *Analytical studies in world music*. New York: Oxford University Press.

Van Dalen, B. 2002. Islamic and Chinese astronomy under the Mongols: A little-known case of transmission. In *From China to Paris: 2000 years transmission of mathematical ideas*, ed. Yvonne Dold-Samplonius et al., 327–56. Stuttgart: Franz Steiner Verlag.

Vogel, K. 1968. *Neun Bucher arithmetischer Technik*. Braunschweig: Vieweg.

Wagner, D. B. 1978. Liu Hui and Tsu Keng-chih on the volume of a sphere. *Chinese Science* 3:59–79.

Wang Ling and J. Needham. 1955. Horner's method in Chinese mathematics: Its origins in the root extraction procedures of the Han dynasty. *Toung Pao* 43:345–88.

MATHEMATICS OF INDIA

Abdi, W. H. 1982. Some works of Indian mathematicians who wrote in Persian. *Ganita Bharati* 4 (2): 6–9.

Almeida, D. F., J. K. John, and A. Zadorozhnyy. 2001. Keralese mathematics: Its possible transmission to Europe and the consequential educational implications. *Journal of Natural Geometry* 20:77–104.

Almeida, D., and G. G. Joseph. 2004. Eurocentrism in the history of mathematics: The case of the Kerala school. *Race and Class* 45 (4): 45–60.

———. 2007. Kerala mathematics and its possible transmission to Europe. *Philosophy of Mathematics Education Journal*, retrieved from www.people.ex.ac.uk/PErnest/pome20/index.htm.

———. 2009. A report of the investigation on the possibility of the transmission of the medieval Kerala mathematics to Europe. In *Medieval Kerala mathematics: Its history and possibility of transmission to Europe*, ed. G. G. Joseph, 257–75. Delhi: B. R. Publishing Corporation.

Alsdorf, A. 1991. The pratyayas: Indian contribution to combinatorics. *Indian Journal of History of Science* 26 (1): 17–61.

Andrews, G. E. 1979. An introduction to Ramanujan's "lost" notebook. *American Mathematics Monthly* 86:89–108.

Aryabhata. 499. *Aryabhatiya of Aryabhata*. Trans. K. S. Shukla and K. V. Sarma. New Delhi: Indian National Science Academy, 1976.

Aryabhata Group. 2002. Transmission of the calculus from Kerala to Europe. In *Proceedings of the International Seminar and Colloquium on 1500 Years of Aryabhateeyam*, 33–48. Kochi: Kerala Sastra Sahitya Parishad.

Bag, A. K. 1976. Madhava's sine and cosine series. *Indian Journal of History of Science* 11 (1): 54–57.

———. 1979. *Mathematics in ancient and medieval India*. Varanasi: Chaukhambha Orientalia.

Bag, A. K., and S. R. Sarma, eds. 2003. *Concept of sunya*. New Delhi: Aryan Books.

Bala, A. 2009. Establishing transmissions: Some methodological issues. In *Kerala mathematics: Its history and possibility of transmission to Europe*, ed. G. G. Joseph, 155–80. Delhi: B. R. Publishing Corporation.

Baldini, U. 1992. *Studi su filosofia e scienza dei gesuiti in Italia, 1540–1632*. Firenze, Bulzoni Editore.

———. 2009. The Jesuit mathematicians in India (1578–1650) as possible intermediaries between European and Indian mathematical traditions. In *Kerala mathematics: Its history and possibility of transmission to Europe*, ed. G. G. Joseph, 277–306. Delhi: B. R. Publishing Corporation.

Bein, R. 2007. Viète's controversy with Clavius over the truly Gregorian calendar. *Archive for History of Exact Sciences* 61 (1): 39–66.

Bernard, H. 1973. *Matteo Ricci's scientific contribution to China*. Westport, CT: Hyperion Press.

Bose, D. M., S. N. Sen, and B. V. Subbarayappa, eds. 1971. *A concise history of science in India*. New Delhi: Indian National Science Academy.

Burnett, C. 2002. Indian numerals in the Mediterranean Basin in the twelfth century, with special reference to the "Eastern Forms." In *From China to Paris: 2000 years transmission of mathematical ideas*, ed. Yvonne Dold-Samplonius et al., 237–87. Stuttgart: Franz Steiner Verlag.

Cantor, M. 1905. Uber die alteste indische Mathematique. *Archiv der Mathematik und Physic* 8:63–92.

Chattopadhyaya, D. 1986. *History of science and technology in ancient India: The beginnings*. Calcutta: Firma KLM PVT.

Datta, B. 1929. The Bakhshali Manuscript. *Bulletin of the Calcutta Mathematical Society* 21:1–60.

———. 1932a. *The science of the Sulbas: A study in early Hindu geometry*. Calcutta: Calcutta University Press.

———. 1932b. On the relation of Mahavira to Sridhara. *Isis* 17 (1): 25–33.

Datta, B., and A. N. Singh. 1962. *History of Hindu mathematics*. 2 vols. Bombay: Asia Publishing House.

Dharampal. 1971. *Indian science and technology in the eighteenth century*. New Delhi: Biblia Impex.

———. 1983. *The beautiful tree*. New Delhi: Biblia Impex.

Duke, D. 2005. The equant in India: The mathematical basis of ancient Indian planetary models. *Archive for History of Exact Sciences* 59:563–76.

———. 2007. The second lunar anomaly in ancient Indian astronomy. *Archive for the History of Exact Sciences* 61:147–57.

Ferroli, D. 1939. *The Jesuits in Malabar*, 1951 edition. Bangalore: King and Co.

Ganguli, S. 1929. Notes on Indian mathematics: A criticism of G. R. Kaye's interpretation. *Isis* 12:132–45.

Gupta, R. C. 1967. Bhaskara I's approximation to the sine. *Indian Journal of History of Science* 2:121–36.

———. 1974a. Sines and cosines of multiple arcs as given by Kamalakara. *Indian Journal of History of Science* 9:143–50.

———. 1974b. Addition and subtraction theorems for the sine and cosine in medieval India. *Indian Journal of History of Science* 9:164–77.

———. 1975. Circumference of the Jambudvipa in Jaina cosmography. *Indian Journal of History of Science* 10 (1): 38–46.

———. 1977. Paramesvara's rule for the circumradius of a cyclic quadrilateral. *Historia Mathematica* 4:67–74.

———. 1978. Indian values of the sinus totus. *Indian Journal of History of Science* 13 (2): 125–43.

———. 1986. On the derivation of Bhaskara I's formula for the Sine. *Ganita Bharati* 8:39–41.

———. 1992. On the remainder term in the Madhava-Leibniz's series. *Ganita Bharati* 14:68–71.

Gurukkal, R. 1992. *The Kerala temple and the early medieval agrarian system.* Sukapuram: Vallathol Vidyapeetham.

Hankel, H. 1874. *Zur Geschichte der Mathematik in Alterthum und Mittelalter.* Leipzig: Teubner.

Hayashi, T. 1995. *The Bakhshali Manuscript: An ancient Indian mathematical treatise.* Groningen: Egbert Forsten.

———. 1997. Aryabhata's rule and table for sine-differences. *Historia Mathematica* 24:396–406.

Hayashi, T., and T. Kusuba. 1998. Twenty-one algebraic normal forms of Citrabhanu. *Historia Mathematica* 25 (1): 1–21.

Hayashi, T., T. Kusuba, and M. Yano. 1989. Indian values for π from Aryabhata's value. *Historia Scientarium* 37:1–16.

———. 1990. The correction of the Madhava series for the circumference of a circle. *Centaurus* 33:149–74.

Heefer, A. 2007. The reception of ancient Indian mathematics by Western historians. In *Ancient Indian leaps in the advent of mathematics,* ed. B. S. Yadev. Basel: Birkhauser.

Hoemle, A.F.R. 1888. The Bakhshali Manuscript. *Indian Antiquary* 17:33–48 and 275–79.

Hunter, G. R. 1934. *The script of Harappa and Mohenjodaro and its connection with other scripts.* London: Kegan Paul, Trenchy, Trabner and Co.

Iannaccone, I. 1998. *Johann Schreck Terrentius.* Napoli: Instituto Universitario Orientale.

Ikeyama, S., and K. Plofker. 2001. The Tithicintamani of Ganesa, a medieval Indian treatise on astronomical tables. *SCIAMVS* 2:251–89.

Jain, L. C. 1973. Set theory in the Jaina school of mathematics. *Indian Journal of History of Science* 8 (1/2): 1–27.

———. 1982. *Exact sciences from Jaina sources.* 2 vols. Jaipur: Rajasthan Prakrit Bharti Sansthan.

Jeganathan, P. 1993. *On the structural reading and the evolution of the Indus script.* Technical Report 220, Department of Statistics, University of Michigan, Ann Arbor, MI.

Joseph, G. G. 1994c. Different ways of knowing: Contrasting styles of argument in Indian and Greek mathematical traditions. In *Mathematics, education and philosophy: An international perspective*, ed. P. Ernest, 194–204. London: Falmer Press.

———. 1995a. The geometry of Vedic altars. In *Nexus: Geometry and architecture*, ed. K. Williams, 97–113. Fucecchio, Italy: Edizione Dell'Erba.

———. 1995b. Cognitive encounters in India during the age of imperialism. *Race and Class* 36 (3): 39–56.

———. 1995c. Transmissions of mathematical knowledge during the medieval period. [In Malayalam.] *Kerala Padanengal* 5:126–40.

———. 1996a. Calculating people: Origins of numeracy in India and the West. In *Challenging ways of knowing in English, math, and science*, ed. D. Baker, J. Clay, and C. Fox, 194–203. London: Falmer Press.

———. 1996b. Mathematics in ancient India. Harappa civilization: Mathematics in metrology. *Mathesis* 12:236–83.

———. 2002a. The enormity of zero. *Revista Brasileira de Historia da Matematica* 2 (4): 155–67.

———. 2002b. Infinite series in Kerala: Background and motivation. In *Proceedings of the International Seminar and Colloquium on 1500 Years of Aryabhateeyam*, 129–35. Kochi: Kerala Sastra Sahitya Parishad.

———. 2009. *A passage to infinity: Medieval Indian mathematics from Kerala and its impact*. New Delhi: Sage Publishers.

Kak, S. C. 1989. Indus writing. *Mankind Quarterly* 30:113–18.

———. 1990. Indus and Brahmi: Further connections. *Cryptologia* 14 (2): 169–83.

———. 1993. Astronomy of the Vedic altars. *Vistas in Astronomy* 36:117–40.

———. 2005. *The Indo-Aryan controversy: Evidence and inference in Indian history*. London and New York: Routledge.

Kanigel, R. 1991. *The man who knew infinity: A life of the genius Ramanujan*. New York: Charles Scribner's Sons.

Kaye, G. R. 1908. Notes on Indian mathematics. *Journal of the Asiatic Society of Bengal* 4:111–141.

———. 1912. The Bakhshali Manuscript. *Journal of the Asiatic Society of Bengal* 8:349–361.

———. 1915. *Indian mathematics*. Calcutta: Amd Simla.

———. 1933. The Bakhshali Manuscript: A study in mediaeval mathematics. *Archaeological Survey of India*, New Imperial Series, 43, parts 1–3.

Keller, A. 2006. *Expounding the mathematical seed*. 2 vols. Berlin: Birkhauser Verlag.

Krishna Tirthaji, B. 1965. *Vedic mathematics*. New Delhi: Motilal Banarsidass.

Kulkarni, R. P. 1971. Geometry as known to the people of the Indus civilisation. *Indian Journal of History of Science* 13:117–24.

———. 1983. *Geometry according to* Sulba Sutra. Pune: Vaidika Samsodhana Mandala.

Kunjunni Raja, K. 1963. Astronomy and mathematics in Kerala. *Adyar Library Bulletin* 27:117–67.

Kuppanna Sastry, S. 1985. *Vedanga Jyotisa of Lagadha*. Ed. K. V. Sarma. New Delhi: Indian National Science Academy.

Madhavan, S. 1991. Origin of Katapyadi system of numerals. *Sri Ravi Varma Samskrta Granthvali Journal* 18 (2): 35–48.

Mainikar, V. B. 1984. Metrology in the Indus civilisation. In *Frontiers of the Indus civilisation*, ed. B. Lal and S. P. Gupta, 35–54. New Delhi: Motilal Banarsidass.

Mallayya, V. M. 2002. Geometrical approach to arithmetic progressions from Nilakanta's *Aryabhatiyabhasya* and Sankara's *Kriyakramakari*. In *Proceedings of the International Seminar and Colloquium on 1500 Years of Aryabhateeyam*, 143–47. Kochi: Kerala Sastra Sahitya Parishad.

———. 2004. An interesting algorithm for computation of sine tables from the *Golasara* of Nilakantha. *Granita Bharati* 26:40–55.

———. 2008. Vatesvara's trigonometric tables and the method. *Ganita Bharati* 30: 61–79.

———. Forthcoming. Trigonometric sines and sine tables in India: A survey. AHRB Project, University of Exeter.

Mallayya, V. M., and G. G. Joseph. 2009a. Indian mathematical tradition: The Kerala dimension. In *Kerala mathematics: History and its possible transmission to Europe*, ed. G. G. Joseph, 35–58. Delhi: B. R. Publishing Corporation.

———. 2009b. Kerala mathematics: Motivation, rationale, and method. In *Kerala mathematics: History and its possible transmission to Europe*, ed. G. G. Joseph, 77–112. Delhi: B. R. Publishing Corporation.

Matilal, B. K. 1985. *Logic, language, and reality: An introduction to Indian philosophical studies*. New Delhi: Motilal Banarsidas.

Parameswaran, S. 1980. Kerala's contribution to mathematics and astronomy. *Journal of Kerala Studies* 7:135–47.

———. 1983. Madhava of Sangamagramma. *Journal of Kerala Studies* 10:185–217.

Pingree, D. 1981. History of mathematical astronomy in India. In *Dictionary of scientific biography*, ed. C. C. Gillespie, vol. 15 (suppl.). New York: Charles Scribner's Sons.

Plofker, K. 2007. Mathematics in India. In *The mathematics of Egypt, Mesopotamia, China, India, and Islam: A sourcebook*, ed. V. Katz, 385–514. Princeton, NJ, and Oxford: Princeton University Press.

———. 2009. *Mathematics in India*. Princeton, NJ: Princeton University Press.

Rajagopal, C. T., and K. Mukanda Marar. 1944. On the quadrature of the circle. *Journal of the Royal Asiatic Society* (Bombay Branch) 20:65–82.

Rajagopal, C. T., and M. S. Rangachari. 1978. On an untapped source of medieval Keralese mathematics. *Archive for History of Exact Sciences* 18:89–102.

———. 1986. On medieval Keralese mathematics. *Archive for History of Exact Sciences* 35:91–99.

Rajagopal, C. T., and T. V. Vedamurthi Iyer. 1952. On the Hindu proof of Gregory's series. *Scripta Mathematica* 18:65–74.

Rajagopal, C. T., and A. Venkataraman. 1949. The sine and cosine power-series in Hindu mathematics. *Journal of the Royal Asiatic Society* (Bengal Branch) 25 (1): 1–13.

Rajagopal, P. 1991. The *Sthanga Sutra* programme in Indian mathematics. *Arhat Vacana* 3 (2): 1–8.

Rajasekhar, P. 2009. Derivation of the infinite series expansion for π as demonstrated in the *Yuktibhasa*. In *Kerala mathematics: History and its possible transmission to Europe*, ed. G. G. Joseph, 113–136. Delhi: B. Publishing Corporation.

Raju, C. K. 2007. *Cultural foundations of mathematics—the nature of mathematical proof and the transmission of the calculus from India to Europe in the 16th century CE*. Patparganj, Delhi: Center for Studies in Civilization (CSC) and Pearson Longman.

Ramanujan, S. 1913/1914. Modular equations and approximations to π. *Quarterly Journal of Pure and Applied Mathematics* 45:350–72.

———. 1985. *Notebooks*. Ed. B. C. Berndt. New York: Springer-Verlag.

Ramasubramanian, K., M. D. Srinivas, and M. S. Sriram. 1994. Heliocentric model of planetary motion in the Kerala school of Indian astronomy. *Current Science* 66 (10): 784–90.

Rao, S. R. 1973. *Lothal and the Indus civilisation*. New York: Asia Publishing Co.

Rizvi, S.A.H. 1983. On trisection of an angle leading to the derivation of a cubic equation and computation of value of sine. *Indian Journal of History of Science* 19 (1): 77–85.

Rodet, L. 1879. Leçons de calcul d'Aryabhata. *Journal Asiatique* 13:393–434.

Roy, R. 1990. The discovery of the series formula for π by Leibniz, Gregory and Nilakantha. *Mathematics Magazine* 63:291–306.

Sarasvati, T. A. 1963. Development of mathematical series in India. *Bulletin of the National Institute of Sciences* 21:320–43.

———. 1979. *Geometry in ancient and medieval India*. Delhi: Motilal Banarsidass.

Sarma, K. V. 1972. *A history of the Kerala school of Hindu astronomy*. Hoshiarpur: Vishveshvaranand Institute.

———. 1986. Some highlights of astronomy and mathematics in medieval India. *Sanskrit and World Culture* 18:595–605.

———. 2008. *Ganita-Yukti-Bhasa of Jysthdeva*. 2 vols. New Delhi: Hindustan Book Agency.

Sarma, K. V., and S. Hariharan. 1991. Yuktibhasa of Jyesthadeva. *Indian Journal of History of Science* 26:186–207.

Sarma, S. R. 2002. Rule of three and its variations in India. In *From China to Paris: 2000 years transmission of mathematical ideas*, ed. Yvonne Dold-Samplonius et al., 133–56. Stuttgart: Franz Steiner Verlag.

———. 2009. Early transmissions of Indian mathematics. In *Kerala mathematics: History and its possible transmission to Europe*, ed. G. G. Joseph, 205–31. Delhi: B. R. Publishing Corporation.

Selenius, C. 1975. Rationale of the Chakravala process of Jayadeva and Bhaskara II. *Historia Mathematica* 2:167–184.

Sen, S. N., and A. K. Bag, ed. and trans. 1983. *The Sulbasutras*. New Delhi: Indian National Science Academy.

Sengupta, P. C. 1927. The Aryabhatiyam, translated into English. *Journal of the Department of Letters* 16:1–56.

Singh, N. 1984. Foundations of logic in ancient India: Linguistics and mathematics. In *Science and technology in Indian culture*, ed. A. Rahman. New Delhi: National Institute of Science, Technology and Development Studies (NISTAD).

———. 1987. *Jain theory of actual infinities and transfinite infinities*. New Delhi: National Institute of Science, Technology and Development Studies (NISTAD).

Singh, P. 1981. Narayana's treatment of net of numbers. *Ganita Bharati* 3: (1–2): 16–31.

———. 1982. Total number of perfect magic squares: Narayana's rule. *Mathematics Education* (Saiwan) 16 (2): 32–37.

———. 1986. Narayana's treatment of magic squares. *Indian Journal of History of Science* 21 (2): 123–30.

———. 1988. The *Ganita Kaumudi* of Narayana Pandita. *Ganita Bharati* 20 (1–4): 25–82.

Srinivasiengar, C. H. 1967. *The history of ancient Indian mathematics*. Calcutta: The World Press Private Ltd.

Staal, F. 1978. The ignorant Brahmin of the Agnicayana. *Annals of the Bhandarkar Oriental Research Institute* (Pune), Diamond Jubilee Number, 337–48.

———. 1955. The Sanskrit of science. *Journal of Indian Philosophy* 23:73–127.

Subbarayappa, V. 1993. Numerical system of the Indus Valley civilisation: A monograph on the Indus script. Mimeo. Indian Council of World Culture, Bangalore.

Thibaut, G. 1875. On the *Sulba-sutra*. *Journal of the Asiatic Society of Bengal* 44:7–75.

Vijayalekshmy, M., and G. G. Joseph. 2009. An intellectual history of medieval Kerala with special reference to mathematics and astronomy. In *Kerala mathematics: History and its possible transmission to Europe*, ed. G. G. Joseph, 59–73. Delhi: B. R. Publishing Corporation.

MATHEMATICS OF THE ISLAMIC WORLD

Aaboe, A. 1954. Al-Kashi's iteration method for sin 1°. *Scripta Mathematica* 20:24–29.

Aballagh, M. 1988. Les fondements des mathématiques à travers le "Raf al-Hijab" d'Ibn al-Banna (1256–1321). Presented at Histoire des mathématiques Arabes, December 1–3, 1986. In *Actes du colloque*, 133–156. Alger: Maison du Livre.

———. 1992. Les fractions entre la théorie et la pratique chez Ibn al-Banna al-Murrakusi (1256–1321). In *Histoire de fractions, fractions d'histoire*, ed. P. Benoit, K. Chemla, and J. Ritter, 247–58. Basel: Birkhauser.

Balty-Guesdon, M. G. 1992. Le Bayt al-hikma de Baghdad. *Arabica* 39 (2): 131–50.

Berggren, J. L. 1983. The correspondence of Abu Sahl al-Kuhi and Abu Ishaq al-Sabi: A translation with commentaries. *Journal for the History of Arabic Science* 7:39–124.

———. 1986. *Episodes in the mathematics of medieval Islam*. New York: Springer-Verlag.

———. 2002. Some ancient and medieval approximations to irrational numbers and their transmission. In *From China to Paris: 2000 years transmission of mathematical ideas*, ed. Yvonne Dold-Samplonius et al., 31–44. Stuttgart: Franz Steiner Verlag.

———. 2007. Mathematics in medieval Islam. In *The mathematics of Egypt, Mesopotamia, China, India, and Islam: A sourcebook*, ed. V. Katz, 515–675. Princeton, NJ, and Oxford: Princeton University Press.

Berggren, J. L., and G. Van Brummelen. 1975. Al-Kuhi's revision of Book I of Euclid's *Elements*. *Historia Mathematica* 32 (4): 426–52.

Brentjes, S. 2007. Khwarizmi: Muhammad ibn Musa al-Khwarizmi. In *The biographical encyclopedia of astronomers, Springer reference*, ed. Thomas Hockey et al., 631–633. New York: Springer.

Daffa, A. al-, and J. J. Stroyls. 1984. *Studies in the exact sciences in medieval Islam*. New York: John Wiley & Sons.

Djebbar, A. 1981. *Enseignment et recherche mathématique dans le Maghreb des XIIIe–XIVe siècles.* Paris: Publications Mathématiques d'Orsay.

———. 1985. *L'Analyse combinatoire au Maghreb: l'exemple de Ibn Munim (XIIe–XIIIe siècles).* Paris: Publications Mathématiques d'Orsay.

———. 1990. Mathématiciens du Maghreb, médiéval (IXe–XVIe siécles): Contribution á létude des activités scientifiques de l'Occident musulman. Thése de Doctorat, Université de Nantes, France.

———. 1997. Mathematics of Africa: The Maghreb. In *Encyclopedia of the history of science, technology, and medicine in non-Western cultures,* ed. H. Selin, 613–16. Dordrecht: Kluwer Academic Publishers.

Gandz, S. 1938. The algebra of inheritance. *Osiris* 5:319–91.

Goldstein, B. R. 1985. *The astronomy of Levi ben Gerson (1288–1344): A critical edition of chapters 1–20 with translation and commentary.* New York: Springer-Verlag.

Gutas, D. 1988. *Greek thought, Arabic culture: The Graeco-Arabic translation movement in Baghdad and early Abbasid society (2nd–4th/8th–10th century).* London: Routledge.

Hodgson, M.G.S. 1961. *The venture of Islam: Conscience and history in a world civilization.* Vol. 1. Chicago: University of Chicago Press.

Joseph, G. G. 1994d. The Arabs. In *Nuffield advanced mathematics: History of mathematics.* London: Longman Education.

Kasir, D. S., ed. 1931. *The algebra of Omar Khayyam.* New York: Teachers' College Press, Columbia University. Reprinted by College Press, New York, 1972.

Katz, V. J. 1993. Using the history of calculus to teach calculus. *Science & Education* 2:243–249.

———. 1995. Ideas of calculus in Islam and India. *Mathematics Magazine* 68:163–174.

Kennedy, E. S. 1964. The Chinese-Uighur calendar as described in the Islamic sources. *Isis* 55:435–443.

King, D. 2000. Mathematical astronomy in Islamic civilisation. In *Astronomy across Cultures: The history of Non-Western astronomy,* ed. H. Selin, 585–613. Dordrecht, Boston & London: Kluwer Academic Publishers.

Langermann, Y. T. 1984. The mathematical writings of Maimonides. *Jewish Quarterly Review* 75:57–65.

———. 1994. Mathematics in medieval Hebrew literature. In *Companion encyclopedia of the history and philosophy of the mathematical sciences,* ed. I. Grattan-Guinness, vol. 1. London: Routledge.

Levy, T. 1996. Hebrew mathematics in the Middle Ages: An assessment. In *Tradition, transmission, transformation: Ancient mathematics in Islamic and Occidental cultures,* ed. F. Jamil Ragep and Sally P. Ragep with Steven J. Livsey. Leiden: E. J. Brill.

———. 1997. Mathematics of the Hebrew people. In *Encyclopedia of the history of science, technology, and medicine in non-Western cultures,* ed. H. Selin, 632–34. Dordrecht: Kluwer Academic Publishers.

Mercier, R. 2004. *Studies on the transmission of medieval mathematical astronomy.* Ashgate: Variorum.

Nasr, S. H. 1968. *Science and civilization in Islam.* Cambridge, MA: Harvard University Press.

Oakes, J. A., and H. M. Alkhateeb. 2007. Simplifying equations in Arabic algebra. *Historia Mathematica* 34 (1): 45–61.

Pingree, D. 1970. The fragments of the works of al Fazari. *Journal of Near Eastern Studies* 29:103–123.

Rashed, R. 1986. *Al-Tusi, Oevres mathématiques: Algèbre et géometrie au XII siècle*. Paris: Les Belles Lettres.

Rosen, F. 1831. *The algebra of Muhamed ben Musa*. London: Oriental Translation Fund.

Rozenfeld, B. A. 1988. *A history of Non-Euclidean geometry: Evolution of the concept of a geometric space*. New York: Springer-Verlag.

Saidan, A. S. 1974. Arithmetic of Abul Wafa. *Isis* 65:367–75.

Saliba, G. 1994. *A history of Arabic astronomy: Planetary theories during the golden age of Islam*. New York: New York University Press.

———. 2007. *Islamic science and the making of the European Renaissance*. Cambridge, MA: MIT Press.

Sarfati, G. B. 1968. *Mathematical terminology in Hebrew scientific literature of the Middle Ages*. Jerusalem: Magnes Press.

Savory, R. M. 1976. *Introduction to Islamic civilisation*. Cambridge: Cambridge University Press.

Sayili, A. 1960. Thabit ibn Qurra's generalisation of the Pythagorean theorem. *Isis* 51:35–37.

Scriba, C. J. 1966. John Wallis's *Treatise on angular sections* and Thabit ibn Qurra's *Generalisations of the Pythagorean theorem*. *Isis* 57:55–66.

Sesiano, J. 1991. Two problems of number theory in Islamic times. *Archives for History of Exact Sciences* 41 (3): 235–38.

———. 2004. *Les carrés magiques dans les pays islamiques*. Lausanne, Switzerland: Presses polytechniques et universitaires romanes.

Yushkevich, A. P. 1964. *Geschichte der Mathematik im Mittelater*. Leipzig. B. G. Teubner (ISL).

Zaimeche, S. 2003. *Aspects of the Islamic influence on science and learning in the Christian West (12th–13th century)*. Manchester, UK: Foundation for Science, Technology and Civilisation.

MATHEMATICS OF OTHER CULTURES

Ascher, M. 1988. Graphs in cultures II. A study in ethnomathematics. *Archive for History of Exact Sciences* 39:75–95.

Ascher, M., and R. Ascher. 1972. Numbers and relations from ancient Andean quipus. *Archive for History of Exact Sciences* 8:288–320.

———. 1981. *Code of the Quipu*. Ann Arbor: University of Michigan Press.

———. 1997. *Mathematics of the Incas: Code of the Quipu*. Mineola, NY: Dover Publications.

Aveni, A. F. 1981. Tropical archeoastronomy. *Science* 213:161–71.

Bernal, M. 1994. Response to Robert Palter. *History of Science* 32:445–468.

Bishop, A. J. 1995. What we can learn from the counting systems research of Dr. Glendon Lean. Keynote address to the History and Pedagogy of Mathematics Conference, Cairns, Queensland, Australia.

Bogoshi, J., N. Naidoo, and J. Webb. 1987. The oldest mathematical artefact. *Mathematical Gazette* 71 (458): 294.

Clark, W. E. 1930. *The Aryabhatiya of Aryabhata, translated into English with notes.* Chicago: University of Chicago Press.

Closs, M. P. 1986. *Native American mathematics.* Austin: University of Texas Press.

———. 1992. *I am a Kahal; my parents were scribes.* Research Reports on Ancient Maya Writing 39. Washington, DC: Center for Maya Research.

———. 1995. Mathematicians and mathematical education in ancient Maya society. Mimeo. Privately circulated.

Crowe, D. 1971. The geometry of African art I: Bakuba art. *Journal of Geometry* 1:169–82.

———. 1975. The geometry of African art II: A catalogue of Benin patterns. *Historia Mathematica* 2:253–71.

———. 1982a. The geometry of African art III: The smoking pipes of Begho. In *The geometric vein: The Coxeter Festschrift,* ed. C. Davis, B. Grunbaum, and F. Sherk, 177–89. New York: Springer-Verlag.

———. 1982b. Symmetry in African art: Ba Shiru. *Journal of African Languages and Literature* 3:57–71.

Day, C. L. 1967. *Quipus and witches' knots.* Lawrence: University of Kansas Press.

de Acosta, José. 1596. *Historia natural moral de las Indias.* Madrid.

de Heinzelin, J. 1962. Ishango. *Scientific American* 206 (June): 105–16.

Eglash, R. 1995. Fractal geometry in African material culture: Symmetry, natural and artificial. *Culture and Science* 6 (1): 174–77.

———. 1997. When math worlds collide: Intention and invention in ethno-mathematics. *Science, Technology & Human Values* 22:79–97.

Gay, J., and M. Cole. 1967. *The new mathematics and an old culture: A study of learning among the Kpelle of Liberia.* New York: Holt, Rinehart and Winston.

Gerdes, P. 1985b. Conditions and strategies for emancipatory mathematics education in underdeveloped countries. *For the Learning of Mathematics* 5 (3): 15–20.

———. 1986. How to recognize hidden geometrical thinking: A contribution to the development of anthropological mathematics. *For the Learning of Mathematics* 6 (2): 10–17.

———. 1988a. On possible uses of traditional Angolan sand drawings in the mathematical classroom. *Educational Studies in Mathematics* 19 (1): 3–22.

———. 1988b. A widespread decorative motif and the Pythagorean theorem. *For the Learning of Mathematics* 8 (1): 35–39.

———. 1988c. On culture, geometrical thinking and mathematics education. *Educational Studies in Mathematics* 19 (3): 123–35.

———. 1990. On mathematical elements in the Tchokwe 'sona' tradition. *For the Learning of Mathematics* 10:31–34.

———. 1991. On ethnomathematical research and symmetry. *Symmetry: Culture and Science* 1:154–70.

———. 1995. *Ethnomathematics and education in Africa.* Stockholm: Stockholms Universitet.

———. 1999. *Geometry from Africa: Mathematical and educational explorations.* Washington, DC: The Mathematical Association of America.

———. 2002. *Sipatsi: Cestaria e geometria na cultura Tonga de Inhambane*. Maputo: Moçambique Editora.

Gerdes, P., and A. Djebbar. 2007. *Mathematics in African history and cultures: An annotated bibliography*. Self Publishing—Lulu.com. http://www.lulu.com/product/paperback/mathematics-in-african-history-and-cultures-an-annotated-bibliography/835410.

Gerdes, P., and J. Fauvel. 1990. African slave and calculating prodigy: Bicentenary of the death of Thomas Fuller. *Historia Mathematica* 17:141–51.

Heath, T. L. 1910. *Diophantus of Alexandria*. Cambridge: Cambridge University Press.

Huylebrouck, D. 2008. The ISShango project. *Journal of Mathematics and the Arts* 2:145–152.

Johnson, R. K. 1997. Administration and language policy in Papua New Guinea. In *New Guinea area languages and language study*. Vol. 3: *Language, culture, society, and the modern world*, ed. S. A. Wurm. Canberra: Australian National University, Research School of Pacific Studies, Department of Linguistics.

Kelley, D. 2000. Mayan astronomy. In *Encyclopedia of astronomy and astrophysics*, ed. Paul Murdin, article no. 1920, 1–4. Bristol. Institute of Physics Publishing.

Lagercrantz, S. 1973. Counting by means of tally sticks or cuts on the body in Africa. *Anthropus* 68:569–88.

Laurencich-Minelli, L., and G. Magli. 2008. A calendar quipu of the early 17th century and its relationship with the Inca astronomy. Preprint arkiv.org/abs/0801.1577, to appear in *Archaeoastronomy*.

Laycock, D. C. 1975. Observations on number systems and semantics. In *New Guinea area languages and language studies*. Vol. 1, *Papua languages and New Guinea linguistic scene*, ed. S. A. Wurm. Canberra: Dept. of Linguistics, Research School of Pacific Studies, The Australian National University.

Lean, G. A. 1996. *Counting systems of Papua New Guinea and Oceania*. 21 vols. Papua New Guinea: PNG Institute of Technology.

Locke, L. L. 1912. The ancient quipu, a Peruvian knot record. *American Anthropologist* 14:325–52.

———. 1923. *The ancient quipu or Peruvian knot record*. New York: American Museum of Natural History.

Mackie, E. W. 1977. *The megalithic builders*. Oxford: Phaidon.

Majumdar, N. K. 1911–12. Aryabhata's rule in relation to indeterminate equations of the first degree. *Bulletin of the Calcutta Mathematical Society* 3:11–12.

Mann, A. 1887. Notes on the numeral system of the Yoruba nation. *Journal of the Anthropological Institute of Great Britain and Ireland* 16:59–64.

Marshack, A. 1972. *The roots of civilisation*. London: Weidenfeld & Nicolson.

Megitt, M. J. 1958. Mae Enga time reckoning and calendar. *Man* 58:74–77.

Mubumbila, M. 1988. *Sur le sentier mystérieux des nombres noirs*. Paris: L'Harmattan.

Murata, T. 1975. Pour une interpretation de la destinée du Wasan—aventure et mésaventure de ces mathématiques. In *Proceedings of the 14th International Congress of the History of Science*, 2:184–99. Tokyo: Science Council of Japan.

Murra, J. V. 1968. An Aymara kingdom in 1567. *Ethnohistory* 15:115–51.

Niles, S. 2007. Considering quipus: Andean knotted string records in analytical context. *Reviews in Anthropology* 36:85–102.

Open University. 1975. *Counting: I. Primitive and more developed counting systems.* AM 289 N1. Milton Keynes: Open University Press.

———. 1976. *Written numbers.* Vol. 2, *History of mathematics: Counting, numerals, and calculation.* Milton Keynes: Open University Press.

Pareja, D. 1986. Pre-Hispanic tools of computation: The quipu and the yupana. [In Spanish.] *Revista Integración Temas de Matemáticas* 4 (1): 37–56.

Plester, V., and D. Huylebrouck. 1999. The Ishango artefact: The missing base 12 link. *Forma* 14 (4): 339–46.

Poma de Ayala, Guaman. 1936. *Nueva corónica y buen gobierno.* Paris: Institute d'ethnologie.

Posner, J. 1982. The development of mathematical knowledge in two West African societies. *Child Development* 53:200–208.

Rambane, D. T., and M. C. Mashige. 2007. The role of mathematics and scientific thought in Africa: A Renaissance perspective. *International Journal of African Renaissance Studies* 2:183–199.

Saxe, G. 1982. Developing forms of arithmetical thought among the Oksapmin of Papua New Guinea. *Development Psychology* 18 (4): 583–94.

Schmandt-Besserat, D. 1987. Oneness, twoness, threeness. *Sciences* 27:44–48.

———. 1992. *Before writing.* 2 vols. Austin: University of Texas Press.

Shirley, L. 1988. Historical and ethnomathematical algorithms for classroom use. Mimeo. Ahamadu Bello University Zaria, Nigeria.

Spinden, H. J. 1924. The reduction of Mayan dates. In *Papers of the Peabody Museum of American Archaeology and Ethnology,* vol. 6, no. 2. Cambridge, MA: Harvard University Press.

Thom, A. 1967. *Megalithic sites in Britain.* Oxford: Oxford University Press.

Tozzer, A. M. 1941. *Land's relacion de las Cosa de Yucatán: A translation with notes.* Papers of the Peabody Museum, vol. 18. Cambridge, MA: Harvard University Press.

Urton, G. 1977. *The social life of numbers: A Quechua ontology of numbers and philosophy of arithmetic.* Austin: University of Texas Press.

Vega, Garcilaso de la. 1966. *The royal commentaries of the Incas.* Trans. H. V. Livermore. Austin: University of Texas Press.

Wassen, H. 1931. The ancient Peruvian abacus. *Comparative Ethnological Studies* 9:191–205.

Zaslavsky, C. 1970. Mathematics of the Yoruba people and of their neighbors in southern Nigeria. *The Two-Year College Mathematics Journal* 1 (2): 76–99.

———. 1973a. *Africa counts: Number and pattern in African cultures.* Boston: Prindle, Weber & Schmidt.

———. 1973b. Mathematics in the study of African culture. *Arithmetic Teacher* 20:532–35.

———. 1991. Women as the first mathematicians. *Women in Mathematics Education Newsletter* 14 (1).

———. 1994. Mathematics in Africa: Explicit and implicit. In *Companion encyclopedia of the history and philosophy of the mathematical sciences,* ed. I. Grattan-Guinness, 1:85–92. London: Routledge.

Name Index

Subject Index